New Directions in Wireless Communications Systems

From Mobile to 5G

New Directions in Wireless Communications Systems

From Mobile to 5G

Edited by
Athanasios G. Kanatas
Konstantina S. Nikita
Panagiotis Mathiopoulos

CRC Press
Taylor & Francis Group
Boca Raton London New York

CRC Press is an imprint of the
Taylor & Francis Group, an **informa** business

CRC Press
Taylor & Francis Group
6000 Broken Sound Parkway NW, Suite 300
Boca Raton, FL 33487-2742

© 2018 by Taylor & Francis Group, LLC
CRC Press is an imprint of Taylor & Francis Group, an Informa business

No claim to original U.S. Government works

Printed on acid-free paper

International Standard Book Number-13: 978-1-4987-8545-7 (Hardback)

Visit the Taylor & Francis Web site at
http://www.taylorandfrancis.com

and the CRC Press Web site at
http://www.crcpress.com

Contents

Preface

We are experiencing the dawn of what is called the *digital age*. The Internet and digital technologies are transforming our world. Wireless communication systems constitute a basic component of the current and the future information society. The *fifth generation* of communication systems, or 5G, will be, in the years to come, the most critical building block of what is referred to as a *digital society*. It is expected that 5G will be a truly converged network environment, because not only mobile communications systems but also other wireless systems and wired networks will coexist and use the same infrastructure.

The vision ahead is that 5G will provide virtually ubiquitous, ultra-fastbroadband, *connectivity* not only to millions of individual users but also to billions of connected objects. Therefore, 5G systems should have the capabilities to (a) serve the data-hungry devices, for example, smartphones and tablets, providing rates of Gbps; (b) enable machine-to-machine (M2M) communications, for example, vehicle-to-vehicle (V2V) networks; (c) allow for interconnectivity of massive devices, for example, sensors and e-health equipment; and (d) support a wide range of applications and sectors. All these expectations set a wide variety of technical requirements for the design of 5G systems, including higher peak and user data rates, extremely low latency and response times, enhanced indoor and outdoor coverage, significantly increased number of devices, reaching the astronomical number of 100 billion, seamless mobility, security and privacy, increased battery lifetime, and improved quality of service (QoS) while maintaining low operational costs.

All the predictions available in the literature lead to the conclusion that beyond 2020, wireless communication systems will be able to support more than 1000 times the traffic volume served today by the current telecommunication systems. This extremely high traffic load is one of the major issues faced by the 5G designers, manufacturers, and researchers, alike. It appears that this challenge will be addressed by a combination of parallel techniques that will use more spectrum more flexibly, realize higher spectral efficiency, and densify cells. Therefore, novel techniques and paradigms should be developed to support such challenges.

In this context, this book addresses diverse key-point issues of the next-generation wireless communication systems and attempts to identify promising solutions. The core of the book deals with the techniques and methods belonging to what is generally referred to as *radio access network*. The increased needs and the users' expectations from the next-generation systems have been based for long on enabling technologies provided by the physical layer. These technologies are mainly developed to combat signal degradations imposed by the wireless channel and to support increased user data rates and improved QoS.

Chapter 1 presents an overview of the wave propagation and modelling of radio channels with characteristics that vary in time and space. Then, the influence of multipath propagation on the signal-distorting characteristics of wireless channels and the resulting effects on digital communications are outlined. Both narrowband and wideband channel modeling is discussed with suggestions for assessing whether a wide-sense stationary (WSS) model is appropriate and for modeling channel processes with nonstationary characteristics and correlated scattering. Spatial channel characterization and MIMO channel models are reviewed and the measurement equipment and techniques are presented. Moreover, in Chapter 2, various channel models are compared, modeling techniques involved (stochastic, ray-tracing, etc.) are analyzed, and the applications of the involved theory in optimizing the cell planning procedures are demonstrated.

The millimeter-wave (mm-Wave) band has been foreseen for the development of 5G networks. A large amount of spectral space, from 28 to 95 GHz, is available to establish such systems providing increased data rates. Hence, the design issues of such networks necessitate the thorough knowledge of mm-Wave propagation characteristics. In Chapter 3, the state-of-the art channel models, derived from the latest measurement campaigns in various propagation scenarios (indoor, outdoor, etc.) are presented. In addition, models that describe the spatial–temporal variations of the mm-Wave channel are explained. Recent advances of MIMO and massive MIMO technology exploitation in mm-Wave propagation are also examined.

Chapter 4 presents the fundamental concepts and basic theoretical tools for the qualitative analysis of fading channels. This section helps the reader to understand the nature and types of fading, the basic characteristics of a wireless channel, and the impact of fading on signal transmission. Fading mitigation techniques are presented, and a complete classification of the existing transmit/receive diversity techniques that exploit the spatial, frequency, time, and the polarization domains, is also provided.

OFDM modulation and OFDM-based transmission schemes (e.g., OFDMA) have dominated the current modern wireless standards. However, despite OFDM's implementation simplicity, some key weaknesses such as the need for extended guard bands leading to reduced spectral efficiency, the increased peak-to-average power ratio, and its inherent sensitivity to frequency offset impairments have motivated in the research of alternative multicarrier or single-carrier modulation techniques based on filter banks. Therefore, Chapter 5 reviews the most important modulation and multicarrier schemes, including OFDM that have become strong candidates for the future generations of wireless standards.

The 5G radio access will be built on both new radio access technologies (RATs) and the evolved existing wireless technologies (LTE, HSPA, GSM, and WiFi). Chapter 6 provides an overview of radio network planning techniques from 2G to 4G. Then it describes the energy-efficient *green* radio network planning concept for heterogeneous 4G wireless networks and presents a cellular layout adaptation method.

The incorporation of MIMO transmission in the modern wireless standards has provided a significant improvement in the achieved spectral efficiency and system capacity. As technology moves toward 5G, the concept of MIMO has been evolved with the definition of complex centralized schemes, such as massive MIMO, or decentralized schemes, that is, network MIMO and coordinated multipoint transmission. Moreover, alternative advanced MIMO schemes reducing the number of required RF chains, enabling full duplex communications, and allowing the use of compact antennas have been considered. Chapter 7 reviews the most important trends and challenges in MIMO transmission with emphasis on their application to 5G systems.

Alternative enabling technologies are also presented. As an example, Chapter 8 introduces the load-controlled parasitic antenna arrays (LC-PAA) that resemble the operation of conventional antenna arrays with many elements, thus boosting the performance of multiantenna wireless communication systems, whereas at the same time providing cost and energy consumption savings and size reduction. Therefore, a novel method is presented that enables one to perform arbitrary channel-dependent precoding with LC-PAAs. The possible application of this technique in point-to-point MIMO, multiuser MIMO (MU-MIMO), coordinated MIMO, and massive MIMO setups is investigated. The design and implementation of various types of LC-PAAs for these frequency bands are also shown.

Spatial modulation (SM) has been recently proposed as a promising transmission concept that reduces the complexity and the cost of multiple-antenna schemes. At the same time, it guarantees high data rates, improved system performance, and energy efficiency. Working principles of SM are presented in Chapter 9 and the advantages and disadvantages of SM-MIMO as compared to the state-of-the-art MIMO communications are also discussed. Various transmission techniques for SM-MIMO, namely space shift keying (SSK), generalized space shift keying (GSSK) and generalized spatial modulation (GSM) are also further discussed. An analytical framework for the performance evaluation of SM-MIMO over fading channels in terms of the average bit error probability (ABEP) is also presented. Some MIMO transmission schemes closely related to the SM-MIMO paradigm, that is, single RF MIMO schemes, the incremental MIMO, and the antenna subset modulation (ASM) schemes, are briefly discussed. Moreover, several applications of SM-MIMO for future 5G wireless communications are presented, including MIMO implementations that exploit the massive MIMO paradigm as well as the combination of both orthogonal frequency division multiplexing (OFDM) and single carrier with SM-MIMO.

Device-to-device (D2D) communication is recognized as one of the technology components of the evolving 5G architecture. The reuse of cellular resources by D2D links that are located randomly inside a macro-cell imposes a cochannel interference to the base station (BS), cellular users and to other D2D receivers. Several aspects of D2D communications are investigated in Chapter 10, such as interference statistics of interferers scattered according to a homogeneous Poisson point process, D2D neighbor discovery based on signal-to-interference-and-noise ratio association metric, D2D link performance in the presence of interference and power control imposed by the BS, and the impact of mobility on the D2D link performance. In addition, a V2V use case is considered and the performance of a multiuser cooperative V2V communication system is studied.

Next, Chapter 11 addresses the virtualization of wireless access in order to provide the required capacity to a set of virtual base stations (VBSs) with diverse requirements, instantiated in a given geographical area. A network architecture is presented, based on a generic network virtualization environment, in which both physical and virtual perspectives are considered and the main stakeholders are taken into account. A new tier of radio resource management (RRM) is proposed for inter-VNets (virtual networks) RRM aiming at transposing the cooperative set of functionalities to the virtualization environment.

Cooperative techniques have significantly contributed toward improving the capacity of the 4G networks and are considered as a basic element of the imminent 5G networks. This topic is addressed in Chapter 12 of this book. Among cooperative techniques, cooperative relaying (CR) has received significant attention from researchers due to the gains that it offers to the network. The techniques that have received a large amount of contributions are the opportunistic relay selection (ORS) and successive relaying (SuR) and several policies, such as successive opportunistic relaying (SOR) with full duplex (FD) operation and SOR for networks with buffer-aided (BA) relays that offer additional degrees of freedom in inter-relay interference (IRI) mitigation. In addition, an overview of interference mitigation techniques designed for relay networks is provided.

Cognitive radio networks (CRNs) have been proposed as a promising solution to cope with the spectral scarcity, and this is the main topic of Chapter 13. In addition, as the technology of CRN can help to unlock the full potential of 5G wireless systems, the 5G key enablers are described and finally the role of CRN in 5G networks is also highlighted. Furthermore, several formulations of different CRN's resource allocation problems are presented based on various mathematical approaches such as optimization techniques, game theory, matching theory, multi-criteria decision-making theory, and machine learning. The basic concepts of each approach are described and an extensive list of scientific works is also presented. Finally, various future research avenues for the application of cognitive systems to 5G and the investigation of novel flexible algorithms in radio resource management are also discussed.

The last chapter (Chapter 14) focuses on transformational wireless technologies for health care, discussing their potential and the challenges that rose. As it is well known, the significant advances in wireless communications, sensing technologies, and sensor data analytics are opening new opportunities in medicine, and are promising to address the unsustainability of current health care provision models. Notably, health care challenges, including rising health care costs, aging populations, and emerging disease threats rank among the most serious concerns in the world. Wireless technology can empower both patients and medical providers by providing round-the-clock health status information. Examples include wireless on-body (wearable, epidermal) and in-body (implantable, ingestible) medical devices that may be used as sensors, actuators, and/or drug delivery devices. Remote diagnosis, vital parameter control, elderly monitoring, and chronic disease management are just some of the examples of applications of wireless technologies. Exploitation of wireless technologies and sensor data analytics in health care can lead to healthier citizens, reduced hospital stays, and lower costs.

This book is dedicated to Philip Constantinou, an inspiring professor, a valuable colleague, and a loyal friend. He was an exceptional professor who was always there for students, who could understand your thoughts and concerns, who was the inspiration, and who gave confidence to the team. A testimony to this was the overwhelming response we received from many colleagues who wanted to contribute to this book because they wanted to join us in celebrating his research and educational achievements in the general field of wireless communications for the past 25 years.

Philip was a kind human being, a generous and a welcoming person open to the community, showing great concern and affection for people with special needs. Before joining the National Technical University of Athens, Athens, Greece, he and his family have lived for many years in Canada, where he was granted the Master of Applied Science by University of Ottawa and started his great telecom career in Telesat in Ottawa, Canada. Then he worked for several years for the Government of Canada in the Department of Communications, while at the same time pursuing his PhD degree at the Carleton University, Ottawa, Canada. Being an international man, he believed in openness, valued collaboration, and always tried to impart the importance of communicating engineering novelty to his students. He was always supportive to the junior communications engineers' ideas and encouraged them to act on their own initiative following their dream. Philip was a workaholic lab enthusiast who loved to deliver working systems. Although he was a zealous and motivated leader, he had a methodic approach regularly reminding to his colleagues two of his favorite phrases: "One step at a time" and "I am always a practical man." His students and his colleagues, alike, owe him a great debt of gratitude for making us feel special, strong, and capable of doing things.

Athanasios (Thanasis) G. Kanatas
Konstantina (Nantia) S. Nikita
Panagiotis (Takis) Mathiopoulos

About the Editors

Athanasios (Thanasis) G. Kanatas is a professor at the Department of Digital Systems and dean of the School of Information and Communication Technologies at the University of Piraeus, Greece. He received his diploma in electrical engineering from the National Technical University of Athens (NTUA), Greece, in 1991, MSc degree in satellite communication engineering from the University of Surrey, Surrey, UK in 1992, and earned his PhD degree in mobile satellite communications from NTUA, Greece in February 1997. From 1993 to 1994 he was with National Documentation Center of the National Research Institute. In 1995, he joined SPACETEC Ltd. as technical project manager for VISA/EMEA VSAT Project in Greece. In 1996, he joined the Mobile Radiocommunications Laboratory as a research associate. From 1999 to 2002, he was with the Institute of Communication and Computer Systems responsible for the technical management of various research projects. In 2000, he became a member of the board of directors of OTESAT S.A. In 2002, he joined the University of Piraeus as an assistant professor. He has published more than 150 papers in international journals and international conference proceedings. He is the author of 6 books in the field of wireless and satellite communications. He has been the technical manager of several European and National R&D projects. His current research interests include the development of new digital techniques for wireless and satellite communications systems, channel characterization, simulation, and modeling for mobile, mobile satellite, and future wireless communication systems, antenna selection and RF preprocessing techniques, new transmission schemes for MIMO systems, V2V communications, and energy efficient techniques for wireless sensor networks.

He has been a senior member of IEEE since 2002. In 1999, he was elected chairman of the Communications Society of the Greek Section of IEEE. He has been a member of the TPC of more than 40 international IEEE conferences. He was a corecipient of two best paper awards for papers published in the International Conference on Advances in Satellite and Space Communications (SPACOMM), Athens, Greece, (2010) and Global Wireless Summit, Wireless VITAE, Atlantic City, United States, (2013).

Konstantina (Nantia) S. Nikita received her diploma in electrical engineering and earned her PhD degree from the National Technical University of Athens (NTUA), as well as MD degree from the Medical School, University of Athens. From 1990 to 1996, she worked as a researcher at the Institute of Communication and Computer Systems. In 1996, she joined the School of Electrical and Computer Engineering, NTUA, as an assistant professor, and since 2005, she serves as a professor at the same school. Moreover, she is an adjunct professor of Biomedical Engineering and Medicine, Keck School of Medicine and the Viterbi School of Engineering, University of Southern California. She has authored or coauthored 165 papers in refereed international journals, 41 chapters in books, and more than 300 papers in international conference proceedings. She is the editor of seven books in English and author of two books in Greek. She holds three patents. She has been the technical manager of several European and National R&D projects. She has been honorary chair/chair of the program/organizing committee of several international conferences, and has served as a keynote/invited speaker at international conferences, symposia, and workshops organized by NATO, WHO, ICNIRP, IEEE, URSI, and so on. She has been the advisor of 27 completed PhD theses, several of which have received various awards. Her current research interests include biomedical telemetry, biological effects and medical applications of radiofrequency electromagnetic fields, biomedical signal and image processing and analysis, simulation of physiological systems, and biomedical informatics. She is an associate editor of the IEEE Transactions on Biomedical Engineering, the IEEE Journal of Biomedical and Health Informatics, the *IEEE Transactions on Antennas and Propagation, the Wiley Bioelectromagnetics,* and *the Journal of Medical and Biological Engineering and Computing.* She has received various honors/awards, with the Bodossakis Foundation Academic Prize (2003) for exceptional achievements in "Theory and Applications of Information Technology in Medicine" being one of them.

She has been a member of the board of directors of the Atomic Energy Commission and of the Hellenic National Academic Recognition and Information Center, as well as a member of the Hellenic National Council of Research and Technology. She has also served as the deputy head of the School of Electrical and Computer Engineering of the NTUA. She is a member of the Hellenic National Ethics Committee, a founding fellow of the European Association of Medical and Biological Engineering and Science (EAMBES), a fellow of the American Institute of Medical and Biological Engineering (AIMBE), and a member of the Technical Chamber of Greece and of the Athens Medical Association. She is also a member of the BHI Technical Committee, the founding chair and ambassador of the IEEE–EMBS, Greece chapter and has served as the vice-chair of the IEEE Greece Section.

Panagiotis (Takis) Mathiopoulos is a professor of telecommunications at the Department of Informatics and Telecommunications, University of Athens, Greece. Prior to that, he was with the Institute for Space Applications and Remote Sensing (ISARS) of the National Observatory of Athens, first as its director (2001–2005) and then as director of research (2006–2014). From 1989 to 2003, he was a faculty member in the Department of Electrical and Computer Engineering, the University of British Columbia (UBC), where he was a professor from 2000 to 2003. From 2008 to 2013, he was appointed as guest professor at the Southwest Jiaotong University, China. He has been also appointed by the Government of People's Republic of China as a senior foreign expert at the School of Information Engineering, Yangzhou University (2014–2016) and by Keio University as a visiting (Global) professor in the Department of Information and Computer Science (2015–2016 and 2017–2018) under the Top Global University Project of the Ministry of Education, Culture, Sports, Science, and Technology (MEXT), Government of Japan.

For the past 25 years, he has been conducting research mainly on the physical layer of digital communication systems for terrestrial and satellite applications, including digital communications over fading and interference environments. He coauthored a paper in GLOBECOM'89 establishing for the first time in the open technical literature the link between maximum likelihood sequence estimation (MLSE) and multiple (or multisymbol) differential detection for the AWGN and fading channels. He is also interested in channel characterization and measurements, modulation and coding techniques, synchronization, SIMO/MIMO, UWB, OFDM, software/cognitive radios, green communications, and 5G. In addition, since 2010, he has been actively involved in research activities in the fields of remote sensing, LiDAR systems, and photogrammetry. In these areas, he has coauthored more than 110 journal papers, mainly published in various IEEE and IET journals, 4 book chapters, and more than 120 conference papers.

Prof. Mathiopoulos has been or currently serves on the editorial board of several archival journals, including the *IET Communications,* and *the IEEE Transactions On Communications* (1993–2005). He has regularly acted as a consultant for various governmental and private organizations. Since 1993, he has served on a regular basis as a scientific advisor and a technical expert for the European Commission (EC). In addition, from 2001 to 2014, he has served as a Greek representative to high level committees in the European Commission (EC) and the European Space Agency (ESA). He has been a member of the TPC of more than 70 international IEEE conferences, as well as TPC vice-chair for the 2006-S IEEE VTC and 2008-F IEEE VTC as well as cochair of FITCE2011. He has delivered numerous invited presentations, including plenary and keynote lectures, and has taught many short courses all over the world. As a faculty member at the ECE of UBC, he was elected as ASI fellow and a Killam research fellow. He was a corecipient of two best paper awards for papers published in the *2nd International Symposium on Communication, Control, and Signal Processing* (2008) and *3rd International Conference on Advances in Satellite and Space Communications* (2011).

Contributors

Georgia E. Athanasiadou
Department of Informatics and
  Telecommunications
University of Peloponnese
Tripoli, Greece

Petros S. Bithas
Department of Digital Systems
University of Piraeus
Piraeus, Greece

Robert Bultitude
Department of Systems and Computer
  Engineering
Carleton University
Ottawa, Ontario, Canada

Luisa Caeiro
ESTSetúbal/INESC-ID
University of Lisbon
Lisbon, Portugal

Filipe Cardoso
ESTSetúbal/INESC-ID
University of Lisbon
Lisbon, Portugal

Theofilos Chrysikos
Wireless Telecommunications
  Laboratory
Department of Electrical and
  Computer Engineering
University of Patras
Patras, Greece

Luis M. Correia
IST/INESC-ID
University of Lisbon
Lisbon, Portugal

George P. Efthymoglou
Department of Digital Systems
University of Piraeus
Piraeus, Greece

Athanasios G. Kanatas
Department of Digital Systems
University of Piraeus
Piraeus, Greece

Vasileios M. Kapinas
Department of Electrical and
  Computer Engineering
Aristotle University of Thessaloniki
Thessaloniki, Greece

George K. Karagiannidis
Department of Electrical and
    Computer Engineering
Aristotle University of Thessaloniki
Thessaloniki, Greece

Konstantinos Karathanasis
School of Electrical and Computer
    Engineering
National Technical University of Athens
Athens, Greece

Stavros Kotsopoulos
Wireless Telecommunications
    Laboratory
Department of Electrical and
    Computer Engineering
University of Patras
Patras, Greece

Konstantinos Maliatsos
Department of Digital Systems
University of Piraeus
Piraeus, Greece

Panagiotis Mathiopoulos
Department of Informatics and
    Telecommunications
University of Athens
Athens, Greece

Nektarios Moraitis
School of Electrical and Computer
    Engineering
National Technical University of Athens
Athens, Greece

Konstantina S. Nikita
School of Electrical and Computer
    Engineering
National Technical University of Athens
Athens, Greece

Nikolaos Nomikos
Department of Information and
    Communication Systems
    Engineering
University of Aegean
Mytilene, Greece

Dimitrios Ntaikos
Broadband Wireless & Sensor
    Networks (B-WiSE) Research
    Laboratory
Athens Information Technologies
Marousi, Greece

Konstantinos Ntougias
Broadband Wireless & Sensor
    Networks (B-WiSE) Research
    Laboratory
Athens Information Technologies
Marousi, Greece

Georgia D. Ntouni
Department of Electrical and
    Computer Engineering
Aristotle University of Thessaloniki
Thessaloniki, Greece

Athanasios D. Panagopoulos
School of Electrical and Computer
    Engineering
National Technical University of Athens
Athens, Greece

Constantinos B. Papadias
Broadband Wireless & Sensor
    Networks (B-WiSE) Research
    Laboratory
Athens Information Technologies
Marousi, Greece

Konstantinos Peppas
Department of Informatics and
  Telecommunications
University of Peloponnese
Tripoli, Greece

Marios I. Poulakis
School of Electrical and Computer
  Engineering
National Technical University of Athens
Athens, Greece

Anargyros J. Roumeliotis
School of Electrical and Computer
  Engineering
National Technical University of Athens
Athens, Greece

George V. Tsoulos
Department of Informatics and
  Telecommunications
University of Peloponnese
Tripoli, Greece

Ioannis Valavanis
Department of Informatics and
  Telecommunications
University of Peloponnese
Tripoli, Greece

Stavroula Vassaki
School of Electrical and Computer
  Engineering
National Technical University of Athens
Athens, Greece

Demosthenes Vouyioukas
Department of Information and
  Communication Systems
  Engineering
University of Aegean
Mytilene, Greece

Dimitra Zarbouti
Department of Informatics and
  Telecommunications
University of Peloponnese
Tripoli, Greece

Chapter 1

Propagation Measurement-Based Wireless Channel Characterization and Modeling

Robert Bultitude

Contents

1.1 Introduction

Any communications link can be modeled as shown in Figure 1.1, the model having three major components: (1) the source, usually referred to as the transmitter; (2) the transmission channel; and (3) the sink, usually referred to as the receiver, or the destination of the transmitted signals.

After final design and implementation of the transmitter, receiver, and antennas (if the communication signals are to be carried via radiowave propagation), any distortion of the transmitted waveform, $s(t)$, on its path from the transmitter to the receiver comes from either noise or the physical characteristics of the radio channel. These characteristics can primarily be categorized as transmission loss, the dispersion (or spreading) of the transmitted signal in time as a result of propagation delays, its dispersion in frequency as a result of Doppler effects, or its dispersion in three-dimensional space. On a static or nonvarying radio channel, which exists any time neither the communications link antennas nor anything in the environment of operation is moving and the physical characteristics of the environment are not changing, such distortion can be corrected using fixed technology, including hardware or software. However, when the transmission characteristics of a radio channel are randomly varying because changes in the environment are random, the correction of distortion becomes challenging. Models, and knowledge of the parameters

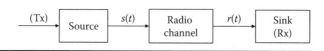

Figure 1.1 Basic model of a communications link.

thereof, for the random variations are thus needed to assess whether the correction of distortion is needed to achieve the desired quality of service on the communication link, and if so, to determine design criteria for distortion correcting hardware and software, and for use in simulation models to test its effectiveness. This chapter focuses on the modeling of randomly varying radio channels, the estimation of model parameters from radio propagation measurements, and the characterization of measured channels based upon model parameters, which is the first step in assessing the requirement for distortion correcting technology.

Following this introduction, this chapter has three main sections. Section 1.2 begins with a discussion of the most fundamental impairment to wireless transmission: transmission loss in an environment where there are no reflectors, scatterers, or obstructions. Then, the subject of multipath propagation when there are objects in the environment of operation that can reflect, scatter, and obstruct radiowaves is introduced and the review of a method for modeling transmission loss in a multipath environment based on radio propagation measurements is discussed. Next, in Section 1.3, the influence of multipath propagation on the signal-distorting characteristics of wireless channels and resulting effects on digital communications are outlined. This is accompanied by a review and discussion of a classical model, applicable to the analysis of single-input-single-output systems, for modeling randomly varying channels as a Gaussian wide-sense-stationary process. It is followed by the general assessment of channel quality, with particular emphasis on methods by which measured channels can be classified in accordance with the subject model and by which model parameters can be estimated from measured data for use in channel modeling and simulation. Section 1.3.6 summarizes an extension from the literature of the classical model to make it applicable for multiple-input multiple-output systems, which are envisaged as a key to the success of 5G systems in meeting anticipated transmission capacity requirements.

Section 1.4 focuses on the data qualification preprocessing of measured data that is necessary for the reduction of errors in analyzing such data for the estimation of model parameters. First, a very general overview of radio channel sounding is given. A method from the literature for the reduction of noise from wideband channel measurements is then reviewed. A discussion follows on the assessment of measured data to determine if modeling of random variations in the time domain using a wide-sense-stationary model is appropriate. Section 1.4.3 is a review of modeling cumulative probability distributions based on estimates thereof from measured data. Then, the estimation of spaced-frequency correlation functions for randomly time varying channels is outlined, with emphasis on how results can be used to verify conditions required for modeling variations in the frequency domain as a wide-sense-stationary process. Section 1.4.5 is a very short note on considerations regarding the assessment of whether a Gaussian model is appropriate and whether assumptions of ergodicity are reasonable.

1.2 Transmission Loss

What is referred to as free-space transmission occurs anytime radiowaves travel between a radiating source and a receiving antenna without interacting with anything, such as physical obstructions, particles, or electric charges (e.g., in an ionospheric plasma). If one considers the transmission of radio waves from a point source, or isotropic antenna, which is a fictitious construct that cannot be realized in the real world, radiation occurs equally in all direction and the radiated waves spread out in spheres of ever-increasing radius, r, with the isotropic antenna at their center. Since the area of the surface of a sphere is given by

$$A_s = 4\pi r^2 \tag{1.1}$$

the power loss (most frequently referred to as propagation loss), at any frequency, as a result of the propagation of radiowaves in free space is inversely proportional to r^2. A practical transmit (Tx) antenna focuses energy in specific directions such that the ratio of power density it radiates in any direction, θ, to that which would be radiated by an isotropic antenna, is given by its directive gain (often referred to as gain), $G_t(\theta)$. Similarly, the ratio of power received by a receive (Rx) antenna from any direction, θ, to the power that would be received by an isotropic antenna, is given by its gain, $G_r(\theta)$. Associated with a receive antenna is an effective area, A_e, which can be thought of as a capture area, over which power incident upon it can be received. This effective area [1] is related to the gain of the antenna in accordance with the equation

$$A_e = \frac{G_r \lambda^2}{4\pi} \tag{1.2}$$

Using Equation 1.1 and knowledge of the focusing ability of a Tx antenna, the power density, S, incident on any surface as a result of the radiation of power, P_t, by the Tx antenna can be computed as

$$S = \frac{E^2}{\eta} = \frac{P_t G_t}{4\pi r^2} \tag{1.3}$$

where E is the magnitude of the electric field and η is the intrinsic impedance of free space [2]. Then, using Friis equation [2], the power received through the Rx antenna is given by

$$P_r = SA_e = \frac{P_t G_t G_r \lambda^2}{(4\pi r)^2} \tag{1.4}$$

and the free-space transmission loss is given by

$$L = \frac{P_r}{P_t} = \frac{G_t G_r \lambda^2}{(4\pi r)^2} \tag{1.5}$$

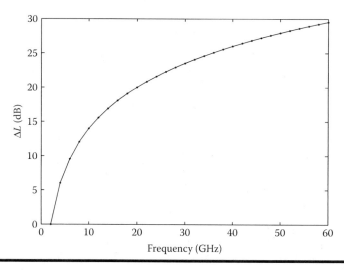

Figure 1.2 The difference of transmission loss between equal gain antennas at the specified frequencies and at 2 GHz.

Note that because the effective area of the Rx antenna is dependent upon wavelength, λ, free-space transmission loss has a direct-quadratic dependence upon λ, and hence, an inverse-quadratic dependence upon the frequency of operation, f. The fact that the frequency dependence of transmission loss is an antenna property rather than a property of the propagation environment is often overlooked. If Equation 1.5 were used to calculate the ratio of transmission loss at two different frequencies f_1 and f_2 over a free space radio link having equal gain antennas at the two frequencies, it is clear that this ratio would be given by $\Delta L = (\lambda_2/\lambda_1)^2$. Figure 1.2 is a plot of this ratio for the case of $f_1 = 2\,\mathrm{GHz}$, which is a ratio of interest to many in the consideration of the design of 5G systems, which are proposed for operation above 6 GHz.

The quadratic relationship, resulting in a decreasing slope of the ratio of transmission loss with respect to that at 2 GHz às frequency increases, is worth noting.

1.2.1 *Multipath Propagation*

In many instances free-space transmission is impossible. This is a result of the fact that as the radiated waves spread out in spheres of ever-increasing radius, they impinge upon obstacles, or encounter particles or, as in the ionosphere, regions where there are concentrated charge densities. In the troposphere, the interaction of radiowaves with obstacles causes reflection, diffraction, and scattering. These phenomena, in turn, generate secondary waves that can interfere with waves that arrive at a Rx antenna directly from the Tx antenna. Such interference is referred to as multipath interference, and it can either enhance or diminish the performance of a radio communication link, depending upon the overall link design.

In the region of an Rx antenna, the electromagnetic fields from the multipath waves add vectorally such that the magnitude and phase of the total electric field, $\vec{E}_{tot}$, can be written as

$$\left|\vec{E}_{tot}\right| = \left[E_1 e^{-j\beta r_1} + E_2 e^{-j\beta r_2} + \cdots + E_n e^{-j\beta r_n}\right] \tag{1.6}$$

where, using Equation 1.3 and assuming single interactions, $E_i = \sqrt{(\eta P_t G_t / 4\pi r_i^2)}\rho_i$, and ρ_i is a complex coefficient resulting from the interaction of the "ith" multipath wave with some obstacle or surface before it is reflected, diffracted, or scattered toward the Rx antenna. If radio path "i" is the direct path between the Tx and Rx antennas, $\rho_i = 1$. The vector sum of the multipath waves can either enhance or diminish received power, depending on the phases of the interfering waves at any specific receive location. This, in turn, can modify the dependence of transmission loss on distance, r, of what must now be referred to as average transmission loss, from the inverse quadratic dependence appropriate for free space, and increases the exponent to which r is raised, or decreases it, depending upon the circumstance. In a reverberation chamber, for example, where all waves that impinge upon the chamber walls are reflected with a reflection coefficient near unity, this exponent is reduced, resulting in lower average transmission loss than that which would occur in free space. This is because energy is prevented from leaving the chamber by its walls. On a radio path between two antennas where there is one reflection point and the angle between the direction of travel of the impinging wave and the reflecting surface is small, the reflection coefficient is near unity, and if the electric field is perpendicular to the plane containing the direction (or Poynting) vectors of both the incident and reflected waves, there is a phase shift of π radians on reflection. In this case, received power can diminish in proportion to r^4 in accordance with the well-known two-ray model [2] at Tx–Rx ranges beyond a breakpoint, after which the direct and reflected waves begin to have an inverse phase relationship and their vector addition is therefore destructive. This distance is a function of the offset of the antennas from the reflecting surface (i.e., antenna heights if the surface is the earth) and the wavelength at the frequency of operation.

Multipath propagation and its effects can be exemplified by considering the simple case, at a frequency of 2.35 GHz, of equal-height antennas in a small empty room of dimensions 5 m long × 3.6 m wide × 4 m high, where the direct wave from the Tx antenna, as well as reflections from the four walls, the floor, and the ceiling are received. A simple simulation using Equation 1.6 and assuming vertical quarter-wavelength monopoles with radiation patterns as in [1] at the Tx and Rx yields a received power pattern along the centerline of the room as shown in Figure 1.3. For this simulation, the transmit power, P_t, was set equal to −7 dBm, transmission line losses were equal to 2.2 dB, the walls were modeled as being

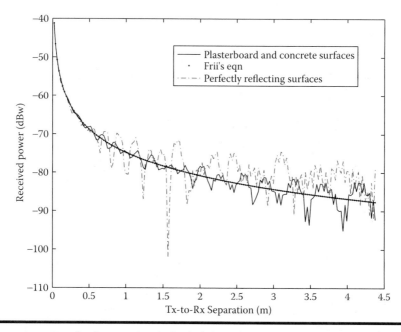

Figure 1.3   Results from a simulation of the received sum of multipath waves as a function of distance along the centerline of a small, empty room.

constructed of plasterboard on a wooden support structure, and the floor and ceiling were modeled as being made of dry concrete. Equations for the reflection coefficients and the material constants were taken from [2]. Results for the case in which the walls and floor are perfectly conducting are also shown in Figure 1.3.

The undulations of the multipath sums as the constituent waves add with different phase relationships as a function of the distance between the Tx and the Rx can clearly be seen. It can also be seen that the average power of the multipath sum for the case of perfectly reflective surfaces decreases less rapidly than the curve given by Friis equation, leading to a model transmission loss is proportional to r raised to an exponent that is less than 2.

1.2.2 Modeling of Transmission Loss in a Multipath Environment

The modeling of transmission loss is something that has received much attention by many. This is an important part of communication system design, since the capacity of any radio link is directly related to the ratio of received signal power to noise power, or the SNR, at the receiver.

The modeling of P_r as a function of Tx–Rx separation, say, from simulations such as above, or from propagation measurements, is best done by expressing P_r

in terms of the logarithm of distance. When plotted, this relationship is a straight line and is easy to model. However, transmission loss, the ratio of P_r-to-P_t is modeled more often than P_r. In a multipath environment, by convention, before such modeling, averaging is conducted to eliminate transmission loss variations that result from the vector addition of multipath waves. This averaging can be conducted over small ranges of time, space, or frequency, since the physics associated with the seemingly random variations that result from the vector addition are the same in all three domains. If the averaging is done in time, motion of either the antennas, or objects in the operating environment is needed to *stir* the component multipath waves (referred to as multipath components, or MPCs) to achieve a variation in time that covers as much as possible as the full dynamic range commensurate with the multipath sum where the measurements are being made. If the averaging is over space, one of the link antennas must be moved in order to achieve the variations of interest. A wideband channel sounding signal [2] facilitates ease in averaging over frequency. The range over which averaging must be conducted is best determined by the observation of the absence of variations in averaged results.

Modeling of the log–log relationship between average transmission loss and Tx–Rx antenna separation (sometimes herein referred to as range) can be effected using a standard least-mean-squared (LMS) error approach. This can be done by considering a system of multiple equations, each for the average transmission loss at a different position along the trajectory of the Rx antenna. The modeling process begins by representing the logarithm of the Tx–Rx separation (d) as

$$x = \log_{10}\left(\frac{d}{d_{\text{ref}}}\right) \tag{1.7}$$

where d_{ref} is an arbitrary short distance from the Tx antenna, in its far field. The mean transmission loss can then be modeled as

$$E[L(x)] = n\log_{10}\left(\frac{d}{d_{\text{ref}}}\right) + L_{\text{ref}} = nx + L_{\text{ref}} \tag{1.8}$$

where L_{ref} represents the path loss at d_{ref}. If measured or simulated path loss samples are ordered as a function of d, one can write the system of equations as

$$L_1 = nx_1 + L_{\text{ref}} + S_1$$

$$L_2 = nx_2 + L_{\text{ref}} + S_2 \tag{1.9}$$

$$\vdots = \vdots \quad \vdots \quad \vdots$$

$$L_M = nx_M + L_{\text{ref}} + S_M$$

where the S_i, $i = 1...M$, represents deviations in propagation loss with respect to its range-dependent mean. This system can be written in matrix form as

$$y = Az + e \qquad (1.10)$$

in which $y = [L_1 \; L_2 \cdots L_M]^T$, $A = \begin{bmatrix} x_1 \; x_2 \cdots x_M \\ 1 \; 1 \cdots 1 \end{bmatrix}^T$, $z = [n \; L_{ref}]^T$, and $e = [S_1 \; S_2 \cdots S_M]^T$.

The best fit model parameters, n and L_{ref}, can be found by regarding the deviations, S_i, as errors, and minimizing their sum, given by

$$\begin{aligned} \left[S_1^2 + S_2^2 \cdots + S_M^2 \right] &= e^T e \\ &= (y - Az)^T (y - Az) \\ &= y^T y - y^T Az - z^T A^T y + z^T A^T Az \end{aligned} \qquad (1.11)$$

Minimization of this expression can be achieved by setting its derivative with respect to the parameter vector z equal to zero, giving

$$y^T A - A^T y + 2A^T Az = 0 \qquad (1.12)$$

or,

$$z = \left(A^T A \right)^{-1} A^T y \qquad (1.13)$$

where $\left(A^T A \right)^{-1} A^T$ is the pseudo-inverse of A.

It should be noted that there are discussions within the research community concerning the specification of d_{ref} in the procedure described in the foregoing. When the model is to be applied where Tx-to-Rx separations are in the range of Tx-to-Rx separations over which the propagation measurements upon which the model is based were made, it is acceptable to use the minimum value of Tx-to-Rx separation that was used during the measurements, and therefore L_{ref} is the associated local average transmission loss, as estimated from the measurements. However, if in applying the model, extensions need to be made to the range of Tx-to-Rx separations used during the measurements in either direction, inaccuracies can increase linearly with the amount of the extension. For this reason, it has been proposed that the model for average transmission loss should be forced to have a transmission loss equal to that of the free-space model at a Tx–Rx antenna separation of 1 m, and equations for such modeling are given in [3]. It is argued that, whereas it does not matter if 1 m happens to be in near field of the Tx antenna since no actual measurements are involved, in most cases there is a low probability that there are obstructions between the Tx and Rx antennas at a separation of 1 m. Friis equation at a range of 1 m is therefore thought to provide a good physical basis for the model. Models from this approach have been tested against measured data at extended ranges and after frequency scaling, which is sometimes used to predict losses in a given environment at frequencies where no measurements in that environment have been reported. Robustness of the model with a 1 m free space intercept is reported to be good [4]. This anchoring of L_{ref}, clearly, fixes the slope of the model, and another anchor that, for the same reasons, has been used by the author at a Tx–Rx separation of 1 wavelength most often produces a model with a different slope. Although the robustness to extensions of models resulting from the latter approach has not been tested, the benefit is that the reference transmission loss, L_{ref}, for transmission between unity gain antennas, is 22.5 dB, regardless of operating frequency. Figure 1.4 shows the results of measurements at 2.35 GHz in a room of the same dimensions as those used in the simulations for Figure 1.3. Each of the measurements was derived from the average insertion loss recorded by a vector-network-analyzer-based measurement system every 2.5 MHz across a 500 MHz bandwidth. The room was furnished as a study room, with three library-type cubicles aligned along each side wall, with a walk space down the center of the room. The Tx antenna was a biconical at 1 m high, in a fixed location at one end of the room, on its centerline. The Rx antenna was a monopole at the same height. In Figure 1.4, unconstrained results from the described LMS error modeling procedure are labeled L_{FI}, where the "FI" denotes a floating transmission loss intercept. The constrained models are labeled L_{FSI}, where "FSI" denotes a free-space

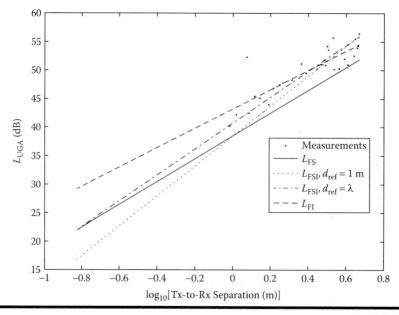

Figure 1.4 **Measurement and modeling results from a propagation experiment in a small, furnished, study room at 2.35 GHz.**

intercept, and in each of these two cases, d_{ref} is given. The ordinate in this figure is labeled L_{UGA}, representing transmission loss between unity gain antennas, which the author has found to be a convenient concept for comparing transmission losses at different frequencies when measurement system antenna gains are different at different frequencies. This is in preference to the term *propagation loss*, as it does not imply the loss is a property of the transmission medium.

The reader should note that Figure 1.4 is merely an example of results from application of the methods described in the text. Nothing should be inferred regarding which model is best from a comparison of results in the single example shown in the figure, as this comparison varies for different propagation scenarios and operating environments.

In addition to modeling of the average transmission loss, variations with respect to the average, referred to as large area, long-term, or shadowing variations, by radio engineers, whether or not there are obstructions on the radio paths between the Tx and Rx,* are often modeled as having a log-normal probability distribution with a specified standard deviation. Such a model can be verified as appropriate, and its standard deviation specified by forming an estimated probability density function (EPDF), or an estimated cumulative probability function (ECDF), for variations in local average transmission loss with respect to a linear model derived as explained in the foregoing. This modeling cannot be done by searching, for example, for the LMS difference between the epdf, or the ECDF and a model of interest. The reason is explained in Section 1.4.3, along with a description of better procedures for the derivation of models for the pdfs and CDFs that characterize variations of the underlying random process from which measurements are sampled. Figure 1.5 shows the results of such modeling, for transmission loss variations with respect to the floating intercept model, L_{FI}.

1.3 Characterization of Multipath Radio Channels for Digital Communications

1.3.1 Introduction

In addition to modifying propagation loss parameters with respect to those applicable to free space, multipath propagation also causes time dispersion on a radio channel, which can spread transmitted digital symbols in time, so that symbols arriving consecutively at a receiver overlap and cause intersymbol interference. This results in the vector sum of received multipath signals, rather than the desired

---

* Both small area and large area variations in received power can result from multipath propagation, in the absence of obstructions on any of the radio paths between the Tx and the Rx, as for example, when there is a direct wave and a ground-reflected wave on a line-of-sight radio link [4,5].

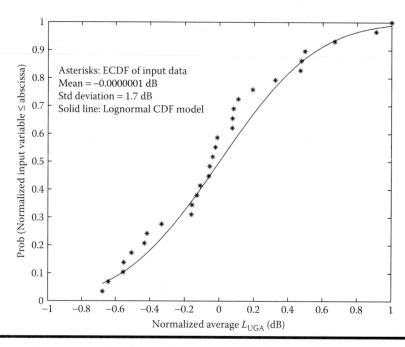

Figure 1.5 Example showing modeling of the distribution of shadow loss variations in a small furnished room at 2.35 GHz.

symbol at sample times, and, when severe enough, causes symbol errors at a rate that cannot be reduced, regardless of SNR at the receiver. Resulting error rates are referred to as irreducible error rates, or error-rate floors.

The time dispersion discussed in the foregoing is caused by the fact that different multipath waves generally arrive at a receiver at different times since the paths over which they travel at the speed of light generally have different lengths. Thus, if it were possible to transmit an impulse from the Tx antenna, a series of short attenuated pulses would arrive at the Rx at different propagation delays, τ. For the real-world small empty room example from Section 1.2.2, such an impulse response, $h(\tau)$, for the propagation channel when the Tx and Rx antennas are separated by 1 m can be plotted as in Figure 1.6.

Six power-scaled pulses are represented in Figure 1.6, whereas the direct wave and six reflections (from walls, floor, and ceiling) were simulated. This is because the reflections from the two-side walls have the same powers and propagation delays, so they are superimposed on each other and cannot be distinguished in the figure.

The impulse response in Figure 1.6 can be Fourier transformed to yield the frequency-domain transfer function, $H(f)$, for the radio channel. This function is plotted in Figure 1.7. Note the variations, from frequency-to-frequency, in the transfer function gain that result from variations in the vector sum of MPCs at different frequencies.

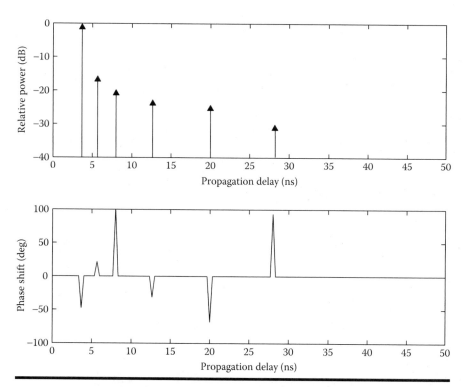

Figure 1.6 **Channel impulse response at a midband frequency of 2 GHz in the small empty room with the Tx and Rx antennas separated by 1 m.**

Given either the impulse response or the transfer function of any radio channel, the output resulting from transmitting any waveform, $s(t)$, on the channel can be calculated: the impulse response can be applied using convolution with the transmitted waveform in the time domain, and the transfer function, through multiplication with the transmitted waveform spectrum, $S(f)$, in the frequency domain. That is

$$r(t) = s(t) * h(\tau) \qquad (1.14)$$

and

$$R(f) = S(f)H(f) \qquad (1.15)$$

While the overlapping of received digital symbols as a result of the multipath echoes in Figure 1.6 can be imagined, its dual in the frequency domain, plotted in Figure 1.7, can easily also be envisaged as resulting in distortion of the transmit signal spectrum. This distortion can be corrected using a fixed equalizer when there are no time variations on the channel. This mitigates multipath effects, resulting in an overall radio channel transfer function that has close to the same gain, and linear phase shift, as a function of frequency over the bandwidth of interest. When multipath

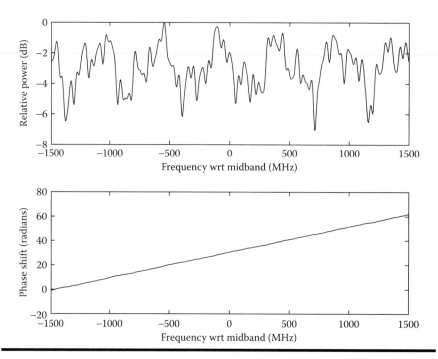

Figure 1.7 **Channel transfer function at a midband frequency of 2 GHz in the small empty room with the Tx and Rx antennas separated by 1 m.**

dispersion changes in time, however, equalization also becomes dynamic and more complex and costly. Engineers are therefore in need of measures by which the severity of changing multipath conditions can be assessed so means can be considered for the avoidance of a requirement for other than very simple dynamic equalizers.

1.3.2 Frequency Correlation Measures for Randomly Varying Multipath Channels

Assessment of the severity of multipath conditions on randomly varying channels can be effected by regarding time variations as a stochastic process, being random primarily as a result of the randomness with which reflecting, diffracting, and scattering phenomena within an operating environment, as well as the transmit and receive antenna locations, change as time progresses. If one considers random time variations of $H(f)$, it is easy to envisage the fact that distortion of transmitted symbols is more severe, the narrower the bandwidth over which the correlation of such variations diminishes. When this co-called correlation bandwidth (B_c) is large, there is greater probability that all frequencies in the transmit signal spectrum vary in unison, maintaining symbol shape in the time domain constant, varying only received power. A channel with this characteristic is referred to as frequency-flat fading channel [6]. When B_c is small, signal spreading occurs in the time domain.

A time-varying channel with a narrow correlation bandwidth is referred to as frequency-selective fading channel [7]. Care should be taken to note that this concept is not useful on nonvarying, or static, radio channels. A static channel can be frequency selective, but frequency correlation is not a useful measure, as the selectivity is fixed over time.

Turning now to a statistical characterization, through consideration of the central limit theorem, and the fact that a sum of MPCs is often comprised of a large number of MPCs, transfer function variations are often modeled as being random and Gaussian [6]. The channel process (CP) can therefore be completely characterized by its mean and autocorrelation. The autocorrelation of variations of $H(f)$ can be written in complex baseband notation as

$$R_H(f, f') = \frac{1}{2} E\left[H(f)H^*(f') \right] \tag{1.16}$$

where the operation $(\cdot)^*$ denotes complex conjugate. Now, if the time-varying equivalent of $H(f)$ is denoted $H(t; f)$, and the CP is ergodic, the ensemble average can be replaced by a time average, and Equation 1.16 becomes

$$R_H(\Delta t; f, f') = \frac{1}{2} \int_{-\infty}^{\infty} H(t; f)H^*(t + \Delta t; f')\, dt \tag{1.17}$$

where the time domain wide-sense stationary (WSS) condition implied by considering only time shifts (Δt) accompanies the ergodicity. Further, for the purpose of assessing distortion at the level of each symbol, it is clear that Δt must equal zero. Then, if the CP is also WSS in the frequency domain, R_H is independent of the specific frequencies, f and f', being only dependent upon their separation, Δf, and identical for all reference frequencies, allowing one to write

$$R_H(0; \Delta f) = \frac{1}{2} \int_{-\infty}^{\infty} H(t; f)H^*(t; f + \Delta f)\, dt \tag{1.18}$$

Finally, if one is analyzing samples from time series recordings of measurements, the integral must be replaced by a finite sum, giving an estimate of R_H that can be written as

$$\hat{R}_H(\Delta f) = \frac{1}{2N} \sum_{i=1}^{N} H_i(f)H_i^*(f + \Delta f) \tag{1.19}$$

It should be noted in applying Equation 1.19 that the time average of $H(f)$ should not be subtracted to give what is sometimes referred to as a covariance estimate. The problem with doing this is that radio channels that would normally exhibit narrowband envelope variations that are Rician-like are automatically analyzed as if they exhibit

Rayleigh fading*, which is concomitant with much narrower correlation bandwidths. On Rician channels, correlation is maintained high across wider bandwidths by the steady component, or specular content, in the received signal, which, after phase synchronization to replicate the operation of a receiver [8], results in a non-zero mean for $H(t; f)$. An automatic assumption that the CP is WSS in the frequency domain, which gives a symmetrical result for R_H as a function of Δf, making it independent of f, should also be avoided. A better approach is to process measured data with the sampled equivalent of Equation 1.17, as discussed in Section 1.4.4; then, if the result is symmetrical, or nearly so, declare that the CP is WSS in the frequency domain, which allows the use of Equation 1.19, and implies uncorrelated scattering [6] in the time domain. This, in turn, simplifies delay-domain analyses, which is discussed in Section 1.3.3. Figure 1.8 shows results from the estimation of R_H using Equation 1.77, which is the sampled equivalent of Equation 1.17, to analyze data measured on radio channels, for which Rayleigh and Rician cumulative distribution functions (CDFs) were found, via the Kolmogorov-Smirnov (KS) hypothesis test, to be good models for narrowband envelope fading. The KS test is discussed in Section 1.4.3.

The frequency offset of the time series plot is a result of not using the spectral line at midband as a reference to avoid d.c. offsets in the measurement system.

The curve in Figure 1.8a, labeled *Time Series Result* was estimated using Equation 1.77 to analyze a recorded time series of 800 channel transfer function estimates (TFEs), with the correlation across a 100 MHz bandwidth with respect to the time series of complex values of the 800 TFEs at midband. The dashed curve is the result of Fourier transforming the channel's average power-delay profile, which is discussed in Section 1.3.3. Two observations can be made. The first is that this frequency correlation function (FCF) estimate is almost symmetrical, indicating WSS characteristics in the frequency domain. The second is that the two curves in the figure match fairly closely. This is because the measured channel was a Rayleigh fading channel, and, as observed, its characteristics were WSS in the frequency domain. These two conditions are necessary to make the Fourier Transform (FT) method valid. In Figure 1.8b, it can be seen that the time series FCF estimate from using Equation 1.78 is reasonably symmetrical, but the fact that the channel on which data for this estimate were measured exhibited Rician, rather than Rayleigh fading, makes the FT method invalid, so the two curves in the figure do not match. It can be seen that the FT result would grossly under-estimate B_C according to any definition.[†] The FT approach, however, is often applied without checking to see if the Rayleigh fading condition is satisfied. This can lead to over-design of communication systems, since an interpretation would be that symbol rates need to be much lower than is actually necessary to minimize equalization requirements.

---

* See Section 1.3.6 for an explanation of Rayleigh and Rician fading.

† In fact, the time series result diminishes only because of decreased SNR at the band edges of the channel sounding system used to measure the data from which the FCFs were estimated. It is considered reasonable to assume that its magnitude would remain very high for extremely large bandwidths otherwise.

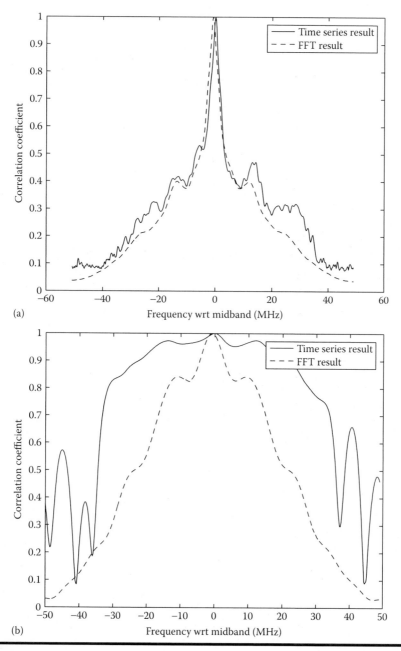

Figure 1.8 **(a) Frequency correlation functions estimated by two different methods from data measured on a Rayleigh fading channel. (b) Frequency correlation functions estimates by two different methods from data measured on a Rician fading channel $K_{Rice} = 9$ dB.**

It was noted that the FCF estimates in Figure 1.8a report correlation with respect to temporal variations at the midband frequency. This is sufficient if the CP is WSS in the frequency domain. However, if this midband result were to be asymmetrical, it would be necessary to estimate the frequency correlation with respect to references at every spectral position across the bandwidth of interest. This results in 3D FCFs, as shown in Figure 1.9, for the examples discussed in the foregoing.

Consider now an ideal frequency-flat fading channel. If such a channel had a bandwidth, B, then the volume (V_{R_H}) defined by its 3D FCF would be equal to B^2. It has been suggested in the literature, therefore [9], that when it is found that frequency-domain WSS characteristics cannot be assumed, the ratio, to B_m^2, of the volume of 3D FCFs estimated from measurements, where B_m is the measurement system bandwidth, could be used to assess channel selectivity in much the same way that R_H is applied when frequency-domain WSS conditions prevail. Whereas B_c is often defined to be the bandwidth at which R_H diminishes to 50%, or 1/e, it might also be considered to be the bandwidth at which V_{R_H}/B_m^2 diminishes to 50%, or 1/e. This would apply if the channel exhibits Rayleigh fading, but is not frequency-domain WSS. If, however, Rican fading is found to be a good model, it is suggested that if the ratio, K, of specular to random power is anything greater than about 0 dB, it is reasonable to assume that the correlation bandwidth is large enough that intersymbol interference is not of concern, provided $B_m \gg B_{sys}$, where B_{sys} is the bandwidth of the communication system being considered for operation on the measured channel.

1.3.3 Delay Dispersion Measures for Multipath Channels

Similar to the way in which a frequency correlation function can be defined as a statistical measure for variations of $H(t; f)$, one can define an autocorrelation function to characterize time variations of $h(t; \tau)$, the time varying equivalent of $h(\tau)$. Again, one begins by taking ensemble averages to give

$$R_h\left(t, t'; \tau, \tau'\right) = \frac{1}{2} E\left[h\left(t; \tau\right) h^*\left(t'; \tau'\right)\right] \tag{1.20}$$

Under ergodicity, the ensemble average can be replaced by a time average to give

$$R_h\left(\Delta t; \tau, \tau'\right) = \frac{1}{2} \int_{-\infty}^{\infty} h\left(t; \tau\right) h^*\left(t + \Delta t; \tau'\right) dt \tag{1.21}$$

where, as before, the time domain WSS condition implied by considering only time shifts (Δt) accompanies the ergodic condition. Further, for the purpose of assessing distortion at the level of each symbol, it is clear that Δt must equal zero. In addition, it can be noted that if the random variations of the channel's impulse response

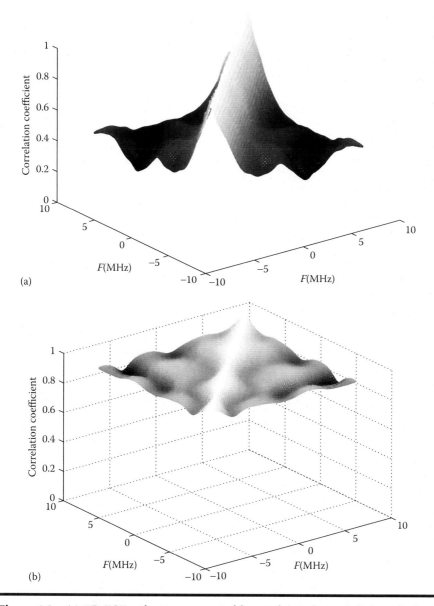

Figure 1.9 (a) 3D FCF estimate, computed by applying the sampled equivalent of Equation 1.17 to estimate an FCF with reference to each sampled frequency across the measurement system bandwidth on the same channel for which Figure 1.8a was estimated. (b) 3D FCF estimate, computed by applying the sampled equivalent of Equation 1.17 to estimate an FCF with reference to each sampled frequency across the measurements system bandwidth on the same channel for which Figure 1.8b was estimated.

at delay τ are uncorrelated with the random variations delay τ', R_h is equal to zero unless $\tau = \tau'$, and one can write

$$R_h\left(0;\tau\right) = \frac{1}{2}\int_{-\infty}^{\infty} h\left(t;\tau\right)h^{*}\left(t;\tau\right)dt \tag{1.22}$$

Finally, in the sampled data case, the integral can be replaced by a sum, giving

$$\hat{R}_h\left(\tau\right) = \frac{1}{2N}\sum_{i=1}^{N}\left|h_i\left(\tau\right)\right|^2 \tag{1.23}$$

Since Equation 1.23 represents the average power on the channel at delay τ, it is widely known as the channel's average power-delay profile, or its delay power spectrum [10], whether or not the condition of uncorrelated variations at different delays, referred to as uncorrelated scattering, prevails and permits its recognition as a true autocorrelation function estimate. If there is uncorrelated scattering, and, in addition, $H(t; f)$ has a zero mean, which results in Rayleigh envelope fading, it can also be shown [6,7,10,11] that

$$R_h\left(\tau\right) = \int_{-\infty}^{\infty} R_H\left(\Delta f\right)e^{j2\pi\Delta f\tau}d\Delta f \tag{1.24}$$

Equation 1.24 is the FT relationship discussed in association with the curve labeled *FFT result* in Figure 1.8. Recall that Figure 1.8b is a demonstration that the Rayleigh fading condition is crucial if Equation 1.24 is to hold, as is the uncorrelated scattering condition. Hence, when analyzing measured data, estimating the distributions of narrowband fading envelopes (see Section 1.4.3) and assessing the symmetry of a time series estimate of frequency correlation (see Section 1.4.4) before applying Equation 1.24 is considered to be prudent. In addition, throughout, the condition of conformance to a time-domain WSS model has been evoked. A discussion of this can be found in Section 1.4.2.

While plots of $\hat{R}_h(\tau)$ are interesting and often foretelling that problems with the reception of multipath echoes can be anticipated, it is desirable to be able to use a single value, like B_c, for characterizing the delay dispersion exhibited by a radio channel. For this, the power-weighted standard deviation of delay with respect to the centroid of $\hat{R}_h(\tau)$ is used.

If the earliest received MPC has delay τ_0, then in discrete form, in the analysis of measured impulse response estimates, each with M samples in delay, the centroid, or power-weighted average excess delay is given by

$$\overline{\tau_w} = \frac{\sum_{i=1}^{M}\hat{R}_h\left(\tau_i\right)\times\tau_i}{\sum_{i=1}^{M}\hat{R}_h\left(\tau_i\right)} - \tau_0 \tag{1.25}$$

The power-weighted standard deviation of excess delays, or second central moment of $h(\tau)$ can then be written as

$$\sigma_\tau = \sqrt{\frac{\sum_{i=1}^{M} \hat{R}_h(\tau_i)\left(\tau_i - \overline{\tau_w} - \tau_0\right)^2}{\sum_{i=1}^{M} \hat{R}_h(\tau_i)}} \tag{1.26}$$

and is frequently referred to as the rms delay spread, or simply delay spread of the channel. As a result of the Fourier transform relationship in Equation 1.24, the relationship $B_c = 1/\sigma_\tau$, which has been studied in [12], is sometimes used for rule-of-thumb calculations. If intersymbol interference is to be avoided, T_S must be significantly* greater [7,10] than σ_τ, or $B_S \ll B_c$, where T_S is the duration and B_S is the bandwidth of digital symbols transmitted over the channel. It should be noted that, because of the zero-mean transfer function and uncorrelated scattering conditions for the validity of Equation 1.24, reciprocal relationship between σ_τ and B_c should only be applied when it is known that a channel exhibits uncorrelated scattering and Rayleigh fading.

Frequently σ_τ, though formally defined for the characterization of fading channels, is applied to characterize static, or nonvarying channels, or channels that cannot be categorized as zero-mean Gaussian wide-sense-stationary uncorrelated-scattering (ZMGWSSUS) channels because one or several of the required conditions do not apply. In this case, it is sometimes useful to use a CDF estimated from measured data (herein referred to as an ECDF) to model static (sometimes referred to as instantaneous) rms delay spread (σ_{τ_s}) values estimated by applying Equation 1.23 in the analysis of measured data, but with $N = 1$ such that there is no averaging over time. Such an ECDF might be used, for example, to characterize changes in σ_{τ_s} as a function of operating scenario, or Tx-to-Rx separation.

1.3.4 Characterization of Frequency Dispersion on Time-Varying Multipath Channels

The Doppler shift in frequency undergone by a signal received from, or by, a moving antenna or that is received via reflection from, or by, diffraction around a moving obstruction, is a function of the relative position, relative direction of motion, and relative speed of the moving antennas or obstructions and the wavelength of the signal. For cases where multiple MPCs are received from different angles of arrival, each MPC could have been modified by a different frequency shift. Similarly, each of the spectral components in a wideband signal would also be modified by a different frequency shift. The result is that the spectrum occupied by the originally transmitted signal is broadened. If a sinusoid were transmitted in an environment where antennas and obstructions are moving for example, a signal with a bandwidth greater than zero would be received. The bandwidth of such

---

* Often taken to be one tenth.

a frequency-spread signal is referred to as the Doppler spread of the channel over which the signal is received, and it can be characterized through the autocorrelation function in Equation 1.17 in a manner similar to that in which delay spread is characterized. If a Fourier transform of Equation 1.17 with respect to Δt is taken, an intermediate function, $S_H(\vartheta; f_i, f_j)$, results [10], where

$$S_H\left(\vartheta; f, f'\right) = \int_{-\infty}^{\infty} R_H\left(\Delta t; f, f'\right)e^{-j2\pi\vartheta\Delta t}d\Delta t \tag{1.27}$$

Then, if there is uncorrelated scattering, the channel is WSS in the frequency domain, and its autocorrelation is dependent only on Δf, so that

$$S_H\left(\vartheta; \Delta f\right) = \int_{-\infty}^{\infty} R_H\left(\Delta t; \Delta f\right)e^{-j2\pi\vartheta\Delta t}d\Delta t \tag{1.28}$$

and for the sinusoidal case,

$$S_H\left(\vartheta\right) = \int_{-\infty}^{\infty} R_H\left(\Delta t\right)e^{-j2\pi\vartheta\Delta t}d\Delta t \tag{1.29}$$

$S_H(\vartheta)$ is referred to as the Doppler power spectrum of the channel. Its power-weighted variance, σ_D, can be estimated in a manner similar to the way in which delay spreads are estimated using Equation 1.26. If uncorrelated scattering conditions prevail, because of the Fourier transform relationship in Equation 1.29, the correlation time of the channel can be approximated as

$$T_c = \frac{1}{\sigma_D} \tag{1.30}$$

T_c is the time duration over which the phase of the channel can be assumed to be approximately constant. If irreducible error rates are to be avoided in digital systems, symbol bandwidths, B_s, must be significantly greater than σ_D (or, $T_S \ll T_c$). Thus, when both delay spread and Doppler spread are considered for time-varying multipath channels, it can be concluded that for the avoidance of error-rate floors, $\sigma_\tau \ll T_S \ll T_c, \sigma_D \ll B_S \ll B_c$. A channel with such characteristics is referred to as a doubly-flat [7,10], or just flat, fading* channel. A channel with characteristics outside these boundaries in either direction is referred to as an overspread fading channel, and this type of channel can be frequency flat and time selective, time flat and frequency selective, or doubly selective.

---

* Recall that these characteristics cannot be defined for non-time-varying channels, since, in the real world, where ensemble averages do not exist, correlation functions cannot be defined for non-time-varying channels. Note also that the adjective "fading" means there are time variations.

If $S_H(\vartheta;\Delta f)$ is inverse Fourier transformed with respect to Δf, another function $S(\vartheta;\tau)$, which is sometimes reported in the literature, results. This function is referred to as the channel's scattering function, and can be written as

$$
\begin{aligned}
S(\vartheta;\tau) &= \int_{-\infty}^{\infty} S_H(\vartheta;\Delta f)e^{+j2\pi\tau\Delta f}d\Delta f \\
&= \int_{\infty}^{\infty}\int_{\infty}^{\infty} R_H(\Delta t;\Delta f)e^{-j2\pi\vartheta\Delta t}e^{+j2\pi\tau\Delta f}d\Delta t d\Delta t
\end{aligned}
\tag{1.31}
$$

It characterizes the average power in the channel as a function of multipath delay and Doppler frequency. Note that because this function is independent of both time and frequency, there is a requirement for a channel to conform to WSS models in both time and frequency if it is to be estimated from measured data.

1.3.5 Characterization of Received Signal Envelope Variations

In addition to the characterization of time variations by the quantification of Doppler spread, there is often interest in determining the extent of time variations of the absolute value, or envelope of a received signal. As for the quantification of Doppler spread, a single frequency is analyzed. This is a worst case because depths of fading decrease in direct proportion to signal bandwidth [14]. Often the target is to determine whether an ECDF for envelope fading matches one of several fading models, to enable the reproduction of experimentally observed envelope variations during wireless system simulations.

The most frequently cited model for envelope variations is the Rayleigh fading model, which is based on consideration of the variations of the resultant (r) of the vector sum of n unit vectors of arbitrary phase and the same frequency, when n is large. While a previous author had shown that this resultant tends to a fixed value, n, Lord Rayleigh showed that r is random, even for very large n [15], and is described by the probability density function (PDF)

$$
p(r) = \frac{2r}{n}e^{-\frac{r^2}{n}}
\tag{1.32}
$$

and through integration, the corresponding CDF is

$$
P[r \le R] = 1 - e^{-\frac{R^2}{n}}
\tag{1.33}
$$

Lord Rayleigh also showed that the expectation of power (time-averaged ac power for an ergodic process) is n. The cited original work goes on to show that the pdf in Equation 1.32 is not altered even if the vectors in the sum have different amplitudes, as long as there is a large number having each amplitude. It is common in modern literature [10] to see a derivation of the Rayleigh pdf based on the assumption that, from central limit theorem considerations, for large n, the resultant of the sum of n waves can be decomposed into the sum of two orthogonal Gaussian processes, with mean zero, and standard deviation σ. In this case, the expectation of power is $2\sigma^2$ and Equations 1.32 and 1.33 are modified through the replacement of n by $2\sigma^2$. Lin [16] showed, however, that if a received signal envelope follows a Rayleigh pdf, it does not necessarily mean that there are a large number of waves, or that the Gaussian decomposition is accurate. The Gaussian–Rayleigh relationship is unique, however, under the conditions that the amplitude and phase of each component wave are statistically independent, and the phase of each wave is randomly and uniformly distributed on $[0, 2\pi]$.

If a received signal reaches the receiver over direct line of sight or via specular reflection, the Rayleigh distribution discussed in the foregoing may not be appropriate. If there is a steady, or coherent, component and a large number of randomly varying MPCs, the situation is analogous to that of a sine wave in Gaussian noise. Writing in terms of noise currents, Rice [17,18] represented this signal combination as

$$I = S\cos(pt) + I_N \tag{1.34}$$

where S and p are constants, and I has an envelope, $r = \sqrt{(S+I_c)^2 + I_s^2}$, where I_c and I_s are independent quadrature zero mean Gaussian components of I_N with equal variance, σ^2, and showed that the pdf for r is given by

$$p(r) = \frac{r}{\sigma^2} \exp\left[-\frac{r^2 + S^2}{2\sigma^2}\right] I_0\left(\frac{rS}{\sigma^2}\right) \tag{1.35}$$

where I_0 is the Bessel function of the first kind, order zero. When applied to characterize multipath propagation-induced random signal envelope fluctuations, this pdf is often written in terms of the ratio, $k = S^2/2\sigma^2$, of the power in the steady component to the power in the random component, and the total received power, $\Omega = S^2 + 2\sigma^2$, giving

$$p(r) = \frac{2(k+1)r}{\Omega} \exp\left(-k - \frac{(k+1)r^2}{\Omega}\right) I_0\left(2\sqrt{\frac{k(k+1)}{\Omega}}\, r\right) \tag{1.36}$$

Often in channel characterization and modeling work, k must be estimated from the measurements. There are several ways in which this can be done. When only received power, or envelope data are available, the method in [19] gives good results. However, this method can sometimes yield complex values as estimates for k when the underlying process is not Rician, or when the estimates of statistical moments are poor. When complex, rather than just power or received signal envelope data are available, the method in [20] can be used. Whichever method is used, results should be verified using hypothesis tests, as discussed in Section 1.4.3.

As the derivation for the Rician envelope pdf is based upon an assumption of quadrature Gaussian signal components and, as reported in a previous paragraph, the Rayleigh pdf can result from such a signal composition, derivations for both pdfs can be found in modern literature [10] that start from the assumption that the signal of interest is composed of the sum of quadrature Gaussian components. From such derivations, it is easy to identify that the Rician pdf becomes identical to the Rayleigh pdf when $k \to 0$. It is also easy to realize that if the envelope of a received signal were reported to vary in accordance with either a Rayleigh or a Rician distribution, such an envelope could be simulated as $r = \sqrt{x^2 + y^2}$, where in the Rayleigh case, x and y are independent Gaussian random variables with standard deviation, σ, and zero mean value, and in the Rician case, x and y are independent Gaussian random variables with standard deviation σ, and mean values, $S \cos\theta$ and $S \sin\theta$, respectively, where θ is any real number.

1.3.6 Characterization of Direction Dispersion on Multipath Radio Channels

With the advent of increased interest in smart antenna and multiple-input multiple-output (MIMO) systems, methods and measures for the quantification of direction dispersion became very important. Direction dispersion is quantified through measures of the statistical properties of random variations of directions from which MPCs arrive at a receive antenna (directions of arrival [DOAs]), and the directions in which the same MPCs departed (directions of departure [DODs]) from the Tx antenna. Knowledge of these directions and their statistical moments is important since they influence the correlation characteristics of the random variations of the complex physical link transfer functions between Tx and Rx antenna pairs, which are critical in determining the capacity of a MIMO system. To understand this, consider the equation for the mutual information, or capacity (C) on a MIMO communication link [21] with multielement antenna arrays at both the Tx and the Rx, when knowledge of the channel transfer function, G, is known only at the Rx, through its estimation, for example, by the processing of a training sequence. This capacity is given by

$$C = \log_2 \det \left[I_{n_R} + \left(\frac{\rho}{n_T} \right) HH^* \right] \text{bits/S/Hz} \tag{1.37}$$

where:

n_T and n_R are the numbers of Tx and Rx antenna elements, respectively

I_{n_R} is the $n_R \times n_R$ identity matrix

ρ is the average signal-to-noise ratio at each receiver branch, the exponent $(\cdot)^*$ represents transpose conjugate

$\boldsymbol{H}$ represents the normalized MIMO channel transfer function matrix which is given as follows:

$$\boldsymbol{H} = \begin{bmatrix} H_{1,1} & \cdots & H_{1,n_T} \\ \vdots & & \vdots \\ H_{n_R,1} & & H_{n_R,n_T} \end{bmatrix} \tag{1.38}$$

where $H_{i,j}$ is the normalized frequency-domain transfer function of the physical link between the ith Rx antenna element and the jth Tx antenna element. Further note that $G(f)$, the frequency domain channel transfer function before normalization, is assumed to represent a narrowband channel, such that it has equal gain across its bandwidth (i.e., $G(f) \rightarrow G$), and it is slowly time-varying such that it is assumed to be nonvarying for the duration of at least one digital transmission frame. The normalization to derive H from G is given by $\sqrt{\hat{P}} \cdot G = \sqrt{P} \cdot H$, where $\hat{P}$ is the total radiated power regardless of the number of Tx antennas and P is the average power at the output of each Rx antenna. It can be shown [21] that, for the case of a frequency-selective channel, assumed to be composed of N contiguous sub-bands of constant gain, the total capacity is the sum of the capacities for the N frequency-flat sub-bands, regardless of the transmission scheme (e.g., OFDM, CDMA, TDMA).

If in such a system, the jth antenna element at the Tx transmits a narrowband complex baseband signal, $u(t)$, the complex baseband signal received over the physical link between the jth Tx antenna element and the ith Rx antenna element can be written as

$$y_{ij}(t) = h_{ij}(\tau) * u(t) \tag{1.39}$$

where h_{ij} is the complex baseband representation of the impulse response of the ijth physical link. With the help of the development in [22,23], and neglecting antenna efficiencies, y_{ij}, can be written in terms of the Tx and Rx antenna element characteristics as

$$y_{ij} = \sum_{l=1}^{L} \left\{ \begin{array}{l} \alpha_l v_{tx_j}\left(\theta_{tx,l},\phi_{tx,l}\right) G_{tx_j}\left(\theta_{tx,l},\phi_{tx,l}\right) \\ v_{rx_i}\left(\theta_{rx,l},\phi_{rx,l}\right) G_{rx_i}\left(\theta_{rx,l},\phi_{rx,l}\right) u\left(t - \tau_l\right) \end{array} \right\} \tag{1.40}$$

where:
α_l is the complex amplitude of the lth MPC
$(\theta_{tx,l},\phi_{tx,l})$ and $(\theta_{rx,l},\phi_{rx,l})$ are its DOD and DOA, respectively
θ and ϕ represent azimuth and elevation angles, respectively

G_{tx} and G_{rx} represent the gains of the Tx and Rx antenna elements, respectively

v_{tx_j} represents the *j*th element of the Tx antenna array steering vector

v_{rx_i} represents the *i*th element of the Rx antenna array steering vector

τ_l is the propagation delay of the *l*th MPC

Then, from inspection of Equations 1.39 and 1.40, and noting the constant gain of H, it can be seen that H_{ij} can be written as

$$H_{ij} = \sum_{l=1}^{L} \alpha_l v_{tx_j} \left(\theta_{tx,l}, \phi_{tx,l} \right) G_{tx_j} \left(\theta_{tx,l}, \phi_{tx,l} \right) v_{rx_i} \left(\theta_{tx,l}, \phi_{tx,l} \right) G_{rx_j} \left(\theta_{tx,l}, \phi_{tx,l} \right) \quad (1.41)$$

It is thus clear that the complex voltages on the antenna elements of the Tx and Rx arrays are dependent upon the DODs and DOAs of the MPCs. It is also clear that DODs and DOAs have significant influence on the correlation matrix, $R_H = HH^*$, which, in turn, can be seen from Equation 1.37 to be critical in the determination of *C*. In fact, it can be shown [24–26] that the magnitude of R_H is inversely proportional to angular dispersion, or the spread in DOAs and DODs. Angular dispersion is therefore measured and modeled often during radio propagation and channel modeling research, since it is a significant factor in the estimation of *C*.

The modeling of directional characteristics on time-varying ZMGWSSUS channels can begin [27] with a straight-forward extension of the expression for such a channel's time-varying impulse response,

$$h(t;\tau) \rightarrow h\left(t;\tau; \theta_{tx}; \phi_{tx}; \theta_{rx}; \phi_{rx} \right) \quad (1.42)$$

However, here, as in [27] for clarity, only one of the angular dimensions, θ_{rx}, is considered, assuming uncorrelated scattering, and that DODs and DOAs are statistically independent, that is,

$$h(t;\tau) \rightarrow h\left(t;\tau; \theta_{rx} \right) \quad (1.43)$$

Just as for the relationship between $h(t;\tau)$ and $H(t;f)$ additional relationships can be derived by Fourier transforms with respect to t, τ, and θ_{rx}. While the FT relationships involving the first two are well known, it is interesting to consider a FT with respect to θ_{rx}. Following [27], this can be realized by consideration of antenna theory [1], where it is known that in the far field, the E field can be related to the aperture distribution by the equation:

$$E(x/\lambda) = \int_{-\infty}^{\infty} E\left[\sin(\theta) \right] e^{-j2\pi(x/\lambda)\sin(\theta)} d\left[\sin(\theta) \right] \quad (1.44)$$

and for a finite aperture of extent "a," Equation 1.44 can be written as

$$E(\theta) = \int_{-a/2\lambda}^{a/2\lambda} E\left(\frac{x}{\lambda}\right) e^{j2\pi(x/\lambda)\sin(\theta)} d\left(\frac{x}{\lambda}\right) \tag{1.45}$$

Thus, for a WSSUS channel, a FT of Equation 1.43 with respect to the angle θ_{rx} leads to function in a domain that can be reasonably referred to as the aperture domain, which is a finite segment from the spatial domain. Then, after consideration of the linearity of the FT, it can be understood that MPCs that arrive with different angles of arrival will superimpose by vector addition, leading to fluctuations in the spatial domain identical to multipath fading in the time domain or frequency selectivity in the frequency domain. In accordance, the relationships discussed earlier for nondirectional time-varying WSSUS channels can be reasonably extended in the case of directional channels. For a WSS channel, one can then recognize the existence of a correlation function given by

$$R_h\left(\Delta t; \tau, \tau'; \theta_{rx}, \theta'_{rx}\right) = E\left[h^*\left(t; \tau; \theta_{rx}\right) h\left(t + \Delta t; \tau'; \theta'_{rx}\right)\right] \tag{1.46}$$

When there is uncorrelated scattering, as before, this expression should go to zero if $\tau \neq \tau'$, $\theta_{rx} \neq \theta'_{rx}$. Then, for $\Delta t = 0$

$$R_h\left(0; \tau; \theta_{rx}\right) = E\left[\left|h\left(\tau; \theta_{rx}\right)\right|^2\right] \tag{1.47}$$

which can reasonably be referred to as a power-delay-DOA profile. Then, integrating with respect to τ gives

$$R_h\left(\theta_{rx}\right) = E\left[\int_{-\infty}^{\infty} \left|h\left(\theta_{rx}\right)\right|^2 d\tau\right] \tag{1.48}$$

If the discrete form, appropriate to samples from measured data is now considered, and, under ergodicity, the expectation is replaced by a time average, Equation 1.48 becomes

$$\hat{R}_h\left(\theta_{rx}\right) = \frac{1}{2NM} \sum_{i=1}^{N} \sum_{j=1}^{M} \left|h_i\left(\theta_{rxj}\right)\right|^2 \tag{1.49}$$

In a manner similar to that in which delay spread is given by Equation 1.26, a DOA spread can now be written as

$$\sigma_{\theta_{rx}} = \sqrt{\frac{\sum_{i=1}^{M} \hat{R}_h\left(\theta_{rxi}\right)\left(\theta_{rxi} - \overline{\theta}_{rxw}\right)^2}{\sum_{i=1}^{M} \hat{R}_h\left(\theta_{rxi}\right)}} \tag{1.50}$$

where:

$$\overline{\theta}_{rx_w} = \frac{\sum_{i=1}^{M} \hat{R}_h\left(\theta_{rx_i}\right) \times \theta_{rx_i}}{\sum_{i=1}^{M} \hat{R}_h\left(\theta_{rx_i}\right)} \tag{1.51}$$

Continuing under the assumption of uncorrelated scattering and the independence of DODs and DOAs, one could write similar equations for $\sigma_{\phi_{rx}}$, $\sigma_{\theta_{tx}}$, and $\sigma_{\phi_{tx}}$. Also note that, considering the FT relationships between the various system functions, these spreads can be considered inversely related to various spatial correlations. That is, the larger the angular spreads, the shorter the associated spatial correlations. Thus, just as delay and Doppler spreads are used in the assessment of the ability of channels to support high data rate digital communications, angular spreads are used to assess the characteristics of the all-critical MIMO matrix correlation function R_H. If the spreads are small, a qualitative assessment would be that capacity gains available on the associated MIMO channel are limited.

As in the case of $\hat{R}_h(\tau)$, after normalization by its integral with respect τ, a normalized experimentally determined estimate, $\hat{R}_h(\theta)$, can be considered as a power-weighted probability density function. The independence of $\hat{R}_h(\theta_{tx})$, $\hat{R}_h(\phi_{tx})$, $\hat{R}_h(\theta_{rx})$, and $\hat{R}_h(\phi_{rx})$ can therefore be assessed by compiling estimates (i.e., time-averaged histograms) of corresponding joint power-weighted probability density functions and comparing them to the products of marginals. If independence cannot be verified, joint angular power spectra must be estimated and modeled [28].

In addition to an analysis of angular power spectra, the degree to which wireless MIMO channels can support spatial multiplexing can also be assessed more directly by an examination of the similarity in the eigenvalues of R_H. This similarity, S_E [29], can be quantified through the ratio of the geometric mean of the eigenvalues to their arithmetic mean, given by

$$S_E = \frac{\left(\prod_{k=1}^{K} \lambda_k\right)^{1/K}}{\frac{1}{K}\sum_{k=1}^{K} \lambda_k} \tag{1.52}$$

where:
K is the number of eigenvalues
λ_k is the kth eigenvalue in the unordered set of eigenvalues

If the eigenvalues are similar, indicating that the eigenmodes over which data are effectively transmitted have similar gains, S_E is close to unity, and the potential for multiplexing gains is high. If the eigenvalues are spread over a wide range of values, indicating the eigenmodes are unbalanced, S_E diminishes, and the potential for multiplexing gains is low.

1.4 Data Qualification Processing prior to the Estimation of Parameters from Measured Data

In the design of digital wireless systems for operation with high-data rates, there is a need to not only know the transmission loss, but also the delay, frequency, and direction dispersion characteristics of the radio channels over which transmissions are planned. These characteristics are often estimated from measurements using test, or channel sounding, transmissions over radio channels typical of the type of channel associated with the wireless system of interest.

To *sound* a radio channel for the estimation of propagation loss, a single tone is often transmitted over a calibrated transmit/receive system, and propagation loss is taken as the difference between the transmitted and received power after other system gains and losses, as, for example, transmission line losses and antenna gains have been accounted for. This works well in free space. However, when MPCs can be received, the received signal is the vector sum of the multipath waves received from all directions, and this vector sum varies over space, time, and frequency, as the MPCs add with different phase relationships in accordance with the lengths of the paths over which they travel. If the Tx or/and the Rx antennas, or anything in the environment, moves, received signal power also varies in time as a result of time-varying vector addition of the component waves. It is generally accepted that when propagation loss is modeled, the effects of such multipath propagation are removed. Then, later in the system design process, account can be made for temporal, frequency, or spatial variations by adding a received power margin in accordance with known statistical distributions of such variations. The effects of the multipath-induced power variations can be mitigated and often almost eliminated by transmitting a very wideband signal so that different multipath components can be resolved in time, thereby avoiding vector addition. The remaining multipath effects can be eliminated by averaging over time, space, or frequency. Experimentally, a good approach is to average over the minimum range in any of these three domains required to effectively eliminate power variations. If the chosen method is to average in time, then a mechanism must be devised to *stir* the mixture of multipath waves so that an appropriate range of powers is received at each Rx antenna location such that averaging is effective. If spatial averaging is chosen, a mechanism must be devised to move at least one of the sounding system antennas over a small area (having dimensions of about a wavelength) repetitively, while the received signal is sampled. To enable averaging over frequency, a wideband sounding signal must be used.

A wideband sounding signal must also be used to resolve multipath components in delay and allow the estimation of DODs, DOAs, and delay dispersion. Ideally, the bandwidth should be infinite, allowing the impulse response of the channel of interest to be measured. This yields exact knowledge of dispersion on the channel as the time interval between when the direct signal or when this is absent, the MPC with the shortest delay, is received over the channel and when the last MPC is received. In practice, however, it is difficult to sound radio channels with impulses because of the peak power limitations of hardware, in addition to hardware bandwidth limitations and radio regulations. In reality, a rule of thumb for channel sounding is that the channel sounding bandwidth should be that bandwidth for which the width of the measurement system back-to-back impulse response is very much less than that of the channel to be sounded at –20 dB relative power. There are alternate definitions of this requirement, but they relate to the bandwidth intended for transmission in the system under design, the specification of which is often one of the objectives of channel sounding.

For the estimation of frequency dispersion, fast (Nyquist) sampling of the received signal resulting from single frequency transmission is required while the Tx and Rx antennas or objects in the environment move at their fastest rates.

There are a number of methods available [2] for conducting wideband channel sounding, including PN sounding, chirp sounding, and swept frequency sounding. The latter method can most easily be implemented [30] for operation over multiple frequency bands using a commercially available vector network analyser (VNA). When a VNA is employed, not only the frequency band to be sounded is selectable, but also the sounding bandwidth up to the maximum allowed by the VNA. Based upon the consideration of multipath delay resolution, which is the reciprocal of the channel sounding bandwidth, antenna bandwidths, and interference avoidance, a transmission bandwidth of 500 MHz is often chosen. A drawback with using the VNA approach is that time series measurements can only be made at a single frequency, and over only short time durations. Thus, frequency dispersion cannot be studied. A second drawback is that transmission range (i.e., channel sounder Tx or Rx separation) is historically limited to the combined lengths of transmission lines that can be run from the VNA to the Tx antenna, and from the Rx antenna to the VNA. The authors of [31] report a method by which the limitation can be overcome using two VNAs, one at the Tx and one at the Rx, but the cited paper reports only narrowband measurements using this system.

The primary objective of using wideband sounding systems is to estimate either the sounded channel impulse response, $h(\tau)$, or its frequency-domain transfer function, $H(f)$. Once one of these functions is estimated from measurements, an estimate of the other can be obtained by appropriate windowing and direct or inverse Fourier transformation. To estimate the time-varying versions, $h(t;\tau)$ and $H(t;f)$, multiple sequential sounding measurements are made. In order to capture Doppler effects so as to enable the estimation of correlation times and Doppler spreads, the frequency of repetition must be at least twice the maximum expected Doppler shift, in accordance with the Nyquist theorem, and is often limited to this value as a result of system constraints such as data storage capacity, or data transfer rates. This is acceptable in most instances, even without post-measurement filtering for interpolation, since often time variations do not have to be reproduced, and it is only the statistics of such variations need to be estimated. Likewise, in order to capture spatial variations, multiple channel soundings must be made while the location of one or both the channel sounder antennas are moved in space, in increments smaller than half the wavelength of the shortest cycles in the anticipated spatial variations.

Regardless of which method is used for channel sounding, and which variations on the channel it is desired to analyze, measured data and estimates made from them are subject to corruption by noise. In addition, if the classical modeling and channel classification approach discussed in Section 1.3 is to be applied, the stochastic characteristics of the time, frequency, and/or spatial series of measurements must conform at least with those of a WSS model. Often this is true only for finite segments in each domain, as either objects in the measurement environment move, or the channel sounder antennas move, or both. Throughout the past several decades, there have been repeated attempts to specify tests that can be conducted in a data qualification stage [32] before final analysis and channel modeling to ensure WSS characteristics prevail [33,34]. At the time of writing, however, the author would label available techniques as contributions in the gradual development of standard methods, with the development of such tests still an area that is open for considerable new research. However, later paragraphs in this section describe some of the techniques that are known to, and have been applied by, the author in data qualification procedures. The following paragraphs begin with the review of a probabilistic method from the literature, for the rejection of noise in impulse response estimates. Methods for assessing whether WSS conditions prevail are discussed next. This is followed by a description of methods for the estimation of frequency correlation functions and envelope fading distributions, the results from which are used in data qualification, as well as for channel modeling. Finally, a note is added with regard to assessing whether random channel variations can be modeled as Gaussian, and whether an assumption of ergodicity is reasonable.

1.4.1 Rejection of Noise from Measured Impulse Response Estimates

Of importance is a method for determining when a specific sample within a channel impulse response estimate (IRE) represents a valid multipath echo and when it is noise. Sometimes researchers simply zero all samples in the envelopes of IREs if their associated powers are below a fixed rejection threshold, as for example, –20 dB relative to the peak power per delay resolution interval in the IRE. An alternate method [35] is to consider that the last n (i.e., 50 or so) samples in IREs are the result of Gaussian noise, rather than multipath echoes, as the associated measurement system should have been designed to enable measurement of echoes with delays greater than the maximum expected. A cumulative distribution function (CDF) for their envelope values therefore has a Rayleigh cumulative distribution function (CDF), and its complimentary CDF is given by

$$P\left(|v| > x\right) = e^{\frac{-x^2}{2\sigma^2}} \qquad (1.53)$$

where σ is the standard deviation of the envelope of the *n* IRE sample values with the greatest delay. This is the probability that the envelope voltage, *x,* is exceeded if it is noise, and in this application, unless noise is rejected, it becomes the probability of false detection (PFD), or the probability that a noise sample is taken as being a signal sample from an MPC. Taking logarithms, at the median $|v|_{50}$, Equation 1.53 can be rearranged to give

$$\sigma = \sqrt{\frac{-|v|_{50}^2}{2\ln(0.5)}} = .8493|v|_{50} \qquad (1.54)$$

Thus, if the median of the last *n* samples in an IRE is estimated, σ can also be estimated. Then, for any desired PFD, the envelope voltage, $|v|_{th}$, that must be exceeded if an IRE sample is to be considered as energy from an MPC can easily be calculated

$$|v|_{th} = \sqrt{2\sigma^2 |\ln(\text{PFD})|} \qquad (1.55)$$

If it is known how many samples are in each IRE, one can easily set the PFD as, for example, the probability that a single sample within, say 2 IREs, will be incorrectly taken to be a sample from an MPC when it is actually noise. Once a value for PFD has been decided upon, a noise rejection threshold can be set using Equation 1.55 to find $|v|_{th}$, such that if the samples in any part of the envelope of an IRE have voltages below this threshold, they are set equal to zero. Figure 1.10 shows an example of an IRE from measurements made with a VNA, before and after probabilistic noise thresholding. In this case, the PFD was set to 1.3×10^{-3}. It can be noted that rather than replacing samples where the envelope of the IRE was below the probabilistically estimated rejection threshold with zeroes, they were replaced with a value equal to 10^{-10} so a logarithm could be taken for plotting purposes.

Sometimes, even after thresholding, there are sample groups that appear after a number of leading samples have been zeroed. It is reported in [36] that if measurement system sampling windows are made long enough so that several impulse response functions can be estimated from data sequentially recorded in each sample window, a further reduction of noise can be made by rejecting those samples that only occur in one of the several IREs estimated during that data window. Based upon similar reasoning, single samples that are high enough above the noise rejection threshold so that two or more samples of the bandlimited IRE envelope should have been recorded* if a valid MPC had been received can be rejected. This is because even an IRE estimated from a single-propagation path measurement system back-to-back connection is usually represented by more than a single sample.

---

* Recall that even isolated MPCs have a time duration of more than one sample in IREs because of the finite bandwidth of any measurement system.

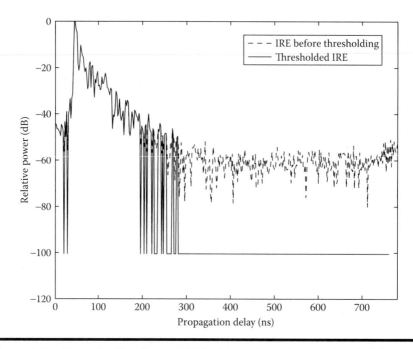

Figure 1.10   Example showing an IRE before noise rejection thresholding, and the same IRE after thresholding.

Once noise is rejected from IREs, all other parameters and functions of interest can be derived from such noise-thresholded IREs, or their Fourier transforms, with better estimation accuracy than that which would prevail without such thresholding.

1.4.2 Assessment of Conformance to a Wide-Sense-Stationary Model

As is well-known, the characteristics of a WSS process include a constant mean, and an autocorrelation function that is independent of the reference time (or location in space or reference frequency, when considering random variations over space or frequency). While it appears initially that verification of these characteristics through the examination of sample mean and autocorrelation function estimates should be straightforward, there are complications. First, the CPs $h(t; \tau)$ and $H(t; f)$ that are to be analyzed are wideband processes. The question, therefore, arises as to whether the widest available bandwidth should be assessed, or whether analyses of narrowband samples of the measured CP are adequate. The second problem is that time or spatial processes within measured data are in series that have finite lengths. Analyses of such series generate sidelobes in autocorrelation estimates. In addition, experience shows that the longer a measured series is, the more probable it is that WSS characteristics degenerate. After long intervals, the averages that are used to characterize WSS processes are no longer constant, and conclusions that are drawn from using them as a basis for the characterization, modeling, or simulation of wireless channels can be misleading. Thus, the series of samples that must be analyzed is often even much shorter than the total length of a measured series, making the sidelobes and any estimation errors even more severe. In fact, reported methods for the assessment of whether either the mean of a sample from a stochastic process is constant, or its autocorrelation is invariant to the index parameter of the process (i.e., time, location, or frequency) involve segmenting the sample series into shorter intervals, hereinafter referred to as subseries, and comparing results from the analysis of sequential subseries.

One method from the literature [32] is an application of the *RUNs* test, which is really for the assessment of the randomness of a data set. For testing whether the mean of a process is constant, it is hypothesized as part of the Runs test that subseries means should be random because of estimation errors. Positive or negative trends, indicative of nonstationary characteristics, can also be detected. The author has used this test to assess the WSS characteristics of a series of frequency domain TFEs, $\hat{H}(t; f)$, as follows. Regardless of the system used for the measurements, TFEs in such a series are made up of a number of frequency-domain samples, herein referred to as *spectral line* samples. There are a number, M, of such samples stretching across the measurement bandwidth, in steps of equal separation, and the spectral lines from which they are sampled are comprised of a number, say *N*, of samples that stretch along the index dimension of the CP, in this case, time. Initial focus can be on the analysis of one such spectral line, representing a narrowband (or CW) signal. If the data are complex, the means of the raw data may be zero due to external factors, like receiver motion [8], which would be mitigated by phase-synchronization in a communication receiver. For the purpose of explanation, therefore, it will be assumed that the envelope of the chosen spectral line is under analysis. To conduct the Runs test, the mean, median, or some other convenient value of the entire series is estimated and used as a reference value. Then, the series is broken up into subseries of equal lengths, and the mean of each of the subseries is estimated. If the sample mean for a particular subseries is greater than the reference value, it is assigned a "+," and if it is smaller than the reference value it is assigned a "−." A run is then defined as a sequence of such results that either contain all positive or all negative assignments. The number of runs, N_R, is then counted. The null hypothesis H_0 for the test[*] is then designated to be the case when the sequence is random, with the alternate hypothesis, H_a, being that the sequence is not random. The test statistic is then computed [37] as

$$Z = \frac{N_R - \overline{N}_R}{\sigma_{N_R}} \tag{1.56}$$

---

[*] Hypothesis testing is summarized in Section 1.4.3.

where $\overline{N}_R$ is the expected number of runs and σ_{N_R} is its expected standard deviation under H_0, with values given by

$$\overline{N}_R = \frac{2n_1 n_2}{n_1 + n_2} + 1 \tag{1.57}$$

and

$$\sigma_{N_R}^2 = \frac{2n_1 n_2 \left(2n_1 n_2 - n_1 - n_2\right)}{\left(n_1 + n_2\right)^2 \left(n_1 + n_2 - 1\right)} \tag{1.58}$$

where n_1 and n_2 represent the number of positive and negative assignments, respectively. The null hypothesis is rejected if

$$|Z| > Z_{\left(1 - \frac{\alpha}{2}\right)} \tag{1.59}$$

where:

α is the significance of the test

Z is taken as standard normal if n_1 and n_2 are large (say > 10)

For smaller values of n_1 and n_2, the (critical) value that Z must be greater than can be found in tables [32,38]. If the test shows that the subseries means are not random, the conclusion should be that the measured series should not be modeled as being WSS. If the overall hypothesis of conformance to a WSS model is rejected, the usual procedure is to shorten the series under analysis and repeat the test until H_0 cannot be rejected. The result from a long series of measurements may be a series of shorter *consistency* intervals of data that can be analyzed under WSS assumptions, and the statistics of results can, in turn, be estimated. For wideband data, this test could be applied for more than one spectral line, and conclusions drawn from the overall results. Rather than applying the runs test to assess randomness, one could also apply the Student's t test [39] to determine the significance of differences in the subseries means.

A search of the literature has failed to identify a hypothesis test that can be applied to determine if autocorrelation function estimates from consecutive subseries in a longer measured data series are really estimates of the same autocorrelation, which would indicate invariance of the autocorrelation with the index parameter of the process. However, on the assumption that sidelobes from autocorrelation function estimates are random, averaging could be applied for their reduction. In this case, autocorrelation estimates centered on different references could be computed and averaged over short windows, with the results from consecutive windows compared in order to draw conclusions. A related procedure would involve the estimation of structure functions [40] and their comparison with autocorrelation estimates. In [40], an important extension to processes that are classified as statistically stationary is reported to have been proposed by Kolmogorov, who noted that, although velocity fields in turbulence are not homogeneous, the difference in velocities at two different locations is homogeneous over wide spatial ranges. That is, if a random function of this type is denoted $f(\mathbf{r})$, the function $|f(\mathbf{r}+\Delta\mathbf{r})-f(\mathbf{r})|$ is homogeneous, even though $f(\mathbf{r})$ is not. Such a process, referred to in [40] as a random process with stationary increments if its random variations are in time, and a locally homogeneous random function if its variations are in space, is character-ized by the fact that $E\left[|f(t+\Delta t)-f(t)|\right]$ is a function only of Δt and the fact that the structure function

$$D_f(\Delta t) = E\left[|f(t+\Delta t)-f(t)|^2\right] \qquad (1.60)$$

is also a function only of Δt. For a stationary process, there is a one-to-one cor-respondence between the autocorrelation function $R_f(\Delta t)$ of the process and its structure function, in accordance with the relationships

$$D_f(\Delta t) = 2R_f(0) - R_f(\Delta t) - R_f^*(\Delta t) \qquad (1.61)$$

$$D_f(\infty) = 2R_f(0) \qquad (1.62)$$

and

$$Re\left[R_f(\Delta)\right] = \frac{1}{2}\left[D_f(\infty) - D_f(\Delta t)\right] \qquad (1.63)$$

Further, it is shown in [40] that the power spectral density, Φ, can be written in terms of $D_f(\Delta t)$ as

$$\Phi(\lambda) = \frac{1}{8\pi\lambda}\int_0^\infty \sin\lambda t\, \frac{d}{d\Delta t}D_f(\Delta t)d\Delta t \qquad (1.64)$$

whereas for a stationary random process

$$R_f(\Delta t) = \int_{-\infty}^\infty \Phi(\lambda)e^{j\lambda\Delta t}d\lambda \qquad (1.65)$$

One can therefore envisage a process to test a WSS assumption, in which the rela-tionships Equations 1.61 through 1.65 are studied using estimates of $D_f(\Delta t)$ and $R_f(\Delta t)$ from measured data.

Approaches that have also been used by the author are more physical based. In these approaches, it is conjectured that the multipath process that causes variations of the channel transfer function needs to be consistent if the CP is to be modeled as WSS process. Here, it is meant by consistent, either that variations in some attribute of the CP are from the same underlying random distribution, or that so-called clusters in its spatial characteristics remain constant (i.e., the same physical clusters of scattering or reflecting objects determine the CP over a consistency interval). To test consistency in terms of randomness, the Two-Sample KS test can be used. This test is conducted in a similar manner to the One-Sample KS test, described in Section 1.4.3, except that the test statistic, D_{max}, is set equal to the maximum difference between two ECDFs, rather than an ECDF and a model. This Two-Sample test is conducted the same way in which the One-Sample KS test is conducted, and the complimentary cumulative probability distribution for D_{max} is given approximately by Equation 1.72, except that the number N_e, that should be used in this equation is given by

$$N_e = \frac{N_1 N_2}{N_1 + N_2} \tag{1.66}$$

where:
   N_1 is the number of independent samples in the first ECDF
   N_2 is the number of independent samples in the second ECDF

A hypothesis test for independence of a data series is described in Section 1.4.3. The Two-Sample KS test is used to determine if two ECDFs represent the same underlying random process. Thus, if ECDFs can be estimated for radio channel parameters and compared for sequentially recorded data using KS tests, it should be possible to determine when changes in a CP take place. However, the requirement for multiple data snapshots whenever high-resolution parameter estimates are made limits time and distance resolutions by smearing change points. In addition, critical values (thresholds for rejection of H_0) for the required KS tests are obscure. This is because such values depend on the number of independent samples being analyzed, but it is not clear how to estimate this number when multiple MPC parameters (e.g., DOAs, DODs delays, and powers) are being estimated each time the CP is sampled. The first of these problems can be overcome by making sliding, or overlapping estimates of the channel parameters. The second can be mitigated using preprocessing to estimate critical values by recognizing and accounting for extraneous parameters that, in addition to true changes in the CP, can lead to distances between compared ECDFs. These factors include finite sample sizes and errors in the processes used to estimate channel parameters. To account for such factors, the first step involves the estimate of multiple ECDFs for channel parameters from multiple data sets recorded under physical conditions wherein it is reasonably certain that there are no true changes in the CP. Critical values are then set equal to the maxima of ECDF distances for such sample sets. Consistency intervals can then be estimated by estimating DOA, DOD, and excess delay ECDFs from consecutive data snapshots. Then, ECDFs associated with reference snapshots are compared with ECDFs associated with previous and following snapshots using the

Two-Sample KS test until H_0 is rejected, thereby defining the limits of a consistency interval. Robustness can be ensured by requiring that consistency is satisfied for all estimated parameters simultaneously. An example of the use of this method for estimating consistency intervals by detecting changes in envelope fading ECDFs can be found in [41]. An example of estimating CIs by monitoring changes in directional data after estimating cluster properties can be found in [23].

For the assessment of whether WSS-like characteristics prevail on MIMO channels, a metric referred to as correlation matrix distance has been applied [42]. This metric was conceived by considering the $n \times 1$ signal vector, $x(t)$ of complex voltages on a MIMO receive antenna array to be a zero-mean stochastic vector process that is fully characterized by the time-dependent correlation matrix

$$R(t) = E\left[x(t)x^H(t)\right] \tag{1.67}$$

and observing that the inner product between this matrix at time t, and time $t = t + \Delta t$ should obey the relationships

$$\langle R(t)R(t+\Delta t)\rangle = tr\left[R(t)R(t+\Delta t)\right] \tag{1.68}$$

and

$$\langle R(t)R(t+\Delta t)\rangle = \|R(\Delta t)\|_2 \|R(t+\Delta t)\|_2 \tag{1.69}$$

where:
$tr(\cdot)$ is the trace operator
$\|\cdot\|_2$ represents a Frobenius norm

The *correlation matrix distance* was then defined as

$$d_{cM}(t, t+\Delta t) = 1 - \frac{tr\left[R(t)R(t+\Delta t)\right]}{\|R(t)\|_2 \|R(t+\Delta t)\|_2} \in [0,1] \tag{1.70}$$

which has a value of zero for equal correlation matrices and unity if they maximally differ. In [42] it is observed that d_{cM} can be calculated for spatial autocorrelations, R at either a Tx antenna array, an Rx antenna array or for a full MIMO channel correlation matrix, given by

$$R_H = E\left[\text{vec}(H)\text{vec}(H)^H\right] \tag{1.71}$$

To apply d_{cM} to find consistency intervals, a reference time, t, or a reference location, would be established, and the time shift, Δt, or a spatial shift, Δx would be incremented until d_{cM} reaches some predefined value near unity, which would define the outer boundary of the consistency interval.

1.4.3 Modeling Estimates for Probability Distribution Functions

To determine if the probability distribution of the underlying random population from which a series of values is sampled can be modeled using a specific pdf or CDF, hypothesis testing is necessary.* In this process, what are referred to as *null* and *alternate* hypotheses, H_0 and H_a, are established. A test statistic, ε, that can be computed in some way by comparison of some characteristic (e.g., mean, epdf, ECDF) of the collection of sampled values to the same characteristic of a model is specified. A distribution for the test statistic, given the null hypothesis is true, is then found either analytically or through simulations. The distributions for the test statistics that are used in many well-known hypothesis tests can be found in the literature. A value ε' for the test statistic is then computed, and the probability $P\left[\varepsilon > \varepsilon' \mid H_0\right]$ that the test statistic is greater than its value as estimated from the data, given H_0 is true, is computed from the distribution for ε. If this probability, referred to as the "*p*" value for the test, is lower than a value, α, referred to as the significance level of the test, H_0 is rejected and the alternate, H_a is concluded to be true. The significance level, α, can be considered to be the probability, given H_0 is true, that H_0 is incorrectly rejected. A commonly-used value for α is 5%. To test whether an analytical pdf is a good model for the pdf of the underlying random population from which a collection of experimental values is sampled, the Chi-square test can be used. To test the appropriateness of a model CDF, the One Sample Kolmogorov-Smirnov test can be used. The accuracy and relative merits of the two tests under different circumstances is discussed in [39,43]. Whereas it is easy to find a good description of the Chi-square test in the literature [39], descriptions of the one-sample KS test are sometimes incomplete, such that it is considered of value to provide a description here.

The One-Sample KS test uses as its test statistic the maximum, D_{max}, of the difference $\left|F_N(x) - F(x)\right|$ between an ECDF, $F_N(x)$ and a model CDF, $F(x)$, which has been found [43,44] to follow a Kolmogorov-Smirnov distribution. The test can either involve rejecting H_0 if D_{max} is greater than a critical value, v, for which $\text{Prob}\left[D_{max} > v\right] = \alpha$, or the *p* value, $\text{Prob}\left[D > D_{max}\right]$ can be found, and H_0 rejected if this value is smaller than α. Although there is no analytic solution for *p*, it is given approximately by the equation [39]

$$p = \text{Prob}\left[D > D_{max}\right] = Q_{KS}\left[\left(\sqrt{N'} + 0.12 + 0.11/\sqrt{N'}\right)D_{max}\right] \quad (1.72)$$

---

* Simple least mean squared fits to pdfs and CDFs estimated from sampled values can lead to significant errors in modeling, as such estimates can vary from sample set-to-sample set. Instead, the probabilistic consideration of functions that truly represent characteristics of the underlying random population, rather than just a sample from that population, are necessary.

where:

$$Q_{KS}(\beta) = 2\sum_{j=1}^{\infty}(-1)^{j-1}e^{-2j^2\beta^2} \tag{1.73}$$

and N' is the number of *independent* values in the sampled data set. Equation 1.72 is strictly applicable only if the parameters of the model distribution, $F(x)$ are completely known, independent of the experimental data. In many instances, however, such parameters are not known and must be estimated from the sampled data[*]. One might imagine then, if the parameters of $F(x)$ are chosen specifically to fit the sampled data, D_{max} has significant probability of being smaller than it might otherwise be, so that p is larger, [43]. The effect is that a decision based on such values is conservative in that the significance level, or the probability of incorrectly rejecting H_0, is lower than for the case when $F(x)$ is completely specified.

The estimation of N' requires the estimation of the time-series lag, δ, at which the sampled data are independent, or so-called serially independent. The value N' is then equal to (N/δ). Alternately, the data can be thinned by repeatedly removing every second value until consecutive values can be declared, as for example, through hypothesis testing, as being from independent underlying populations, after which N' is taken to be the number of remaining data samples. In general, the determination of serial independence seems, from a study of the literature, to be difficult and a number of fairly complicated papers on the subject exist. However, for the purposes of the current application, three approaches can be taken. The first of these is to plot the sampled data in pairs, separated by n-sample lags, in a scatter plot, with the first sample from each pair on the ordinate, say, and the second sample on the abscissa. The approximate lag, δ, between pairs of samples that are serially independent can then be taken as the lag at which there is a significant scatter in the resulting plot. Although an absence of dependence is indicated by an absence of correlation only for Gaussian processes strictly, another approximate method for estimating N' would be to divide N by the sample lag at which an estimate of the autocorrelation function for the data set diminishes to some small value, say 1/e. Finally, the problem of determining serial independence in the residuals of regression models and autoregressive integrated moving average models has been studied rigorously by several researchers [37,38], and resulting hypothesis tests appear to be applicable to the problem under discussion. In particular, a hypothesis test published by Ljung and Box [38] can be applied using the following steps:

---

[*] Papers from literature on statistics recommend that parameters for $F(x)$ should be estimated using maximum likelihood methods. However, in practice, the method of moments is often used, wherein, for example, means are estimated by using sample means, and standard deviations are estimated using sample standard deviations.

First, H_0 is specified to be the hypothesis that the sampled data are independently distributed, whereas H_a is taken to be the hypothesis that the data are not independently distributed. The test statistics is then estimated as [38]

$$Q = n(n+2)\sum_{k=1}^{h}\frac{\hat{\rho}_k^2}{n-k} \qquad (1.74)$$

where:

n is the total number of data samples in the current test for serial independence
$\hat{\rho}_k$ is the sample autocorrelation at lag k
h is the number of lags within the n samples at which is being tested

Under H_0 the statistic Q follows a $\chi^2_{(h)}$ distribution. For significance level α, the critical region for rejection of the hypothesis of randomness is $Q > \chi^2_{1-\alpha,h}$, which is the $1-\alpha$ quantile of the chi-squared distribution with h degrees of freedom.

1.4.4 Estimation of Frequency Correlation Functions and the Confirmation of Uncorrelated Scattering

The most often used method by which FCFs are derived from measured impulse response estimates is through the FT relationship in Equation 1.24. However, recall that the validity of Equation 1.24 is based upon the modeling of a time-varying wireless channel as a ZMGWSSUS stochastic process. The conditions required for such modeling are not always met. An alternate [8] approach that avoids a requirement for most of the ZMGWSSUS conditions is to directly cross correlate spectral lines in a series of TFEs. Let a time series of TFEs derived by FT from a time series of noise-thresholded IREs be denoted as $\dot{H}(t_i; f_j)$, where "i" orders the time-domain samples from 1 to N, and "j" orders the spectral lines across the bandwidth of the measurement system from 1 to M. As a first step, a spectral line at a frequency (f_{ref}) near midband* can be chosen as a reference, and the time variations at each spectral line can be cross correlated against this reference to give

$$\hat{R}_H(f_{ref},f_j) = \frac{\hat{\gamma}_H(f_{ref},f_j)}{\sqrt{\hat{\gamma}_H(f_{ref},f_{ref})\hat{\gamma}_H(f_j,f_j)}} \qquad (1.75)$$

where:

$$\hat{\gamma}_H(f_j,f_k) = \frac{1}{N}\sum_{i=1}^{N}\dot{H}_i(f_j)\dot{H}_i^*(f_k) \qquad (1.76)$$

---

* The midband component itself is not a good choice, since it contains any dc offsets in the measurement system.

The symmetry of $\hat{R}_H(f_{\text{ref}}, f)$ can then be assessed. If it is symmetrical about f_{ref}, the condition of uncorrelated scattering can be considered to hold. If in addition, an analysis of envelope fading as outlined in Section 1.4.3 shows that the Rayleigh distribution is a good model, $\hat{R}_H(\Delta f)$, from Equation 1.24 can be applied to yield approximately the same result as $\hat{R}_H(f_{\text{ref}}, f)$, with any differences being that the FT in Equation 1.64 automatically applies averaging for different values of f_{ref}. If an analysis of envelope fading shows that the Rayleigh pdf is not a good model, and, in particular, if it shows that a Rician distribution is a good model, the time series result $\hat{R}_H(f_{\text{ref}}, f)$ should be accepted as a more accurate characterization of frequency correlation than $\hat{R}_H(\Delta f)$. This is because, in this case, it is likely that $H(t; f)$, after phase synchronization, would have a nonzero mean along the "t" dimension, at any frequency, f, hence violating the zero-mean assumption in the channel modeling. This consideration is important, for example, when estimating B_c as the bandwidth over which correlation drops to some prespecified value. If $\hat{R}_H(f_{\text{ref}}, f)$ is not symmetrical, then frequency correlation likely changes for different reference spectral lines. In this case, $\hat{R}_H(f_{\text{ref}}, f)$ should be estimated for a reasonable large number of reference spectral lines in the midband region of $\hat{H}(f)$, where the channel sounder signal spectrum is flat.[*] Then, the ratio of its volume, $V_{\hat{R}_H(f_{\text{ref}}, f)}$, to the square of the measurement bandwidth, B_m, can be used to assess B_c, as outlined at the end of Section 1.3.2.

1.4.5 Considerations Related to Assumptions of Gaussianity and Ergodicity

It should be noted [8] that neither the analysis of measured data, (e.g., single spectral lines in a time, frequency, or spatial series of values for $\hat{H}(f)$), to determine if complex quadrature components can be modeled as Gaussian nor a determination of whether narrowband envelope variations can be modeled as Rayleigh or Rician, can be used to determine if an assumption of Gaussianity is justified. This is because such modeling inherently assumes ergodicity, which may not be a valid assumption. A better approach, particularly with the availability of high-resolution processing results, is to asses Gaussianity through the central limit theorem. Then, if the measured data also conform to a WSS model, ergodicity can be considered to be a reasonable assumption over WSS intervals. As verification, an analysis of fading distributions can then be conducted. The author knows of no other way, except by intuitive reasoning, that an assumption of ergodicity can be justified.

---

[*] Contoured sounding spectra, such as that produced by pseudo-noise channel sounders, can be accounted for as in [8].

Acknowledgements

The author would like to gratefully acknowledge the help of Dr. Athanasios Kanatas and Dr. Ghassan Dahman for their work in reviewing the original manuscript for this chapter and recommending many modifications and improvements.

References

1. Kraus, J.D., *Antennas*, McGraw-Hill: New York, 1988.
2. Salous, S., *Radio Propagation Measurement and Channel Modelling*, John Wiley & Sons: Hoboken, NJ, 2013.
3. Sun, S. et al., Investigation of the prediction accuracy, sensitivity, and parameter stability of large-scale propagation path loss models for 5G wireless communications, *IEEE Transactions on Vehicular Technology*, 65, 5, 2843–2860, 2016.
4. Bryant, G.H., Bultitude, R.J.C., and Neve, M.J., A spatial field model for mobile radio, *Wave Propagation and Remote Sensing: Proceedings of the 8th URSI Commission F Triennial Open Symposium*, University of Aveiro, Aveiro, Portugal, September, 1998, pp. 220–223.
5. Bultitude, R.J.C., Schenk, T.C.W., Op den Kamp, N.A.A., and Adnani, N., A propagation-measurement-based evaluation of channel characteristics and models pertinent to the expansion of mobile radio systems beyond 2 GHz, *IEEE Transactions on Vehicular Technology*, 56, 2, 382–388, 2007.
6. Bello, P.A., Characterization of time-variant linear channels, *IEEE Transactions on Communications Systems*, 11, 4, 360–393, 1963.
7. Bello, P.A. and Nelin, B.D., The effect of frequency selective fading on the binary error probabilities of incoherent ant differentially coherent matched filter receivers, *IEEE Transactions on Communications Systems*, CS-11, 170–186, 1963.
8. Bultitude, R.J.C., Estimating frequency correlation functions from propagation measurements on fading radio channels: A critical review, *IEEE Journal on Selected Areas in Communications*, 20, 6, 1133–1143, 2002.
9. Bultitude, R.J.C. and Leslie, A.W., Propagation measurement-based probability of error predictions for digital lands mobile radio, *IEEE Transactions on Vehicular Technology*, 46, 3, 717–729, 1997.
10. Proakis, J.D., *Digital Communications*, Chapter 7, McGraw-Hill: New York, 1983.
11. Bultitude, R.J.C., A study of coherence bandwidth measurements for frequency-selective radio channels, *Proceedings of the 33rd IEEE Vehicular Technology Conference*, NJ, May, 1983, pp. 269–278.
12. Fleury, B.H., New bounds for the variation of mean-square-continuous wide-sense-stationary processes, *IEEE Transactions on Information Theory*, 41, 849–852, 1995.
13. Bello, P.A. and Nelin, B.D., Optimization of subchannel data rate in FDM-SSB transmission over selectively fading media, *IEEE Transactions on Communications Systems*, CS-12, 1, 46–53, 1964.
14. Bultitude, R.J.C., Hahn, R.F., and Davies, R.J., Propagation considerations for the design of indoor broadband communications systems at EHF, *IEEE Transactions on Vehicular Technology*, 47, 1, 235–245, 1998.

15. Lord R., On the resultant of a large number of vibrations of the same pitch and arbitrary phase, *Philosophical Magazine and Journal of Science*, 5th Series, 10, 60, 73–78, 1880.

16. Lin, S.H., Statistical behaviour of a fading signal, *Bell System Technical Journal*, 50, 10, 3211–3270, 1971.

17. Rice, S.O., Mathematical analysis of random noise, *Bell System Technical Journal*, 23, 3, 282–332, 1944.

18. Rice, S.O., Statistical properties of a sine wave plus random noise, *Bell System Technical Journal*, 27, 1, 109–157, 1948.

19. Greenstein, L.J., Michelson, D.G., and Erceg, V., Moment-method estimation of the Ricean K factor, *IEEE Communications Letters*, 3, 6, 175–176, 1999.

20. Baddour, K.E. and Willink, T.J., Improved estimation of the Ricean K-factor from I/Q fading channel samples, *IEEE Transactions on Wireless Communications*, 7, 12, 5051–5057, 2009.

21. Foscini, G.J. and Gans, M.J., On the limits of wireless communications in a fading environment when using multiple antennas, *Wireless Personal Communications*, 6, 3, 311–335, 1998.

22. Suvikunnas, P. et al., Evaluation of the performance of multiantenna terminals using a new approach, *IEEE Transactions on Instrumentation and Measurements*, 55, 5, 1804–1813, 2006.

23. Dahman, G.S., Multi-Antenna mobile radio channels: Modelling and system performance predictions, PhD Thesis Dissertation, Carleton University Department of Systems and Computer Engineering, 2010.

24. Zhao, G. and Loyka, S., On multipath clustering in angular domain and its impact on channel capacity and diversity gain, *Proceedings of the European Conference on Antennas and Propagation*, Nice, France, November 6–10, 2006.

25. Vaughan, R., Spaced directive antennas for mobile communication by the Fourier transform method, *IEEE Transactions on Antennas and Propagation*, 48, 7, 1025–1032, 2000.

26. Fleury, B., First- and second-order characterization of direction dispersion and space selectivity in the radio channel, *IEEE Transactions on Information Theory*, 46, 6, 2027–2744, 2000.

27. Kattenbach, R., Statistical modeling of small-scale fading in directional radio channels, *IEEE Journal on Selected Areas in Communications*, 20, 3, 584–592, 2002.

28. Betlehem, T., Lamahewa, T.A., and Abhayapala, T.D., Dependence of MIMO system performance on the joint properties of angular power, *IEEE 2006 International Symposium on Information Theory (ISIT 2006)*, Seattle, Washington, DC, July 9–14, 2006.

29. Suvikunnas, P. et al., Comparison of MIMO antenna configurations: Methods and experimental results, *IEEE Transactions on Vehicular Technology*, 57, 2, 1021–1031, 2008.

30. Tholl, D., Fattouche, M., Bultitude, R.J.C., Melancon, P., and Zaghloul, H., A niques, *IEEE Transactions on antennas and propagation*, 41, 4, 515–517, 1993.

31. Molina-Garcia-Pardo, J-M., Rodriguez, J-V., and Juan-Llacer, L., MIMO channel sounder based on two network analyzers, *IEEE Transactions on Instrumentation and Measurements*, 57, 9, 2052–2058, 2008.

32. Bendat, J.S. and Piersol, J.D., *Measurement and Analysis of Random Data*, John Wiley & Sons: New York, 1966.

33. Willink, T.J., Wide-sense stationarity of mobile radio channels, *IEEE Transactions on Vehicular Technology*, 57, 2, 204–214, 2008.
34. Clarke, R.H., A statistical theory of mobile radio reception, *Bell System Technical Journal*, 47, 6, 957–1000, 1968.
35. Rappaport, T.S. and Seidel, S., 900 MHz multipath propagation measurements for U.S. digital cellular radiotelephone, CTIA Advanced Digital Cellular Radiotelephone Panel Report, January 19, 1989.
36. Sousa, E.S., Yovanovic, V.M., and Daigneault, C., Delay spread measurements for the mobile radio channel in Toronto, *IEEE Transactions on Vehicular Technology*, 43, 837–847, 1994.
37. Durbin, J. and Watson, G.S., Testing for serial correlation in least squares regression. III, *Biometrika*, 58, 1, 1–19, 1971.
38. Ljung, G.M. and Box, G.E.P., On a measure of lack of fit in time series models, *Biometrika*, 65, 2, 297–303, 1978.
39. Press, W.H., Teukolsky, S.A., Vetterling, W.T., and Flannery, B.P., *Numerical Recipes in C*, Cambridge University Press: New York, 1992.
40. Ishimaru, A., *Wave Propagation and Scattering in Random Media*, Vol. 2, Multiple Scattering, Turbulence, Rough Surfaces, and Remote Sensing, Academic Press: New York, 1978, Appendix B: Structure Functions, pp. 518–523.
41. Levin, G. and Bultitude, R.J.C., Modelling, measurement, and analysis of fast fading on narrowband relay channels, *Proceedings on Personal, Indoor and Mobile Radio Communications 2011*, Toronto, Canada, September 11–14, 2011.
42. Herdin, M. and Bonek, E., A MIMO correlation matrix-based metric for characterizing non-stationarity, *IST Mobile and Wireless Communications Summit*, Lyon, France, June, 2004.
43. Massey, F.J., The Kolmogorov-Smirnov test for goodness of fit, *American Statistical Association Journal*, 46, 68–78, 1951.
44. Dixon, W.J. and Massey, F.J., *Introduction to Statistical Analysis*, 3rd ed., McGraw-Hill: New York, 1969.

Chapter 2

RF Channel Modeling for 5G Systems

Theofilos Chrysikos and Stavros Kotsopoulos

Contents

2.1 Introduction: From LTE to 5G

The technology of 4G, namely LTE and LTE-A [1], has unleashed an unprecedented growth of connected devices and mobile data rates reaching up to 100 and 300 Mbps, respectively (theoretical values). This paves the way for the standardization, implementation, and launching of the 5th-generation (5G) technology and services [2], as the number of connected Internet of things (IoT) is estimated to reach 50 billion by 2020 [3], while the mobile data traffic is expected to grow to 24.3 Exabytes per month by 2019 [4].

The number of LTE networks deployed per region was studied in [5]. In Europe, after a slow start, the adoption of LTE has significantly accelerated, with 124 commercial networks up and running. At the end of Q2-2014, there were more than 280 million LTE subscribers worldwide (with a large portion being in Unites States).

LTE networks are deployed in a broad range of frequency bands. In Europe, most of the LTE deployments are happening in new 4G FDD bands such as 800 MHz (LTE band 20) and 2.6 GHz (LTE band 7). Many deployments are also on-going in 1800 MHz, which was previously allocated for GSM. A few countries (especially in Scandinavia at present) are also deploying LTE at 450 MHz. Today, LTE can be deployed in a large range of spectrum bands starting from 450 MHz up to 3.8 GHz [5].

It is suggested that a new generation of 5G standards may be introduced approximately in the early 2020s [6]. The 5G will provide not only higher data rates, as this is already in the work in progress phase of current backhaul technologies for the congested LTE/LTE-A networks (particularly in urban areas with dense population and various obstacles) [7], but will also introduce a whole new concept of design and implementation toward the IoT, by allowing the interconnection of mobile and handheld devices with vehicles, automated equipment, and paving the way for machine-to-machine communications [8].

The 5G technology will satisfy key requirements such as: (a) massive number of connected devices, (b) very high link reliability, and (c) low latency and real-time operation [9]. Various services and applications will be made feasible and available through 5G: smart houses and smart offices, as well as smart cities on a greater urban scalability. In addition, telemetry and e-health applications will provide higher reliability and will allow for the development of both telemedicine services in remote *white* areas as well as the further progress in body area networks (BANs) with both in-body and on-body sensors.

Figure 2.1 presents some of the key applications supported and enhanced by 5G.

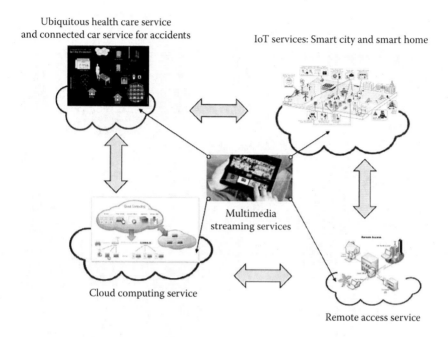

Figure 2.1   5G services.

2.2 5G below 6 GHz

2.2.1 The Move toward mm-Wave Communications

The spectrum employed for 5G is one of the key issues of discussion, research, even debate amidst industry and academia. Recent works argue in favor of dedicating frequencies above 30 GHz (EHF band) for the deployment of both the *core* and the backhaul 5G network [10]. More specifically, potential bands for 5G spectrum are: the 28 GHz band (27.5–29.5 GHz), the 38 GHz band (36–40 GHz), the 60 GHz band (57–64 GHz), the E-band (71–76 GHz and 81–86 GHz) and finally, the W band (92–95 GHz) [11].

The frequency band of 28 GHz is considered as a possible candidate for the core 5G network, whereas small cell solutions for 5G (the equivalent of today's backhaul for LTE networks) can operate in the mm-Wave band above 70 GHz. In addition, indoor 5G can be deployed in the EHF band as well, utilizing the directional, O_2-attenuated spectrum around 60 GHz [12–14].

Channel modeling for the mm-Wave band has been thoroughly investigated in a number of recent published works [15–26]: investigating the increased distance-dependent attenuation in the EHF band, the intrinsic channel characteristics due to atmosphere and oxygen absorption, as well as rain attenuation (i.e., at the 70 GHz channel), and the directivity that is an outcome of all the aforementioned losses.

Obstacle losses, by nature frequency-dependent, will also come into the forefront as critical parameters for the calculation of excess path loss and the physical limitations of the EHF channel will serve as a basis for the service limitations when a massive number of users need to be simultaneously connected within a network providing robust and seamless broadband services over wireless links in metropolitan urban areas.

In [15], a path loss investigation for the 28 GHz channel for a densely populated, heavily shadowed urban area (Manhattan, New York City), as a validation scheme for the physical limitations of a potential core 5G network, has been conducted. As already mentioned, the indoor 60 GHz channel can be employed for upcoming 5G *femto* infrastructure that will allow the design and implementation of smart houses. In [12], a comparative study between the current 2.4 GHz channel and the proposed 60 GHz scheme was conducted.

2.2.2 Launching 5G in Microwave

Propagation validation and channel modeling for the mm-Wave band has gathered a lot of scientific interest with recent projects investigating the potential of implementing 5G directly in the higher EHF spectrum [27–35]. These publications conduct both simulation and experimental measurements and provide solutions for hardware equipment as well, that is, channel sounders, in order to study the impact of small-scale fading (multipath).

In addition, as stated in [36], the FCC allocated approximately 11 GHz for flexible, mobile, and fixed use wireless broadband, comprising 7 GHz of unlicensed spectrum from 64 to 71 GHz and 3.85 GHz of licensed spectrum, designated as a new *upper microwave flexible use* service, in three bands:

- 27.5–28.35 GHz
- 37–38.6 GHz
- 38.6–40 GHz

The move toward the mm-Wave spectrum reveals a lot of potential for bandwidth, throughput, and capacity growth for the 5G upcoming technology. It is, however, a band that still needs to be thoroughly investigated and its specific characteristics are yet to be fully addressed. Atmosphere effects and attenuation due to irregularities need to be considered. In addition, providers will have to face the cost of implementing their 5G equipment from scratch; therefore, Capital and Operational Expenditure (Capex & Opex, respectively) need to be carefully planned before moving to the EHF electronics and hardware.

A possible transitional step toward mm-Wave 5G might be to launch the technology in the currently utilized microwave spectrum. In 37, a possible scheme of transition toward 5G via the existing LTE networks is considered, according to which the launching of 5G is possible via an upgraded migration from the existing LTE core in the following years leading up to 2020.

A significant milestone in the migration from LTE/LTE-A to 5G is the microwave backhaul provided for the present congested 4G networks [38]. The sub-6 GHz band provides backhaul support in both the licensed and the unlicensed spectrum, and various solutions have been proposed in order to mitigate user traffic and provide total downlink bit rates up to 500 Mbps. Mobile operators and wireless technology providers have already begun the transition toward 5G via the existing mobile/cellular broadband infrastructure [39].

Launching 5G in the microwave spectrum requires investigation concerning the candidate carrier frequencies, as well as the possibility of employing cognitive radio to utilize additional spectrum in the sub-6 GHz band. Capacity and congestion issues will remain critical since many of these frequencies are already *overcrowded* and the service limitations have already put boundaries on the 4G technology. However, as a transitional step until mm-Wave band equipment and logistics fall into reasonable costs for the operators, it is a viable and feasible solution. Therefore, the key aspects of backhaul support for 4G/4G+ need to be briefly discussed.

2.2.3 *LTE/LTE-A Backhaul*

Backhaul for LTE/LTE-A was developed in order to mitigate congestion especially in urban shadowed areas where user traffic and severe attenuation decreased the performance of wireless broadband services [40–42]. As shown in [43,44], the theoretical bounds of 100 and 300 Mbps for 4G and 4G+, respectively, are approximately met when high SNR requirements are met, if *macro cells* are considered (the core LTE network). Small cells can operate as backhaul structures, intervening either in *cell-edge* or in selected locations within an urban deployment scheme with antennas located as high as lamp posts or street signs [45,46].

Current backhaul providers have investigated the demands in both point-to-point and point-to-Multipoint LTE services. Open issues, confirmed and emerging trends have been discussed thoroughly in [46]. More specifically, confirmed trends include small-cell backhaul requirements different from macro-cell backhaul, a higher (%) of backhaul TCO in small cells than in macro cells, the emergence of solutions specific to small—cell deployments, coexistence of multiple solutions (Fiber and Wireless, LOS and NLOS), and a complexity from a multiventor, multitechnology backhaul approach. Emerging trends include increased flexibility in available backhaul solutions, partnerships among vendors, more Wi-Fi and 3G small cells, new support for indoor small cells, serious consideration of the 3.5 GHz channel for small-cell access, a blurring line between LOS and NLOS, and vendor focus on shorter installation process.

Even though fiber backhaul is also supported by specific providers, its cost and complexity, related both to the actual construction of the fiber infrastructure as well as legal and regulatory issues, render it as a less popular solution amidst providers and clients. Wireless backhaul for outdoor urban congested LTE networks has increased in both popularity and demand and newer solutions exploit both the sub-6 GHz band and new, higher frequencies of interest in the upper bound of the SHF band and the EHF band.

Cost is a critical issue when providing a backhaul solution for congested LTE cells, which are considered as *macro-cells* operating at the microwave frequencies established by each 4G provider. Backhaul solutions can either consider line-of-sight (LOS) or non-line-of-sight (NLOS) scenarios, even though there is a blurring line between the two conditions, with the emergence of obstructed-line-of-sight (OLOS) as a site-general method for outdoor propagation topologies [47].

The impact of backhaul on the percentage of total equipment cost for small cells and macro-cells differs significantly. Naturally, the cost increases as we move to licensed bands and the LOS scenario (PTP) demands more resources. When macro-cells are considered, the cost decreases as small cells are essentially implemented for backhaul support of LTE networks. If fiber is not available as an existing infrastructure or if legal or regulatory issues arise, then wireless backhaul in the sub-6 GHz band should be investigated, for licensed frequencies. If such spectrum is not available or is insufficient for the needs of the congested *macro* LTE network, then mm-Wave backhaul should be selected as the best solution, if LOS is available and reliable. Otherwise, if relay hop cannot cover for the LOS requirements, then unlicensed spectrum in the sub-6 GHz should be sought. If none of the wireless backhaul solutions work, then the only solution would be to seek fiber-optic backhaul regardless of the cost.

The methodology described in the small-cell backhaul decision tree was investigated and implemented for simulation purposes in [48], as a mathematical validation of a web-based, techno-economic tool for PPDR services over a dedicated LTE security network, deployed in an urban environment at 700 MHz. The total cost was calculated over a six-year period for various channel conditions (NLOS) and for variable urban surface coverage. The results are shown in Figure 2.2.

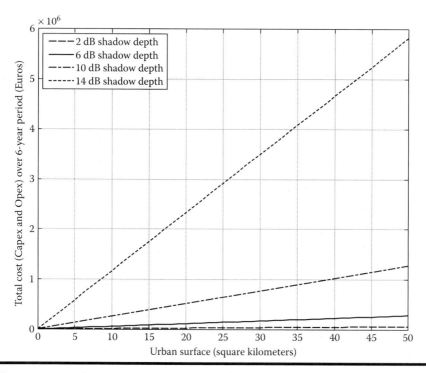

Figure 2.2 Total cost (Capex and Opex) over a six-year period. (From Chrysikos, T. et al., In *TEMU 2016 International Conference on Telecommunications & Multimedia*, Heraklion, Greece, July 25–27, 2016.)

2.3 Path Loss Models in the Sub-6 GHz Band

2.3.1 Path Loss Modeling for Microwave Propagation

Path loss modeling involves the calculation of all attenuation losses in a logarithmic formula that incorporates both losses from distance-dependent attenuation due to free space propagation as well as losses from all various mechanisms influencing the signal trajectory, such as reflection, scattering, and diffraction [49]. In addition, path loss modeling accounts for the losses due to various intrinsic irregularities as well as losses due to atmospheric phenomena [50].

The fundamental reference path loss model is the Free Space model which is a logarithmic expression of the Friis formula, assuming an ideal propagation scenario with no obstacles, or terrain geographic characteristics, and no antenna heights [51]. According to path loss modeling theory, the average received power at a given measurement location is provided by Parsons [52]:

$$P_r(\mathrm{dBm}) = P_t(\mathrm{dBm}) + K(\mathrm{dB}) - 10n \log_{10}\left(\frac{d}{d_0}\right) \qquad (2.1)$$

where:

d is expressed in meters (T-R separation)

d_0 is the reference distance which in indoor propagation schemes equals 1 m

n is the path loss exponent (set to 2 for the Free Space model)

K is the reference path loss at 1 m, which equals –40 dB for 2.4 GHz (802.11 g protocol)

Average path loss according to the Free Space model is given by [52]:

$$P_L = 32.45 + 20\log_{10}\left[f(\text{MHz})\right] + 20\log_{10}\left[d(\text{km})\right] \tag{2.2}$$

where:

f is the carrier frequency of the transmitted signal (MHz)

d is the T-R separation (in km)

In both equations, the T-R separation distance has been calculated in order to incorporate the transmitter and receiver antenna heights.

The idealistic propagation assumptions of the Free Space model render it reliable only for LOS cases. In the original Free Space formula, the path loss exponent is set to $n = 2$, which is in accordance with the inverse-square law derived out of the Friis equation (ideal propagation of electromagnetic wave into free space).

The large-scale variations of the received power have been attributed to losses by obstacles of proportions significantly larger than the signal wavelength, which remain constant over a time scale of seconds or minutes (large-scale fading). The shadowing deviation, or shadow depth, expresses the excess path loss, defined by Jakes as "the difference (in decibels) between the computed value of the received signal strength in free space and the actual measured value of the local mean received signal" [53].

These fluctuations have been proven to follow the log-normal distribution [54]. From the aforementioned discussion it is more than apparent that the accurate prediction of received power fluctuations around a mean value is of critical importance in order to achieve a robust network planning in terms of Outage Probability calculation.

Shadow fading losses can be incorporated in the logarithmic path loss formula with a Gaussian variable. This is the mathematical expression of the Log-Distance path loss model [55]:

$$P_L = P_L(d_0) + 10n\log_{10}\left(\frac{d}{d_0}\right) + X_\sigma \tag{2.3}$$

$$X_\sigma = z \times \sigma(\text{dB}) = 1.645 \times \sigma(\text{dB}) \tag{2.4}$$

where:

$P(d_0)$ is the reference path loss at 1 m from the transmitter

n is the path loss exponent

X_σ is a zero-mean Gaussian variable (dB) that expresses the losses due to (log-normal) shadowing

The parameter z stands for the percentage of coverage probability [54]. For best case scenarios with a coverage probability of 98%, z equals 2. In sub-optimal, realistic schemes, as the one considered in this work, z equals 1.645 for a coverage probability of 95% [49].

The log-Distance model attempts to incorporate all losses other than distance-dependent attenuation due to free space propagation in one single variable. This demands a very precise calculation of the shadow depth, particularly in complex, obstacle-dense outdoor propagation topologies. Such a method has been developed and will be discussed next. It is, however, imperative to be able to take into consideration the antenna heights as well as density of building structure discretely and not in a single, all-inclusive variable, in order to investigate the impact of these different phenomena independently. This has led to the development and validation of certain empirical path loss models, where the mathematical formula is derived from extensive on-site measurements, either in urban and suburban or in open, rural areas [55].

The Hata model [56] is a mathematical expression of an empirical model first developed by Okumura in the 1960s [57]. Whereas the Okumura model was based on curves obtained from extensive measurements in urban areas in Japan, the Hata model, developed in 1980, allowed for an elaborate mathematical logarithmic formula [56]:

$$P_L(\text{dB}) = 69.55 + 26.16 \log_{10} f(\text{MHz}) - 13.82 \log_{10} h_t$$
$$- a(h_{re}) + (44.9 - 6.55 \log_{10} h_t) \log_{10} d(\text{km}) \tag{2.5}$$

where $a(h_{re})$ is a correction factor for the receiving antenna height, based on the topology and the channel characteristics.

The original Hata model is distance-bound (1–20 km) and frequency-bound (150–1500 MHz). Various extensions have been suggested, with an original 2 GHz bound [54]. The Hata model was validated for frequencies beyond the 2 GHz in [47].

Another empirical path loss model suitable for validation of RF propagation in heavily shadowed urban areas is the Walfish-Ikegami, an elaborate path loss model originally developed for cellular bands (800 MHz–2 GHz limitation) with a 5 km upper distance-bound [58,59].

In the case of an urban LOS scenario, the Free Space model is employed for $d < 20$ m, and beyond that the following formula applies:

$$P_L = 42.6 + 20 \log_{10} f(\text{MHz}) + 26 \log_{10} d(\text{km}) \tag{2.6}$$

In the NLOS case:

$$P_L(\text{dB}) = L_o(\text{dB}) + L_{\text{rts}}(\text{dB}) + L_{\text{msd}}(\text{dB}) \tag{2.7}$$

where:

L_o represents free space loss and is provided by Equation 2.1

L_{rts} is a correction factor representing diffraction and scattering from rooftop to street

L_{msd} represents multiscreen diffraction due to urban rows of buildings

These terms vary with street width, building height and separation, and angle of incidence.

$$L_{rts}(\text{dB}) = -16.9 - 10\log_{10} w + 10\log_{10} f(\text{MHz})$$
$$+ 20\log_{10}(h_{roof} - h_{re}) + L_{ori} \tag{2.8}$$

$$L_{msd}(\text{dB}) = L_{bsh} + K_a + K_d \log_{10} d(\text{km})$$
$$+ K_f \log_{10} f(\text{MHz}) - 9\log_{10} b \tag{2.9}$$

where:

w is the average street width

b the average building separation

h_{roof} the building height

h_{re} the receiving antenna height, all expressed in meters

These models have been developed especially for urban and suburban shadowed areas. A number of RF models has been developed for open, rural areas, such as the Longley-Rice and the Egli model, where scarce number of users over extended coverage distances is assumed [49,54].

Indoor propagation presents an even more complex environment in terms of both attenuation losses and the number of phenomena that need to be considered and investigated. The increased number of obstacles of various types and materials whose dimensions are comparable to the signal wavelength cause signal scattering, which adds to the complexity of the wireless channel characterization. Further attenuation is caused by signal penetration of walls and floors, as well as from human intervention, also known as *body shadowing* [60].

The ITU indoor path loss model is described by the following formula [49]:

$$P_L = 20\log_{10}(f) + 10n\log_{10}(d) + Lf(n) - 28 \text{ dB} \tag{2.10}$$

where:

f is the carrier frequency expressed in MHz

n is the path loss exponent

$Lf(n)$ is the floor penetration factor

For same floor measurements, $Lf(n) = 0$. ITU specifications [49] provide a number of values for the slope factor (path loss exponent) for different carrier frequencies.

These original specifications have, however, been proven to be inaccurate for the 2.4 GHz frequency band for all indoor propagation topologies. Numerical adjustments have been provided for the slope factor values for each topology, and for the multiple floors measurements in the office topology, the floor penetration factor has also been corrected [61].

The Multi-Wall-Floor (MWF) model is a path loss model whose formula is provided by Lott and Forkel [62]:

$$L = L_0 + 10n\log_{10}(d) + \sum_{i=1}^{I}\sum_{k=1}^{K_{wi}} L_{wik} + \sum_{j=1}^{J}\sum_{k=1}^{K_{fj}} L_{fjk} \tag{2.11}$$

where:
I, J is the number of types of walls and floors
L_{wik} is the attenuation due to kth traversed wall type i
L_{fjk} is the attenuation due to kth traversed floor type j
K_{wi} is the number of walls type i
K_{fj} is the number of floors type j

The mathematical expression assumes two categories of losses: losses due to free space propagation and losses caused by all the various types of walls and floor that may come into the path of the propagated signal. Moreover, the MWF model considers different losses for different types of materials, at a given frequency of transmission.

The MWF model also takes into account the decreasing penetration loss of walls and floors of the same material as their number increases [62].

Attenuation over distance (dB/m) is another key parameter employed for the calculation of excess path loss in wireless propagation. Instead of incorporating all excess path loss in a single shadow variable, the attenuation over distance considers a rate of attenuation as the distance linearly increases and also provides such a metric for the characterization of the propagation topology.

Attenuation over distance can be calculated via the Linear Attenuation Model (LAM), also known as the Devasirvatham model [63]:

$$P_L(\text{dB}) = P_{L0}(\text{dB}) + 10n\log_{10}(d) + ad \tag{2.12}$$

where:
$P_L(\text{dB})$ stands for the average path loss (dB)
$P_{L0}(\text{dB})$ stands for the frequency-dependent reference path loss (path loss at 1 m distance from the transmitter)
n is the path loss exponent that expresses the rate of attenuation losses
a is the attenuation over distance (dB/m)
d is the transmitter-receiver distance (T-R separation), in meters

When the total EIRP is known (in dBm), and the local mean values of received signal power can be measured, then it is possible to characterize the wireless channel for the topology in question by calculating the attenuation over distance (in dB/m) by the following formula:

$$a = \frac{\text{EIRP(dBm)} - P_r(\text{dBm}) - 10n \log_{10} d - P_{L0}(\text{dB})}{d} \quad (2.13)$$

2.3.2 The 3.5 GHz Channel

Another frequency of interest is the 3.5 GHz channel. Originally considered as a WiMax frequency, the 3.5 GHz channel has also been featured in scientific works [64–66], which investigate the variations of the signal amplitude and phase in both indoor and outdoor scenarios.

The 3.5 GHz band has emerged as a promising candidate, in the long term, for the dedicated use of small cells [46]. Offering a presently underutilized spectrum, the 3.5 GHz channel remains a matter of debate amidst providers, as to whether it can be more suitably employed as an access or a backhaul frequency, offering advantages and disadvantages for both scenarios.

In [67], the fundamental path loss models were employed for the validation of their predictions, in terms of received signal power and compatibility of the local mean values of the signal strength with the log-normal distribution, for a selected, obstacle-dense, indoor propagation topology. The results, depicted in Tables 2.1 and 2.2, allowed for the first time the calculation of obstacle losses for this specific frequency. Comparison with losses measured for the 2.4 GHz confirmed the frequency-dependent nature of obstacle attenuation and opened the way for immediate future work in the 3.5 GHz channel including more indoor topologies, as hinted in [68], as well as outdoor propagation scenarios.

Table 2.1 Frequency-Dependent Obstacle Losses

Number	2.4 GHz (dB)	3.5 GHz (dB)
1	7–8	9
2	5–6	7
3	3–4	5
4	1–2	2
5 and more	1–2	2

Source: Chrysikos, T. et al., In *Wireless Telecommunications Symposium 2015 (WTS 2015)*, New York, April 15–17, 2015.

Table 2.2 MWF Performance Based on Obstacle Losses

RF Model	Mean Error (%)
MWF (2.4 GHz)	3.63
MWF (3.5 GHz)	2.99

Source: Chrysikos, T. et al., In *Wireless Telecommunications Symposium 2015* (*WTS 2015*), New York, April 15–17, 2015.

2.3.3 The 5–6 GHz Band

The 802.11p protocol has been designed for vehicular communications and applied mostly in outdoor highway scenarios that can potentially include roadside units (RSUs) [69–70]. A series of works has investigated the robustness of protocols designed and developed in order to support vehicle-to-vehicle (V2V) or vehicle-to-infrastructure (V2I) communications that can contain warning messages or periodical signal information about road traffic, weather updates, and safety information.

Operating in the 5.89 GHz frequency, these protocols opened the way for research that incorporates studies of channel characterization along with routing techniques for highway and vehicular scenarios [71–77]. These works assumed a propagation topology outside the boundaries of urban and suburban topologies. Essentially an open, rural-area scenario, the dominant attenuation loss in the highway environment was due to free space propagation, and the excess path loss was confined to scattering losses, diffraction, and shadow losses due to the height of vehicles such as trucks and buses. As vehicular communications continued to grow, the need for further channel modeling and RF validation in the 5–6 GHz band provided further results, as shown in [78,79].

In [80], the deployment of a roadside network was investigated for V2I communications. The findings of this work have been extended to include LTE-A and a possible migration to 5G for a city scenario instead of a highway topology. The process of moving the 5–6 GHz band from the open highway to the urban scenario is already beginning to garner scientific and academic interest. A thorough survey in [81] showcased the dynamics of channel modeling for this spectrum at urban propagation topologies.

In [46], the backhaul solutions for the congestion of 4G/4G+ networks included a scenario in the 5–6 GHz for NLOS backhaul by commercially available products. Vehicular communications and roadside networking will continue to grow as we move toward the 5G deployment. Therefore, further research in this band will certainly enhance the findings concerning path loss modeling and will contribute to the spectrum potential for 5G.

2.3.4 All-Inclusive Frequency Range Models

Recent published works [82–86] have studied, as part of a broader research, channel models with a very large band spectrum of applicability. From 500 MHz up to 80 GHz, 3D channel models have been validated and their prediction reliability is suggested as an input to 5G deployment. As the 700 MHz band is reserved for safety solutions and security networks in Europe [5] and the mm-Wave frequencies such as 73 and 80 GHz have been suggested as potential backhaul for 5G [15], channel models that have been validated for the whole spectrum of UHF, SHF, and EHF bands are of critical importance.

A major modeling work was conducted in [85] across a wide range of selected frequencies: 2, 5.6, 10.25, 28.5, 39.3, and 73.5 GHz. Two sets of data processing were employed: the first set was the measurement set, which was used to compute the parameters of the RF models and the second set was the prediction set where the shadow fading standard deviation was computed at different distances, frequencies, and environments (dB). The measured data was from 2 to 28 GHz, whereas the ray tracing technique provided data sets up to 73 GHz.

2.4 Large-Scale Fading for 5G

Research in both theoretical and experimental works has confirmed that the large-scale variations of the received power follow the log-normal distribution [53,54]. Hence, the logarithmic values of the local mean power (dBw or dBm) comply with the Gaussian distribution. The probability density function (PDF) of the Gaussian distribution for the logarithmic values of the received power values is derived by Rappaport [54]:

$$p(x) = \frac{1}{\sigma\sqrt{2\pi}} e^{-\frac{(x-\bar{x})^2}{2\sigma^2}} \tag{2.14}$$

where:

x is the received power (logarithmic value) in each measurement location (local mean strength)

$\bar{x}$ is the average received power (logarithmic value) for all measurement locations (median value of the received power overall the topology in question)

σ is the standard deviation of the shadowing losses (in dB)

Given a pool of experimentally obtained i received power values corresponding to respective measurement locations in a propagation topology, the mean value and standard deviation of the distribution of these values can be calculated by Seybold and Parsons [49,52]:

$$\bar{x} = \frac{1}{n}\sum_{i=1}^{n} x_i \tag{2.15}$$

$$\sigma^2 = \frac{1}{n}\sum_{i=1}^{n}(x_i - \bar{x})^2 \tag{2.16}$$

In order to provide unbiased results, a significantly large number of i samples are required [53]; therefore, the measurements need to be extensive and reliable, averaging the small-scale phenomena with sufficient local mean samples [54].

The Log-Distance model is the reference method that incorporates shadow fading losses in its mathematical formula and provides predictions for the excess path loss. It requires, however, a precise value assignment for both the path loss exponent and the log-normal shadowing variable in order to provide reliable estimations. Expressing all the various attenuation losses in a single variable can compromise the reliability of the predictions.

A reliable alternative solution for estimating the excess path loss lies in the MWF model [62] that calculates the losses due to obstacles, walls, and floors notably, and the decreasing penetration as the number of obstacles of a given type and material increases.

Therefore, the losses (in decibels) caused by walls and floors and expressed by $\sum_{i=1}^{I}\sum_{k=1}^{K_{wi}}L_{wik} + \sum_{j=1}^{J}\sum_{k=1}^{K_{fj}}L_{fjk}$ in the MWF model formula, are Gaussian. This has provided the basis of a novel empirical method for the calculation of shadowing deviation [87].

$$\sigma\ (\mathrm{dB}) = \frac{\displaystyle\sum_{i=1}^{I}\sum_{k=1}^{K_{wi}}L_{wik} + \sum_{j=1}^{J}\sum_{k=1}^{K_{fj}}L_{fjk}}{z} \tag{2.17}$$

Thus, the shadowing deviation (in dB) is directly calculated from the losses caused by obstacles along the signal propagation path. This method does not require extensive RF measurements; it only requires limited measurements near the obstacles in order to obtain the respective penetration losses. The robustness of this method has been validated in [87] for the 2.4 GHz channel and in 67 for the 3.5 GHz channel.

An alternative method for calculating the excess path loss lies in the LAM that incorporates all losses in the attenuation over distance metric (in dB/m). Comparing the shadow depth, as provided from the obstacle losses, to the attenuation over distance has provided with some a new formula for calculating the linear attenuation metric [88]:

$$a = \frac{z}{d}\sigma(\mathrm{dB}) \tag{2.18}$$

The mean error of this method is provided in Table 2.3.

In the channel models covering the whole range from microwave to the mm-Wave band, the shadow fading deviation has been calculated from prediction sets with input values derived from measured sets over distance, frequency, and different topologies [85].

Table 2.3 Mean Error for Calculation of Attenuation over Distance Directly from the Shadow Depth, Compared to the Original LAM Formula

Propagation Topologies	Mean Error (%)
Office—same floor	18.63
Office—same floor (LOS locations omitted)	5.78
Office—multiple floor	3.43
Library	14.71
Library (LOS locations omitted)	2.53

Source: Chrysikos, T. and Kotsopoulos, S., In *IEEE Vehicular Technology Conference (VTC Fall)*, 2012.

2.5 Comparison of Modeling Approaches: Conclusions and Future Work

Both the microwave spectrum (below 6 GHz) and the mm-Wave band have demonstrated a wide variety of path loss models, empirical and semi-empirical, stochastic and deterministic, with varied reliability and scalability performance for different propagation topologies and channel characteristics. Measurements have always been the criterion with which each model is validated in terms of relative mean error and error deviation (usually in dB).

Microwave models will remain relevant since the sub-6 GHz may very well be the launching stage of 5G, migrating from the backhaul extension of 4G/4G+. Models such as Walfish-Ikegami and even the modified Hata model will still function as reference models for urban and suburban topologies, and researchers will investigate the potential of extending these models beyond their original spectrum and distance limitations, as proven in a first step in [47].

Concerning the 5–6 GHz band, the models described in Section 2.3.3 provide predictions of the received signal strength employing both geometric stochastic and deterministic methods, as well as non-geometric calculations. The shift from the open, rural environment of conventional highway scenarios toward the urban and suburban road network converges with the migration of V2V, V2I communications and roadside networks toward the 5G era of low-latency vehicular broadband communications.

The mm-Wave band provides a potential for bit rates up to 10 Gbps (burst peak rate) and average rates of 1 Gbps, with opportunities for bandwidth up to 1 GHz. The phenomena of very severe directivity and loss of omni-directional transmission in the higher frequencies, that is, 60, 73, and 80 GHz is mitigated by the potential for beam forming and massive MIMO. There are, however, cost issues with regard to massive production and deployment of hardware electronics in those frequencies as well as the need to launch the access network of 5G in a lower EHF frequency, that is, 39 GHz, and maintain the higher frequencies for 5G backhaul. Moreover, research has to deal with on-going propagation and attenuation issues in the mm-Wave band (atmospheric absorption, path loss irregularities, etc). In addition, the lack of channel sounders for the EHF band needs to be dealt with, in order to provide metrics for the multipath (small-scale) fading.

Large-scale fading both in the microwave and mm-Wave band is still an issue of on-going interest and research. Whereas the site-generic rule of the log-normal distribution remains (with mathematical closed-form approximations alternatively provided by the Gamma distribution), there are challenges regarding the derivation of the shadow depth from obstacle losses, which are frequency-dependent and therefore demand a more focused approach when predictions are calculated in a wide range of frequencies. Predictions in distance need to be averaged for eliminating biased results and predictions over different topologies test the limits of site-generic methodology. Attenuation over distance can be employed to describe the excess path loss in a mathematical approximation and dB/m, as well as devices per square kilometer, will remain a parameter of interest in wireless communications.

As academia and industry moves toward the launching of 5G, immediate future work lies in the further optimization and investigation of LTE-A backhaul, and addressing issues related with the possible migration toward 5G as an early phase through the microwave backhaul solutions currently provided for the 4G+ congested networks in urban shadowed areas.

Regulatory and legal issues related to spectrum allocation and licensing in the sub-6 GHz band will have to be resolved, and immediate future work arises for path loss modeling in the 3.5 GHz as potential 5G microwave frequency. In addition, vehicular and roadside applications and scenarios for the urban environments will be addressed in the short-term future.

The time scope for launching 5G in 2020 is ambitious, yet feasible. Research internationally faces challenges and prospects as 5G strives to surpass all previous mobile broadband networks and lead us in the new era of wireless communications.

References

1. LTE; Evolved universal terrestrial radio access (E-UTRA); Base Station (BS) radio transmission and reception (3GPP TS36.104 version 11.8.2. R11).
2. Samsung 5G Vision. Available at http://www.samsung.com/global/business-images/insights/2015/Samsung-5G-Vision-0.pdf (accessed July 18, 2017).
3. UMTS, Mobile traffic forecasts 2010-2020 Report, UMTS Forum, January 2011.
4. Cisco, Cisco Visual Networking Index: Global Mobile Data Traffic Forecast Update: 2013–2018, Cisco, February 2014.
5. SALUS FP7 EU Project. Available at https://www.sec-salus.eu/ (accessed July 18, 2017).
6. X. Li, A. Gani, R. Salleh, and O. Zakaria. The future of mobile wireless communication networks. In *International Conference on Communication Software and Networks*, pp. 554–557, IEEE, Macau, China.

7. A. K. Pachauri and O. Singh. 5G technology–redefining wireless communication in upcoming years. *International Journal of Computer Science and Management Research*, 1, 1, 12–19.

8. 5G research centre gets major funding grant. Available at http://www.bbc.co.uk/news/technology-19871065 (accessed January 15, 2014).

9. 5GrEEn—towards green 5G mobile networks. Available at http://www.eitictlabs.eu/innovationareas/networking-solutions-for-future-media/5green-towardsgreen-5g-mobile-networks/ (accessed January 15, 2014).

10. K. Haneda et al. 5G 3GPP-like channel models for outdoor urban microcellular and macrocellular environments, In *2016 IEEE 83rd Vehicular Technology Conference (VTC 2016-Spring)*, May 2016, Nanjing, China.

11. C. Kourogiorgas, N. Moraitis, and A. D. Panagopoulos. Radio channel modeling and propagation prediction for 5G mobile communication systems, *Handbook of Research on Next Generation Mobile Communication Systems, IGI Global*, pp. 1–30, 2016. doi:10.4018/978-1-4666-8732-5.ch001.

12. J. Kyröläinen, P. Kyösti, J. Meinilä, V. Nurmela, L. Raschkowski, A. Roivainen, and J. Ylitalo. Channel modelling for 5th generation mobile communications, In *Proceedings on 8th European Conference on Antennas and Propagation (EuCap 2014)*, pp. 219–223, April 2014, The Hague, The Netherlands.

13. M. Fallgren and B. Timus (Eds.). Future radio access scenarios, requirements and KPIs. Deliverable D1.1, V1.0, ICT-317669, METIS Project, May 1, 2013. Available at http://www.metis2020.com (accessed July 18, 2017).

14. H. Zhao et al. 28 GHz millimeter wave cellular communication measurements for reflection and penetration loss in and around buildings in New York City, In *2013 IEEE International Conference on Communications (ICC)*, pp. 5163–5167, June 2013, Budapest, Hungary.

15. T. S. Rappaport et al. Millimeter wave mobile communications for 5G cellular: It will work!, *IEEE Access*, 1, 335–349, 2013.

16. T. S. Rappaport, R. W. Heath, Jr. R. C. Daniels, and J. N. Murdock. *Millimeter Wave Wireless Communications*. Prentice Hall, Upper Saddle River, NJ, 2015.

17. T. S. Rappaport, G. R. MacCartney, Jr. M. K. Samimi, and S. Sun. Wideband millimeter-wave propagation measurements and channel models for future wireless communication system design, *IEEE Transactions on Communications*, 63, 9, 3029–3056, 2015. Available at http://ieeexplore.ieee.org/stamp/stamp.jsp?arnumber=7109864 (accessed July 18, 2017).

18. G. R. MacCartney, Jr. T. S. Rappaport, S. Sun, and S. Deng. Indoor office wideband millimeter-wave propagation measurements and channel models at 28 and 73 GHz for ultra-dense 5G wireless networks, *IEEE Access*, 3, 2388–2424, 2015. Available at http://ieeexplore.ieee.org/xpl/articleDetails.jsp?arnumber=7289335 (accessed July 18, 2017).

19. M. K. Samimi, T. S. Rappaport, and Jr. G. R. MacCartney. Probabilistic omnidirectional path loss models for millimeter-wave outdoor communications, *IEEE Wireless Communications Letters*, 4, 4, 357–360, 2015. Available at http://ieeexplore.ieee.org/document/7070688/?arnumber=7070688 (accessed July 18, 2017).

20. M. K. Samimi and T. S. Rappaport. 3-d statistical channel model for millimeter-wave outdoor communications, In *2015 IEEE International Conference on Communications (ICC)*, June 2015, London, UK. Available at http://arxiv.org/abs/1503.05619 (accessed July 18, 2017).

21. G. R. MacCartney, Jr. J. Zhang, S. Nie, and T. S. Rappaport. Path loss models for 5G millimeter wave propagation channels in urban microcells, In *2013 IEEE Global Communications Conference (GLOBECOM)*, pp. 3948–3953, December 2013.

22. S. Sun et al. Propagation path loss models for 5G urban micro- and macro-cellular scenarios, In *2016 IEEE 83rd Vehicular Technology Conference (VTC2016-Spring)*, May 2016, Nanjing, China. Available at http://arxiv.org/abs/1511.07311 (accessed July 18, 2017).

23. C. Larsson et al. Polarisation characteristics of propagation paths in indoor 70 GHz channels, In *8th European Conference on Antennas and Propagation (EuCAP 2014)*, pp. 3301–3304, April 2014, The Hague, The Netherlands.

24. Semann et al. 3-D statistical channel model for millimeter-wave outdoor communications, In *2014 IEEE Global Communications Conference (GLOBECOM) Workshop*, pp. 393–398, December 2014.

25. S. Sun et al. Path loss, shadow fading, and line-of-sight probability models for 5G urban macro-cellular scenarios, In *2015 IEEE Global Communications Conference, Exhibition & Industry Forum (GLOBECOM) Workshop*, December 2015.

26. S. Piersanti, L. Annoni, and D. Cassioli. Millimeter waves channel measurements and path loss models, In *2012 IEEE International Conference on Communications (ICC)*, pp. 4552–4556, June 2012.

27. ETSI, New ETSI group on millimetre wave transmission starts work, Technical Report. Available at http://www.etsi.org/news-events/news/866-2015-01-press-new-etsigroup-on-millimetre-wave-transmission-starts-work (accessed July 18, 2017).

28. METIS, Mobile and wireless communications enablers for the twenty-twenty information society, EU 7th Framework Programme, Available at http://www.metis2020.com (accessed July 18, 2017).

29. A. Osseiran et al. Scenarios for the 5G mobile and wireless communications: The vision of the METIS project, Accepted for publication in *IEEE Communications Magazine*; *Feature Topic on 5G Wireless Communication Systems: Prospects and Challenges*.

30. Department of Electronics, Computer Science and Robotics of the University of Bologna, Pervasive mobile and ambient wireless communications: COST Action 2100. Available at http://www.cost2100.org/ (accessed July 18, 2017).

31. R. Verdone and A. Zanella (Eds.). *Pervasive Mobile and Ambient Wireless Communications: COST Action 2100*. Springer, London, UK, 2012. Available at http://www.springer.com/la/book/9781447123149 (accessed July 18, 2017).

32. http://www.ic1004.org/.

33. NIST, 5G mmWave channel model alliance. Available at https://www.nist.gov/ctl/5g-mmwave-channel-model-alliance (accessed July 18, 2017).

34. MiWEBA, Channel modeling and characterization, Technical Report MiWEBA, Deliverable D5.1, June 2014. Available at http://www.miweba.eu/wp-content/uploads/2014/07/MiWEBA_D5.1_v1.011.pdf (accessed July 18, 2017).

35. mmMagic, mmMAGIC: Millimetre-wave based mobile radio access network for fifth generation integrated communications. Available at https://5g-ppp.eu/mmmagic/ (accessed July 18, 2017).

36. G. Lerude. FCC allocates nearly 11 GHz of spectrum above 24 GHz for 5G, *Microwave Journal*, July 14, 2016. Available at http://www.microwavejournal.com/articles/26798-fcc-allocates-nearly-11-ghz-of-spectrum-above-24-ghz-for-5g (accessed July 18, 2017).

37. Ericsson microwave towards 2020 report. Available at https://www.ericsson.com/res/docs/2015/microwave-2020-report.pdf (accessed July 18, 2017).

38. Microwave Backhaul, *Ericsson Technology Review*, 93, 1, 2016. Available at https://www.ericsson.com/res/thecompany/docs/publications/ericsson_review/2016/etr-multiband-booster-bachhaul.pdf (accessed July 18, 2017).

39. Ericsson, TeliaSonera and Ericsson go 5G, *Microwave Journal*, January 22, 2016. Available at http://www.microwavejournal.com/articles/25825-teliasonera-and-ericsson-go-5g (accessed July 18, 2017).

40. 3GPP, Study on 3D channel model for LTE, Technical Report 3GPP 36.873 (V12.2.0), July 2015.

41. ITU-R, Guidelines for evaluation of radio interface technologies for IMT-Advanced. Technical Report ITU-R Rep M.2135-1.

42. Guidelines for evaluation of radio interface technologies for IMT-Advanced, International Telecommunication Union (ITU), Geneva, Switzerland, Report ITU-R M.2135-1, 12/2009.

43. P. Vieira, M. P. Queluz, and A. Rodrigues. LTE multi antenna bit rate expectation for urban macro-cell networks, In *7th Ibero-American Congress on Sensors* (*IBERSENSOR 2010*), November 9–11, 2010, Lisbon, Portugal.

44. P. Vieira, M. P. Queluz, and A. Rodrigues. LTE spectral efficiency using spatial multiplexing MIMO for macro-cells, In *2nd IEEE International Conference on Signal Processing and Communication Systems* (*ICSPCS 2008*), December 15–17, 2008, Gold Coast, Australia.

45. T. Giles, J. Markendahl, J. Zander, P. Zetterberg, P. Karlsson, G. Malmgren, and J. Nilsson. Cost drivers and deployment scenarios for future broadband wireless networks—key research problems and directions for research, In *IEEE Vehicular Technology Conference* (*VTC-Spring 2004*), May 17–19, 2004, Milan, Italy.

46. P. Monica, H. Lance, and F. Rayal. Small-cell backhaul: Industry trends and market overview, Senza Fili Consulting, Available at www.senzafiliconsulting.com.

47. T. Chrysikos and S. Kotsopoulos. Site-specific validation of the path loss models and large-scale fading characterization of large-scale fading for a complex urban propagation topology at 2.4 GHz, In *The 2013 IAENG International Conference on Communication Systems and Applications* (*IMECS 2013*), March 13–15, 2013, Hong Kong, China.

48. T. Chrysikos, P. Galiotos, T. Dagiuklas, and S. Kotsopoulos. Techno-economic analysis for the deployment of PPDR services over 4G/4G+ Networks, In *TEMU 2016 International Conference on Telecommunications & Multimedia*, July 25–27, 2016, Heraklion, Greece.

49. J. Seybold. *Introduction to RF Propagation*. Wiley Interscience, Hoboken, NJ, 2005.

50. A. Aguiar and J. Gross. Wireless channel models, Technical Report, TKN, Berlin, 2003.

51. A. Goldsmith. *Wireless Communications*. Cambridge University Press, Cambridge, UK, 2005.

52. J. D. Parsons. *The Mobile Radio Propagation Channel*. Wiley Interscience, Hoboken, NJ, 2000.

53. W. C. Jakes (Ed.). *Microwave Mobile Communications*. Wiley Interscience, New York, 1974.

54. T. Rappaport. *Wireless Communications: Principles & Practice*. Prentice Hall, Upper Saddle River, NJ, 1999.

55. J. Andersen, T. S. Rappaport, and S. Yoshida. Propagation measurements and models for wireless communications channels, *IEEE Communications Magazine*, 3, 1, 42–49, 1995.

56. M. Hata. Propagation measurements and models for wireless communications channels, *IEEE Transactions on Vehicular Technology*, 29, 3, 317–325, 1980.

57. Y. Okumura, E. Ohmori, T. Kawano, and K. Fukuda. Field strength and its variability in VHF and UHF Land-Mobile radio service, In *Review of the Electrical Communication Laboratory*, 16, 9–10, 825–873, 1968.

58. F. Ikegami, S. Yoshida, T. Takeuchi, and M. Umehira. Propagation factors controlling mean field strength on urban streets, In *IEEE Transactions on Antennas & Propagation*, AP-32, 822–829, 1984.

59. J. Walfish and H. L. Bertoni. A theoretical model of UHF propagation in urban environment, In *IEEE Transactions on Antennas & Propagation*, AP-36, 1788–1796, 1988.
60. R. Mathur, M. Klepal, A. McGibney, and D. Pesch. Influence of people shadowing on bit error rate of IEEE 802.11 2.4 GHz channel, In *1st International Symposium on Wireless Communication Systems (ISWCS 2004)*, pp. 448–452, September 20–22, 2004, Port-Louis, Mauritius.
61. T. Chrysikos, G. Georgopoulos, and S. Kotsopoulos. Site-specific validation of ITU indoor path loss model at 2.4 GHz, In *4th IEEE Workshop on Advanced Experimental Activities on Wireless Networks and Systems*, June 19, 2009, Kos Island, Greece.
62. M. Lott and I. Forkel. A multi wall and floor model for indoor radio propagation, In *IEEE Vehicular Technology Conference (VTC 2001-Spring)*, May 6–9, 2001, Rhodes Island, Greece.
63. D. M. J Devasirvatham, C. Banerjee, R. R. Murray, and D. A. Rappaport. Four-frequency radiowave propagation measurements of the indoor environment in a large metropolitan commercial building, In *IEEE Global Telecommunication Conference (GLOBECOM 1991)*, 2, pp. 1282–1286, December 2–5, 1991, Phoenix, AZ.
64. K. W. Cheung, J. H. M. Sau, and R. D. Murch. A new empirical model for indoor propagation prediction, In *IEEE Transactions on Vehicular Technology*, 47, 3, 996–1001, 1998.
65. C. Oestges, P. Castiglione, and N. Czink. Empirical modeling of nomadic peer-to-peer networks in office environment, In *IEEE Vehicular Technology Conference (VTC 2011-Spring)*, May 15–18, 2011, Budapest, Hungary.
66. D. W. Matolak, Q. Wu, and I. Sen. 5 GHz band vehicle-to-vehicle channels: Models for multiple values of channel bandwidth, In *IEEE Transactions on Vehicular Technology*, 59, 5, 2620–2625, 2010.
67. T. Chrysikos, C. Papadakos, and S. Kotsopoulos. Wireless channel measurements and modeling for an office topology at 3.5 GHz, In *Wireless Telecommunications Symposium 2015 (WTS 2015)*, April 15–17, 2015, New York.
68. G. Xu et al. Measurement-based wireless network planning, monitoring, and reconfiguration solution for robust radio communications in indoor factories, *IET Science, Measurement & Technology*, 10, 4, 2016.
69. IEEE 802.11p, Draft standard for information technology—telecommunications and information exchange between systems—local and metropolitan area networks—specific requirements: Wireless access in vehicular environments, *IEEE P802.11p/D9.0*, September 2009.
70. IEEE Std 802.11p-2010, IEEE Standard for Information technology–Local and metropolitan area networks–Specific requirements–Part 11: Wireless LAN Medium Access Control (MAC) and Physical Layer (PHY) Specifications Amendment 6: Wireless Access in Vehicular Environments.
71. C. S. Lin, B. C. Chen, and J. C. Lin. Field test and performance improvement in IEEE 802.11p V2R/R2V environments, In *IEEE International Conference on Communications, IEEE Vehicular Networks & Applications Workshop*, May 2010, Cape Town, South Africa.
72. M. Wellens, B. Westphal, and P. Mahonen. Performance evaluation of IEEE 802.11-based WLANs in vehicular scenarios, In *IEEE 65th Vehicular Technology Conference*, pp. 1167–1171, April 2007.
73. L. Cheng, B. E. Henty, R. Cooper, and D. D. Stancil. A measurement study of time-scaled 802.11a waveforms over the mobile-to-mobile vehicular channel at 5.9 GHz, *IEEE Communications Magazine*, 46, 84–91, May 2008.

74. D. Jiang and L. Delgrossi. IEEE 802.11p: Towards an international standard for wireless access in vehicular environments, In *IEEE 67th Vehicular Technology Conference*, pp. 2036–2040, May 2008.

75. D. Carona, A. Serrador, P. Mar, R. Abreu, N. Ferreira, T. Meireles, J. Matos, and J. Lopes. A 802.11p prototype implementation, In *Intelligent Vehicles Symposium (IV)*, 2010 IEEE, June 2010, La Jolla, CA.

76. K. Y. Ho, P. C. Kang, C. H. Hsu, and C. H. Lin. Implementation of WAVE/DSRC devices for vehicular communications, In *Computer Communication Control and Automation (3CA)*, 2010 International Symposium, 2, May 2010.

77. A. Paier, R. Tresch, A. Alonso, D. Smely, P. Meckel, Y. Zhou, and N. Czink. Average downstream performance of measured IEEE 802.11p infrastructure-to-vehicle links, In *2010 IEEE International Conference on Communications (ICC) Workshops*, pp. 1–5, May 23–27, 2010, Cape Town, South Africa.

78. D. W. Matolak. Channel modeling for vehicle-to-vehicle communications, In *IEEE Communications Magazine (special section on Automotive Networking)*, 46, 5, 76–83, 2008.

79. D. W. Matolak, Q. Wu, and I. Sen. 5 GHz band vehicle-to-vehicle channels: Models for multiple values of channel bandwidth, In *IEEE Transactions on Vehicular Technology*, 59, 5, 2620–2625, 2010.

80. G. Charalampopoulos, T. Dagiuklas, and T. Chrysikos. V2I applications in highways: How RSU dimensioning can improve service delivery, In *2016 23rd International Conference on Telecommunications (ICT)*, May 16–18. 2016, Thessaloniki, Greece.

81. W. Viriyasitavat et al. Vehicular communications: Survey and challenges of channel and propagation models vehicular technology magazine, *IEEE Vehicular Technology Magazine*, 10, 2, 55–66, 2015.

82. P. Kyösti et al. WINNER II channel models, IST-4-027756 WINNER II Deliverable D1.1.2 v.1.2.4.2.2008. Available at https://cept.org/files/8339/winner2%20-%20 final%20report.pdf (accessed July 18, 2017).

83. Aalto University, BUPT, CMCC, Nokia, NTT DOCOMO, New York University, Ericsson, Qualcomm, Huawei, Samsung, Intel, University of Bristol, KT Corporation, University of Southern California, 5G Channel model for bands up to 100 GHz, Technical Report, December 6, 2015. Available at http://www.5gworkshops.com/5G_ Channel_Model_for_bands_up_to100_GHz(2015-12-6).pdf (accessed July 18, 2017).

84. I. Rodriguez et al. Radio propagation into modern buildings: Attenuation measurements in the range from 800 MHz to 18 GHz, In *2014 IEEE 80th Vehicular Technology Conference (VTC Fall)*, September 14–17, 2014, Vancouver, Canada.

85. T. Thomas et al. A prediction study of path loss models from 2–73.5 GHz in an urban-macro environment, In *2016 IEEE VTC-Spring 2016*, pp. 1–5, May 2016. Available at http://arxiv.org/abs/1512.01585.

86. H. C. Nguyen et al. An empirical study of urban macro propagation at 10, 18, and 28 GHz, In *2016 IEEE VTC-Spring 2016*, pp. 1–5, May 2016.

87. T. Chrysikos, G. Georgopoulos, and S. Kotsopoulos. Empirical calculation of shadowing deviation for complex indoor propagation topologies at 2.4 GHz. *ICUMT*, pp. 1–6, 2009.

88. T. Chrysikos and S. Kotsopoulos. Characterization of large-scale fading for the 2.4 GHz channel in obstacle-dense indoor propagation topologies. In *IEEE Vehicular Technology Conference (VTC Fall)*, 2012, Quebec, Canada.

Chapter 3

Advances in Millimeter Wave Propagation for 5G Mobile Communication Systems

Nektarios Moraitis

Contents

3.1 Introduction

It is projected that wireless data traffic will increase about 1000 fold by 2020, and likely more than 10,000 fold by the year 2030 (Muirhead et al. 2015). This tremendous increase in wireless capacity could be realized through an increase in performance, spectrum availability, and massive densification of small cells (Ghosh et al. 2014). These requirements have been envisioned for the research and development of the 5th-generation (5G) wireless communication networks. These networks will have to support high capacity and massive connectivity with an increasingly diverse set of services, applications, and users, such as machine-to-machine (M2M) and Internet of things (IoT), to enable communications anywhere, anytime, and by anything (Boccardi et al. 2014). The wireless spectrum below 6 GHz is already congested and will not be enough to meet these future requirements. Therefore, there has been growing interest in moving up frequency into millimeter-wave (mm-Wave) bands, between 30 and 300 GHz, where a tremendous amount of bandwidth is available. Furthermore, the advancement of semiconductor technology has made mm-Wave cellular systems feasible (Rappaport et al. 2011). Candidate frequency bands for the deployment of 5G networks are: the 28 GHz band (27.5–29.5 GHz), the 38 GHz band (36–40 GHz), the 60 GHz band (57–64 GHz), the E-band (71–76 GHz and 81–86 GHz), and finally, the W-band (92–95 GHz) (Ghosh et al. 2014). The communication scenarios of the 5G networks include outdoor as well as indoor users, which will be realized on the basis of the aforementioned mm-Wave frequency bands, as shown in Figure 3.1. Therefore, various communication links

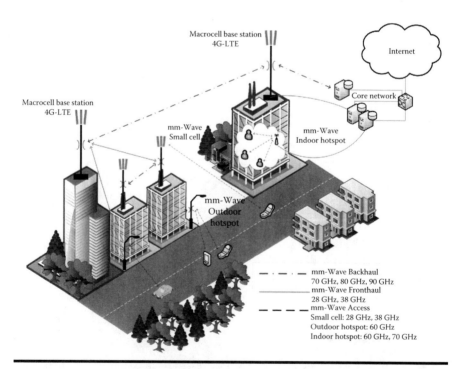

Figure 3.1 A prospective network deployment ushering the 5G era. The wireless links exploit the mm-Wave bands to realize all the communication scenarios including outdoor and indoor users.

are foreseen that include: (1) links from macro-to-macro cell base stations, that is, backhaul links; (2) fronthaul links between a macro-cell base station and the station of a small cell; and (3) radio access links for the end users (outdoor-to-outdoor and indoor-to-indoor), as also indicated in Figure 3.1.

More specifically, for an outdoor scenario and for access or fronthaul links, the frequencies of 28 or 38 GHz will be employed for the system operation, while for the indoor users (implementing the so called *femto-cellular* hotspots) the 60 or 70 GHz band will be considered. The 60 GHz band can be also utilized for short-range outdoor hotspots (up to 50 m). Finally, for the backhauling network architecture, the most probable frequency bands to be used are between 70 and 90 GHz.

The development of cellular networks in the mm-Wave band imposes significant challenges, which necessitate a rigorous assessment. Therefore, it is vital to understand the propagation channel characteristics across all the proposed mm-Wave frequencies in order to perform a precise and consistent system design. The propagation scenarios and models would incorporate fixed and mobile access users, indoor and outdoor environments, as well as line-of-sight (LOS) and non-line-of-sight (NLOS) links. Due to the higher free-space loss at mm-Wave signals in respect to the conventional systems currently operating up to 3 GHz, proportional increase in antenna gains with appropriate beam forming will be required. Another important issue is that mm-Wave signals can be severely susceptible to shadowing resulting in outages, rapidly varying channel conditions and discontinuous connectivity (Akdeniz et al. 2014). This is more concerning especially in urban environments where NLOS links will be also deployed.

The first research studies and measurement results for mm-Wave frequencies are reported in Xu et al. (2000), Rappaport et al. (2013a), and Rappaport et al. (2013b), and have shown that outdoor access links up to 200 m are viable and will be able to provide the appropriate bandwidth for the realization of the next-generation cellular systems. The fact that the cell size decreases, assists in the improvement of system's throughput and capacity, as the spatial frequency reuse will be lower. The propagation characteristics in the mm-Wave band include the fading due to local environment and the atmospheric effects. The latter imposes significant challenges to the signal transmission, which due to the very small wavelength introduces severe attenuation and link outages, especially in the upper part of the mm-Wave region. Aside from the rain attenuation, the atmospheric effects incorporate the oxygen and water vapor absorption, fog, and amplitude signal scintillation due to turbulence. The local fading effects include multipath propagation, as well as the shadowing effects and free-space loss. Furthermore, the vegetation attenuation and the building penetration are very important in the design of such links because the introduced attenuation is much higher compared to the frequencies up to 3 GHz. Therefore, all of the aforementioned mm-Wave propagation characteristics and effects are vital to be assessed and modeled, leading toward a reliable network design.

In this chapter, the current propagation trends of the mm-Wave frequencies toward 5G mobile communication systems are thoroughly studied and presented.

State-of-the art channel models, derived from the latest measurement campaigns, concerning various propagation scenarios and architectures are assessed. This chapter will mainly emphasize on channel models that describe the fading effects due to local environment, and less on the fading phenomena due to the atmosphere. Therefore, attention will be given especially in models that characterize the large and small scale fading effects of the mm-Wave channel, taking into account alternative propagation characteristics, and conditions. In addition, models that describe the spatial and temporal variations of the mm-Wave channel will be provided. The recent advances of multiple-input multiple-output (MIMO) and massive MIMO technology exploitation in mm-Wave propagation will be also investigated. Finally, future directions regarding the propagation and channel model prediction for 5G millimeter wave mobile communication systems will be proposed.

3.2 Advances in Millimeter-Wave Propagation

3.2.1 Overview of Research Activities in mm-Wave Propagation

There are several studies and measurement campaigns in the mm-Wave frequencies for wireless communications systems, dating back to 2000 that give valuable information for the future design challenges. In Xu et al. (2000), a wide-band measurement campaign has been performed on three outdoor point-to-point (P2P) links at 38 GHz. The purpose was to examine the effects of various weather events such as rain and hail on the behavior of the mm-Wave link. Multipath statistics showed that while very few multipath components were detected in clear, dry weather, multipath components were increased during rain events. It is also found that Rician K-factor (direct-to-scattered power ratio) depends on the point rainfall rate according to the linear relationship $K = 16.88 - 0.04R$, where K is the Rician factor (in decibels) and R is the rain rate (in mm/h). It can be observed that when the rain rate increases, either the power of the LOS component decreases, either the scattered power increases, or both. However, two possible explanations have been proposed, one is the increase of multipath due to the scattering on hydrometeors or due to the changes in the electrical properties of surfaces due to rain drops. Apparently, the proposed relationship combines the local multipath with the atmospheric effects and is usable for determining system outage probabilities.

Propagation measurements using steerable antennas were performed in Rappaport et al. (2012, 2013a), also at 38 GHz, considering an outdoor urban cellular, and pedestrian walkway scenario. The specific experiment uses a variety of elevated transmitters that represent typical 5G base-station locations at heights of two or more stories above ground level, and various ground-level receiver locations.

It is found that the elevated transmitters at heights of two to eight stories above ground require 60° of scanning (up to ±30° off-boresight) in the azimuth direction to cover nearly all possible NLOS links. The Rx antenna, however, would benefit from larger scanning freedom. A NLOS link is rarely preferred over a LOS or partially obstructed LOS link, as NLOS links tend to have 10–50 dB more path loss and higher expected root-mean-square (RMS) delay spread. Nevertheless, when the LOS component is completely blocked by a building or other shadowing objects, the results showed that a reflection, scattered, or diffraction path may still have sufficient power to be received, although at a lower level. Hence, for mm-Wave systems using beam-steering antenna arrays, in case of a blocked LOS path, the beam will be steered until the strongest NLOS link is identified. Almost all of the LOS links have near free-space path loss and no RMS delay spread (mean values on the order of 1 ns), whereas NLOS present higher RMS delays (mean values between 13.7 and 14.8 ns). Another interesting result is that many distinct paths can be formed in NLOS and LOS channels using narrow-beam antennas. By picking the best combination of Tx and Rx antenna pointing angles at any location, path loss and RMS delay spread can be substantially reduced. However, while NLOS paths may be formed in mm-Wave channels, they will require equalization and will have greater propagation latency, higher power consumption, and lower data rates than LOS channels (Rappaport et al. 2012). It is also important to note that for both the high and low base-station Tx locations, no outages were observed for all random measurement locations within a 200 m cell radius, suggesting that mm-Wave systems could work best in dense urban environments with small cell deployments within that range.

Extensive mm-Wave measurement results, for cellular 5G systems, are presented in Azar et al. (2013), Samimi et al. (2013), and Rappaport et al. (2013b), where outdoor point-to-multipoint measurements were carried out in a dense urban environment using steerable antennas at 28 GHz. In a dense urban environment such as New York, the path loss exponent was found 2.55 for LOS locations, but increases up to 5.76 in NLOS cases. Multipath delay spread is found to be much larger in dense cities, due to the highly reflective nature of the dense urban environment. The measured data show that a large number of resolvable multipath components exist in both LOS and NLOS environments, with observed multipath excess delay spreads as great as 1388.4 ns and 753.5 ns, respectively (Azar et al. 2013). From the angular resolved measurements it is found that dense cities are a multipath-rich environment, and that an average of 2.5 signal lobes exists at any Rx location. Each lobe has an average angle spread of 40.3°, and an RMS angle spread of 7.8°. The small scale fading is negligible and about ±2 dB about the mean power level over a 10 λ distance (Samimi et al. 2013). The widely diverse spatial channels observed at any particular location suggest that mm-Wave mobile communication systems with electrically steerable antennas could exploit resolvable multipath components to create reliable links for cell sizes on the order of 200 m. In addition, outdoor materials exhibit higher penetration losses in urban buildings and signals cannot propagate through, thus indoor networks will be totally isolated from outdoors, which advocates the installation of data showers, repeaters, and access points for handoffs at entrances of commercial and residential buildings (Rappaport et al. 2013b). Additional measurements at 28 GHz were performed in an urban environment in Daejeon, Korea, for NLOS conditions, where steerable antennas were used (Hur et al. 2015). After post-processing the angular received power, the omnidirectional measurements were synthesized for mm-Wave channel modeling. The NLOS path loss exponent was found 3.53, whereas from the spatio-temporal analysis the channel can be modeled with an average number of 3.62 clusters, with an RMS and subpath delay spread values of 22.29 and 5.45 ns, respectively. Finally, similar types of measurements at 28 GHz were carried out in Ko et al. (2016). Two different environments were examined. An open space stadium (having LOS conditions) and an urban ski-resort town (preserving NLOS characteristics). The measurement system utilizes steerable antennas at both Tx and Rx, but the omnidirectional results were synthesized for further processing. The spatio-temporal analysis showed that the mm-Wave LOS channel is modeled with 2.69 clusters and an RMS delay spread of 2.75 ns, whereas the NLOS links can be characterized by 5.33 clusters, and an RMS delay spread of 60.41 ns.

The license-free band at 60 GHz (providing a spectrum width of up to 7 GHz) has been considered promising, foreseen either for short range indoor communications, or for short-range outdoor cellular mobile multi-Gbps systems (Rappaport et al. 2012). Comparative measurements between 38 and 60 GHz, regarding an urban peer-to-peer scenario, revealed that the 60 GHz channel experiences notably more path loss, smaller delay spreads, and fewer unique antenna angles for creating a link (Rappaport et al. 2012). The mean RMS delay spread was found 23.6 and 7.4 ns at 38 and 60 GHz, respectively, considering NLOS links. However, very low mean values (much lower than 10 ns) were observed in LOS cases at both frequencies. Apparently, the very small wavelength of the signals at 60 GHz band constitutes the propagation quasi-optical in nature with most of the signal power received through the LOS, and first- and second-order reflections (Maltsev et al. 2010). There is an extensive research especially for indoor 60 GHz radio channels where a *femto-cellular* architecture will be adopted, providing high data rate services for densely populated buildings, carrying many times more traffic than current systems (Anderson and Rappaport 2004). Indoor cell sizes are on the order of 50 m, therefore, the oxygen attenuation has no impact and can be neglected (Moraitis and Constantinou 2004). Likewise, reliable outdoor access links will be created for similar cell radii.

Polarization measurements in an indoor conference room at 60 GHz channel have shown that a mismatch of polarization characteristics of Tx and Rx antennas can result in a large degradation of the received signal power between 10 and 20 dB (Maltsev et al. 2010). Furthermore, measurements in the exact same environment (Maltsev et al. 2009), using steerable antennas, have shown that the 60 GHz channel is formed by five distinct clusters. One path (LOS or a reflected path) constitutes a single cluster, and the scattered rays inside one path comprise an intra-cluster structure, which statistical characteristics include the average number of rays, ray arrival rate, and a power decay rate. Actually, the intra-cluster parameters of a single cluster define the delay spread or the frequency selective characteristics of the propagation channel.

Comparative indoor channel measurements at 60 and 70 GHz in four different environments, using a steerable antenna only at the Tx, were carried out in Haneda et al. (2015) where the spatio-temporal characteristics of the channels were derived. The results showed that the dominance of the LOS and specular propagation mechanisms is more apparent in open indoor environments such as large shopping mall and railway station, than in office environments. In addition, 70 GHz radio channels suffer from higher losses of both specular and diffuse components than at 60 GHz and show faster power decay as the propagation delay increases.

Indoor wideband measurements at 60 GHz, using directional antennas, showed that the RMS delay spread varies from 12.34 to 15.04 ns in corridors and from 12.56 to 21.09 ns inside laboratory environments (Moraitis and Constantinou 2006). Another convenient way is to describe the mm-Wave channel as a tapped delay line. Tapped delay line models for 60-GHz channels are proposed in Sawaya and Clavier (2003) and Moraitis and Constantinou (2006), where the channel can be represented with six or four taps, respectively. In both cases the proposed tap delay line models are derived on the basis that the channel preserves wide-sense stationary uncorrelated scattering (WSSUS) characteristics. Furthermore, from wideband measurements in Xu et al. (2002), conducted in different indoor office environments at 60 GHz, it was found that when the receiver antenna is pointing into different directions, the delay spread is in the range of 7.5–76.46 ns for hallways and 10.89–41.01 ns for rooms. When the receiver antenna is fixed, the delay spread is significantly smaller. Typical ranges are from 4.6 to 47.3 ns in the hallways and from 4.7 to 33.9 ns in rooms. The angular distribution of multipath power is briefly characterized by angular spread. Angular spread results in Xu et al. (2002) showed dependency on the examined environment. Measurement results show that the angular spread ranges from 0.36 to 0.78 for hallway measurements, 0.62–0.84 for room measurements, from 0.63 to 0.81 for indoor NLOS propagation, and 0.12–0.49 for outdoor measurements with few nearby obstructions. These measurement results show a clear increase of the angular spread from the least cluttered outdoor environments to obstructed NLOS indoor environments. Additional angular spread measurements were conducted in Moraitis et al. (2011), where for room LOS conditions (either laboratory or residence) the angular spread takes large values indicating that multipath comes from different directions. This is attributed to the heavy cluttered environment where lot furniture is present in the propagation channel. In residential environments the angular spread was found between 0.81 and 0.89 in LOS locations, and between 0.41 and 0.45 in NLOS locations, respectively (Moraitis et al. 2011). Nevertheless, sophisticated measurement and modeling of angle-of-arrival (AOA) is required to allow optimization of beam steering requirements and algorithms.

Indoor and outdoor measurements at 73-GHz band were performed in MacCartney et al. (2015a, 2015b) and Rappaport et al. (2015). Increased values of path loss exponent were observed in indoor environments at 73 GHz, having values of 3.1 and 5.7, for LOS and NLOS conditions, respectively. On the other hand, very low values of RMS delay spread were obtained, where the mean values do not exceed 12.1 ns in both LOS and NLOS cases. The low delay spread values can be attributed to the increased path loss in the specific frequency band. Outdoor 73 GHz experiments in Manhattan, New York, showed a path loss exponent of 2.3 and 4.7, for LOS and NLOS links, respectively. The results corroborate the fact that when directional antennas are used, the LOS channel provides virtually no delay spread values. Maximum RMS delay spread values in LOS cases were found between 1.2 and 1.5 ns, whereas in NLOS urban links the maximum observed delays are between 30.3 and 48.5 ns. The aforementioned delay values are smaller than these observed in 28 GHz band, due to the more reflective environment of Manhattan and greater energy being scattered at lower mm-Wave frequencies. Wavelengths at 73 GHz are smaller than 28 GHz, giving rise to more diffuse scattering during propagation, which results in weaker path not detectable at the Rx (Rappaport et al. 2015).

Special care, in designing indoor mm-Wave systems, should be taken in case of a human presence and motion in the wireless channel. Both transmission and diffraction effects are very small especially in the 60 and 70 GHz bands due to its quasi-optical nature, a human body constitutes a practically insurmountable obstacle and produces attenuation values around 22 dB when the human is inside the direct LOS path (Moraitis and Constantinou 2004). Measurements in Colloge et al. (2004) showed that when the direct path is shadowed by a person, the attenuation generally increases by more than 20 dB, for a median duration of about 100 ms for an activity of 1–5 persons and 300 ms for 11–15 persons. Globally, the channel is *unavailable* for about 1% or 2% of the time in the presence of one to five persons. In addition, from experiments in Cassioli and Rendevski (2014), for the 54–59 GHz and 61–66 GHz bands, respectively, was observed that the induced shadowing by human bodies varied between 12 and 25 dB, whereas was statistically modeled by the truncated second order Gaussian mixture model (GMM) (Jacob et al. 2011). The GMM was developed on the basis of a ray tracing procedure for the human blockage, where the human body is represented by a multiple knife-edge diffraction model (Jacob et al. 2011). The experimental results from the latter work manifest that the influence of a person induced shadowing phenomena that may exceed 35 dB. The RMS delay spread increases from 2.6 ns in LOS situations to 6.5 ns when human blockage occurs.

Finally, the presence of many persons moving at various speeds up to 1 m/s, results in a Doppler spread of about 800 Hz and, equivalently, a coherence time of about 1.25 ms (Smulders 2009). This is the absolute lower limit that might be approached when omnidirectional antennas are used. Measurements in Moraitis and Constantinou (2004), where people walked along the LOS path while not obstructing it, with a maximum walking speed of 1.7 m/s, revealed a minimum coherence time of 32 ms which is, indeed, much higher than the aforementioned limit. The channel dynamics preserved slow fading behavior characterized as quasi-wide sense stationary.

To conclude, taking human mobility into consideration, mm-Wave links are intermittent. Therefore, maintaining a reliable link for delay-sensitive applications would be a challenging task for mm-Wave communications. Applying appropriate beamforming techniques in indoor environments alternative paths could be used to preserve a consistent connection when the LOS path is blocked by human presence.

3.2.2 Material Characteristics in mm-Wave Frequencies

A consistent planning and design of future mm-Wave wireless networks requires a thorough knowledge of the common building material characteristics such as reflection and penetration loss (induced attenuation). Initial reflection and penetration measurements were conducted at the frequency of 38 GHz (Xu et al. 2000). The results revealed a 25.5 dB penetration loss through a double-pane, tempered, and tinted window glass. Furthermore, controlled reflection measurements on glass and brick surfaces showed that the specular reflection can increase as much as 6.8 dB when the surface is wet, which constitutes wet exterior surfaces highly reflective. Significant penetration losses of up to 17 dB were also observed through a dense canopy of an oak tree, corroborating vegetation attenuation as a crucial parameter for a proper link budget analysis, averting outdoor access link outages. Attenuation measurements for various material types were carried out in Moraitis and Constantinou (2004) for an indoor office environment at 60 GHz. The results showed that the induced attenuation depends significantly on the thickness of the material. Indoor soft partitions such as plywood panels or simple glass, introduce a 6 and 3.5 dB attenuation, respectively, which renders NLOS propagation feasible for very short distances (up to 10–15 m, and up to four intervening partitions). For greater distances and higher number of intervening partitions the introduced isolation makes intra-room communication impossible, giving rise to the single-cell-per-room network configuration (Anderson and Rappaport 2004). On the other hand, exterior materials such as brick walls showed a high attenuation of up to 48 dB, constituting indoor-to-outdoor links beyond reality, isolating indoor mm-Wave hotspots. In addition, reflection and penetration loss measurements at 28 GHz, for common building materials, were conducted in Rappaport et al. (2013b). Comparing the results for indoor and outdoor materials, the latter presents a higher reflection coefficient of 0.986 for tinted glass, and 0.815 for concrete, as compared to clear nontinted glass and drywall, which have lower reflection coefficients of 0.740 and 0.704, respectively. The highly reflective external building materials can enhance outdoor signal coverage, and allow a wide range of possible angles-of-arrival to create Tx–Rx links in outdoor environments. The results showed that the typical exterior surface of urban buildings such as tinted glass or brick pillars exhibits high penetration loss that could reach 28.3 and 40.1 dB, respectively. This indicates that building penetration of mm-Waves will be extremely difficult for outdoor transmitters, thus providing sufficient isolation between outdoor and indoor networks. On the contrary, typical indoor materials at 28 GHz such as clear

Table 3.1   Measured Penetration Loss of Common Building Materials in mm-Wave Frequencies

Reference	Frequency (GHz)	Location	Material Description	Thickness (cm)	Penetration Loss (dB)
Rappaport et al. (2013b)	28	Exterior	Tinted glass	3.8	40.1
			Brick pillar	185.4	28.3
		Interior	Clear glass	<1.3	3.9
			Tinted glass	<1.3	24.5
			Wall	38.1	6.8
Xu et al. (2000)	38	Exterior	Double-pane, tampered tinted glass	–	25.5
Moraitis and Constantinou (2004)	60	Interior	Double glass	1.5	4.5
			Simple glass	0.5	3.5
			Whiteboard	1.5	11.6
			Plywood panels	0.5	6
		Exterior	Brick wall with plasterboard	23	48

nontinted glass and drywall only have 3.6 and 6.8 dB of losses, respectively, which are relatively low. Therefore, the high penetration loss through outdoor building materials and low attenuation through indoor materials suggest that RF energy can be contained in intended areas within buildings, which reduces cochannel inter-cell interference, yet making outdoor-to-indoor penetration difficult (Rappaport et al. 2013b). Table 3.1 summarizes the penetration losses for various material types derived by measurement campaigns in the mm-Wave frequencies.

The aforementioned measurement results revealed that is infeasible to achieve indoor coverage from outdoor base stations, as buildings are almost impenetrable from outdoor access links. In that case, relays and repeaters will likely be used to achieve indoor-to-outdoor coverage, or else outdoor mobile users will need to hand-off into the indoor network (perhaps unlicensed spectrum or reused mm-Wave spectrum) as a user enters a building. Furthermore, indoor-to-outdoor penetration is also impractical in mm-Wave frequencies, isolating effectively indoor hotspots. In general, the lower penetration loss of indoor materials in conjunction with the reflective and high loss outdoor materials (brick walls and glass) helps reduce interference between indoor and outdoor mm-Wave networks, suggesting a high frequency reuse.

3.2.3 Large Scale Channel Models in mm-Wave Frequencies

For link budget and interference calculations, path loss as a function of propagation distance is fundamental for system coverage and interference analysis and design (Rappaport et al. 2015). In order to predict the path loss, various empirical models have been proposed and will be described in the following. The first model is the single slope or close-in (CI) free-space reference distance path loss model that incorporates a path loss exponent which has physical relevance, as path loss is tied to the free-space loss at a specific reference distance (usually 1 m is convenient and practical at mm-Wave frequencies) (Ghosh et al. 2014). The modeled path loss (in decibels) is given by

$$PL^{CI}(f,d) = PL_{FS}(f,d_0) + 10\bar{n} \log_{10}\left(\frac{d}{d_0}\right) + X_\sigma \tag{3.1}$$

where:
  d_0 is the close-in free-space reference distance
  $PL_{FS}(f,d_0)$ is the free-space path loss
  $\bar{n}$ is the path loss exponent
  d is the distance between transmitter (Tx) and receiver (Rx)

X_σ is the shadow fading term, which is a zero-mean Gaussian random variable with a given standard deviation, σ (i.e., the shadow factor in decibels) that represents the large-scale fluctuations originating from shadowing by large obstructions in the wireless channel (Rappaport 2002). In Equation 3.1 the free-space path loss at a reference distance d_0 is given by

$$PL_{FS}(f,d_0) = 20\log_{10}(f) + 20\log_{10}\left(\frac{4\pi d_0 \cdot 10^9}{c}\right) \tag{3.2}$$

where:
  c is the speed of light
  f is the frequency, scaled in GHz

The CI model is a fundamental large scale path loss model, whereas $\bar{n}$ is derived from a best fit in a minimum mean square error (MMSE) sense over all the measured data. Table 3.2 summarizes the path loss exponent and shadowing factor results for the CI model, derived from a wide range of measurement campaigns, considering all the candidate frequency bands for the next-generation mm-Wave networks. As it is evident the path loss exponent is highly dependable on the measurement environment and propagation condition, where in the majority of the LOS links the exponent is found close to 2.0. On the other hand, in NLOS situations the path loss exponent can reach up to 5.76 in dense urban environments. The provided omnidirectional data, synthesized from the results of the rotating Tx,

Table 3.2 Typical Parameters of the Close-in Path Loss Model Collected from a Wide Range of Measurement Campaigns, Considering Alternative Environment Types and Operational mm-Wave Frequencies

Reference	Frequency (GHz)	Environment	Antenna Type Tx/Rx	$\bar{n}$		σ (dB)	
				LOS	NLOS	LOS	NLOS
Azar et al. (2013)	28	Dense urban (Brooklyn/Manhattan-NYC)	Dir/Dir NB/NB Fixed/Fixed	2.55	5.76	8.66	9.02
Kim et al. (2015)	28	Urban low-rise (Gyanpyung, Daejeon)	Dir/Dir NB/NB Fixed/Fixed	2.06	3.52	4.03	8.87
		Urban very high-rise (Gangnam, Seoul)	Dir/Dir NB/NB Fixed/Fixed	2.21	2.99	3.74	4.6
Hur et al. (2015)	28	Typical urban (Daejeon, S. Korea)	Omni/Omni[a] Fixed/Fixed	–	3.53	–	6.69
Ko et al. (2016)	28	Open square stadium (Pyeongchang ski resort)	Omni/Omni[a] Fixed/Fixed	2.09	–	3.36	–
MacCartney et al. (2015b)	28	Dense urban (Manhattan-NYC)	Omni/Omni[a] Fixed/Fixed	2.1	3.4	3.6	9.7

(Continued)

Table 3.2 (Continued) Typical Parameters of the Close-in Path Loss Model Collected from a Wide Range of Measurement Campaigns, Considering Alternative Environment Types and Operational mm-Wave Frequencies

Reference	Frequency (GHz)	Environment	Antenna Type Tx/Rx	$\bar{n}$		σ (dB)	
				LOS	NLOS	LOS	NLOS
Rappaport et al. (2012)	38	Pedestrian walkway (University of Texas campus)	Dir/Dir NB/NB Fixed/Fixed	2.0	4.57	3.79	11.72
		Light urban campus (University of Texas) Cellular channel	Dir/Dir NB/NB Fixed/Fixed	2.20	3.88	10.3	14.6
Rappaport et al. (2013a)	38	Light urban campus (University of Texas) Cellular channel	Dir/Dir NB/WB Fixed/Fixed	1.9	2.8	3.5	10.3
MacCartney et al. (2015b)	38	Light urban campus (University of Texas)	Omni/Omni[a] Fixed/Fixed	1.85	2.5	3.8	8.8
Rappaport et al. (2012)	60	Pedestrian walkway (University of Texas campus) Peer-to-peer channel	Dir/Dir NB/NB Fixed/Fixed	2.25	4.22	2.0	10.12
Ben-Dor et al. (2011)	60	Inter-vehicle (Urban parking lot) Vehicular channel	Dir/Dir NB/NB Fixed/Fixed	2.5	5.4	3.5	14.8
Moraitis and Constantinou (2004)	60	Indoor office	Dir/Dir NB/NB Fixed/Fixed	1.8	–	1.13	–

(Continued)

Table 3.2 (Continued) Typical Parameters of the Close-in Path Loss Model Collected from a Wide Range of Measurement Campaigns, Considering Alternative Environment Types and Operational mm-Wave Frequencies

Reference	Frequency (GHz)	Environment	Antenna Type Tx/Rx	$\bar{n}$ LOS	$\bar{n}$ NLOS	σ (dB) LOS	σ (dB) NLOS
Moraitis and Panagopoulos (2015)	60	Indoor corridor	Dir/Dir NB/NB Fixed/Mobile	1.95	–	3.05	–
		Indoor laboratory		2.35	–	2.55	–
		Empty residence		2.07	–	3.17	–
		Furnished residence		1.95	2.85	4.05	4.25
Rappaport et al. (2015)	73	Dense urban (Manhattan-NYC)	Dir/Dir NB/NB Fixed/Fixed	2.3	4.7	6.1	12.6
MacCartney et al. (2015b)	73	Dense urban (Manhattan-NYC)	Omni/Omni[a] Fixed/Fixed	2.0	3.4	4.8	7.9
MacCartney et al. (2015a)	73	Typical indoor office (MetroTech Center-NYC)	Dir/Dir NB/NB Fixed/Fixed	3.1	5.7	16.8	16.7
			Omni/Omni[a] Fixed/Fixed	2.4	3.8	12.0	12.9

Note: Dir stands for directional antenna, Omni denotes omnidirectional antenna, and finally, NB, and WB indicate narrow-beam and wide-beam antennas, respectively.

[a] The omnidirectional results are derived after synthesizing the received power of the rotating directional antennas.

Rx antennas, can also be utilized through the use of 3D ray-tracing algorithms and network capacity simulators.

An alternative path loss model is the floating intercept (FI), or alpha-beta model, used in the WINNER II and 3GPP standards. The FI model has no physical reference, but simply fits the best line to the measured data. This model has the following form (Sun et al. 2016a):

$$PL^{\mathrm{FI}}(d) = \alpha + 10\beta \log_{10}(d) + X_\sigma \tag{3.3}$$

where:

α is the intercept in decibels
β is the slope, both determined via a MMSE fit to the measured data
X_σ is the shadow fading term

In Equation 3.3, β cannot be assumed to be the same as the path loss exponent $\bar{n}$ because it is floating, has no physical meaning or frequency dependence, and simply serves as a value of slope that provides the best fit to the data. The advantage of this model is that minimizes the standard deviation (in other words the mean square error fit to the data) but its drawback is that different researchers cannot directly derive intuitive information from the two model parameters. Both CI and FI models are usable to characterize mm-Wave channels either in LOS or NLOS propagation conditions. Nevertheless, for path loss channel modeling, it is preferable to use the CI model as it has a simpler form and a physical meaning, as well as the benefit of having an agreed-upon standard that is usable with reasonable accuracy across many environments, scenarios, and frequency bands.

Table 3.3 lists the parameters of the FI model (α, β, and σ) obtained after a MMSE fit to the measured data, considering alternative measurement campaigns, mm-Wave frequencies and propagation environments. As previously mentioned, parameter β has no physical meaning, and according to the results introduced in Table 3.3, cannot be correlated with path loss exponent.

A multifrequency three-parameter model that aims to predict large-scale path loss as a function of both frequency and distance is the alpha-beta-gamma (ABG) model. The predicted path loss in decibels is expressed as follows (MacCartney et al. 2015a):

$$PL^{\mathrm{ABG}}(f,d) = 10\alpha \log_{10}\left(\frac{d}{d_0}\right) + \beta + 10\gamma \log_{10}(f) + X_\sigma \tag{3.4}$$

where:

α and γ are coefficients that describe the distance and frequency dependence on path loss
β is an optimized offset parameter that is devoid of physical meaning
f is the frequency in GHz
X_σ is the shadow fading term

Table 3.3 Typical Parameters of the Floating Intercept Path Loss Model Derived from Various Measurement Campaigns, Considering Alternative Environment Types and Operational mm-Wave Frequencies

Reference	Frequency (GHz)	Environment	Antenna Type Tx/Rx	LOS α β	LOS σ (dB)	NLOS α β	NLOS σ (dB)
Nguyen et al. (2016)	28	Urban residential district (Vestby, Aalborg)	Dir, WB/Omni Fixed/Mobile	61.5 2.1	4.2	63.8 2.5	6.4
Ko et al. (2016)	28	Typical urban (Alpensia resort town)	Omni/Omni[a] Fixed/Fixed	–	–	3.96 52.15	5.69
Rappaport et al. (2015)	28	Dense urban (Manhattan-NYC)	Dir/Dir NB/NB Fixed/Fixed	45.3 2.9	0.04	57.6 4.7	10.0
Rappaport et al. (2015)	38	Light urban campus (University of Texas) Cellular channel	Dir/Dir NB/WB Fixed/Fixed	71.1 1.6	3.75	116.9 0.68	7.8
Rappaport et al. (2015)	73	Dense urban (Manhattan-NYC)	Dir/Dir NB/NB Fixed/Fixed	127.9 –1.2	4.6	118.2 2.1	11.3
MacCartney et al. (2015a)	73	Typical indoor office (MetroTech Center-NYC)	Dir/Dir NB/NB Fixed/Fixed	94.5 0.7	15.8	117.8 1.3	11.7
			Omni/Omni[a] Fixed/Fixed	86.3 0.8	11.3	88.1 2.2	12.1

Note: Dir stands for directional antenna, Omni denotes omnidirectional antenna, and finally NB, and WB indicate narrow-beam and wide-beam antennas, respectively.

[a] The omnidirectional results are derived after synthesizing the received power of the rotating directional antennas.

The ABG model is an extension of the FI model for multiple frequencies, and reverts to the FI model if only a single frequency is used (when $\gamma = 0$ or 2). The ABG model is solved in a MMSE sense by simultaneously solving for α, β, and γ.

Table 3.4 summarizes the basic parameters of the multifrequency ABG model, considering alternative environments and propagation conditions for a wide frequency range. According to Sun et al. (2016b), the ABG model presents a difference in shadowing factor between −0.1 and 1.2 dB, compared to the CI model, which indicates the good approximation that succeeds.

A recent path loss model also suitable for multifrequency modeling is the close-in free-space reference distance with frequency-dependent path loss exponent (CIF) path loss model. The CIF model is considered as a general form of CI, and is given, in decibels, by (Sun et al. 2016a):

$$PL^{\mathrm{CIF}}(f,d) = PL_{FS}(f,d_0) + 10n\left[1 + b\left(\frac{f - f_0}{f_0}\right)\right]\log_{10}\left(\frac{d}{d_0}\right) + X_\sigma \qquad (3.5)$$

where:

n denotes the distance dependency of path loss (i.e., the path loss exponent)

b is an intuitive model-fitting parameter that characterizes the slope of linear frequency dependency of path loss

$PL_{FS}(f, d_0)$ is the free-space path loss given by Equation 3.2

f_0 is a fixed reference frequency that serves as the balancing point of the linear frequency dependency of n, and is based on the weighted average of all frequencies represented by the model

X_σ is the shadow fading term

If data from only one frequency are used, then $b = 0$, and the CIF model simplifies to the classic CI model. The parameter f_0, expressed GHz, is given by Sun et al. 2016a:

$$f_0 = \frac{\displaystyle\sum_{k=1}^{K} f_k N_k}{\displaystyle\sum_{k=1}^{K} N_k} \qquad (3.6)$$

where f_0 is the weighted frequency average of all measurements for each specific environment and antenna scenario, found by summing up, over all frequencies, the number of measurements N_k at a particular frequency and antenna scenario, multiplied by the corresponding frequency f_k, and dividing that sum by the entire number of measurements $\sum_{k=1}^{K} N_k$ taken over all frequencies for that specific environment and antenna scenario. Table 3.5 presents the typical parameters of the

Table 3.4 Typical Parameters of the Multifrequency ABG Path Loss Model Derived from Various Measurement Campaigns, Considering Alternative Environment Types

Reference	Frequency (GHz)	Environment	Antenna Type Tx/Rx	LOS		NLOS	
				α β γ	σ (dB)	α β γ	σ (dB)
Sun et al. (2016a)	28 and 73	Dense urban (Manhattan-NYC)	Dir/Dir NB/NB Fixed/Fixed	1.0 55.0 1.7	4.3	2.8 46.7 1.9	8.4
MacCartney et al. (2015a)	28 and 73	Typical indoor office (MetroTech Center-NYC)	Dir/Dir NB/NB Fixed/Fixed	0.7 47.1 2.5	14.2	1.9 27.5 4.5	11.0
			Omni/Omni[a] Fixed/Fixed	1.1 17.7 3.5	9.5	2.9 4.5 4.1	11.6
Sun et al. (2016b)	2–38	Urban macrocell	Dir/Dir NB/NB Fixed/Fixed	1.9 35.8 1.9	2.4	3.5 13.6 2.4	5.3
	2.9–73	Urban microcell	Dir/Dir NB/NB Fixed/Fixed	–	–	2.8 31.4 2.7	6.8
	2.9–73	Indoor office hotspot		1.6 32.9 1.8	4.5	3.9 19.0 2.1	7.9

Note: Dir stands for directional antenna, Omni denotes omnidirectional antenna, and finally NB, and WB indicate narrow-beam and wide-beam antennas, respectively.

[a] The omnidirectional results are derived after synthesizing the received power of the rotating directional antennas.

Table 3.5 Typical Parameters of the CIF Path Loss Model Derived from Various Measurement Campaigns, Considering Alternative Environment Types

Reference	Frequency (GHz)	Environment	Antenna Type Tx/Rx	LOS n β	LOS σ (dB)	NLOS n β	NLOS σ (dB)
Sun et al. (2016a)	28 and 73	Dense urban (Manhattan-NYC)	Dir/Dir NB/NB Fixed/Fixed	2.0 −0.06	4.4	3.4 0.0	8.4
MacCartney et al. (2015a)	28 and 73	Typical indoor office (MetroTech Center-NYC)	Dir/Dir NB/NB Fixed/Fixed	3.0 0.07	15.2	5.2 0.21	14.5
			Omni/Omni[a] Fixed/Fixed	2.1 0.32	9.9	3.4 0.22	11.9
Sun et al. (2016b)	2–38	Urban macrocell	Dir/Dir NB/NB Fixed/Fixed	2.0 −0.014	2.4	2.9 −0.002	5.7
	2.9–73	Urban microcell	Fixed/Fixed	–	–	3.2 0.076	7.1
	2.9–73	Indoor office hotspot		1.5 −0.102	4.4	3.1 −0.001	8.3

Note: Dir stands for directional antenna, Omni denotes omnidirectional antenna, and finally NB, and WB indicate narrow-beam and wide-beam antennas, respectively.

[a] The omnidirectional results are derived after synthesizing the received power of the rotating directional antennas.

CIF model, considering alternative environments and propagation conditions for a wide frequency range. In the majority of the cases, parameter β in CIF model was found almost zero. In cases where β is slightly positive, indicates that path loss increases with frequency beyond the first meter of free-space propagation (Sun et al. 2016b).

The CI and CIF models incorporate a close-in, free-space anchor point (d_0) that guarantees that the path loss model (regardless of transmit power) always has a physical connection to the transmitted power over distance. On the other hand, FI and ABG models use a floating constant based on a fit to the measured data, without consideration for the close-in, free-space propagation that always occurs in practice near an antenna out in the open. This implies that particular measured path loss values could greatly affect FI and ABG model parameters, as there is not a physical anchor to assure that close-in free-space transmission occurs in the first meter of propagation from the transmit antenna (Sun et al. 2016b).

Coverage distances in 5G networks will be much shorter when utilizing mm-Wave frequencies. Furthermore, in future small cells, base stations are likely to be mounted closer to obstructions (Rappaport et al. 2013b; 2015). Consequently, the $d_0 = 1$ m reference distance, used in CI and CIF models, is a proposed standard that ties the true transmitted power or path loss to a convenient close-in distance. Standardizing to a reference distance of 1 m makes comparisons of measurements and models simple and provides a standard definition for the path loss exponent, while enabling perception and quick evaluation of large-scale path loss. According to Rappaport et al. (2015), emerging mm-Wave mobile systems will have very few, or no users at all, within a few meters of the transmit antenna, as the base stations will be mounted on rooftops, lampposts, or indoor ceilings. Users in the near field will have strong signals or will be power controlled compared to typical users much farther from the Tx such that any path loss error in the near field will be negligible, and so much smaller than the dynamic range of signals experienced by users in a commercial system.

Another path loss model that is applicable for NLOS street canyon environments is provided by ITU-R, Recommendation P.1411-8 (ITU 2015). The specific model considers that the dominant NLOS obstruction in street grid is the corner transition and the path loss is calculated according to:

$$PL_{\text{NLOS}}(x_1, x_2; L_{\text{corner}}, d_{\text{corner}}, \beta) = PL_{\text{LOS}}(x_1) + L_c + L_{\text{att}} \tag{3.7}$$

where:

x_1 and x_2 are distance parameters denoting the distance between the Tx and the corner and that between the corner and Rx, respectively

PL_{LOS} is the LOS path loss given by Equation 3.1

The parameters L_c and L_{att} are expressed as follows (ITU 2015):

$$L_c = \begin{cases} \dfrac{L_{corner} \log_{10}\left(x_2 - \dfrac{w_1}{2}\right)}{\log_{10}\left(1 + d_{corner}\right)}, & \dfrac{w_1}{2} + 1 < x_2 \leq \dfrac{w_1}{2} + 1 + d_{corner} \\[4ex] L_{corner}, & x_2 > \dfrac{w_1}{2} + 1 + d_{corner} \end{cases} \tag{3.8}$$

$$L_{att} = \begin{cases} 10\beta \log_{10}\left(\dfrac{x_1 + x_2}{x_1 + \dfrac{w_1}{2} + d_{corner}}\right), & x_2 > \dfrac{w_1}{2} + 1 + d_{corner} \\[4ex] 0, & \text{otherwise} \end{cases} \tag{3.9}$$

and w_1 is the street width at the transmitter's side. From Equations 3.8 and 3.9, the NLOS path loss is parameterized by L_{corner}, d_{corner}, and β, which values, for the frequency band below 16 GHz, are suggested as 20 dB, 30 m, and 6, respectively, in an urban environment (ITU 2015). This model was developed, based on measurements at frequency range between 2 and 16 GHz, and has been used in many applications as a standard. An extension of ITU-R P.1411-8 model up to 28 GHz is presented in Lee et al. (2016). Measurements were conducted at 28 GHz, in an urban district of Seoul, which is an area composed of rectangular flat street grids with skyscrapers and 54 m wide streets. Fitting the path loss model parameters, L_{corner}, d_{corner}, and β to the measured data, minimizing the mean square error (MSE), the predictions were found to be in a good agreement with measured data for $L_{corner} = 24.59$ dB, $d_{corner} = 30$ m, and $\beta = 3.65$. The MSE was found 4.83 dB, verifying the good applicability of the ITU-R NLOS model at 28 GHz.

Finally, a combined path loss model that is applicable in both LOS and NLOS situations is presented in Moraitis and Panagopoulos (2015), where measurements were performed in different indoor environments at 60 GHz (including offices, hallways, and residences). The specific model takes into account the additional loss induced by the intervening partitions between the Tx and Rx. The path loss expressed in decibels is given by

$$PL(d) = \underbrace{PL_{FS}\left(d_0\right) + 10\bar{n}\log_{10}\left(\frac{d}{d_0}\right)}_{\text{LOS}} + \underbrace{X_\sigma + \sum_i k_i F_i}_{\text{NLOS}} \tag{3.10}$$

where:

$\bar{n}$ is the path loss exponent

$PL_{FS}(d_0)$ is the free-space path loss given by Equation 3.2

X_σ is the shadow fading term

The NLOS part of the expression considers the number, k_i, of the intervening partitions, as well as the attenuation factor, F_i, of the specific partition (penetration loss of the material). Specific values of material penetration losses can be selected from Table 3.1. In LOS situations it is plausible that $k_i = 0$ in Equation 3.10. On the other hand, in NLOS conditions the model is applied considering the excess loss introduced by the intervening partitions and fitting the model in a MMSE sense to find the path loss exponent. For example in NLOS cases, in the residential environment, the excess losses were between 24 and 44 dB (Moraitis and Panagopoulos 2015). In addition, it is interesting to note the differences in the shadowing factor between furnish and unfurnished residences where in the latter case increased values were observed. In general, the NLOS version of Equation 3.10 fits very well to the measured data obtaining standard deviations on the order of 4 dB.

3.2.4 Small Scale Channel Models in mm-Wave Frequencies

Small-scale fading represents the fluctuations in the received signal over short, sub-wavelength receiver distances and is physically explained by the coherent phasor sum of many random multipath components (MPCs) arriving within the measurement system resolution (Rappaport 2002). Actually, the sum of the MPCs constitutes the channel impulse response (CIR). It is a common assumption that CIRs obtained from appropriate wideband measurement systems are typically ray-like and can be expressed as a temporal discrete multipath model as follows:

$$h(\tau;t) = \sum_{m=1}^{M} a_m(t) e^{j\theta_m(t)} \delta\left[\tau - \tau_m(t)\right] \tag{3.11}$$

where:

M is the total number of rays

a_m and θ_m are the amplitude and phase of the mth ray

τ_m is the delay of the mth ray

$\delta(\cdot)$ denotes the Dirac delta function

In Equation 3.11, a_m is a statistical complex expression of the path gain. The phase θ_m follows a uniform distributed random variable over $[0, 2\pi)$, whereas the amplitude follows a log-normal distributed random variable with standard deviation σ. Note that τ represents the excess delay, whereas the t-dependence represents the changes with time of the very structure of the impulse response. Whenever, $\tau = 0$, then Equation 3.11 simplifies to a narrowband channel representation. The environment where the majority of mm-Wave measurement campaigns were conducted (Table 3.1) is changing much slower than the transmission rate, so that the Doppler effects can be neglected, that is, the time dependence in Equation 3.11 can be disregarded and the parameters can be treated as virtually time-invariant random variables. Therefore, the channel can be considered as physically stationary, exhibiting wide-sense stationary (WSS) characteristics. The channel representation given by Equation 3.11 can be regarded as single-cluster time-of-arrival model. In order to determine the average power for a ray at excess time τ the average power delay profile (APDP) is modeled as a function of excess delay that consists of a direct ray, and an exponentially decaying part (linear in dB). The APDP or average path power gain can be expressed as (Smulders 2009):

$$\overline{|a_m|^2} = \Omega_0 e^{-\tau_m/\gamma} \tag{3.12}$$

where:

Ω_0 stands for the average path power gain of the first arrived component, that is, $\overline{|a_1|^2}$

γ denotes the and ray power-decay time constant

The empirical frequency distributions of the ray inter-arrival time approximate very well the exponential distribution, which implies that the ray arrival process approximates a Poisson process (the probability density function [PDF] of ray inter-arrivals is exponential).

A well-known discrete impulse response model that takes ray-clusters into account is the Saleh-Valenzuela (SV) model and can be regarded as multicluster time-of-arrival model. The CIR based on the SV-model is expressed as follows (Saleh and Valenzuela 1987):

$$h(\tau;t) = \sum_{l=1}^{L} \sum_{m=1}^{M_l} a_{l,m}(t) e^{j\theta_{l,m}(t)} \delta\left[\tau - T_l - \tau_{l,m}(t)\right] \tag{3.13}$$

where:

L is the number of clusters

M_l is the number of rays in the lth cluster

$a_{l,m}$ and $\theta_{l,m}$ are the amplitude and phase of the mth ray in the lth cluster

T_l is the arrival time of the first ray of the lth cluster

$\tau_{l,m}$ is the delay of the mth ray within the lth cluster relative to T_l

The SV model has been considered as a static channel model so the aforementioned parameters can be taken as time-invariant. According to the model the cluster decays exponentially and also the rays within each cluster have an exponential decay. An example of a multicluster time-of-arrival model representation is shown in Figure 3.2. The average path power gain, that is, $\overline{|a_{l,m}|^2}$, is characterized by (Shoji et al. 2009):

$$\overline{|a_{l,m}|^2} = \Omega_0 e^{-T_l/\Gamma} \cdot e^{-\tau_{l,m}/\gamma} \tag{3.14}$$

where:

Γ and γ denote the cluster and ray power-decay time constants
Ω_0 stands for the average path power gain of the 1st ray in the 1st cluster, that is, $|a_{1,1}|^2$

In case of $L = 1$ (one cluster) then $T_l = 0$, and Equation 3.13 simplifies to the single-cluster form given by Equation 3.11. The cluster arrival rate as well as the ray arrival rate can be modeled as a Poisson process, so it is assumed that all clusters have the same ray arrival rate. In Equation 3.14, T_l and $\tau_{l,m}$ are characterized by using probability function of the event x under a condition y, that is, $p(x|y)$, according to Shoji et al. (2009):

$$p(T_l \mid T_{l-1}) = \Lambda \exp[-\Lambda(T_l - T_{l-1})], \quad \text{where } l > 0 \tag{3.15}$$

$$p(\tau_{l,m} \mid \tau_{l,m-1}) = \lambda \exp[-\lambda(\tau_{l,m} - \tau_{l,m-1})], \quad \text{where } m > 0 \tag{3.16}$$

where Λ and λ denote the cluster ray and ray arrival rates constants, respectively.

In addition to spread and clustering in the time domain also spread and clustering in the angular domain has been observed (Spenser et al. 2000; Xu et al. 2002). Therefore, in case where antenna directivity is concerned the SV model has been extended by (Spenser et al. 2000) to include the AOA statistics as follows:

$$h(\tau;t) = \sum_{l=1}^{L} \sum_{m=1}^{M_l} a_{l,m}(t) e^{j\theta_{l,m}(t)} \delta[\tau - T_l - \tau_{l,m}(t)] \delta[\varphi - \Psi_l - \psi_{l,m}(t)] \tag{3.17}$$

where Ψ_l and $\psi_{l,m}$ represent the angle of arrival in the azimuth plane of each cluster and the angle of arrival of the mth ray related in the lth cluster, respectively. The average path power gain, that is, $\overline{|a_{l,m}|^2}$, is characterized by (Shoji et al. 2009):

$$\overline{|a_{l,m}|^2} = \sqrt{G_r(0, \Psi_l + \psi_{l,m})} \Omega_0 e^{-T_l/\Gamma} \cdot e^{-\tau_{l,m}/\gamma} \tag{3.18}$$

where $G_r(\theta_r, \varphi_r)$ stands for the normalized antenna gain for the angle of arrival along the vertical axis, θ_r, and horizontal axis, φ_r (i.e., elevation and azimuth). The angular distribution of the rays within a cluster is usually modeled by the Laplacian distribution (Smulders 2009):

$$p(\psi_{l,m}) = \frac{1}{\sqrt{2}\sigma_\varphi} e^{-\sqrt{2}\psi_{l,m}/\sigma_\varphi} \tag{3.19}$$

where σ_φ is the angular-spread of rays in the lth cluster. Alternatively, the angular distribution of the paths can also be approximated by the uniform distribution (Haneda et al. 2015).

Novel communication systems in the mm-Wave bands will require high gain directional and steerable antennas on both Tx and Rx so as to compensate increased path loss and create viable links with appropriate beamforming. Therefore, due to directionality and steerability double-directional channel models have to be considered that include both ray departing and arriving angular characteristics. A cluster-based form of the double-directional omnidirectional CIR is commonly used to represent the radio propagation channel between a Tx and Rx, and can be expressed as (Samimi et al. 2015):

$$h(\tau;t;\vec{\Theta},\vec{\Phi}) = \sum_{l=1}^{L}\sum_{m=1}^{M_l} a_{l,m}(t)e^{j\theta_{l,m}(t)}\delta\big[\tau - T_l - \tau_{l,m}(t)\big]$$

$$\delta\big[\vec{\Theta} - \vec{\Theta}_{l,m}(t)\big]\delta\big[\vec{\Phi} - \vec{\Phi}_{l,m}(t)\big]$$

(3.20)

where $\vec{\Theta} = (\theta,\varphi)_{Tx}$ and $\vec{\Phi} = (\theta,\varphi)_{Rx}$ are the vectors of azimuth/elevation angle-of-departure (AOD) and angle-of-arrival (AOA), respectively. In addition, $\vec{\Theta}_{l,m}$ and $\vec{\Phi}_{l,m}$ are the vectors of azimuth/elevation AOD and AOA, respectively, of the mth ray contained in the lth cluster. The omnidirectional CIR can further be partitioned to yield directional PDPs at a desired Tx–Rx unique pointing angle, and for arbitrary Tx and Rx antenna patters we get (Samimi et al. 2015):

$$h(\tau;t;\overrightarrow{\Theta_d},\overrightarrow{\Phi_d}) = \sum_{l=1}^{L}\sum_{m=1}^{M_l} a_{l,m}(t)e^{j\theta_{l,m}(t)}\delta\big[\tau - T_l - \tau_{l,m}(t)\big]$$

$$g_{Tx}\big[\overrightarrow{\Theta_d} - \vec{\Theta}_{l,m}(t)\big]g_{Rx}\big[\overrightarrow{\Phi_d} - \vec{\Phi}_{l,m}(t)\big]$$

(3.21)

where:

$(\overrightarrow{\Theta_d},\overrightarrow{\Phi_d})$ are the desired Tx–Rx antenna pointing angles

$g_{Tx}(\vec{\Theta})$ and $g_{Rx}(\vec{\Phi})$ are the arbitrary 3D (azimuth and elevation) Tx and Rx complex amplitude antenna patterns of multielement antenna arrays, respectively

Consequently, the small scale fading of the mm-Wave channel can be described by its spatial and temporal characteristics based on Equations 3.20 and 3.21. The spatial lobes model the phenomenon of spatial directionality and have been used successfully to model the 2D and 3D spatial channels. Both 3GPP and WINNER II channel models do not distinguish a spatial cluster from a time cluster. They represent spatial clusters of MPCs by assigning one group of traveling MPCs to one random AOA. From field measurements with high directional antennas it is observed

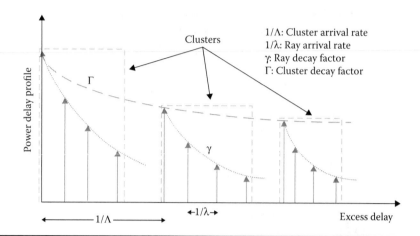

Figure 3.2   Indicative example of a multicluster time-of-arrival channel model.

that multiple groups of multipath clusters can arrive at the same unique pointing angle in azimuth and elevation but at different time delays, and hence, the spatial clusters are generalized into spatial lobes, where a spatial lobe can receive more than one traveling cluster (Rappaport et al. 2015). The multipath cluster is defined as a group of MPCs within a PDP for a specific Tx and Rx antenna spatial angle such that a cluster represents signal energy travelling closely together in time and angular space (Samimi et al. 2013). Intra-cluster statistics revealed that time clusters are subdivided into smaller sub-path components, whose power levels exhibit (on average) an exponential fall-off behavior as shown in Figure 3.2. The temporal statistics include: (1) the number of clusters in a PDP, (2) the number of MPCs within a cluster, (3) the inter-arrival distributions within the cluster (extraction of $1/\lambda$), (4) the cluster multipath amplitude distribution (extraction of γ), (5) the cluster internal (sub-path) delay, (6) the cluster inter-arrival distribution in a PDP (extraction of $1/\Lambda$), (7) the PDP multipath amplitude distribution (extraction of Γ), and (8) the RMS delay spread (RDS) and mean excess delay of the PDP. Directional spatial statistics can be used to model the spatial directionality at both Tx and Rx and include: (1) the mean direction of arrival/departure of a lobe (AOA and AOD), (2) the angle spread (AS) denoting the angle span of a lobe above a specific threshold, (3) the RMS AS (standard deviation of AS) indicating the angle span of a lobe in which most power is received, and (4) the number of lobes (spatial directions) for a particular Rx location. Additional information about the definition and formulation of the temporal and spatial parameters can be found in Xu et al. (2002) and Samimi et al. (2013).

All the aforementioned small-scale parameters are usually obtained using high-resolution joint delay-angle multipath parameter extraction procedures, such as SAGE (Fleury et al. 1999) and KPowerMeans algorithms (Czink et al. 2006a). The latter provides an optimum assignment of MPCs into joint delay-angle clusters given a desired number of clusters from multidimensional CIR data, that is, Equations 3.20 and 3.21, using the mathematically based, power-weighted multipath component distance (MCD) metric (Czink et al. 2006b). The optimum number of clusters is then determined from two optimal criteria, the Calinksi-Harabasz and the Davies-Bouldin indices (Czink et al. 2006b).

The most significant parameter that characterizes the frequency selectivity (time dispersion) of the channel is RDS. Additional useful parameters are the total number of clusters, L, as well as the AS. Table 3.6 summarizes typical values of the RDS, AS and L, derived by different measurement campaigns, representative of various environment types. It is observed that all the listed parameters depend highly on the environment type, the frequency and propagation condition.

According to Table 3.6, measurements across all bands consistently show that when directional antennas are used, the LOS channel exhibits virtually no delay spread (the RDS is the width of the channel sounder's impulse response). For example, RDS values in LOS cases are very low (<15 ns) indicating that the mm-Wave link is realized mainly by the direct component that contains the majority of the transmitted signal energy. On the other hand, in NLOS cases additional paths are captured with greater delays increasing the RDS. According to Rappaport et al. (2013a), mean RDS also increases as the antenna angle is pointed away from LOS angle (i.e., boresight misalignment) between Tx and Rx. Similar findings were also observed in Rappaport et al. (2015) where RDS values greater than 50 ns can occur. From the 28 and 38 GHz measurements was found that in both cases 90% of the RDS values are below 50 and 40 ns, respectively.

It was also found that mean RDS decreases with increasing Tx and Rx separation distance (Rappaport et al. 2013a). This can be attributed to the fact that a stronger received signal is caused by one or few strong MPCs arriving at different specific angles. These strong MPCs dominate the RDS and motivate the use of mm-Wave cellular where directional low path loss links can carry very high data rates with small RDS. However, when the Tx and Rx antennas are steered away from each other at relatively close distance (e.g., <200 m), strong LOS and other strong multipath components are less likely, and RDS becomes much greater as MPCs arrive from many different scattering and reflection mechanisms. At distances beyond several hundred meters, the number of receivable MPCs reduces because of the propagations loss, thus creating fewer detectable MPCs and smaller RDS. Because RDS increases and becomes more variable as Tx and Rx antennas are pointed away from boresight, future mobile devices at a particular location should prefer a link using relatively small off-boresight antenna pointing angles (lower than ±30°). Finally, reaching the cell limits where the Tx, Rx separation may reach nearly a kilometer, the measured RDS values are very low, which requires less equalization. The reduced power and latency for equalization of these cell-edge links can be put to use in other processing areas, such as additional error coding for this lower signal-to-noise ratio (SNR) case (Rappaport et al. 2013a).

Table 3.6 Representative Values of RMS Delay Spread (RDS), Angular Spread (AS), and Number of Clusters, L, from Different Measurement Campaigns, Considering Alternative Environment Types and Operational mm-Wave Frequencies. The AS Is Either for the Arrival (ASA) or the Departure (ASD)

Reference	Frequency (GHz)	Environment	Antenna Type Tx/Rx	RDS (ns) LOS	RDS (ns) NLOS	AS (°)	L
Azar et al. (2013)	28	Dense urban (Brooklyn/ Manhattan-NYC)	Dir/Dir NB/NB Fixed/Fixed	0.878	47.2	ASA: 40.3	4.7
Kim et al. (2015)	28	Urban low-rise (Gyanpyung, Daejeon)	Dir/Dir NB/NB Fixed/Fixed	<10	24-60	ASA: 2.5-32	–
		Urban very high-rise (Gangnam, Seoul)	Dir/Dir NB/NB Fixed/Fixed	<10	55-105	ASA: 2.5-27	–
Hur et al. (2015)	28	Typical urban (Daejeon, S. Korea)	Omni/Omni[a] Fixed/Fixed	–	22.29	ASD: 8 ASA: 31	3.6
Ko et al. (2016)	28	Typical urban (Pyeongchang resort town)	Omni/Omni[a] Fixed/Fixed	–	60.41	ASD: 7 ASA: 43	5.3

(Continued)

Table 3.6 (Continued) Representative Values of RMS Delay Spread (RDS), Angular Spread (AS), and Number of Clusters, L, from Different Measurement Campaigns, Considering Alternative Environment Types and Operational mm-Wave Frequencies. The AS Is Either for the Arrival (ASA) or the Departure (ASD)

Reference	Frequency (GHz)	Environment	Antenna Type Tx/Rx	RDS (ns) LOS	RDS (ns) NLOS	AS (°)	L
Rappaport et al. (2012)	38	Pedestrian walkway (University of Texas campus)	Dir/Dir NB/NB Fixed/Fixed	1.2	23.6	–	–
		Light urban campus (University of Texas) Cellular channel	Dir/Dir NB/NB Fixed/Fixed	1.1	12.2	–	–
Rappaport et al. (2013a)	38	Light urban campus (University of Texas) Cellular channel	Dir/Dir NB/WB Fixed/Fixed	1.1	13.7	–	–
Rappaport et al. (2012)	60	Pedestrian walkway (University of Texas campus) Peer-to-peer channel	Dir/Dir NB/NB Fixed/Fixed	0.8	7.4	–	–
Ben-Dor et al. (2011)	60	Inter-vehicle (Urban parking lot) Vehicular channel	Dir/Dir NB/NB Fixed/Fixed	<0.9	12.3	–	–

(Continued)

Table 3.6 (Continued) Representative Values of RMS Delay Spread (RDS), Angular Spread (AS), and Number of Clusters, L, from Different Measurement Campaigns, Considering Alternative Environment Types and Operational mm-Wave Frequencies. The AS Is Either for the Arrival (ASA) or the Departure (ASD)

Reference	Frequency (GHz)	Environment	Antenna Type Tx/Rx	RDS (ns)		AS (°)	L
				LOS	NLOS		
Moraitis and Constantinou (2006) Moraitis et al. (2011)	60	Indoor corridor	Dir/Dir NB/NB Fixed/Fixed	13.36	–	ASA: 17.6	–
		Indoor laboratory		13.64	20.52	ASA: 25	–
Rappaport et al. (2015)	73	Dense urban (Manhattan-NYC)	Dir/Dir NB/NB Fixed/Fixed	1.2–1.5	5.6–7.1	ASA: 3.7	2.9–3.3
MacCartney et al. (2015a)	73	Typical indoor office (MetroTech Center-NYC)	Dir/Dir NB/NB Fixed/Fixed	12.1	10.7	–	–

Note: Dir stands for directional antenna, Omni denotes omnidirectional antenna, and finally, NB, and WB indicate narrow-beam and wide-beam antennas, respectively.

a The omnidirectional results are derived after synthesizing the received power of the rotating directional antennas.

It is interesting to note the very low RDS values at NLOS situations obtained from 73 GHz measurements. Similar to the 28 and 38 GHz bands, at 73 GHz the RDS values decrease as the Tx, Rx separation distance increases, but the dense urban measurements do not decrease as rapidly, because of the scattering and reflections from large buildings (Rappaport et al. 2015). The mean NLOS RDS values at 38 and 73 GHz are smaller than at 28 GHz, due to the more reflective environment of New York and greater energy being scattered at lower mm-Wave frequencies (where the first meter of free-space path loss is lower). Wavelengths at 73 GHz are extremely small, giving rise to more diffuse scattering during propagation which results in weaker paths not detectable at the Rx. Time dispersion characteristics for indoor measurements at 73 GHz, using directional antennas, showed that 90% of RDS values are less than 30 ns (MacCartney et al. 2015a). In both 28 and 73 GHz measurements, for NLOS situations, the number of MPCs detected at any unique pointing angle combination between Tx and Rx, follow a uniform distribution over all Tx, Rx separation distances, and decreases with distance. Overall, there are more resolvable MPCs as a function of Tx, Rx separation distance in NLOS environments at 28 GHz compared to 73 GHz, attributed to stronger signals and the larger wavelengths at 28 GHz, which allow the signal to reflect more and scatter less. The 73 GHz signals have a smaller wavelength and a higher possibility of getting caught in tiny building cracks and rough surfaces, leading to diffusion.

Considering the cluster number, it is found that more clusters are formed in lower mm-Wave bands in the same examined environment (e.g., measurements at 28 GHz yield on average 5 clusters compared to the 3 clusters derived from measurements at 73 GHz in New York). Omnidirectional results from measurements at 28 GHz in Ko et al. (2016) and Hur et al. (2015) yield 5 and 4 clusters, respectively, in a typical urban environment.

From the AS results it is observed that the spatial multipath distribution was found to be highly correlated with the site-specific environment, the operational frequency, and the Tx, Rx antenna beamwidths. It is interesting to note that high AS values (comparable to outdoor urban results) are obtained in indoor environments (Moraitis et al. 2011), where the highly reflective indoor surfaces with fully furnished offices accounts for this observation. In addition, very low AS values are presented in 73 GHz measurements, due to the narrower Tx antenna and the increased environmental scattering that dampens energy over a wider field of view (Rappaport et al. 2015). The LOS AS are on average slightly larger than the NLOS AS, suggesting that energy arrives in narrow lobes at the Rx, while being more distributed over space and stronger in LOS environments. Comparing 28 and 73 GHz measured AS results, the former exhibits larger values, designating that 28 GHz propagation is spatially prominent (i.e., strong MPCs come from a large number of angles at lower mm-Wave).

There are significant differences in the spatial properties of LOS and NLOS mm-Wave channels, where in NLOS environments more sharply defined power angle spectrum was observed, in contrast with the great azimuthal sparsity of power encountered in LOS situations (Rappaport et al. 2015).

Measurements in Kim et al. (2015) revealed that the directional RDS exhibited strong dependency on the antenna beamwidths, where wider beamwidths presented larger RDS values. The AS values showed similar correlation. The wider beamwidth the antenna has, the higher AS values are yielded. Based on the previous observations, empirical prediction models were produced that estimate for directional RDS and AS from a given beamwidth of antenna. The analytical models as function of antenna beamwidth, θ, can be written as

$$RDS(\theta) = a \cdot \log_{10}(\theta) \tag{3.22}$$

$$AS(\theta) = \alpha \cdot \theta^{\beta} \tag{3.23}$$

where the beamwidth range is $0° \leq \theta \leq 120°$. Specific values for the parameters a, and β can be found in Kim et al. (2015).

The results listed in Table 3.6 provide an insight into systems that systematically search for the strongest Tx and Rx antenna pointing angles that lead to low RDS values (so that simple equalization methods can be used). The temporal statistics for these strong directional beams may help define the channel matrix used to describe LOS and NLOS mm-Wave channels. Numerous antenna elements at both link ends will be used to increase the received SNR or reduce interference via beam combining or beamforming techniques, by combining multipath from many arriving angles and excess delays. It is expected that up to hundreds of miniature on-chip electrically steerable antennas will be used in mm-Wave devices to find the strongest multipath AOA at the Rx so as to improve SNR and to extend coverage distances (Rappaport et al. 2015). A simple algorithm to find the best beam directions that can simultaneously minimize both RDS and path loss (finding the best paths that simultaneously have both strong SNR and very small multipath time dispersion) is suggested in Sun et al. (2014). By selecting a beam with both low RDS and low path loss, relatively high SNR can be achieved at the Rx using directional antennas without complicated equalization, meaning that low latency single carrier (wideband) modulations may be a viable candidate for future mm-Wave wireless communications systems (Ghosh et al. 2014).

Algorithmic procedures that describe the cluster and subpath generation, based on the presented CIR models and spatio-temporal parameters that will help to simulate the measured channel, fall out of the scope of this chapter. Extensive algorithmic approaches can be found in Akdeniz et al. (2014), Haneda et al. (2015), Samimi et al. (2015), and Hur et al. (2016), where the simulated results presented great consistency with the measured values. In the aforementioned work, extensive and detailed values of the spatial and temporal channel parameters are listed, which are useful for developing accurate channel simulators. Representative spatio-temporal channel parameters that are suitable for convenient CIR generation are summarized in Table 3.7, derived by two different measurement campaigns at 28 GHz.

Table 3.7   Indicative Spatial and Temporal Parameters Useful for CIR Generation

Parameter	NLOS (Pyeongchang) (Ko et al. 2016)	NLOS (Daejeon) (Hur et al. 2015)
Number of clusters	5	4
Delay distribution	Exponential ($\lambda = 0.0098$)	Exponential ($\lambda = 0.021$)
RMS delay spread	60 ns	22 ns
Subpath delay distribution	Exponential ($\lambda = 0.0195$)	Exponential ($\lambda = 0.064$)
Subpath delay spread	16 ns	6 ns
Angle distribution	Laplacian	Laplacian
AOD spread (ASD)	7°	8°
AOA spread (ASA)	43°	31°
Subpath ASD	5°	6°
Subpath ASA	11°	16°

Up to now all the presented measurements consider a fixed Tx and Rx terminal. It is also interesting to introduce Doppler effects in the measurements, which induces time-selective fading in the channel, especially for a fast moving terminal in mm-Wave frequencies. For example pedestrians or people inside buildings walking at a speed of 1 m/s induce a maximum Doppler spread of 933 Hz, at 70 GHz. This value may reach up to 6.5 kHz for a slow moving vehicle at 50 km/h. A generalized model that incorporates large- and small-scale fading is proposed in Moraitis and Panagopoulos (2015), where measurements between a fixed Tx and a moving Rx were carried out at 60 GHz in various indoor environments. The Rx was moving at a constant speed of 0.5 m/s with no people being present during the measurements. The proposed model is an extension of Equation 3.10, where the small-scale effects are introduced as well:

$$PL(d) = \underbrace{\underbrace{PL_{FS}(d_0) + 10\bar{n}\log_{10}\left(\frac{d}{d_0}\right)}_{LOS} + X_\sigma + \underbrace{\sum_i k_i F_i}_{NLOS}}_{\text{Large Scale}} + \overbrace{Y\left(\frac{f_m \Delta t}{\lambda}\right)}^{\text{Small Scale}} \quad (3.24)$$

The last term, Y, is the space/time variability due to multipath, which is characterized in linear units as Rice or Rayleigh distributed. In addition, the characterization of the Doppler spectra is important for the determination of the time-variance of the channel. Toward the modeling process of the Doppler spectra it is found that all the power spectral densities (PSDs) for all the LOS runs can be modeled as follows:

$$S_{\text{LOS}}(f) = \left(\frac{1}{K+1}\right)^2 \frac{\bar{P}_r}{16\pi f_m}$$

$$x \left\{ K\left(\sqrt{1-\left(\frac{f}{2f_m}\right)^2}\right) + \frac{K\pi}{\sqrt{1-\left(f/2f_m\right)^2}} + K^2\pi^2 f_m\delta(f) \right\} \tag{3.25}$$

where:

K(·) is the complete elliptic integral of the first kind
$\bar{P}_r$ is the average received power
f_m is the Doppler shift
K is the Rician K-factor

The Doppler spectrum given by Equation 3.25 is valid for $-2f_m \leq f \leq 2f_m$. For NLOS runs the Doppler spectrum can be modeled as the Jake's classic U-shape form:

$$S_{\text{NLOS}}(f) = \frac{\bar{P}_r}{2\pi f_m}\left[1-\left(\frac{f}{2f_m}\right)^2\right]^{-1/2} \tag{3.26}$$

which is also valid for $-2f_m \leq f \leq 2f_m$. Therefore, in Y, the induced time variations are modeled from Equations 3.25 and 3.26, for LOS and NLOS situations, respectively. According to Moraitis and Panagopoulos (2015), the small-scale variations in LOS trajectories follow the Rician distribution with a K-factor ranging between 6.9 and 8.5 dB, whereas in NLOS cases the fading statistics correspond very well with the Rayleigh curves. For all the measured runs, the spectrum width was found between 390 and 420 Hz. Hence, the fading at any measured run and environment type is extremely slow, having a very narrow bandwidth, much narrower than that of practical transmitted signals, verifying the stationary of the channel. It is concluded that the main cause of the temporal variations is the movement of the receiver, as there are no moving scatterers in the channel. Finally, the average coherence time of the channel is found 2.6 ms, which is 0.26 times the ratio between the wavelength and the speed of the mobile terminal. For the envisioned high data rate applications at 60 GHz the calculated coherence time of 2.6 ms is very high compared to the symbol duration. Hence, the channel exhibits slow fading characteristics and can be regarded as static. Nevertheless, the induced slow fading can be mitigated by packet sizing and appropriate coding over the coherence time of the channel.

3.2.5 Atmospheric Effects

A very challenging task in designing mm-Wave systems is the atmospheric effects that due to the small wavelength of the signals exhibit severe attenuation. Apart from the free-space loss that is significantly increased, oxygen and water absorption may cause excess loss especially in the region of 60 GHz. Furthermore, mm-Wave links that will operate outdoors are also limited by the rain, where the droplets both scatter and absorb the incident waves, leading consequently to signal attenuation and link outages.

First, taking into account the effects due to atmospheric gases, oxygen molecules and water vapor mainly contribute to excess attenuation in the mm-Wave frequencies. Figure 3.3 presents the specific attenuation in dB/km due to atmospheric gases, for frequencies between 10 and 300 GHz. The curves have been derived from the attenuation models provided in ITU (2012a). There is peak of oxygen attenuation at 60 GHz that reaches 11.6 dB/km, which renders long-range outdoor links very difficult to deploy. For frequencies below and above 60 GHz (i.e.,

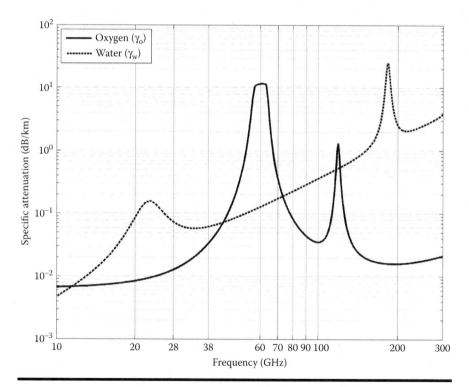

Figure 3.3   Specific attenuation due to oxygen molecules and water vapor of the atmosphere.

28–38 GHz and 70–90 GHz), there are regions which can be exploited for outdoor signal transmission. For example, at 28 or 38 GHz, the additive oxygen and water vapor attenuation is lower than 0.1 dB/km, whereas in the frequency region between 70 and 90 GHz the attenuation will not exceed 0.6 dB/km, as shown in Figure 3.3. Therefore, according to Figure 3.3, atmospheric attenuation for outdoor small cells is virtually negligible. As urban small cells will be designed for inter-site distances within 200 m (for backhaul, fronthaul, and access links), air attenuation will be of little concern (Ghosh et al. 2014). In case of outdoor and indoor access links at 60 GHz, which will be deployed within a range of up to 50 m, the impact of air attenuation is also insignificant (Moraitis and Constantinou 2004).

Second, apart from the oxygen and water absorption, rain attenuation has a detrimental effect on the link quality in the mm-Wave frequencies. For example, the averaged attenuation in a hailstorm was measured to be 25.7 dB over a 605 m path at the frequency of 38 GHz (Xu et al. 2000). According to DeLange et al. (1975), a rain-fading margin of 40–42 dB would be required to meet a reliability objective of less than 5 min. outage per year (a probability of 10^{-5}) on a 1-km path at the frequency of 60 GHz. Other experiments at 60 GHz have shown worst case attenuations (depending on the time availability) between 5.5 and 12.5 dB, and between 6 and 28.5 dB, on a 250 m and 850 m path, respectively (Timms et al. 2005). Therefore, mm-Wave signals, especially in regions above 60 GHz, will be severely affected by rainfall, limiting the range of access or backhaul links. A power-law expression between point rainfall rate and specific rain attenuation has been proposed in Olsen et al. (1978), and in conjunction with parameters defined in ITU (2005), the specific rain attenuation can be calculated, taking into account various frequencies and rainfall rates (in mm/h). Based on these expressions, Figure 3.4 presents the induced rain attenuation for a 1 km path as a function of operating frequency and different rainfall rates. The latter parameter depends directly on the local climatological conditions. For high availability systems, in which the outage probability must be less than 0.01% of time, the rainfall rate may exceed very high values (Kourogiorgas et al. 2016).

According to Figure 3.4, at a heavy rain rate of 10 mm/h, a 1 km link, experiences a rain attenuation of 2 dB at 28 and 2.5 dB at 38 GHz. Even with a very heavy rainfall of 20 mm/h, rain attenuation will not exceed 5 dB. Therefore, access or fronthaul links at 28 or 38 GHz will be totally feasible to be deployed within a radius of 200 m, as also proposed in Rappaport et al. (2013b). On the other hand, outdoor hotspot access links at 60 GHz will have to be limited within a range of 50 m in order to minimize outages and provide consistent services. In general, proper link design (e.g., with varying gain antennas) could account for rain margin in future mobile mm-Wave cellular systems (Azar et al. 2013).

Finally, other atmospheric effects include hydrometeors that constitute fog, as well as amplitude signal scintillation due to turbulence; factors that could also deteriorate outdoor link's performance. The attenuation due to fog depends on the liquid water density of the fog and the operating frequency of the link. According to ITU (2012b), the specific attenuation for a light fog is less than 5 dB/km for frequencies

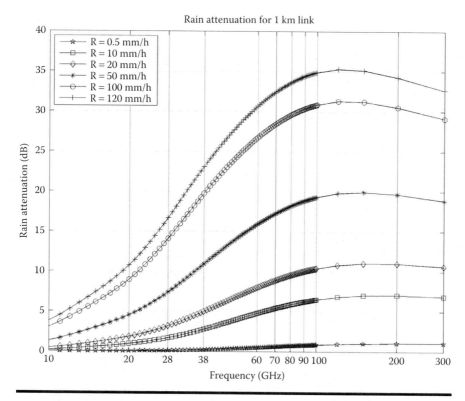

Figure 3.4   Rain attenuation versus frequency for a 1-km link. Different rainfall rates have been considered.

between 20 and 90 GHz and can be neglected for small mm-Wave cells. The impact of atmospheric turbulence is significant, and if not considered properly, amplitude scintillation may affect the performance of a low-power margin communication system. Nevertheless, measurements in Cheffena (2010), for long-range links (up to 5.6 km) at 42 GHz, have shown a small signal variation, which constitutes the impact of turbulence insignificant for short-range outdoor mm-Wave links, and can be neglected as well.

To conclude, the most vital parameter among all the atmospheric effects, for the proper design of next-generation mm-Wave networks, is rain attenuation. Consequently, local rainfall patterns must be considered to ensure a reliable link operation; most telecommunications carriers require links to display 99.999% availability (i.e., downtime of less than 5 min/year). Therefore, it is important to model the first-order statistics of the various losses, as well as the development of channel models, in order to proceed to end-to-end simulations. On the one hand, first-order statistics in the case of atmospheric effects are crucial for the design of system margins and the ability to consider such margins taking into account the costs and the consumed power (Kourogiorgas et al. 2016).

3.3 Future Trends in Millimeter-Wave Propagation

Millimeter-wave propagation has been acknowledged as an ultimate research direction for the deployment of next-generation cellular systems inaugurating 5G networks. The main challenging task to develop such networks is the unique propagation conditions encountered in mm-Wave frequencies between 28 and 90 GHz. The specific frequency span gives the capability to allocate ultra-wideband channels (much greater than 500 MHz) providing Gb/s rates to the users. As described in Section 3.2, mm-Wave channel experiences very high path loss that in practice necessitates the existence of LOS links to achieve the aforementioned rates. Measurement campaigns have shown very low time-dispersion phenomena in LOS situations (RDS lower than 10 ns). On the other hand, enough detectable MPCs occur in NLOS conditions (increasing RDS), even at greater Tx, Rx separations, regardless of the fact that these components have relatively low powers. Nevertheless, frequency selectivity has to be considered due to the wider Rx bandwidth and several strongly localized rays can be found in the angular domain. In case of mobile users, high Doppler spread is expected even in slow mobility due to the small wavelength. Apparently, the implemented physical layer transmission techniques must be able to operate under these extreme propagation conditions.

Millimeter-wave cellular can be realized with high-gain steerable antennas, which are necessary to compensate for the high propagation loss and to direct energy toward optimal directions that can exploit multipath and successfully complete a link. Measurements have shown that NLOS links are also feasible with appropriate beam steering, orienting Tx and Rx antennas at the best pointing angle and taking advantage of the spatially separated multipath. Phased antenna arrays or switched-beam antenna arrays can exploit beam steering to increase link gain. Future smart antenna arrays, possibly on-chip, will algorithmically determine the optimal AOA and AOD. One possible beam steering method is to use narrowband pilot tones that enable the prediction of the spatial location of multipath based on narrowband envelope cross-correlations (Samimi et al. 2013). However, in order to achieve highly directional patterns, large antenna arrays are necessary.

In mm-Wave communications due to the small wavelength, the antenna size can be extremely small and a large number of array elements can be packed into a small area at both link ends, giving rise to mm-Wave MIMO concept. The application of MIMO configuration in the mm-Wave band has been theoretically studied in Moraitis and Constantinou (2007) and Arvanitis et al. (2007) for an indoor radio channel at 60 GHz, where a MIMO channel model was proposed applying uniform linear (ULAs) and uniform rectangular arrays (URAs). The results showed that there is a significant fading reduction in the channel, as well as a capacity enhancement at the downlink that could reach up to 6–7 b/s/Hz (having a 64 × 64 MIMO configuration with URA antennas and 10 dB SNR). Similar theoretical simulations in Akdeniz et al. (2014) have shown spectral efficiencies between 2.6 and 3.34 b/s/Hz, and between 2.6 and 3.0 b/s/Hz, at 28 and 73 GHz, respectively, depending on the antenna elements. In addition, MISO antenna configurations were adopted in Muirhead et al. (2015), where a densely deployed wireless network based on 25 apartment block was used as the basis of the simulation model. Each base station was equipped with a beamformed ULA with 8 elements, and a single omnidirectional antenna at the user terminal (UE). The median achieved capacity at 72 GHz was 9 b/s/Hz for the MISO configuration instead of 4.5 b/s/Hz when two omnidirectional antennas were used at both Tx and Rx (SISO). A comparative study in channel capacities between 2.1 and 2.6 GHz with 20 MHz RF bandwidth, and the 28 and 38 GHz bands with 1 GHz RF bandwidth was carried out in Sulyman et al. (2014). The results revealed that the capacity for 5G systems is over 20 times that of today's 4th-generation (4G) systems (3.8 Gb/s in 5G systems against lower than 0.5 Gb/s for 4G, for a 10 dB carrier-to-interference ratio).

Massive MIMO is also acknowledged as an emerging technology for 5G networks. The Massive MIMO and mm-Wave technologies provide vital means to resolve many technical challenges of the future wireless networks, and they can be seamlessly integrated with the current networks and access technologies. The concept of Massive MIMO is based on the use of a very large number of antenna elements to multiplex messages for several users. The main principle of operation is that the Massive MIMO antennas can focus the radiated energy toward desired directions while minimizing intra- and inter-cell interference, and enhancing significantly the spectral and energy efficiency. In a rich scattering environment, these performance gains can be achieved with simple beamforming strategies, such as maximum ratio transmission (MRT) or zero forcing (ZF) (Bogale and Le 2016). Another critical issue in mm-Wave MIMO is to cope with the existing hardware constraints due to the high bandwidth and operational frequencies. Therefore, there is the need for hybrid precoding, which is a combination of analog and digital processing. Analog beamforming solutions are mainly based on controlling the phase of the signal transmitted by each antenna, but their performance is relatively low. In order to achieve larger gains, hybrid analog/digital processing strategies are proposed. The use of precoding techniques on an antenna array can be used to form a beam toward the receiver or even multiple beams in a multiuser scenario. The aim of precoding is to suppress the cochannel interference between the users, which communicates at the same frequency. More specifically, in precoding techniques, the symbols that are to be transmitted are processed in order to minimize the interference, for example, maximizing the signal-to-noise-interference ratio (SNIR) (Kourogiorgas et al. 2016). A common problem that arises is the channel state information (CSI) at the base stations, especially in case of Massive MIMO where the number of elements of the channel matrix is very large. The tasks of channel estimation and precoder extraction for the large channel matrix become cumbersome. Efficient ways to exploit correlations and sparsity in order to decompose the complicated task must be developed and evaluated. Moreover, it is unrealistic to assume perfect CSI to the transmitter, therefore the effects of partial

CSI in the Massive MIMO transmitter must be considered. Assuming that all the cells were served by a single base station, as the CSI would be known at this base station, the intra-cell interference could be resolved. However, as there are many base stations there would be inter-cell interference with the fundamental problem that the one base station does not have the appropriate knowledge of CSI of the adjacent cells. The problem may be solved through the coordination of the different base stations. It must be noted that the interference between the different cells strongly depends on the frequency reuse factor.

A variation of a large-scale MIMO system is the distributed MIMO architecture, unlike Massive MIMO and its centralized architecture. In distributed MIMO it is assumed that MIMO technology is implemented by different transmitters (each one equipped with a MIMO antenna). The distributed MIMO nodes transmit synchronized signals to maximize spectral efficiency and system capacity through interference alignment. The role of mm-Wave networks in a distributed MIMO system can be twofold. It can be used to implement directive multipoint backhaul connections that are used to coordinate the distributed MIMO nodes, or as the radio access network technology from the base stations to the UE. In distributed MIMO systems, interference is turned from foe to friend, as long as the transmitters are all coordinated and synchronized. The development of proper coordination, synchronization, and precoding transmission scheme for distributed MIMO in mm-Wave frequencies is of great scientific value, as interference coordination may benefit from the directional beams and the propagation conditions of such system.

An additional problem that has to be taken into account is a proper adaptive modulation and coding scheme selection in order to enhance system efficiency. The mm-Wave channel is highly variable between LOS and NLOS situation, which necessitates the ability to extend modulation in each case (either to higher or lower order schemes). High order modulation schemes shall be used in case of high SNIR. On the contrary, low modulation schemes may be used (codes with low coding rates or symbol repetition) in order to support lower values of SNIR. In addition, due to the time-variant nature of the mm-Wave channel, special care should be given to properly select single carrier or multicarrier modulations. For example, OFDM is an appealing solution for spectral efficiency and equalization complexity, but its implementation in mm-Wave frequencies is complicated due to the sensitivity in high peak to average power ratio, frequency and phase shifts. Single carrier modulations do not present such difficulties, however the estimation and equalization complexity increase significantly and the spectral efficiency is also reduced. Therefore, the investigation of other single carrier or multicarrier schemes is demanding for the development of mm-Wave networks in order to define the best approach in terms of throughput, power efficiency, and hardware complexity.

As mentioned in Section 3.2, user mobility can affect significantly mm-Wave transmission because of the induced Doppler effects and its increased influence due to the small wavelength. Communications in the mm-Wave band may be used in vehicular applications (vehicle-to-vehicle, small range radars, etc.). Therefore, it is critical to develop and assess Doppler spread and frequency offset estimation and compensation techniques in order to cope with the time-variant nature of the radio channel. The developed algorithms should be robust and able to track the frequency shifts through changes in direction and speed of movement, especially when using OFDM modulations, due to its sensitivity in phase noise and frequency offsets.

However, many additional measurements are necessary to cover all the environments and scenarios of interest, and to develop full channel models for standards development. Therefore, based on the state of the art research, thoroughly described in Section 3.2, new challenges in measurement campaigns are emerging for the development of future mm-Wave wireless networks. In terms of channel characterization and modeling, novel research activity should be focused on outdoor and indoor channel measurements including additional environment types, scenarios and operational frequencies. Outdoor channel measurements should be performed for fixed and mobile scenarios. Additional measurements are required to cover the gap in some frequency bands, especially at 38 and 60 GHz for outdoor access links. In case of fixed wireless links the 60 GHz band will be mostly investigated for short paths in order to avoid the high oxygen absorption and rain attenuation. For mobility scenarios the investigation of 28 and 38 GHz bands will be more sensible, due to lower atmospheric and rain attenuation, and lower Doppler effects. The majority of the current research is oriented in measurements in urban environments; therefore, alternative types should be examined such as light urban, suburban, and pedestrian. Access links through vegetation is also necessary to be taken into account and investigated. In addition, long-periodic measurements is vital to be performed for various weather conditions, as the effects due to local environment change, for example, in case of rain or hail. Therefore, concurrent measurements of path attenuation, rain intensity, and atmospheric attenuation should be carried out, for fixed/mobile access at the bands considered, in order to investigate the large-scale attenuation. Finally, additional channel measurements should be performed at the 70 GHz band and above, which are candidate operational frequencies for network backhauling.

Indoor measurements should be considered for both fixed and mobile access at 60 and 70 GHz, realizing the femto-cellular architecture of high-speed access hotspots. Extensive measurements are required, especially at the 70 GHz band, where additional space for the development of spatio-temporal channel models still exists. Furthermore, alternative environments have to be examined such as offices, conference rooms, and finally residences, where the current measurement attempts are scarce. Special attention should be given in residences where the deployment of mm-Wave access links is a very challenging attempt, as the specific environment is unique with highly reflective surfaces. It is also vital to investigate the effect of human motion within the wireless channel. Multipath and shadowing phenomena should be carefully assessed and taken into account for designing a proper channel model. Finally, attenuation measurements for indoor partitions are also necessary including various types of materials, in order to determine the expanse of the indoor cells between adjacent rooms.

A remarkable trend in future outdoor and indoor measurement attempts will be the utilization of sophisticated MIMO channel sounders, incorporating several antenna elements at both Tx and Rx (or even large-scale arrays), with view of recording all the valuable time- and angular-resolved factors, characterizing statistically the mm-Wave channel. Polarized MIMO measurements are also valuable so as to examine the polarization multiplexing capabilities of the channel. A requirement for the development and evaluation of mm-Wave physical layer algorithms is the development of a suitable channel modeling tool. Exploitation of radio channel measurements in the mm-Wave bands can be performed with the definition of a channel model based on stochastic and/or geometric characterization of the measurement results. The general purpose is to model statistically the channel, especially in multiantenna scenarios (MIMO). It is crucial to develop models that will be able to characterize the small- and large-scale effects, to calculate the first- and second-order statistics of the wireless channel, and to evaluate its spatial and temporal dispersion. In respect, the channel models must incorporate those attenuation factors that fade and scatter the signal due to atmospheric phenomena, and vegetation. Consequently, the development of theoretical space-time scattering models due to local environment is required, including the modeling of the mobile or fixed MIMO channel according to LOS or NLOS availability. The discrimination of the propagation conditions is important due to the different natures of the propagation types and their impact on the performance of a MIMO system. The narrowband and wideband models must determine the positions of the random scatterers utilizing statistical distributions for different environments. In addition, the developed models will be able to evaluate the space-time correlation between the MIMO subchannels, to estimate the spatial multiplexing efficiency, and to calculate the channel capacity.

Aside from the local propagation environment, the attenuation due to atmospheric effects must be modeled through statistical methods in which the physical properties of the phenomena must be incorporated. For the latter, meteorological measurements are obligatory. Novel analytical expressions of the first-order statistics of atmospheric attenuation will be derived through the measurements for its accurate prediction in the various environments and conditions. Moreover, the spatial and temporal variations of rain attenuation will be investigated, for the adaptive techniques and spatial MIMO channel, as well as the cross-correlation in the temporal and spatial domain so as to implement adaptive techniques and spatial diversity, simultaneously, for alternative operational mm-Wave frequencies.

The final objective for the future deployment of 5G networks is a unified mm-Wave MIMO channel model that could distinguish different propagation environments (e.g., indoor, outdoor), conditions (LOS, NLOS, or both), and scenarios (fixed or mobile). This model will be able to separate the space and frequency domains of the channel and include the space-time correlation of the slow and fast fading, the dual-polarized MIMO radio channel, and finally the spatial MIMO channel. The ultimate target is to develop an empirical model for the mm-Wave MIMO channel that will combine both local environment and atmospheric effects.

3.4 Conclusion

In this chapter, the current advances of the mm-Wave propagation toward the realization of the future 5G mobile wireless networks were thoroughly studied and presented. State-of-the art channel models, derived from the latest available measurement campaigns, concerning various propagation environments and conditions were assessed. The envisaged 5G systems will utilize mm-Wave bands for outdoor and indoor communication, alternative cell types (e.g., macro-cells, small-cells and femto-cells), and different links (e.g., backhaul, fronthaul and access radio links). The large-scale path loss models and multipath time dispersion characteristics presented here will be important for mm-Wave channel modeling and may assist in the deployment of future mm-Wave systems. The presented large-scale path loss models were assessed from extensive measurement campaigns, and new multifrequency path loss models were introduced for use across the entire mm-Wave spectrum. In addition, models and statistics that describe the spatial characteristics of the mm-Wave were also provided.

It is very important to develop models that will incorporate local environment effects (e.g., multipath and path loss) and atmospheric effects. These models will be based on measurements across the entire candidate mm-Wave spectrum investigating additional operational frequencies, propagation scenarios, and environment types. It is also necessary to perform MIMO channel measurements with view of recording all the valuable time- and angular-resolved factors, characterizing statistically the mm-Wave channel. To conclude, the ultimate goal is to develop an accurate empirical model for the mm-Wave MIMO channel that will incorporate various propagation conditions and combine local multipath and atmospheric effects. The development of accurate mm-Wave channel models is a demanding but challenging task for the scientific community, which will support the implementation and pave the way for the 5G mobile wireless networks.

References

Akdeniz, M. R., Liu, Y., Samimi, M. K. et al. Millimeter wave channel modeling and cellular capacity evaluation. *IEEE Journal on Selected Areas in Communications*, 32(6), (2014): 1164–1179.

Anderson, C. R. and Rappaport, T. S. In-building wideband partition loss measurements at 2.5 and 60 GHz. *IEEE Transactions on Wireless Communications*, 3(3), (2004): 922–928.

Arvanitis, A., Anagnostou, G., Moraitis, N. et al. Capacity study of a multiple element antenna configuration in an indoor wireless channel at 60 GHz. In *Proceedings of 65th IEEE Vehicular Technology Conference (VTC)*, Dublin, Ireland, (2007): pp. 609–613.

Azar, Y., Wong, G. N., Wang, K. et al. 28 GHz propagation measurements for outdoor cellular communications using steerable beam antennas in New York City. In *Proceedings of IEEE International Conference on Communications*, Budapest, Hungary, (2013): pp. 5143–5147.

Ben-Dor, E., Rappaport, T. S., Qiao, Y. et al. Millimeter-wave 60 GHz outdoor and vehicle AOA propagation measurements using a broadband channel sounder. In *Proceedings of the Global Communications Conference, (GLOBECOM)*, Houston, TX, (2011): pp. 1–5.

Boccardi, F., Heath, R. W., Lozano, A. et al. Five disruptive technology directions for 5G. *IEEE Communications Magazine*, 52(2), (2014): 74–80.

Bogale, T. E. and Le, L. B. Massive MIMO and mmWave for 5G wireless HetNet. *IEEE Vehicular Technology Magazine*, 11(1), (2016): 64–75.

Cassioli, D. and Rendevski, N. A statistical model for the shadowing induced by human bodies in the proximity of a mmWaves radio link. In *Proceedings of IEEE International Conference on Communications (ICC)*, Sydney, Australia, (2014): pp. 14–19.

Cheffena, M. Measurement analysis of amplitude scintillation for terrestrial line-of-sight links at 43 GHz. *IEEE Transactions on Antennas and Propagation*, 58(6), (2010): 2021–2027.

Collonge, S., Zaharia, G., and Zein, G. E. Influence of the human activity on wide-band characteristics of the 60 GHz indoor radio channel. *IEEE Transactions on Wireless Communications*, 3(6), (2004): 2396–2406.

Czink, N., Cera, P., Salo, J. et al. A framework for automatic clustering of parametric MIMO channel data including path Powers. In *Proceedings IEEE 64th Vehicular Technology Conference (VTC-Fall)*, Quebec, Canada, (2006a): pp. 1–5.

Czink, N., Cera, P., Salo, J. et al. Improving clustering performance using multipath component distance. *Electronics Letters*, 42(1), (2006b): 33–45.

DeLange, O. E., Dietrich, A. F., and Hogg, D. C. An experiment on propagation of 60-GHz waves through rain. *The Bell System Technical Journal*, 54(1), (1975): 165–176.

Fleury, B. H., Tschudin, M., Heddergott, R. et al. Channel parameter estimation in mobile radio environments using the SAGE algorithm. *IEEE Journal on Selected Areas in Communications*, 17(3), (1999): 434–450.

Ghosh, A., Thomas, T. A., Cudak, M. C. et al. Millimeter-wave enhanced local area systems: A high-data-rate approach for future wireless networks. *IEEE Journal on Selected Areas in Communications*, 32(6), (2014): 1152–1163.

Haneda, K., Järveläinen, J., Karttunen, A. et al. A statistical spatio-temporal radio channel model for large indoor environments at 60 and 70 GHz. *IEEE Transactions on Antennas and Propagation*, 63(6), (2015): 2694–2704.

Hur, S., Baek, S., Kim, B. et al. Proposal on millimeter-wave channel modeling for 5G cellular system. *IEEE Journal on Selected Topics in Signal Processing*, 10(3), (2016): 454–469.

Hur, S., Cho, Y. J., Kim, T. et al. Wideband spatial channel model in an urban cellular environments at 28 GHz. In *Proceedings of 9th European Conference on Antennas and Propagation (EuCAP)*, Lisbon, Portugal, (2015): pp. 1–5.

ITU-R. P.1411-8. Propagation data and prediction methods for the planning of short-range outdoor radiocommunication systems and radio local area networks in the frequency range 300 MHz to 100 GHz. International Telecommunication Union. Geneva, 2015.

ITU-R. P.676-9. Attenuation by atmospheric gases. International Telecommunication Union. Geneva, 2012a.

ITU-R. P.838-3. Specific attenuation model for rain for use in prediction methods. International Telecommunication Union. Geneva, 2005.

ITU-R. P.840-5. Attenuation due to fogs and cloud. International Telecommunication Union. Geneva, 2012b.

Jacob, M., Priebe, S., Maltsev, A. et al. A ray tracing based stochastic human blockage model for the IEEE802.11 and 60 GHz channel model. In *Proceedings of 5th European Conference on Antennas and Propagation (EuCAP)*, Rome, Italy, (2011): pp. 3084–3088.

Kim, M. D., Liang, J., Yoon, Y. K. et al. 28 GHz path loss measurements in urban environments using wideband channel sounder. In *Proceedings of IEEE International Symposium on Antennas and Propagation*, Vancouver, Canada, (2015): pp. 1798–1799.

Ko, J., Hur, S., Lee, S. et al. 28 GHz channel measurements and modeling in a ski resort town in Pyeongchang for 5G cellular network systems. In *Proceedings of 10th European Conference on Antennas and Propagation (EuCAP)*, Davos, Switzerland, (2016): pp. 1–5.

Kourogiorgas, C., Moraitis, N., and Panagopoulos, A. D. Radio channel modeling and propagation prediction for 5G mobile communication systems. *Handbook of Research on Next Generation Mobile Communication Systems*, IGI Global, Hershey, PA, (2016): pp. 1–30.

Lee, J., Kim, M. D., Liang, J. et al. Frequency range extension of the ITU-R NLOS path loss models applicable for urban street environments with 28 GHz measurements. In *Proceedings of 10th European Conference on Antennas and Propagation (EuCAP)*, Davos, Switzerland, (2016): pp. 1–5.

MacCartney, G. R. Jr., Rappaport, T. S., Sun, S. et al. Indoor office wideband millimeter-wave propagation measurements and channel models at 28 and 73 GHz for ultra-dense 5G wireless networks. *IEEE Access*, 3, (2015a): 2388–2424.

MacCartney, G. R. Jr., Rappaport, T. S., Samimi, M. K. et al. Millimeter-wave omnidirectional path loss data for small cell 5G modeling. *IEEE Access*, 3, (2015b): 1573–1580.

Maltsev, A., Maslennikov, R., Sevastyanov, A. et al. Experimental investigation of 60 GHz WLAN systems in office environment, *IEEE Journal in Selected Areas in Communications*, 27(8), (2009): 1488–1499.

Maltsev, A., Perahia, E., Maslennikov, R. et al. Impact of polarization characteristics on 60-GHz indoor radio communication systems. *IEEE Antennas and Wireless Propagation Letters*, 9, (2010): 413–416.

Moraitis, N. and Constantinou, P. Indoor channel measurements and characterization at 60 GHz for wireless local area network applications. *IEEE Transactions on Antennas and Propagation*, 52(12), (2004): 3180–3189.

Moraitis, N. and Constantinou, P. Measurements and characterization of wideband indoor radio channel at 60 GHz. *IEEE Transactions on Wireless Communications*, 5(4), (2006): 880–889.

Moraitis, N. and Constantinou, P. Indoor channel capacity evaluation utilizing ULA and URA antennas in the millimeter wave band. In *Proceedings of 18th IEEE International Symposium on Personal, Indoor and Mobile Radio Communications (PIMRC)*, Athens, Greece, (2007): pp. 1–5.

Moraitis, N. and Panagopoulos, A. D. Millimeter wave channel measurements and modeling for indoor femtocell applications. In *Proceedings of 9th European Conference on Antennas and Propagation (EuCAP)*, Lisbon, Portugal, (2015): pp. 1–6.

Moraitis, N., Vouyioukas, D., and Constantinou, P. Indoor angular profile measurements and channel characterization at the millimetre-wave band. In *Proceedings of 5th European Conference on Antennas and Propagation (EuCAP)*, Rome, Italy, (2011): 155–159.

Muirhead, D., Imran, M. A., and Arshad, K. Insights and approaches for low-complexity 5G small-cell base-station design for indoor dense networks. *IEEE Access*, 3, (2015): 1562–1572.

Nguyen, H. C., Rodriguez, I., Sørensen, T. B. et al. An empirical study of urban macro propagation at 10, 18 and 28 GHz. In *Proceedings of IEEE 83rd Vehicular Technology Conference (VTC Spring)*, Nanjing, China, May 2016.

Olsen, R. L., Rogers, D. V., and Hodge, D. B. The aRb relation in the calculation of rain attenuation. *IEEE Transactions on Antennas and Propagation*, 26(2), (1978): 318–329.

Rappaport, T. S. *Wireless Communications: Principles and Practice*, 2nd ed. Upper Saddle River, NJ: Prentice Hall, (2002).

Rappaport, T. S., Ben-Dor, E., Murdock, J. N. et al. 38 GHz and 60 GHz angle-dependent propagation for cellular & peer to peer wireless communications. In *Proceedings of IEEE International Conference on Communications (ICC)*, Ottawa, Canada, (2012): pp. 4568–4573.

Rappaport, T. S., Gutierrez, F. Jr., Ben-Dor, E. et al. Broadband millimeter wave propagation measurements and models using adaptive beam antennas for outdoor urban cellular communications. *IEEE Transactions on Antennas and Propagation*, 61(4), (2013a): 1850–1859.

Rappaport, T. S., MacCartney, G. R. Jr., Samimi, M. K. et al. Wideband millimeter-wave propagation measurements and channel models for future wireless communication system design. *IEEE Transactions on Communications*, 63(9), (2015): 3029–3056.

Rappaport, T. S., Murdock, J. N., and Gutierrez, F. State of the art in 60-GHz integrated circuits and systems for wireless communications. *Proceedings of the IEEE*, 99(8), (2011): 1390–1436.

Rappaport, T. S., Sun, S., Mayzus, R. et al. Millimeter wave mobile communications for 5G cellular: It will work! *IEEE Access*, 1, (2013b): 335–349.

Saleh, A. A. M. and Valenzuela, R. A. A statistical model for indoor multipath propagation. *IEEE Journal on Selected Areas in Communications*, 5(2), (1987): 128–137.

Samimi, M. and Rapparort, T. S. Statistical channel model with multi-frequency and arbitrary antenna beamwidth for millimeter-wave outdoor communications. In *Proceedings of the Global Communications Conference (GLOBECOM)*, San Diego, CA, (2015): pp. 1–7.

Samimi, M., Wang, K., Azar, Y. et al. 28 GHz angle of arrival and angle of departure analysis for outdoor cellular communications using steerable-beam antennas in New York City. In *Proceedings of 77th IEEE Vehicular Technology Conference (VTC)*, Dresden, Germany, (2013): pp. 1–6.

Sawaya, W. and Clavier, L. Simulation of DS-CDMA on the LOS multipath 60 GHz channel and performance with RAKE receiver. In *Proceedings of 14th IEEE Symposium on Personal Indoor Mobile Radio Communications (PIMRC)*, Beijing, China, (2003): pp. 1232–1236.

Shoji, Y., Sawada, H., Choi, C. S. et al. A modified SV-model suitable for line-of-sight desktop usage of millimeter-wave WPAN systems. *IEEE Transactions on Antennas and Propagation*, 57(10), (2009): 2940–2948.

Smulders, P. F. M. Statistical characterization of 60-GHz indoor radio channels. *IEEE Transactions on Antennas and Propagation*, 57(10), (2009): 2820–2829.

Spenser, Q., Jeffs, B., Jensen, M. et al. Modeling the statistical time and angle of arrival characteristics of an indoor multipath channel. *IEEE Journal on Selected Areas in Communications*, 18(3), (2000): 347–360.

Sulyman, A. I., Nassar, A. T., Samimi, M. et al. Radio propagation path loss models for 5G cellular networks in the 28 GHz and 38 GHz millimeter-wave bands. *IEEE Communications Magazine*, 52(9), (2014): 78–86.

Sun, S., MacCartney, G. R. Jr., and Rappaport, T. S. Millimeter-wave distance-dependent large-scale propagation measurements and path loss models for outdoor and indoor 5G systems. In *Proceedings of 10th European Conference on Antennas and Propagation (EuCAP)*, Davos, Switzerland, (2016a): pp. 1–5.

Sun, S., Rappaport, T. S., Heath, R. W. Jr. et al. MIMO for millimeter-wave wireless communications: Beamforming, spatial multiplexing, or both? *IEEE Communications Magazine*, 52(12), (2014): 110–121.

Sun, S., Rappaport, T. S., Thomas, T. A. et al. Investigation of prediction accuracy, sensitivity, and parameter stability of large-scale propagation path loss models for 5G wireless communications. *IEEE Transactions on Vehicular Technology*, 65(5), (2016b): 2843–2860.

Timms, G., Kvičera, V., and Grábner, M. 60 GHz band propagation experiments on terrestrial paths in Sydney and Praha. *Radioengineering*, 14(4), (2005): 27–32.

Xu, H., Kukshya, V., and Rappaport, T. S. Spatial and temporal characteristics of 60-GHz indoor channels. *IEEE Journal on Selected Areas in Communications*, 20(3), (2002): 620–630.

Xu, H., Rappaport, T. S., Boyle, R. J. et al. Measurements and models for 38-GHz point-to-multipoint radiowave propagation. *IEEE Journal on Selected Areas in Communications*, 18(3), (2000): 310–321.

Chapter 4

Basics on the Theory of Fading Channels and Diversity

Vasileios M. Kapinas, Georgia D. Ntouni, and George K. Karagiannidis

Contents

Abbreviations

2D	two-dimensional
AF	amount of fading
AM	amplitude modulation
AWGN	additive white Gaussian noise
BPSK	binary phase-shift keying
CDMA	code division multiple access
CIR	channel impulse response
CSI	channel state information
DoF	degree of freedom
DS-SS	direct-sequence spread spectrum
FDM	frequency division multiplexing
FH-SS	frequency-hopping spread spectrum
i.i.d.	independent and identically distributed
IQ	in-phase/quadrature
ISI	intersymbol interference
LOS	line-of-sight
LTI	linear time-invariant
LTV	linear time-variant
MIMO	multiple-input multiple-output
MISO	multiple-input single-output
MLD	maximum likelihood detection
MLSE	maximum likelihood sequence estimation
MRC	maximum ratio combining

MUSA	multiple unit steerable antenna
OFDM	orthogonal frequency division multiplexing
PDF	probability density function
PRS	pseudo-random sequence
PSD	power spectral density
QAM	quadrature amplitude modulation
QoS	quality-of-service
QPSK	quadrature phase-shift keying
QSFF	quasi-static flat fading
RF	radio frequency
RV	random variable
Rx	receiver
RxD	receive diversity
SAR	synthetic aperture radar
SER	symbol error rate
SIMO	single-input multiple-output
SISO	single-input single-output
SNR	signal-to-noise ratio
Tx	transmitter
TxD	transmit diversity
WSS	wide-sense stationary

In modern communication systems, the sufficient knowledge of the channel behavior and its induced impairments on the transmitted signal are vital for the sophisticated design of advanced mitigation techniques. Fading is a complicated phenomenon that can severely affect the signal propagation in fixed and mobile wireless communication systems. However, its stochastic and highly varying nature makes it very difficult to model the associated channels precisely. Research efforts over the years have led to various statistical models for fading channels, which depend on the radio propagation environment and the communication scenario under study.

The negative impact of fading on signal transmission can be summarized in two major effects: the distortion of the signal due to intersymbol interference (ISI) and the signal-to-noise ratio (SNR) penalty in the error performance compared to the additive white Gaussian noise (AWGN) channel. These problems need to be efficiently tackled in order to guarantee robust communication with high availability. Diversity schemes can improve the transmission reliability by proper utilization of multiple communication channels with different characteristics. Interestingly, diversity has been proved to be one of the most common and efficient techniques for combating the detrimental effects of fading, interference, and error bursts for over a century now in the history of wireless systems.

The main purpose of this chapter is to introduce the fundamental concepts of fading and diversity, mainly in a qualitative way. A more detailed presentation of these subjects can be found in many textbooks, articles and dissertations in the literature, such as [1–10] and references therein. This chapter also provides a detailed historical investigation on the roots of the fading phenomenon and the evolution of the various diversity techniques since the beginning of the wireless communications era. To this end, profound evidence of critical works, either not known to the research community or even uncited in the context of wireless communications, are given. Finally, more than 15 different pure concepts of the diversity are recorded here for the reader's ease, serving as a high-level taxonomy of the versatile diversity techniques that can be employed, either alone or in combination, by wireless communication systems to improve their error performance.

The organization of the chapter is as follows. The first half provides the basic theoretical tools for the analysis of fading, including its nature and types as well as the most commonly used statistical models. The impact of fading on signal transmission is also analyzed, focusing on time and frequency selectivity. In the second half, the chapter elaborates on fading mitigation and continues with a complete high-level classification of the diversity techniques proposed in the entire period of the twentieth century, which exploit various dimensions or degrees of freedom (DoF)s, such as space, frequency, time, and polarization, among others.

Throughout this chapter, various notations are used. Other than those described where read for the first time, some common math symbols adopted here include: (1) $j = \sqrt{-1}$ (imaginary unit), (2) $\mathrm{Re}\{z\}$, $\mathrm{Im}\{z\}$ (real and imaginary parts of complex number z), (3) $|z|$, $\angle z$ (amplitude, $|z| = a$, and phase, $\angle z = \theta$, of complex $z = a\exp(j\theta)$), (4) $\mathbb{C}, \mathbb{R}, \mathbb{Z}$ (sets of complex, real, and integer numbers), (5) $\mathbb{N}, \mathbb{N}_0$ (sets of natural and non-negative integer numbers), (6) $\mathbb{R}^+, \mathbb{R}_0^+$ (sets of positive real and non-negative real numbers), (7) $\mathrm{E}\{\cdot\}$ (statistical expectation operator), (8) $\mu_z = \mathrm{E}\{z\}$, $\mathrm{var}\{z\} = \mathrm{E}\{|z - \mu_z|^2\}$ (mean and variance of RV $z \in \mathbb{C}$), (9) $\forall$ (for each), and (10) $x \mapsto T(x)$ (mapping from x to $T(x)$, also defining the mapping rule T).

4.1 Nature and Types of Fading

Consider a communication channel between a source S and a destination D, where the former intends to deliver an information message to the latter through the available wireless medium. Typically, not a single radio propagation path will exist between S and D due to the presence of various objects and structures surrounding either or both. These obstacles provide alternative paths for the electromagnetic wave propagation through different mechanisms that can generally be attributed to reflection, diffraction and scattering. As a consequence, radio waves travel along diverse paths and arrive at D from several directions with various delays and after experiencing different attenuations. Thus, the *multipath* nature of the medium introduces time spread in the information-bearing signal transmitted through the radio channel. In addition, *time variations* in the structure of the medium or existence of relative motion between S and D induce extra phase shift on the multipath components due to the Doppler effect.

Taking into account the characteristics of the time-variant multipath channel, the received signal $r_l(t)$ at D can be considered as a superposition of a large number of phasors,[*] let us say L_p, with time-varying amplitudes $a_l(t)$ and phases $\theta_l(t)$ that appear to be unpredictable to the user(s) of the channel

$$r_l(t) = \sum_{l=1}^{L_p} a_l(t) \exp\left[j\theta_l(t)\right] \tag{4.1}$$

where:

$\theta_l(t) = -2\pi f_c \tau_l(t)$

$\tau_l(t)$ is the related to the lth path propagation delay

f_c is the carrier frequency

For the sake of clarity, noiseless transmission of an unmodulated carrier is initially considered, while the complete version of the received signal, $r_l(t)$, will be treated in Section 4.3.2. In addition, the model in Equation 4.1 describes a discrete multipath channel, whereas in the case of a continuum of multipath components (i.e., *diffused multipath*), the sum has to be replaced by an integral.

4.1.1 Small and Large Scale Fading

The multipath channel model described earlier implies that the received signal $r_l(t)$ may be viewed as a complex-valued stochastic process with random amplitude and phase in both time and space domain, the latter defined by the set of all possible locations of D. In a more intuitive approach, the associated multipath components may add constructively or destructively at different time realizations, so that the composite signal received by D may experience distortion, strength fluctuation or both. This is the result of *signal fading* caused by the channel impulse response (CIR), that can be classified into major categories based on the nature of the physical medium, the distance between S and D, and the transmitted symbol rate.[†]

From the definition of fading, it is clear that when a received signal experiences fading during transmission, both its envelope (i.e., amplitude) and phase fluctuate over time. However, in most communication system scenarios, the receiver (Rx) is able to perform *coherent detection*, that is to reconstruct the carrier with perfect knowledge of the phase and frequency, and proceed further with complex conjugate demodulation. Therefore, it can generally be assumed that the phase variation

---

[*] The subscript ℓ in $r_l(t)$ indicates the equivalent lowpass signal of the real bandpass signal $r_b(t)$, also called the complex envelope of $r_b(t)$. For more details, see Section 4.3.

[†] We generally assume, excluding the spread spectrum systems, that the transmitted symbol rate ρ_s is approximately equal to the inverse of the signal bandwidth W, namely $\rho_s \approx 1/W$.

due to fading does not affect the error performance of the system, since the channel phase can be perfectly tracked at the Rx. It follows that performance analysis of digital communication systems mainly requires the knowledge of the fading envelope statistics, such as those given in Section 4.1.2, that widely define the type or distribution of the fading process.

Simply put, the rapid fluctuations of the instantaneous received signal strength over small travel distances or short time intervals, caused by the time-variant multipath channel,* result in the so-called *envelope fading* or more officially termed *small-scale fading* effects. This type of fading is relatively fast and is therefore responsible for the short-term signal variations. It is assumed to be a wide-sense stationary random process in time.† Depending on the nature of the radio propagation environment, there are different models describing the statistical behavior of the multipath fading envelope, the most common being the Rayleigh (1880), Nakagami-*q* or Hoyt (1947), Nakagami-*n* or Rice (1948), and Nakagami-*m* (1960). In the context of mobile communications, small-scale fading can result in signal power variations of up to 30–40 dB for a mobile movement of just a fraction of wavelength, while higher speeds can cause the mobile to pass through several fades in a small period of time.

However, as the Rx moves away from the transmitter (Tx) over much longer distances, the average signal power will gradually decrease as a result of the *large-scale fading* effects. Definitely, the most typical representative of channel impairments that fall within this category is the free-space *path loss*, a factor involved in the well-known Friis equation, for the case where there is a clear and unobstructed line-of-sight (LOS) path between the Tx and Rx. Nevertheless, this is a rather simplistic model since, in a real mobile radio channel, propagation is generally neither free space nor LOS. To this end, generalized measurement-based path loss models exist in the literature, where the path loss exponent can take values up to 4, as opposed to 2 for the free-space model. In addition, several empirical models have been obtained by curve fitting experimental data, two of the most popular being the Okumura-Hata (1968, 1980) and Lee (1985). Finally, in typical mobile radio channels, the mean signal level (or *local mean*) experiences slow variations over distances of several tens of wavelengths due to the presence of large buildings, hilly terrain and foliage. This phenomenon is known as shadow fading or *shadowing* and is usually modeled as a multiplicative and, generally, slowly time-varying random process. Experimental observations have confirmed that the shadow fades follow a Lognormal distribution for various outdoor and indoor environments (see, for instance, [11] and references therein).

---

* Even the spatial variations in the received signal due to the mobility of D can be observed by the latter as temporal variations while moving through the multipath field.
† It is recalled that a wide-sense stationary random process has the property that its mean value is constant and its autocorrelation function is invariant to any time shift.

4.1.2 Statistics and Modeling of Fading

It has already been discussed that, in a multipath fading channel, delayed versions of the transmitted signal can be received by reflection, diffraction and/or scattering, where each component corresponds to an appropriate channel *fading coefficient*. As will be seen in Section 4.3, fading has a multiplicative effect on the transmitted signal (see for instance Equation 4.53), while, for narrowband systems, the faded signal can be simply modeled by the product of a single fading coefficient and the complex envelope of the transmitted signal (see Equation 4.62). Under these assumptions, the envelope and phase of the noiseless received signal are actually the envelope and phase of the linearly modulated transmitted symbol by the fading coefficient. However, it has been proved that knowing the statistics of just the fading envelope is sufficient for the performance analysis of wireless communication systems [3, Sec. 2.1.1].[*] Therefore, in the sequel of this section, the envelope statistics of the most popular fading channels will be outlined. Note that, over the years, several models have been adopted to describe the statistical behavior of the fading envelope, each with different flexibility and complexity determined by the nature of the radio propagation environment.

4.1.2.1 Rayleigh Fading

Rayleigh fading is the most commonly used statistical model in wireless communications. Interestingly, Lord Rayleigh (John William Strutt) derived and applied this distribution to the theory of sound [12] long before the understanding of multipath signal reception. In a Rayleigh channel there exists no LOS component and the fading coefficient, h, is modeled as a zero-mean circularly-symmetric complex Gaussian random variable (RV) with variance $\text{var}\{h\} = \text{E}\{|h|^2\}$, that is, $h \sim CN(0, \text{E}\{|h|^2\})$.[†] The probability density function (PDF) of the fading envelope $|h| = a$ equals

$$f_a(a) = \frac{2a}{\Omega} \exp\left(-\frac{a^2}{\Omega}\right), \quad a \geq 0 \tag{4.2}$$

---

[*] Recall also the discussion held in Section 4.1.1 about the coherent detection capability of communication systems.

[†] The symbol C in $CN(\cdot,\cdot)$ denotes that random samples (here snapshots of the fading channel) are both circularly-symmetric and complex. Additionally, it is known that circularly-symmetric complex jointly-Gaussian random vectors are completely determined by their covariance matrix [13], which for the (complex) scalar case reduces to the variance.

where $\Omega = \mathrm{E}\{a^2\}$ is the *average fading power*. It shall be noted here that, if needed, the phase $\angle h = \theta$, in this case, is typically following a uniform distribution in $[0, 2\pi)$, that is, $\theta \sim \mathcal{U}(0, 2\pi)$ and is considered to be statistically independent of a.

Assuming now that E_s is the symbol energy and N_0 the variance of the AWGN, the instantaneous SNR per symbol, $\gamma = a^2 (E_s/N_0)$, is exponentially distributed according to

$$f_\gamma(\gamma) = \frac{1}{\mathrm{E}\{\gamma\}} \exp\left(-\frac{\gamma}{\mathrm{E}\{\gamma\}}\right), \quad \gamma \geq 0 \tag{4.3}$$

In order to measure the severity of the experienced fading, Charash [14] introduced a metric called *amount of fading* (AF), which, according to [3, Sec. 1.1.4], can be used as an alternative performance criterion in the more general context of systems with arbitrary combining techniques and channel statistics. The AF metric is quantitatively expressed by

$$\mathrm{AF} = \frac{\mathrm{var}\{\gamma\}}{\mathrm{E}\{\gamma\}^2} = \frac{\mathrm{E}\{\gamma^2\} - \mathrm{E}\{\gamma\}^2}{\mathrm{E}\{\gamma\}^2} \tag{4.4}$$

It is seen from Equation 4.4 that the calculation of AF requires the computation of the first two moments (i.e., mean and variance) of γ, which for the Rayleigh fading case can be determined by [3, eq. (2.9)]

$$\mathrm{E}\{\gamma^k\} = \Gamma(1 + k)\mathrm{E}\{\gamma\}^k, \quad k \in \mathbb{N} \tag{4.5}$$

where $\Gamma(z) = \int_0^\infty t^{z-1} \exp(-t)\,dt$, $z \in \mathbb{R}^+$ stands for the Euler gamma function. Hence, $\mathrm{AF}_{\mathrm{Rayleigh}} = 1$.

4.1.2.2 Rician (or Nakagami-n) Fading

If there exists a strong LOS component along with many random weaker ones, the channel fading coefficient has a non-zero mean value. In this case, the fading envelope follows the Rice distribution that was first derived by Rice [15]. Its basic parameter K, called *Rician factor*, denotes the average power ratio of the LOS component to all the other multipath components, where increasing K implies decreasing fading severity conditions. In the telecommunications literature, the Rice distribution is also known as Nakagami-*n* distribution with PDF given by [16, eq. (50)]

$$f_a(a) = \frac{2(1+n^2)\exp(-n^2)a}{\Omega} \exp\left(-\frac{(1+n^2)a^2}{\Omega}\right) I_0\left(2na\sqrt{\frac{1+n^2}{\Omega}}\right), \quad a \geq 0 \tag{4.6}$$

where $I_0(\cdot)$ is the zero-order modified Bessel function of the first kind defined in [17, eq. (9.6.3)] and $n \geq 0$ is the Nakagami-*n* fading parameter which is related to the Rician factor by $K = n^2$. Interestingly, for $n = K = 0$, Equation 4.6 reduces to the Rayleigh PDF in Equation 4.2, while for $n = K \to \infty$ there exists no fading at all but the LOS component.

In Rician fading, the instantaneous SNR per symbol follows noncentral chi-square distribution

$$f_\gamma(\gamma) = \frac{(1 + n^2)\exp(-n^2)}{E\{\gamma\}} \exp\left(-\frac{(1 + n^2)\gamma}{E\{\gamma\}}\right) I_0\left(2n\sqrt{\frac{(1 + n^2)\gamma}{E\{\gamma\}}}\right), \quad \gamma \geq 0 \quad (4.7)$$

while the moments of γ are given by [3, eq. (2.18)]

$$E\{\gamma^k\} = \frac{\Gamma(1 + k)}{(1 + n^2)^k} M(-k, 1, -n^2) E\{\gamma\}^k, \quad k \in \mathbb{N} \quad (4.8)$$

where $M(\cdot, \cdot, \cdot)$ stands for the Kummer's confluent hypergeometric function [17, eq. (13.1.2)].

Therefore, with the aid of Equation 4.8, the AF in Rician fading can be computed from Equation 4.4 as

$$\mathrm{AF_{Rician}} = \frac{1 + 2n^2}{(1 + n^2)^2} \quad (4.9)$$

From Equation 4.9, it can be validated that as $n \to \infty$ then $\mathrm{AF_{Rician}} \to 0$, which corresponds to no-fading channel conditions.

4.1.2.3 Nakagami-m Fading

A more general fading distribution that can be well-fitted to experimental data is the Nakagami-*m* distribution suggested in 1960 by Nakagami. In this case, the PDF of the fading envelope is given by [16]

$$f_a(a) = \frac{2m^m a^{2m-1}}{\Omega^m \Gamma(m)} \exp\left(-\frac{ma^2}{\Omega}\right), \quad a \geq 0 \quad (4.10)$$

where $m \geq 0.5$ is the Nakagami-*m* fading parameter. Nakagami-*m* reduces to Rayleigh when $m = 1$, approximates Rician fading when $m = \left[(K + 1)^2 / (2K + 1)\right]$ and results to no fading when $m \to \infty$. The PDF of the instantaneous SNR per symbol is a Gamma distribution given by

$$f_\gamma(\gamma) = \frac{m^m \gamma^{m-1}}{E\{\gamma\}^m \Gamma(m)} \exp\left(-\frac{m\gamma}{E\{\gamma\}}\right), \quad \gamma \geq 0 \quad (4.11)$$

The moments of γ are given by [3, eq. (2.23)]

$$E\{\gamma^k\} = \frac{\Gamma(m+k)}{m^k\Gamma(m)}E\{\gamma\}^k, \quad k \in \mathbb{N} \tag{4.12}$$

Again, with the aid of Equation 4.12, the AF in Nakagami-*m* fading can be computed from Equation 4.4 as

$$AF_{Nakagami} = \frac{1}{m} \tag{4.13}$$

Given that $AF \in [0,2]$, Nakagami-*m* can model worse fading conditions than Rayleigh. Particularly, the channel with the most severe fading conditions, corresponding to $AF = 2$, is referred to as one-sided Gaussian fading and occurs when either $\text{var}\{\text{Re}\{b\}\} = 0$ or $\text{var}\{\text{Im}\{b\}\} = 0$.

4.1.2.4 Hoyt (or Nakagami-q) Fading

The Hoyt model was originally proposed by Hoyt [18] and was further investigated by Nakagami (1960) in his effort to approximate Nakagami-*m* for certain values of *m*. Apart from cellular mobile radio systems, this type of fading channel is common in satellite links affected by strong ionospheric scintillation. The PDF of the fading envelope is given by [16, eq. (52)]

$$f_a(a) = \frac{(1+q^2)a}{q\Omega}\exp\left(-\frac{(1+q^2)^2 a^2}{4q^2\Omega}\right)I_0\left(\frac{(1-q^4)a^2}{4q^2\Omega}\right), \quad a \geq 0 \tag{4.14}$$

where $0 \leq q \leq 1$ is the Nakagami-*q* fading parameter. This distribution can be approximated by the Nakagami-*m* after substituting $m = (1+q^2)^2/[2(1+2q^4)]$.

Regarding the instantaneous SNR per symbol, its PDF reads

$$f_\gamma(\gamma) = \frac{1+q^2}{2qE\{\gamma\}}\exp\left(-\frac{(1+q^2)^2\gamma}{4q^2E\{\gamma\}}\right)I_0\left(\frac{(1-q^4)\gamma}{4q^2E\{\gamma\}}\right), \quad \gamma \geq 0 \tag{4.15}$$

In this case, the moments of γ are given by [3, eq. (2.13)]

$$E\{\gamma^k\} = \Gamma(1+k)F\left(-\frac{k-1}{2}, -\frac{k}{2}; 1; \left(\frac{1-q^2}{1+q^2}\right)^2\right)E\{\gamma\}^k, \quad k \in \mathbb{N} \tag{4.16}$$

where $F(\cdot,\cdot;\cdot;\cdot)$ is the Gauss hypergeometric function [17, eq. (15.1.1)], [19, eq. (9.100)].

Finally, the AF in Hoyt fading can be computed with the aid of Equations 4.4 and 4.16 as

$$AF_{Hyot} = \frac{2(1+q^4)}{(1+q^2)^2} \tag{4.17}$$

taking values within the interval [1,2].

4.1.2.5 *Generalized Gamma (or* α − μ*) Fading*

The generalized Gamma distribution was originally proposed by Stacy [20] as a purely mathematical model in which some statistical properties of a generalized version of the Gamma distribution were investigated. In this sense, it was connected neither with any specific application nor with any physical modeling of any given phenomenon. Generalized Gamma was recently revisited by Yacoub (2007), under the name α − μ distribution, with its parameters being directly associated with the physical properties of the propagation medium of a fading channel. The generalized Gamma or α − μ distribution includes several other distributions as special cases, such as the Gamma, Nakagami-*m*, Exponential, Weibull, one-sided Gaussian, and Rayleigh [21], while fading conditions for values of $m < 0.5$ can also be modeled. The PDF of the fading envelope is [22, eq. (1)]

$$f_a(a) = \frac{\beta m^m a^{m\beta-1}}{\Omega_p^m \Gamma(m)} \exp\left(-\frac{ma^\beta}{\Omega_p}\right), \quad a \geq 0 \tag{4.18}$$

where:
$\beta > 0$ is a parameter that yields the best fit to empirical measurements
$m > 0$ is the fading parameter
Ω_p is a power-scaling parameter given by $\Omega_p = \mathrm{E}\{a^\beta\}$

Denoting by $\tau = \left[\Gamma(m)/\Gamma(m + 2/\beta)\right]$, the PDF of the instantaneous SNR per symbol is given by [23, eq. (1)]

$$f_\gamma(\gamma) = \frac{\beta \gamma^{\frac{m\beta}{2}-1}}{2\Gamma(m)(\tau\mathrm{E}\{\gamma\})^{\frac{m\beta}{2}}} \exp\left[-\left(\frac{\gamma}{\tau\mathrm{E}\{\gamma\}}\right)^{\frac{\beta}{2}}\right], \quad \gamma \geq 0 \tag{4.19}$$

The moments of γ in this case are given by [23, eq. (3)]

$$\mathrm{E}\{\gamma^k\} = \frac{\Gamma[m + (2k/\beta)]}{\Gamma(m)}(\tau\mathrm{E}\{\gamma\})^k, \quad k \in \mathbb{N} \tag{4.20}$$

while the AF in generalized Gamma fading can be calculated from Equations 4.4 and 4.20 as

$$\mathrm{AF}_{\text{gen. Gamma}} = \frac{\Gamma[m + (4/\beta)]\Gamma(m)}{\Gamma^2[m + (2/\beta)]} - 1 \tag{4.21}$$

As mentioned earlier, generalized Gamma can describe a great variety of small-scale fading and shadowing conditions by adopting different values for the two parameters β and m. Indicatively, for $\beta = 2$ and $m = 1$ the Rayleigh fading channel model is obtained, while the Nakagami-*m* can be derived for $\beta = 2$ and $m > 0.5$.

4.1.2.6 Lognormal Shadowing

In the case of large-scale fading and assuming that the multipath effects are somehow eliminated, shadowing is the only phenomenon affecting the performance of the communication system. As mentioned in Section 4.1.1, empirical measurements have confirmed that the shadow fades can be modeled by a Lognormal distribution. In this case, the PDF of the instantaneous SNR per symbol, γ, is given by [3, eq. (2.53)]

$$f_\gamma(\gamma) = \frac{\xi}{\sigma\sqrt{2\pi}\gamma} \exp\left(-\frac{(10\log\gamma - \mu)^2}{2\sigma^2}\right) \tag{4.22}$$

where $\xi = (10/\ln 10)$ and μ (in dB), σ (in dB) are the mean and standard deviation of RV $10\log\gamma$, respectively.

The moments of γ in this case are given by [3, eq. (2.55)]

$$E\{\gamma^k\} = \exp\left(\frac{k\mu}{\xi} + \frac{k^2\sigma^2}{2\xi^2}\right), \quad k \in \mathbb{N} \tag{4.23}$$

while the AF yields the expression

$$\mathrm{AF}_{\mathrm{Lognormal}} = \exp\left(\frac{\sigma^2}{\xi^2}\right) - 1 \tag{4.24}$$

In practical situations, the AF in Lognormal shadowing is bounded by a number that exceeds the maximal AF exhibited by the multipath PDF studied in the previous sections by several orders of magnitude [3, Sec. 2.2.2].

4.1.2.7 Generalized-K Composite Fading/Shadowing

Shadowing results in random rather than deterministic average received power levels, while in many communication systems it has to be encountered along with small-scale fading. The combined effect of small and large-scale fading can be modeled by various distributions, which are derived by the combination of one distribution for the fading envelope and another one for the local mean.

The Generalized-K distribution was introduced by Shankar [24] as a generalized fading/shadowing channel model, combining the Nakagami-m fading and Gamma shadowing models. It is general enough to include as special cases (or approximate well) other distributions. For this composite channel, the PDF of the fading envelope is given by [25, eq. (1)]

$$f_a(a) = \frac{4m^{\frac{k+m}{2}}}{\Gamma(m)\Gamma(k)\Omega_p^{\frac{k+m}{2}}} a^{k+m-1} K_{k-m}\left(2a\sqrt{\frac{m}{\Omega_p}}\right), \quad a \geq 0 \tag{4.25}$$

where k, m are the two shaping parameters, $\Omega_p = \mathrm{E}\{a^2\}/k$, and $K_{k-m}(\cdot)$ is the $(k-m)$th order modified Bessel function of the second kind [19, eq. (8.407.1)]. One can see that, for $m=1$, Equation 4.26 reduces to the K distribution, thus approaching the Rayleigh-Lognormal composite fading/shadowing channel model. In addition, as $k \to \infty$ it approximates Nakagami-m, while, for $m, k \to \infty$, the Generalized-K fading channel approaches AWGN (i.e., no-fading case).

The PDF of the instantaneous SNR per symbol is given by [25, eq. (2)]

$$f_\gamma(\gamma) = 2\left(\frac{km}{\mathrm{E}\{\gamma\}}\right)^{\frac{k+m}{2}} \frac{\gamma^{\frac{k+m-2}{2}}}{\Gamma(k)\Gamma(m)} K_{k-m}\left(2\sqrt{\frac{km}{\mathrm{E}\{\gamma\}}\gamma}\right), \quad \gamma \geq 0 \qquad (4.26)$$

In addition, the moments of γ for the Generalized-K distribution are given by [25, eq. (5)]

$$\mathrm{E}\{\gamma^n\} = \frac{\Gamma(k+n)\Gamma(m+n)}{\Gamma(k)\Gamma(m)}\left(\frac{\mathrm{E}\{\gamma\}}{km}\right)^n, \quad n \in \mathbb{N} \qquad (4.27)$$

Finally, the AF in Generalized-K fading can be directly computed from [24, eq. (8)]

$$\mathrm{AF}_{\mathrm{gen.}\text{-}K} = \frac{1}{k} + \frac{1}{m} + \frac{1}{km} \qquad (4.28)$$

where it is easy to see that (1) low values of k and m correspond to severe fading, (2) values of $k \to \infty$ render Equation 4.28 identical to 4.13, that is, shadowing vanishes giving rise to Nakagami-m fading only, and (3) values of $k, m \to \infty$ give $\mathrm{AF}_{\mathrm{gen.}\text{-}K} = 0$, which corresponds to the ideal case of no fading at all or equivalently to the AWGN channel.

4.1.3 Historical Roots of Fading

The fluctuations in the received signal strength over time and distance were noticed very soon after the first transatlantic transmission, accomplished by Marconi in 1901. Long ago in 1902, Marconi drew attention to the remarkable difference in the strength of signals received by day and by night, and his observations have been repeated and recorded by many other observers [26]. It is very interesting to reproduce below some selected excerpts from the weekly published issues of the "Electrical World and Engineer" periodical of that period, which reveal that the signal intensity variation in long distance transmissions was very early reported by experts and not, though any kind of explanation for this paradox was unable to be provided by that time:

Atmospheric conditions assuredly play an important part here, and it is a common experience of transatlantic travel to find the wireless not working satisfactorily, while at times remarkable results are achieved. At moderate range most of the difficulties seem to disappear.

—*Elec. World Eng.*, vol. XL, no. 20, Nov. 15, 1902;
under the column "Long distance wireless telegraphy"

These long-distance feats are tantalizing in their uncertainty. Messages might be received on one day and the next day, for no apparent reason, might fail to have any accurate effect on the receiving apparatus.... It is thought that long distance experiments are useless for practical purposes.

—*Elec. World Eng.*, vol. XL, no. 20, Nov. 15, 1902;
under the column "Telegraphy, telephony and signals"

In the wireless service on the transatlantic steamers there has been beautiful success in picking up passing neighbors at long range, but there have also been lugubrious failures in reaching the land stations at very moderate distances.

—*Elec. World Eng.*, vol. XL, no. 23, Dec. 6, 1902;
under the column "Round the world by wireless"

However, it seems that it was not before 1910s that first in-depth discussions on this phenomenon and possible explanations appear in the literature. In one of these very early works published in 1911, Taylor uses the term *freak communications* (or *freak signals*) to refer to variations in the range of communication (or signal strength) occasionally occurred by day, which are much more pronounced during the night hours [27]. His broader explanation attributes this phenomenon to "variations in the transmitting efficiency of the atmosphere" adding that "perturbances in this respect exhibit almost the same characteristics of irregularity as atmospheric impulses." In order to highlight the importance of this phenomenon, Taylor remarks that "the sudden veiling or obscuring of distant signals together with the equally sudden 'opening out' of signals observed whilst listening on the receiver at a wireless station during the prevalence of this phenomenon is very impressive." Nevertheless, Commander Loring had some doubts about the explanation given previously [28]:

The prevalence, or otherwise, of atmospherics does not appear to bear any relation to the appearance of freak signals. At any rate, atmospherics have been reported by reliable observers to be least noticeable at periods of freak ranges.

—Commander Loring (1911); comment on [27]. Reproduced by
permission of the Institution of Engineering & Technology.

In the following years, systematic observations and recordings took place with regard to variations of the radio range or signal strength in several wireless links. Indicatively, Marriott [29] presented a record of the most interesting range variations observed at the Manhattan Beach station of the United Wireless Telegraph Company, located at Coney Island in New York City. His measurements indicated that there were more *freaks* in the month of August than in the winter months, namely the ranges showed a greater percentage variation in the summer than in the winter. Marchant (1915) summarized measurements made mainly between stations in Liverpool and Paris regarding fluctuations in the strength of wireless signals during sunset and night hours, taking into account the weather conditions over the period of one year. Among others, Marchant concluded that (1) the *sunset effect*, (i.e., the strengthening of the signal just after sunset) occurs about 3/4 hr after the actual time of sunset and varies with the weather conditions, whereas under rainy conditions it is much less marked and (2) the freak signals are relatively great during night and occur within the space of a few minutes, while greater increases in signal strength can be observed after the cessation of rain either at the transmitting or receiving station [26]. In the same context, Austin (1915) studied the seasonal variation in strength of signals sent from radio stations in the Philadelphia and Norfolk navy yards via observations made for over two years. His findings, among others, experimentally verified that winter signals in general are stronger than those of summer, especially when the transmission takes place overland, while, "contrary to the ideas previously held, there seems to be no very marked connection between rainfall and the transmission of the signals" [30].

From the aforementioned published works, it can be easily observed that during the first two decades of the wireless telegraphy era,* there was considerable confusion in understanding the actual effects of fading on signal transmission and even more difficulties in providing accurate explanations for its genesis. With this in mind, the terms *freak communications/signals/ranges* were actually related to the long ranges that could be covered by low-power shortwave radio transmitters at certain times and places, a phenomenon which is well-known today as *skywave* or *skip* propagation. However, in these early works, the so-called *freaks* were sometimes unintentionally confused with the large-scale fading effects.† Very interestingly, it was in 1913 that the radio pioneer (and engineer of the Federal Telegraph Co.) De Forest officially coined the term *fading* for the first time to describe the fact that during a station-to-station transmission from Los Angeles to San Francisco, being 350 miles apart, two waves differing only 5% in length (3260 and 3100 m, respectively) were received with great variations in their intensity over certain time intervals, sometimes observing even total extinction of the one or the other but very rarely of both simultaneously [34].‡ In order to give some

---

* Inventor and entrepreneur Marconi had actually established radio communication links spanning as far as the English Channel (la Manche) since the early 1890s [31].

† Note that, the first consolidated results on "shadowing" appear in the literature several years later, in [32,33].

‡ This first occurrence of the term "fading" in the literature was also reported by Burrows [35].

explanation for this paradox, De Forest adopted an existing speculative theory about the possible interference of the direct wave with the reflected ones from the upper conducting layer of the atmosphere, thus giving the birth to multipath fading reasoning. The next parts from this seminal work are very illuminative.

> The duration of this fading effect is often several hours after nightfall; then it suddenly vanishes and thereafter both waves have their normal intensity. This alteration of intensity is sometimes for one wave, and sometimes for the other, and rarely for both. Under such conditions there are acting at the receiving stations two trains of waves, which have travelled over paths of unequal lengths or which have travelled with unequal velocities. Consequently, there will be a phase displacement between them and interference at certain localities… it would account for the fact that the Marconi transatlantic stations can operate sometimes with a few kilowatts and sometimes require 125 to 600 kilowatts.

—De Forest. © 1913 IEEE. Reprinted, with permission, from [34].

In 1916, Taylor and Blatterman published a very interesting paper related to nocturnal transmission experiments carried out by radio stations in University of North Dakota and Washington University in St. Louis as an attempt to test the interference theory of fading effects. In this work, the term *swinging* appears for the first time to denote slow variations in signal strength. Particularly, they stated that there seem to be two kinds of fluctuations in nocturnal overland transmission; a rapid fading and a slow swinging in signal strength [36]. In the discussion section of the same paper, some plausible explanations for these two channel oddities are also provided.

> The first (refers to *fading*) may be due to changes, in the nature of interference effects. These could be local at the sender or at the receiver, or they might be caused by rather sharp surfaces of discontinuity almost anywhere between the stations... The second (refers to *swinging*) or slower effect may be due to refracting masses of moving ionized air in the path of transmission, producing at times a lens-like concentration and at other times a dispersive effect.

—Taylor and Blatterman. © 1916 IEEE. Reprinted, with
permission, from [36].

In the coming years, the understanding of the cause and effects of fading or swinging on radio signal transmission proved itself to be one of the most elusive problems faced by both radio engineers and amateurs. As the radio art was

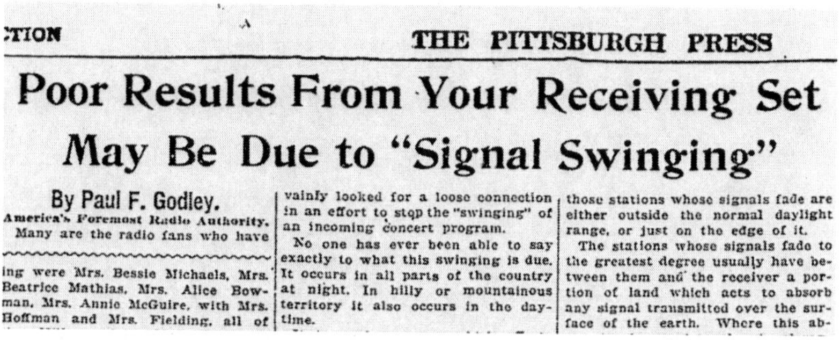

:TION   THE PITTSBURGH PRESS

Poor Results From Your Receiving Set
May Be Due to "Signal Swinging"

By Paul F. Godley.
America's Foremost Radio Authority.
Many are the radio fans who have

ing were Mrs. Bessie Michaels, Mrs. Beatrice Mathias, Mrs. Alice Bowman, Mrs. Annie McGuire, with Mrs. Hoffman and Mrs. Fielding, all of

vainly looked for a loose connection in an effort to stop the "swinging" of an incoming concert program.
No one has ever been able to say exactly to what this swinging is due. It occurs in all parts of the country at night. In hilly or mountainous territory it also occurs in the daytime.

those stations whose signals fade are either outside the normal daylight range, or just on the edge of it. The stations whose signals fade to the greatest degree usually have between them and the receiver a portion of land which acts to absorb any signal transmitted over the surface of the earth. Where this ab-

Figure 4.1   Part from a news article on *signal swinging. The Pittsburgh Press*, June 25, 1922.

progressively moving from spark telegraphy into continuous wave telegraphy and radio telephone broadcasting, any empirical or scientific contribution on the peculiarities of the wireless transmission medium and the possible ways to improve the message quality at the receiving end became matters of common interest. As a consequence, these issues occupied much space in the academic and public press, an example for the latter given in Figure 4.1 that depicts the headline and part of the body of a 1922 article on signal swinging, published in a major daily afternoon newspaper. During that period, another term, that is, *soaring*, appears in the literature to describe the radio signal variations in the same context. Particularly, Pickard [37] mentions that the terms fading, swinging, and soaring were commonly used interchangeably to refer to the large amplitude short period variations of radio signals.

In the meantime, radio amateurs were among the first to perform large-scale experiments on fading and long-distance shortwave communication even from the early 1910s [31], thus having already obtained valuable experience. However, due to the plethora of variables involved in radio transmission, any conclusions regarding fading could be considered safe only after being corroborated by a large number of observers. To this end, Whittemore and Kruse from the U.S. National Bureau of Standards suggested a giant initiative involving extensive experiments by radio amateurs. The so-called *Fading Tests* took place during 1920–1921 with 243 receiving stations from North America producing thousands of records [31]. The results, after being analyzed by the Bureau of Standards, were finally published in [38], mainly validating the long-standing belief that fading at short wavelengths is more serious than at long ones. However, historical investigations, carried out by Yeang (2004) revealed that "the claims associated with atmospheric conditions did not lead to a theory of fading. Nor was the conclusion very credible to the experiments at the bureau" [31].

4.2 Impact of Fading on Signal Transmission

4.2.1 Multipath Spread and Frequency Selectivity

It has been mentioned that multipath causes small-scale fading due to interference of two or more versions of the transmitted signal that arrive at the Rx at slightly different times. The detrimental effects of multipath can be better realized through the *delay power spectrum* or *multipath intensity profile* of the channel, which gives the average power output of the channel as a function of the time delay τ. Particularly, the range of τ values over which the delay power spectrum is essentially nonzero is defined as the *multipath spread* T_m of the channel. Therefore, if the signaling interval T is smaller than T_m, then the channel introduces ISI since the replicas of the neighboring transmitted signals interfere with the one received over the current interval. Equivalently, the Fourier transform of the multipath intensity profile defines the *spaced-frequency correlation function*, which provides a measure of the correlation between two samples of the channel response taken at different frequencies. A critical parameter of this function is the *coherence bandwidth* B_c of the channel that indicates the frequency range over which the fading process is highly correlated.[*]

Therefore, if the signal bandwidth W is greater than B_c, then the spectral components of the transmitted signal are affected by different amplitude gains and phase shifts across the band. Under these conditions, that is, $W > B_c$ or equivalently $T < T_m$, the channel is said to be *frequency-selective* and the signal is severely distorted. However, if the signaling interval is selected to be greater than T_m, then the channel introduces a negligible amount of ISI. Likewise, if the bandwidth is smaller than B_c, all the frequency components of the signal undergo the same attenuation and phase shift in transmission through the channel. In this case, that is, $W < B_c$ or equivalently $T > T_m$, the fading is said to be *frequency-nonselective* or *flat* and the multipath components of the channel are not resolvable, thus having a time-varying multiplicative effect on the transmitted signal.

4.2.2 Doppler Spread and Time Selectivity

The rapidity with which the CIR is changing with time is related to the *Doppler power spectrum* that gives the power profile as a function of the Doppler frequency λ. Particularly, the range of λ values over which the Doppler power spectrum is essentially nonzero is defined as the *Doppler spread* B_d of the channel.[†] Since the time variations in the CIR are evidenced as a Doppler broadening, it follows that a large value of B_d characterizes a rapidly changing channel. Equivalently, the inverse Fourier transform of the Doppler power spectrum defines the *spaced-time correlation*

---

[*] The coherence bandwidth of the channel is related to the multipath spread by $B_c \approx 1/T_m$.

[†] The Doppler spread of the channel is sometimes called *fading rate*.

function, which quantifies the correlation between two samples of the channel response taken at different time instants. A critical parameter of this function is the *coherence time* T_c of the channel that indicates the period of time over which the fading process is highly correlated. Since the reciprocal of B_d is a measure of the channel coherence time, the T_c parameter gives strong evidence of the rate at which the channel attenuation and phase shift change over time due to the nonstatic nature of the environment.

Therefore, a channel with a large Doppler spread compared to the signal bandwidth is also characterized by a small coherence time with respect to the signaling interval. In this case, that is, $W < B_d$ or equivalently $T > T_c$, the transmitted signal undergoes *time-selective* or *fast* fading. In the dual case, if T is selected to be smaller than the coherence time, that is, $W > B_d$ or equivalently $T < T_c$, the channel attenuation and phase shift can be considered fixed during at least one signaling interval, thus giving rise to *time-nonselective* or *slow* fading.

4.2.3 Characterization of Wireless Channels

The adverse effects of signal fading, if not mitigated, can significantly degrade the system's performance, resulting in nonreliable or even no communication during deep fades, a phenomenon called *outage*. Before referring to the versatile ways to tackle the problem of fading, it would be very useful to first classify the fading channels that arise in practice and highlight the different impact they have on signal transmission. Following a top–down approach, small-scale and large-scale fading are the main phenomena involved in the wireless channel propagation. Although they are both experienced by the user(s) of the channel as temporal and spatial fluctuations in the received power, the degradation mechanisms behind them are different, as explained in Section 4.1. From now on, we will focus on small-scale fading and refer to it simply as fading.

In Sections 4.2.1 and 4.2.2, it has been mentioned that fading channels can be characterized by their level of selectivity in frequency and time through the normalized parameters T_m/T and B_d/W, respectively. While the former indicates the dispersion degree of the signal in time due to multipath delays, the latter provides a measure of the signal dispersion in frequency due to Doppler shifts. Interestingly, a simple yet meaningful factor, which encompasses both spread parameters T_m, B_d to characterize a single fading channel, is the *spread factor* given by $T_m B_d \approx (B_c T_c)^{-1}$, where intuitively the $B_c T_c$ quantity reveals the overall coherence of the channel. The value of this factor is critical for the design of a digital communication system as it reflects the ease at which channel estimation and coherent demodulation can be performed, thus affecting the detection performance especially under adaptive transmission. By definition, a channel is said to be *underspread* if $T_m B_d < 1$ and *overspread* if $T_m B_d > 1$. This implies that, in the first case, the CIR can be easily measured, while in the second one, the same process becomes extremely difficult and unreliable, if not impossible.

Taking all these into account, a fading channel can be generally characterized as:

1. Slow flat (or *nondispersive*)
   a. If and only if $T_m < T < T_c$ and $B_d < W < B_c$
   b. If $W \approx 1/T$, then $T_m B_d < 1 \rightarrow$ underspread channel
2. Slow frequency-selective (or *time-dispersive*)
   a. If and only if $T_m > T < T_c$ and $B_d < W > B_c$
   b. If $\rho_s \downarrow$ and B_d is small enough $\rightarrow$ more flat channel
3. Time-selective flat (or *frequency-dispersive*)
   a. If and only if $T_m < T > T_c$ and $B_d > W < B_c$
   b. If $\rho_s \uparrow$ and T_m is small enough $\rightarrow$ more slow channel
4. Doubly-selective (or *doubly-dispersive*)
   a. If and only if $T_m > T > T_c$ and $B_d > W > B_c$
   b. If $W \approx 1/T$, then $T_m B_d > 1 \rightarrow$ overspread channel

The necessary and sufficient conditions per channel case illustrated earlier give evidence that, from a system-level point of view, the transmission characteristics can be chosen in such a way so that the dispersive nature of the channel can be moderated. To this end, in a frequency-selective fading channel, if the symbol rate ρ_s is decreased, the channel becomes more flat.* Correspondingly, in a time-selective fading channel, if ρ_s is increased, the channel varies more slowly. However, the design of more realistic fading mitigation techniques comes up as a requirement in radio applications and their consideration should definitely take into account the different degradation mechanisms involved in each channel type.

For instance, in the presence of slow flat fading with no LOS path, the error probability is asymptotically inversely proportional to the SNR. For a given quality of service (QoS) threshold, this implies a significant loss in terms of SNR compared to the AWGN channel, where the error probability decreases exponentially with the SNR. For visualization purposes, the error performance degradation of binary phase-shift keying (BPSK) and quadrature phase-shift keying (QPSK) systems due to the presence of nondispersive fading is shown in Figures 4.2 and 4.3. All systems are single-input single-output (SISO) and employ maximum likelihood detection (MLD) at the Rx. In many cases, the comparison with respect to the ideal AWGN (i.e., no-fading) channel reveals huge SNR penalties, which for typical target error rates span several tens of dB. The standard methodology and related techniques used to increase the received SNR in wireless communication systems suffering from severe fading are provided in Sections 4.4.2 and 4.4.4.

---

* In this direction, *narrowband* signal transmission, that is, $W \ll f_c$, is almost synonymous with flat fading conditions.

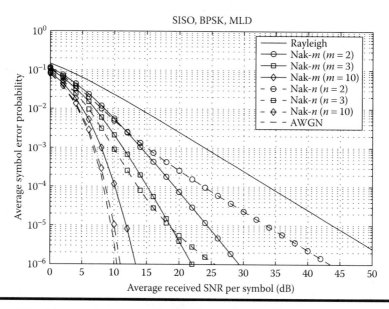

Figure 4.2 **Error performance of a BPSK communication system operating over various slow flat fading channels (i.e., Nakagami-*m*, Nakagami-*n*) and comparison with the ideal AWGN case.**

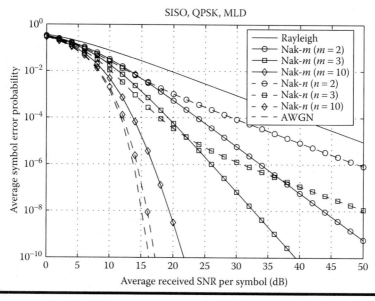

Figure 4.3 **Error performance of a QPSK communication system operating over various slow flat fading channels (i.e., Nakagami-*m*, Nakagami-*n*) and comparison with the ideal AWGN case.**

Even worse, in the case of transmission through a frequency-selective and/or fast fading channel, the system performance can exhibit an irreducible error-rate level, called *error floor*, due to the significant degradation induced by the ISI and/or Doppler spread.[*] In such a scenario, no amount of SNR can help achieve the desired level of performance, unless some forms of mitigation are first employed to reduce or even eliminate the signal distortion. In Section 4.4.1, a list of robust transmission techniques is given for the elimination of ISI and the associated performance error floor.

4.3 Formulation of Signal Transmission over Fading Channels

As our analysis does not depend on the carrier frequency f_c, it is mathematically convenient to make use of the *complex baseband representation* of bandpass signals and channels. To begin with, if $s_b(t)$ is the continuous-time linearly modulated narrowband bandpass signal to be transmitted over the wireless medium, then its general form can be written as [39, Sec. 2.5.2]

$$s_b(t) = \alpha(t)\cos(2\pi f_c t + \phi(t)) \tag{4.29}$$

with $\alpha(t)$ and $\phi(t)$ being the envelope and phase of $s_b(t)$, whereas its canonical form follows as

$$s_b(t) = s_I(t)\cos(2\pi f_c t) - s_Q(t)\sin(2\pi f_c t) \tag{4.30}$$

In Equation 4.30, $s_I(t) = \alpha(t)\cos\phi(t)$ and $s_Q(t) = \alpha(t)\sin\phi(t)$ denote the in-phase/quadrature (IQ) components of the real-valued bandpass signal $s_b(t)$, defining the equivalent lowpass (or baseband) signal $s_\ell(t)$ as

$$s_\ell(t) = s_I(t) + js_Q(t) = \alpha(t)\exp\big(j\phi(t)\big) \tag{4.31}$$

For M-ary digital modulation, both $s_I(t)$ and $s_Q(t)$ take values from predefined finite discrete sets with the additional restriction that eligible $(s_I(t), s_Q(t))$-tuples must belong to a two-dimensional (2D) constellation of size M that satisfies particular design criteria. Then, each constellation point is associated with a distinct $\log_2(M)$-bit length sequence according to a lookup table. In this sense, modulation of the binary information message can be envisaged either through $s_b(t)$ or its complex envelope $s_\ell(t)$ due to their interconnection through the relation

$$s_b(t) = \mathrm{Re}\{s_\ell(t)\exp(j2\pi f_c t)\} \tag{4.32}$$

---

[*] The varying Doppler shifts on different multipath components result in random frequency modulation.

The corresponding spectrum relation is derived by taking the Fourier transform of $s_b(t)$ in Equation 4.32

$$S_b(f) = \mathcal{F}_t\{s_b(t)\} = \frac{1}{2}\int_{-\infty}^{\infty} s_\ell(t)\exp\left(j2\pi f_c t\right)\exp\left(-j2\pi f t\right)dt$$

$$+ \frac{1}{2}\int_{-\infty}^{\infty} s_\ell^*(t)\exp\left(-j2\pi f_c t\right)\exp\left(-j2\pi f t\right)dt \qquad (4.33)$$

$$= \frac{1}{2}\left[S_\ell(f - f_c) + S_\ell^*(-f - f_c)\right]$$

where $S_b(f)$ and $S_\ell(f)$ are the bandpass and lowpass signal spectra, respectively.

Let us now assume that $s_b(t)$ passes through a continuous-time linear time-invariant (LTI) bandpass filter with impulse response $g_b(t)$ that yields the following representation with respect to its complex envelope $g_\ell(t)$ [40]*

$$g_b(t) = 2\text{Re}\{g_\ell(t)\exp\left(j2\pi f_c t\right)\} \qquad (4.34)$$

If $G_b(f)$ and $G_\ell(f)$ are the bandpass and equivalent lowpass frequency responses, respectively, then

$$G_b(f) = \mathcal{F}_t\{g_b(t)\} = G_\ell(f - f_c) + G_\ell^*(-f - f_c) \qquad (4.35)$$

Thus, the output of the bandpass system in the time and frequency domains is given by

$$x_b(t) = s_b(t) * g_b(t) \Rightarrow X_b(f) = S_b(f)G_b(f) \qquad (4.36)$$

By substituting Equations 4.33 and 4.35 into the right-hand side of Equation 4.36, we obtain

$$X_b(f) = \frac{1}{2}\left[S_\ell(f - f_c)G_\ell(f - f_c) + S_\ell^*(-f - f_c)G_\ell^*(-f - f_c)\right.$$

$$\left. + S_\ell(f - f_c)G_\ell^*(-f - f_c) + S_\ell^*(-f - f_c)G_\ell(f - f_c)\right] \qquad (4.37)$$

However, for narrowband signal transmission, where the frequency content of $s_b(t)$ is concentrated in a narrow band around f_c, that is, $W \ll f_c$, it is easy to see that [1, Sec. 4.1.3]

$$S_\ell(f - f_c)G_\ell^*(-f - f_c) = 0 \quad \text{and} \quad S_\ell^*(-f - f_c)G_\ell(f - f_c) = 0 \qquad (4.38)$$

---

* According to Tranter et al. [40, Sec. 4.2.1], the factor 2 in Equation 4.34 preserves the filter passband gain.

Therefore, by taking the inverse Fourier transform of $X_b(f)$ in Equation 4.37, it is straightforward to show that

$$x_b(t) = \mathcal{F}_f^{-1}\{X_b(f)\} = \text{Re}\{(s_\ell(t) * g_\ell(t))\exp(j2\pi f_c t)\} = \text{Re}\{x_\ell(t)\exp(j2\pi f_c t)\} \quad (4.39)$$

which finally defines the input-output relation of the equivalent lowpass system in the time domain as

$$x_\ell(t) = s_\ell(t) * g_\ell(t) \quad (4.40)$$

4.3.1 Time-Variant Multipath Channel Model

Consider next that the bandpass signal $x_b(t)$ is transmitted through a time-variant multipath channel with L_p distinct paths. Normally, the number of propagation paths will be a function of time, that is, $t \mapsto L_p(t)$. However, the assumption $L_p(t) \approx \text{E}\{L_p(t)\} \approx L_p, \forall t$, which implies a constant ensemble and time average number of paths over all signaling intervals, does not affect the generality of our analysis, while it simplifies the derivation of the final system model.

In this case, the faded signal $y_b(t)$ at the channel output consists of the superposition of L_p randomly attenuated and delayed copies of the original signal $x_b(t)$, mathematically formulated as

$$y_b(t) = \sum_{l=1}^{L_p} a_l(t)x_b[t - \tau_l(t)] \quad (4.41)$$

with $a_l(t)$ and $\tau_l(t)$ being the time-varying attenuation and propagation delay induced by the lth path, respectively. By substituting the right-hand side of Equation 4.39 into 4.41, we obtain successively

$$y_b(t) = \sum_{l=1}^{L_p} a_l(t)\text{Re}\{x_\ell(t - \tau_l(t))\exp[j2\pi f_c(t - \tau_l(t))]\}$$

$$\quad (4.42)$$

$$= \text{Re}\{\sum_{l=1}^{L_p} a_l(t)\exp(j\theta_l(t))x_\ell(t - \tau_l(t))\exp(j2\pi f_c t)\}$$

where $\theta_l(t) = -2\pi f_c \tau_l(t)$ is the phase shift associated with the $\tau_l(t)$ delay. By simple inspection, we get the complex envelope $y_\ell(t)$ of the faded signal expressed in terms of the baseband transmitted signal[*]

$$y_\ell(t) = \sum_{l=1}^{L_p} a_l(t)\exp(j\theta_l(t))x_\ell(t - \tau_l(t)) \quad (4.43)$$

---

[*] Equation 4.43 reduces to 4.1 for an unmodulated carrier with no filtering, that is, $s_\ell(t) = 1$, $g_\ell(t) = \delta(t)$, $\forall t$.

In order to proceed with the interpretation of Equation 4.43, it is necessary to invoke some elements from the theory of linear time-variant (LTV) filters. Particularly, if $x_\ell(t)$ is the input to an LTV filter with impulse response $h_\ell(t,\theta)$, where t and θ are the observation and impulse input time instants, respectively, then the input–output relation is defined by the general superposition (or convolution) integral [41, Sec. 6.2]

$$x_\ell(t) * h_\ell(t,\theta) \triangleq \int_{-\infty}^{\infty} h_\ell(t,\theta)x_\ell(\theta)d\theta = \int_{-\infty}^{\infty} h_\ell(t,t-\tau)x_\ell(t-\tau)d\tau \qquad (4.44)$$

where the second integral form is simply derived from the first one by introducing the variable $\tau = t - \theta$ that physically corresponds to the elapsed time of the filter (or age of the input). For the case of an LTI system it holds that $h_\ell(t,\theta) = h_\ell(t+a,\theta+a), \forall a \in \mathbb{R}$. Hence, by setting $a = -\theta$ we get $h_\ell(t,\theta) = h_\ell(t-\theta,0) \triangleq h_\ell(t-\theta) = h_\ell(\tau)$, which reduces Equation 4.44 to the standard convolution integral $\int_{-\infty}^{\infty} h_\ell(\tau)x_\ell(t-\tau)d\tau$ [42, Sec. 1.3].

Although expressions in Equation 4.44 are general enough to handle both LTI and LTV systems, the impulse response $h_\ell(t,\theta)$ does not involve directly the parameter τ, which becomes a drawback when one wants to define the frequency response of the LTV system through the Fourier transform of $h_\ell(t,\theta)$. For this reason, Kailath [43] introduced the modified impulse response $h_\ell(\tau;t)$ of an LTV system, which satisfies the following relations with respect to $h_\ell(t,\theta)$ [43], [41, Sec. 6.2.1]

$$h_\ell(\tau;t) = h_\ell(t,t-\tau) \quad \text{and} \quad h_\ell(t,\theta) = h_\ell(t-\theta;t) \qquad (4.45)$$

Based on these relations, the time-varying convolution integral of Equation 4.44 may be alternatively defined as

$$x_\ell(t) * h_\ell(\tau;t) \triangleq \int_{-\infty}^{\infty} h_\ell(\tau;t)x_\ell(t-\tau)d\tau = \int_{-\infty}^{\infty} h_\ell(t-\theta;t)x_\ell(\theta)d\theta \qquad (4.46)$$

The LTV nature of the filter implies that its impulse response generally changes as a function of both the time t at which the response is observed and the time $t - \tau$ at which the impulse is applied. For an LTI filter the impulse response depends only upon the time difference $t - (t-\tau) = \tau$, regardless of when the impulse is applied or the response is observed, since by adopting the modified version of the impulse response for an LTI system, we take $h_\ell(\tau;t) \triangleq h_\ell(\tau) = h_\ell(t-\theta)$, which for a reference input time $\theta = 0$ gives also $h(t)$ [42, Sec. 1.3]. Thus, considering the faded signal $y_\ell(t)$ in Equation 4.43 as the output of the LTV filter in Equation 4.46, the impulse response of the multipath radio channel $h_\ell(\tau;t)$ may be expressed in terms of $a_l(t)$, $\theta_l(t)$, and $\tau_l(t)$ as [1, Sec. 1.3], [40, Sec. 14.4]*

$$h_\ell(\tau;t) = \sum_{l=1}^{L_p} h_l(t)\delta[\tau - \tau_l(t)] \qquad (4.47)$$

---

* There is a typo in [40, eq. (14.21)]; the quantity $\delta[t - \tau_n(t)]$ should be changed to $\delta[\tau - \tau_n(t)]$.

where $h_l(t) = a_l(t)\exp\left[j\theta_l(t)\right]$ represents the generally time-varying complex coefficient of the lth path. Equation 4.47 is well known in the literature as the *tapped delay line model* for the representation of time-varying frequency-selective channels. In this context, $h_l(t)$, $l = 1, 2, \ldots, L_p$, are the associated complex tap coefficients (or weights). It is interesting to note that, by substituting Equation 4.47 into 4.46 and taking into account Equation 4.43, we validate the input-output relation

$$y_\ell(t) = x_\ell(t) * h_\ell(\tau; t) \tag{4.48}$$

Before proceeding, we need to revisit Equation 4.40 to discuss about the lowpass impulse response $g_\ell(t)$ of the LTI filter. In the system under study, the wireless channel is considered to be band-limited to W, namely its time-varying lowpass frequency response, which is defined now as $H_\ell(f; t) = \mathcal{F}_\tau\{h_\ell(\tau; t)\}$, should be zero for $|f| > W/2, \forall t.^*$ This implies that the transmitted signal $x_\ell(t)$ shall initially be limited to $W/2$, which is accomplished through the transmit pulse-shaping filter with impulse response $p_T(t)$. Hence, the output of the filter is given by replacing $g_\ell(t) = p_T(t)$ into Equation 4.40, that is, $x_\ell(t) = s_\ell(t) * p_T(t)$, which permits the faded signal in Equation 4.40 to be written in the following form

$$y_\ell(t) = s_\ell(t) * p_T(t) * h_\ell(\tau; t) \tag{4.49}$$

In Equation 4.49, the associative property of convolution has been invoked. It would be useful to mention here that the associative and distributive properties still hold for LTV systems, in contrast to the commutativity that is generally valid only for LTI systems [44]. Also, LTV systems satisfy the property of homogeneity of degree one [41, Sec. 6.3.1].

4.3.2 Received Signal in the Presence of Noise

In addition to the fading channel impairments, the Rx front end introduces another source of signal degradation. This is the noise generated by electronic components and devices due to the thermal agitation of electrons, which can be closely modeled by a stationary additive white Gaussian stochastic process with zero mean and flat power spectral density (PSD), a.k.a. power density spectrum, equal to $N_0/4$ W/Hz for all frequencies $f.^\dagger$ We need to mention here that, other types of additive noise include atmospheric noise, cosmic noise, man-made noise, and interference coming from other communication systems [1, Ch. 1]. Nevertheless, in this chapter, only the thermal noise is taken into account, which according to the central limit

---

* Note that the bandwidth of the bandpass signal is twice the bandwidth of the equivalent baseband.
† The PSD of thermal noise is denoted here as $N_0/4$, instead of $N_0/2$, for power normalization purposes.

theorem is Gaussian distributed. In addition, thermal noise is simply characterized as *stationary*, as a wide-sense stationary (WSS) Gaussian stochastic process is strictly stationary as well.

As a white stochastic process cannot be expressed in terms of IQ components due to its wideband character, it is convenient to consider the noise process at the output of an ideal bandpass filter with a completely flat passband that includes the frequencies satisfying the inequalities $|f - f_c| \leq W/2$ and $|f + f_c| \leq W/2$. Particularly, the PSD of the bandpass white noise $\eta_b(t)$ can be written in closed form as $S_{\eta_b}(f) = N_0/4\left[\Pi\left((f - f_c)/W\right) + \Pi\left((f + f_c)/W\right)\right]$. Thus, the signal $z_\ell(t)$ at the Rx input can be modeled as the faded signal $y_\ell(t)$ subject to the additive complex random signal $\eta_\ell(t)$

$$z_\ell(t) = y_\ell(t) + \eta_\ell(t) \tag{4.50}$$

In Equation 4.50, the sample function $\eta_\ell(t)$ of the equivalent lowpass noise process can now be expressed in terms of its IQ components as $\eta_\ell(t) = \eta_I(t) + j\eta_Q(t)$. Following the analysis in [1], it is straightforward to show that both $\eta_I(t)$ and $\eta_Q(t)$ are sample functions of a zero-mean stationary white Gaussian stochastic process with PSD functions given by $S_{\eta_I}(f) = S_{\eta_Q}(f) = N_0/2$ for $|f| \leq W/2; = 0$ for $|f| > W/2$, or more compact by $S_{\eta_I}(f) = S_{\eta_Q}(f) = (N_0/2)\Pi(f/W)$. Furthermore, $\eta_I(t), \eta_Q(t)$ are jointly-stationary random signals. As a Gaussian process is completely characterized by its first two moments, it would also be useful to employ the Wiener-Khinchin theorem,* in order to calculate the associated autocorrelation functions as $\mathcal{R}_{\eta_I}(\tau) = \mathcal{F}_f^{-1}\{S_{\eta_I}(f)\} = \mathcal{F}_f^{-1}\{S_{\eta_Q}(f)\} = \mathcal{R}_{\eta_Q}(\tau) = (N_0/2)W\mathrm{sinc}(W\tau)$. Finally, given that the cross-correlation function $\mathcal{R}_{\eta_Q \eta_I}(\tau)$ is zero, the zero-mean IQ components of $\eta_\ell(t)$ are uncorrelated, and since they are Gaussian, they are also statistically independent [45, Sec. 7.1].

The next operation involves the convolution of $z_\ell(t)$ with the LTI impulse response $p_R(t)$ of the receive filter. Taking into account Equations 4.49 and 4.50, and by invoking successively the distributive and associative properties of convolution, the filtered version of $z_\ell(t)$ can easily be derived as

$$r_\ell(t) = z_\ell(t) * p_R(t) = s_\ell(t) * \hbar_\ell(\tau;t) + n_\ell(t) \tag{4.51}$$

where:

$\hbar_\ell(\tau;t) = p_T(t) * h_\ell(\tau;t) * p_R(t)$ is the compound channel impulse response
$n_\ell(t) = \eta_\ell(t) * p_R(t)$ is the lowpass noise signal at the output of the receive filter

Thus, given that the output of the baseband IQ modulator can be expressed in the form [46, Sec. 1.2]

---

* According to this theorem, the PSD and autocorrelation functions of a WSS process form a Fourier transform pair.

$$s_\ell(t) = \sum_{k=-\infty}^{\infty} s_I(kT)\delta(t-kT) + j\sum_{k=-\infty}^{\infty} s_Q(kT)\delta(t-kT) = \sum_{k=-\infty}^{\infty} s_\ell(kT)\delta(t-kT) \quad (4.52)$$

the output of the receive filter is derived by substituting Equation 4.52 into 4.51, recalling the distributive property of convolution, and applying the integral formulation of Equation 4.46 for its calculation, finally yielding

$$r_\ell(t) = \sum_{k=-\infty}^{\infty} s_\ell(kT)\hbar_\ell(t-kT;t) + n_\ell(t) \quad (4.53)$$

It turns out that analysis can proceed with the establishment of the complex baseband system model instead of the original bandpass one by employing the equivalent lowpass signals and channels. In the sequel, the subscript ℓ will be dropped from all complex envelopes for notational simplicity.

4.3.3 Waveform and Discrete Channel Models

We can elaborate more on Equation 4.53 by adopting the quasi-static flat fading (QSFF) channel that was firstly adopted by Foschini and Gans (1998) in their revolutionary work [47] and is very popular in the context of multiple-input multiple-output (MIMO) and especially space-time block coding systems. This channel model corresponds to a special case of a slow fading environment, where the fading coefficients $h_l(t)$, $l = 1,2,\ldots,L_p$ are assumed to be constant over a frame of $N \in \mathbb{N}$ successive time intervals changing independently from frame to frame, implying that the channel coherence time T_c spans N signaling intervals, that is, $T_c = NT$. In this case, inequality $T_c \geq T$ is fulfilled, thus rendering the fading channel time-nonselective.

Under these channel conditions, it is convenient to consider the associated with the mth transmitted block impulse response of the multipath radio channel, which enables Equation 4.47 to be written in terms of τ only as

$$h^{(m)}(\tau;t) \triangleq h^{(m)}(\tau) \triangleq \sum_{l=1}^{L_p} h_l^{(m)}\delta(\tau - \tau_l) \quad (4.54)$$

where the fading coefficient $h_l^{(m)}$ is constant within the time interval $[(m-1)NT, mNT]$, $m \in \mathbb{Z}$, that is the duration (or length) of the mth block. Following the same notation, the associated with the mth transmitted block compound channel impulse response is reduced now to $\hbar^{(m)}(\tau;t) \triangleq \hbar^{(m)}(\tau)$, which obviously corresponds to the impulse response of an equivalent filter consisting of three LTI filters in cascade. Having this in mind, we can now proceed by considering the mth block-related version of Equation 4.53

$$r^{(m)}(t) = \sum_{k=-\infty}^{\infty} s(kT)\tilde{h}^{(m)}(t-kT) + n^{(m)}(t)$$

(4.55)

$$= \sum_{k=-\infty}^{\infty} s(kT)\Big(h^{(m)}(t-kT) * p(t)\Big) + n^{(m)}(t)$$

where $p(t) = p_T(t) * p_R(t)$ is the nonrelated to block index m overall impulse response of the combined transmit-receive filter.[*] Taking into account Equation 4.54, Equation 4.55 can be further simplified to

$$r^{(m)}(t) = \sum_{l=1}^{L_p} h_l^{(m)} \sum_{k=-\infty}^{\infty} s(kT)p(t-kT-\tau_l) + n^{(m)}(t), \quad (m-1)NT \le t \le mNT \quad (4.56)$$

This last formulation of the signal at the output of the receive filter is very useful for the analysis of MIMO systems, since it incorporates the analog portion of the system in a single chain, comprising the filter/modulator at the Tx, the noisy fading channel, and the demodulator/filter at the Rx. Since it represents the signal waveform relation from the output of the baseband IQ modulator to the input of the Rx decoder, it is also called *waveform channel model*. As described in [40, Sec. 11.2], an equalizer could also be part of the waveform channel. Nevertheless, the equalization procedure has been intentionally omitted from Equation 4.56, since mitigation of ISI caused by the time-dispersive nature of the channel is not our target in this section.

Consequently, analysis will proceed with the replacement of the waveform-based channel model by a discrete-time one, assuming either perfect equalization or flat fading channel conditions. We highlight that, flat fading is a valid consideration under the narrowband signal transmission scenario that has been adopted here (see Section 4.2), whereas for the case of a frequency-selective channel, orthogonal frequency division multiplexing (OFDM) can be used to convert a broadband channel into a set of parallel narrowband (i.e., flat) subchannels (see Section 4.4). Under this consideration, Equation 4.56 may be further simplified by setting $L_p = 1$ and defining $h_1^{(m)} \triangleq h^{(m)}$ and $\tau_1 \triangleq \tau$, which finally takes the form

$$r^{(m)}(t) = h^{(m)} \sum_{k=-\infty}^{\infty} s(kT)p(t-kT-\tau) + n^{(m)}(t), \quad (m-1)NT \le t \le mNT \quad (4.57)$$

The *discrete channel model* can be easily derived by sampling the filter output at the symbol rate, namely at time instants $t_v = vT + \hat{\tau}, v \in \mathbb{N}$, where $\hat{\tau}$ is the propagation delay estimate provided by the timing recovery loop of the Rx. Obviously, assuming

---

[*] In Equation 4.55, the commutative property of convolution has been utilized, which holds for LTI systems.

perfect symbol synchronization, the timing error $\varepsilon = \hat{\tau} - \tau$ becomes zero, and under causality constraints, that is, $k, \nu \in \mathbb{N}$, the sampled version of Equation 4.57 is

$$r^{(m)}(\nu) = h^{(m)} \sum_{k=1}^{\nu} s(k) p(\nu - k) + n^{(m)}(\nu) \tag{4.58}$$

where in Equation 4.58 we have introduced the notations $r^{(m)}(t_\nu) \triangleq r^{(m)}(\nu)$ and $n^{(m)}(t_\nu) \triangleq n^{(m)}(\nu)$ as well as $s(kT) \triangleq s(k)$ and $p(qT) \triangleq p(q)$, with $q \in \mathbb{N}_0$. We may also write Equation 4.58 in the more meaningful form

$$r^{(m)}(\nu) = p(0) h^{(m)} s(\nu) + \upsilon^{(m)}(\nu) + n^{(m)}(\nu) \tag{4.59}$$

with $s(\nu)$ being the desired symbol at the νth sampling instant and $\upsilon^{(m)}(\nu) = h^{(m)} \sum_{k=1}^{\nu-1} s(k) p(\nu - k)$ representing the ISI produced from symbols transmitted prior to $s(\nu)$. At this point, a few words have to be spent on the discrete-time impulse response $p(q)$ of the combined transmit-receive filter. From communication theory, it is known that the optimal choice of $p(q)$ is the one that satisfies the Nyquist condition for zero ISI, which is simply expressed as $p(q) = 1$ for $q = 0$; $= 0$ for $q \neq 0$ [1]. In this case, $\upsilon^{(m)}(\nu) = 0, \forall \nu$, and Equation 4.59 reduces to the more convenient to handle expression

$$r^{(m)}(\nu) = h^{(m)} s(\nu) + n^{(m)}(\nu) \tag{4.60}$$

Considering now the transmission of the mth block, which takes place within the time interval $[(m-1)NT, mNT]$, the associated global sampling index ν is restricted to the set $\{(m-1)N + \kappa : \kappa \in \mathbb{N}; \kappa \leq N\}$. Evidently, the next set of relations hold an equivalent to Equation 4.60 meaning within a block

$$\left. \begin{array}{l} r(\kappa) = h s(\kappa) + n(\kappa) \\[2mm] r_\kappa = h s_\kappa + n_\kappa \end{array} \right\}, \quad \kappa = 1, 2, \ldots, N \tag{4.61}$$

where the new block-based sampling index κ takes now the same values for all m. Interestingly, for $N = 1$, which is a special case of the QSFF channel, and after dropping subscript κ for notational convenience, the following scalar input–output relationship is obtained

$$r = h s + n \tag{4.62}$$

The communication system model in Equation 4.62 describes the well-known complex baseband representation of the open-loop SISO scheme for narrowband digital signal transmission over slow fading channels with coherent reception and no transceiver impairments. The multiplicative fading coefficient $h = a \exp(j\theta)$ represents the independent and identically distributed (i.i.d.) snapshots of the flat fading channel, where the statistics of its amplitude (or fading envelope) $|h| = a$ and phase $\angle h = \theta$ both depend on the nature of the radio propagation environment (for details on the statistics of various flat fading channels, see Section 4.1.2).

4.4 Fading Mitigation Techniques and Diversity

In the system model described by Equation 4.62, N has been interpreted as the number of successive time slots for the single-antenna transmission of N consecutive digital symbols, all experiencing the same fading channel conditions due to the QSFF model assumed. However, this transmission strategy is very vulnerable to fading, as each information symbol realizes just a single instance of the channel. To make it more clear, during a block transmission under unfavorable channel conditions (e.g., deep fading), a whole set of N information symbols is very likely to be erroneously detected. Severe fading also causes failure with large probability in the symbol detection of SISO systems defined by Equation 4.62, although in this case only a single symbol is corrupted by the associated fading coefficient. However, these open-loop configurations can bring benefits in terms of reliability if extra dimensions are introduced and ingeniously utilized along with the spatial one.

In order to give evidence of such a capability, suppose that every individual information symbol utilizes L independent channel snapshots for its transmission through the wireless channel. In this case, there are apparently L new degrees of freedom available in the system. However, a critical question that arises is how these can be efficiently exploited toward more reliable communication. Frankly speaking, there is not a unique methodology for accomplishing this goal. Instead, the happy medium is rather a compromise among the available resources (e.g., bandwidth, power, space) and the practical limitations imposed by the intended application.

In Section 4.2, we have already discussed the two major effects of fading on signal transmission, namely (a) signal distortion mainly due to ISI and (b) SNR penalty in the error performance with respect to the ideal AWGN case. In the sequel of this chapter, efficient ways to tackle these problems will be discussed, starting from the latter.

4.4.1 Compensation of Intersymbol Interference

The ISI caused by a time-dispersive channel can be efficiently compensated by equalization, OFDM, and spread spectrum techniques. In more detail, the process of equalization provides a means to gather the dispersed symbol energy back into its original time interval. To do so, the equalizer adjusts the balance between frequency components, thus approximating an inverse filter of the channel. The target for the combination of the channel and equalizer filter is to provide a flat composite received frequency response and linear phase.

One of the most effective nonlinear equalization techniques is the maximum likelihood sequence estimation (MLSE), first proposed by Forney [48], that is typically implemented using the Viterbi (1967) decoding algorithm. However, in practical scenarios (e.g., mobile radio channels), optimal equalization techniques cannot always be used due to constraints on processing power. In these cases, OFDM is an attractive alternative that achieves the same goal by demultiplexing a data sequence into several streams, transmitted in parallel on different subcarriers and with sufficiently small symbol rates with respect to B_c, so as to consider the subchannels as flat. In addition, the residual ISI can be completely eliminated by the insertion of a cyclic prefix or guard interval, which is always larger than T_m, between successive OFDM symbols. It is interesting to note here that Cimini (1985) proposed and described the use of OFDM in mobile communications. However, the fundamental principle of OFDM is credited to Chang [49] who studied the simultaneous transmission of several messages through a linear bandlimited medium at a maximum data rate without interference.

Spread spectrum techniques can also compensate for the ISI by a different mechanism which is based on bandwidth expansion, namely the transmitted signal is spread over a much larger bandwidth than the minimum required (i.e., the bit data rate). This can be achieved by the introduction of a pseudo-random pattern in the transmitted signal that is later removed at the demodulator to recover the original data sequence without ISI. There are two basic implementations of this technology; the direct-sequence spread spectrum (DS-SS) and the frequency-hopping spread spectrum (FH-SS). According to Scholtz (1982), frequency hopping was a recognized anti-jamming concept during the early 1940s, while the direct sequence followed several years later.

In more detail, in DS-SS systems, a pseudo-random sequence (PRS) of pulses (or chips) is impressed on the original data sequence expanding the allocated bandwidth by a factor equal to the number of chips per information bit, called *processing gain*.[*] As the spread signal bandwidth is approximately equal to the chip rate, multipath components delayed by more than the chip interval can be suppressed by correlating the signal with the same synchronized PRS at the Rx. DS-SS is the key technology behind code division multiple access (CDMA), introduced by Qualcomm, Inc. in 1989, that allows several users to share the same band simultaneously by using distinct PRS codes among different pairs.

In a different way, in a FH-SS system, the total spread bandwidth is divided into a large number of contiguous frequency slots. In any signaling interval, one or more frequency slots are pseudo-randomly selected, according to the PRS, to be occupied by the transmitted signal. Obviously, during this interval, the Rx must be set to the same PRS in order to be tuned to the proper frequency band(s) and hop to the next scheduled slot(s) before the arrival of multipath components.[†]

---

[*] Equivalently, a spread spectrum system is characterized by a chip rate that is much larger than the bit data rate.

[†] The hopping rate of a FH-SS system must be at least equal to the symbol rate.

Nevertheless, in the presence of doubly-selective fading, multipath mitigation strategies alone cannot compensate for the error floor if not assisted by adaptive channel estimation and robust modulation schemes, such as noncoherent, partially coherent, and pilot symbol assisted modulation,[*] among other techniques.

4.4.2 *The Concept of Diversity Gain*

The next step toward reliable communication involves examining ways to reduce the fading-induced penalty on the SNR. This can be efficiently achieved through *diversity*, a classic and well-known concept that has been used for several decades. However, the term *diversity* is rather general and comprises the vast majority of fading mitigation techniques. Diversity is based on the notion that the detector is more prone to errors when the channel is in a deep fade. Therefore, if the Rx is supplied with multiple versions of the same information-bearing signal transmitted over independent fading channels, then the probability that all signals have experienced deep fading will be considerably reduced. In other words, if L is the number of independent fading channels utilized for the transmission of the same signal and p is the probability that any signal will fade below some critical value, then p^L is the probability that all replicas will fade below that critical value.

From another standpoint, diversity has the effect of steepening the rate of descent of the error probability $P_e(\bar{\gamma})$ versus the average SNR, $\bar{\gamma}$. Particularly, in the high SNR regime, the term *diversity order* is used to define the negative slope of $P_e(\bar{\gamma})$ curve on a log–log scale, namely

$$d = -\lim_{\bar{\gamma}\to\infty} \frac{\log P_e(\bar{\gamma})}{\log \bar{\gamma}} \qquad (4.63)$$

By definition, a system is said to achieve *full diversity* if the diversity order d is equal to the number of independent channel realizations (i.e., $d = L$). In this context, a full-diversity scheme is able to exploit all the available DoFs in the direction of decreasing the error probability for a given average SNR and spectral efficiency, while a system with no diversity is characterized by $d = 1$. In order to reveal the intuition behind Equation 4.63, let assume an $1 \times N$ single-input multiple-output (SIMO) maximum ratio combining (MRC) system employing a 16th order quadrature amplitude modulation (QAM) constellation and MLD at the Rx. The fading channel is considered to be slow flat Rayleigh with unity average fading power, that is, $\Omega = 1$ (see Section 4.1.2.1). The average symbol error rate (SER) performance of the system with respect to the average SNR per symbol is shown in Figure 4.4 for various values of N, namely for $N = 1, 2, \ldots, 10$. We first select a relatively high SNR value, for example 50 dB, and measure the corresponding θ_N angles for all N. Then, the slope of each SER curve at SNR equal to 50 dB is given by definition from $-\theta_N$.

---

[*] Analysis of pilot symbol assisted modulation was initially provided by Cavers (1991) for flat Rayleigh fading.

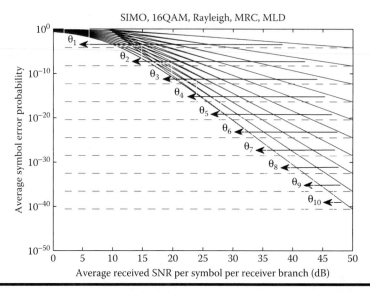

Figure 4.4 **Graphical illustration of *d* for a SIMO MRC system with one-ten antennas at the Rx.**

The graphically-based values for the diversity order can be derived by making use of Equation 4.63 or equivalently by $d_{gr} = \tan(-\text{slope})$. The calculated values are shown in Table 4.1 and obviously they are very close to the expected ones, that is, $d_{th} = N$, which stem from the theory of MIMO systems (note that SIMO MRC are full-diversity schemes).

Table 4.1   Calculation of *d* with the Aid of Figure 4.4

N	θ_N (deg)	Slope (deg)	d_{gr}	d_{th}
1	44.9970	−44.9970	0.9999	1
2	63.4324	−63.4324	1.9998	2
3	71.5631	−71.5631	2.9997	3
4	75.9622	−75.9622	3.9995	4
5	78.6888	−78.6888	4.9994	5
6	80.5366	−80.5366	5.9993	6
7	81.8689	−81.8689	6.9992	7
8	82.8741	−82.8741	7.9990	8
9	83.6590	−83.6590	8.9989	9
10	84.2887	−84.2887	9.9988	10

The benefit of diversity on the system error performance can also be quantified through the related term *diversity gain*, which determines the improvement in the average received SNR obtained at a given error probability using some form of diversity. Very often, the terms diversity order and diversity gain are used interchangeably in the literature to define the increase in the slope of the error rate. However, the distinct definitions coined previously for the two terms are more common [50]. Interestingly, in the high SNR limit, the error probability can be approximated by $P_e(\bar{\gamma}) \approx (G_c \bar{\gamma})^{-d}$, where G_c is the *coding gain* of the system that determines the average SNR shift (in dB) of the $P_e(\bar{\gamma})$ curve relative to a benchmark transmission scheme [51].

In Figures 4.5 and 4.6, the performance of SIMO MRC communication systems in terms of average SER is illustrated for several values of the diversity order d. In all cases, QAM is employed at the Tx and MLD at the Rx, while the fading channel is assumed to be slow flat Rayleigh. Given that SIMO MRC are full-diversity schemes, the diversity order is always equal to the number of deployed antennas at the Rx. It can be easily observed that, for a giver error probability, increasing the diversity order implies substantial diversity or equivalently SNR gains, which are translated to essential savings in the required transmitted power. This great advantage has been exploited for several decades now in the uplink of wireless networks where the design of the Tx is often subject to fabrication and/or power limitations.

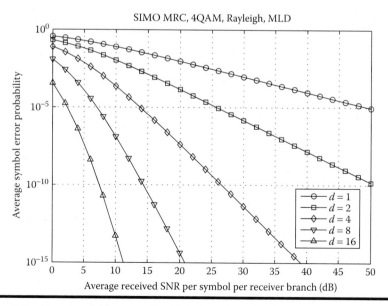

Figure 4.5 **Error performance of a 4QAM communication system operating over a slow flat Rayleigh fading channel for different values of the diversity order d.**

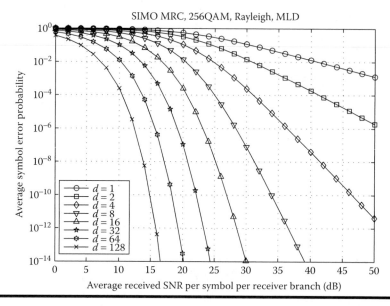

Figure 4.6 Error performance of a 256QAM communication system operating over a slow flat Rayleigh fading channel for different values of the diversity order d.

4.4.3 Tracing the Roots of Diversity

It is very common in the literature to trace the roots of diversity back to 1940s or even early 1950s, probably due to the plethora of diversity receivers proposed during that period, whereas just a few works point out related techniques emerged in 1930s in the context of wireless telegraphy; just to mention the seminal work of Beverage and Peterson (1931) that gives implementation details of some commercial transceiver systems employing three different principles of diversity [52]. An early rack installation of diversity equipment is shown in Figure 4.7a, which illustrates a double rack holding diversity apparatus for two circuits. In 1931, the Riverhead receiving station of R.C.A. Communications, Inc., contained twenty such racks, providing facilities for forty circuits in total. These racks were arranged back to back facing aisles with four circuits per aisle. Ordinarily one man was taking care of the circuits in an aisle. Figure 4.7b presents the end view of an aisle.

However, the diversity concept was known several years before. Particularly, Eckersley [53] gives a detailed diagram, reproduced in Figure 4.8, of an Rx apparatus employing one horizontal and six vertical polarized aerials placed in a line at right angles to the station ray about 12 ft above the surface of the earth. These were connected by phasing coils consisting of vertical lecher wires approximately (1/4)λ in length, λ being the wavelength. The Rx was coupled through a cable to the central lecher wire. This special arrangement was used in a series of tests carried out during

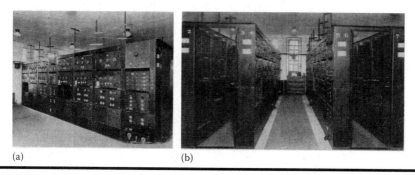

Figure 4.7 Diversity receiving system of R.C.A. Communications, Inc. © 1931 IEEE. (a) Diversity Rx installation. (b) Aisle of four complete diversity sets. (Reprinted with permission from Beverage, H.H. and Peterson, H.O., *Proc. IRE*, 19, 531–561, © 1931 IEEE.)

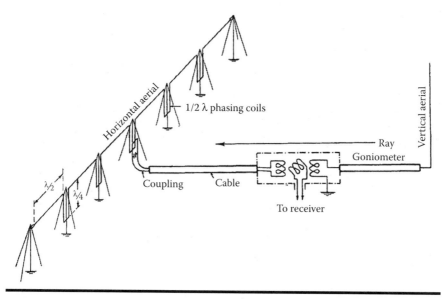

Figure 4.8 The Rx apparatus of Eckersley (1929) for testing polarization diversity. (From Eckersley, T.L., *J. IEEE*, 67, 992–1029, 1929. Reproduced by permission of the Institution of Engineering & Technology.)

the winter of 1927–1928 to compare the relative fading on the two types of aerials, aiming to design a polarization diversity system (see Section 4.4.4.4) through particular combination of the rectified currents coming from the two differently polarized aerials. It is of great interest that, in the same paper, the term *diversity* appears for the first time in the literature to describe one of the very first receive diversity (RxD) techniques, based on *spaced aerials*. The following excerpt reveals that diversity systems had already been quite common among radio experts by that time:

> The fading on two aerials spaced many wave-lengths apart should be very different, and it is actually found to be so. "Diversity" systems of this type, as they are called in America, are in use both here and in that country, and show considerable gain in leveling the fading.
>
> **—Eckersley (1929) [53].**

Furthermore, in 1928, another form of diversity, that is time diversity (see Section 4.4.4.3), was conceived by Verdan and Loiseau, inventors of the antiparasitic device, which was capable of improving the link performance of the Baudot telegraph system subject to atmospheric disturbances [54]. This apparatus, of the later called Baudot-Verdan system, and the associated Verdan principle that calls for retransmission of the same letter with certain delays, are described in a related paper by Harrison [55]. Also, it is almost unknown in the context of wireless communications that, a directional receiving aerial system, patented by Adcock [56] for the purpose of eliminating the errors on closed-coil direction-finders due to down-coming waves, has probably served as a basis for the design of primitive transmit diversity (TxD) systems, such as the rotating-beacon Tx scheme of Smith-Rose. As shown in the schematic diagram in Figure 4.9, the Adcock system may be regarded simply as a pair of spaced vertical aerials with the currents flowing in opposite phase [57].

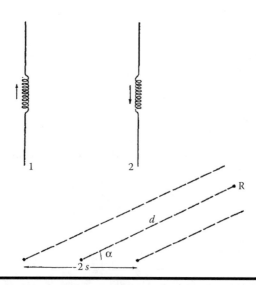

Figure 4.9 Schematic diagram of the Adcock aerial system given by Smith-Rose. (From Smith-Rose, R.L., *J. IEEE*, 66, 270–279, 1928. Reproduced by permission of the Institution of Engineering & Technology.)

Finally, a very interesting spatial TxD system was presented by Hansell (1955), affiliated with R.C.A. Laboratories. In more detail, a wireless system employing two Txs, connected to spaced antennas, and a single Rx antenna was found to be equivalent to a conventional RxD system in terms of overcoming the effects of fading, given that the two Txs are separated in frequency by a very small amount of offset [58]. According to Hansell, the proposed system could be especially suitable for shore to ship transmission because of the impracticability of providing spaced receiving antennas on ships. The paper concludes that it is more cost-effective to improve the signal reception by the addition of a second low-power Tx rather than to increase the power of the single Tx in order to obtain the same performance. Particularly, to do so, the single Tx would require 16–1000 times more power than each of the two Txs. The closing paragraph of [58] is worth being highlighted:

> There is another important advantage that is obtained by transmitter diversity compared to receiver diversity. The receiving equipment is... smaller in size and also easier to operate. This may be quite important, especially for mobile application where quite often space and operating personnel are at a premium.

> **—Hansell (1955) [58].**

Surprisingly, the adopted approach and the discussion made in this early work are very consistent with the revolution that took place some decades later in the context of spatial TxD systems, especially through the use of space-time coding and other MIMO techniques.

Before closing this section, it would be a great omission not to mention the Canadian-born inventor Fessenden, whose technological achievements definitely assisted the understanding of fading effects on signal transmission and their mitigation via the development of proper diversity techniques. In this context, if Marconi is considered to be the pioneer of wireless transmission over long distances, then Fessenden is credited as the first to demonstrate the transmission of voice and music by radio [59]. The works in [60,61] refer to Fessenden as a genius and mathematician, being the father of amplitude modulation (AM) radio and a primary pioneer of radio as we know it today. In November 1899, at the meeting of the American Institute of Electrical Engineers, Fessenden presented a paper on the "possibilities of wireless telegraphy" [62], and soon afterwards, in December 1900, he transmitted, for the first time in the wireless history, intelligible words by electromagnetic waves [60]. The transmission took place on Cobb Island, Maryland, over a distance of 1600 m using a mechanical spark interrupter [59]. However, the fact that his voice was badly distorted and accompanied by a loud noise, convinced him that he needed a "continuously acting, proportional indicating receiver" (Fessenden's words) [60]. For this reason, he advocated for continuous-wave against the spark systems utilized by Marconi [62,63]. In January 1906, the first two-way transatlantic radio telegraphy transmission took place between Fessenden's stations at Brant Rock, Massachusetts and Machrihanish Scotland [61]. Interestingly, Fessenden's concept of continuous-wave radio signals and his development of the heterodyne principle are considered as his greatest contributions to radio technology [61].

4.4.4 Classification of Diversity Techniques

As already discussed, a critical issue in the design of digital communication systems is the provision for high immunity levels to fading by efficient deployment of diversity techniques. As will be shortly described, there is not a single technique that can guarantee this goal. Primarily, diversity systems can be classified according to the nature of fading they are intended to mitigate. Following the discussion in Section 4.2, we focus on *microdiversity* techniques that are used to combat the short-term (i.e., small-scale) fading effects.[*] Before proceeding with a brief summary of them, it would be important to note that practical diversity systems usually combine more than one diversity techniques for better exploitation of the available DoFs of the system.

4.4.4.1 Spatial Diversity

Spatial diversity, also known as *antenna diversity*,[†] is the most commonly used form to obtain several copies of the same information-bearing signal by deploying multiple antennas at the Rx or/and Tx side. The redundant signals are then skillfully combined in order to increase the total average SNR. The term *site diversity* is sometimes used as an alternative to spatial diversity. However, site diversity is a macrodiversity technique that fits better in the context of satellite communications systems operating in Ka band and above, where two or more ground stations are linked together to provide more robust downlink signal reception against rain fading.

Traditionally, the most popular schemes achieving diversity over space have been the SIMO systems, which are usually equipped with linear combiners that require partial or full channel state information (CSI) only at the Rx side. Though the first variants of RxD techniques date back to 1920s, they still offer one of the greatest potentials for radio link performance improvement to many of the current and future wireless technologies. The dual case of SIMO are the multiple-input

---

[*] In analogy, *macrodiversity* can be employed to mitigate the effects of long-term (i.e., large-scale) fading.

[†] More generally, antenna diversity can refer to either of the three techniques, spatial, polarization, or pattern diversity.

single-output (MISO) systems that can obtain TxD under some more sophisticated architecture, usually requiring signal processing and some level of CSI awareness at the Tx through a dedicated feedback link between the two sides of the communication link.

In the more general case, a MIMO system can offer diversity and coding gains, with the challenge being to satisfy the complexity constraints and transmitted power limitations put on the design. However, all configurations require sufficient separation between antenna elements on the same side to ensure independent fading conditions over all channels and realize the diversity advantage. Not very often, the minimum distance between antennas that yield uncorrelated fading is called *coherence space*.

4.4.4.2 Frequency Diversity

In *frequency diversity*, the same information-bearing signal is transmitted on L carriers, where the separation between successive carriers must equal or exceed B_c. Obviously, this approach is very well suited to multicarrier systems working in frequency division multiplexing (FDM) mode. One of the very first applications of this technique in wireless telegraphy is described by Beverage and Peterson [52]. However, the major disadvantages of similar systems are generally the non-efficient bandwidth usage and the requirement for receivers with multiple radio frequency (RF) chains. A more refined version of multicarrier operation to achieve the same target is the combination of OFDM with channel coding and frequency interleaving in the presence of time-dispersive channels.*

In a broader sense, frequency diversity is an intrinsic characteristic of wideband channels, that is, channels with a bandwidth greater than B_c. Particularly, wideband transmission can provide the receiver with several independently fading signal replicas, thus achieving frequency diversity of order $d = W/B_c$, given that the induced ISI is efficiently mitigated. A possible approach to deal with this issue is single-carrier transmission with channel equalization. Alternatively, FH-SS technology utilizes bandwidth expansion to exploit the inherent frequency diversity of the channel. Olofsson et al. [64] showed that FH-SS also introduces *interference diversity*, since different signals interfering with the desired one at different times can result in an averaging effect, which improves performance in system level.

Finally, a DS-SS system employing a *RAKE receiver* is capable of resolving multipath components at different time delays, a phenomenon also called *multipath diversity*. The RAKE receiver, proposed by Price and Green [65], uses several correlators or *fingers*, each assigned to a different multipath component, to separate and coherently combine all the delayed replicas of the wideband signal.

---

* Frequency interleaving is the mapping of the coded bits to the subcarriers.

4.4.4.3 Time Diversity

Time diversity exploits the frequency-dispersive nature of time-varying wireless channels to provide more reliable communication. This can be easily accomplished by using multiple time slots separated by at least the coherence time, T_c, of the channel to transmit the same data symbols. Particularly, every symbol is repeated L times and then interleaved before being transmitted over the wireless channel, so that consecutive symbols experience independent fading conditions.[*] Obviously, this technique can be viewed as repetition coding of the information sequence, with the length L of the code defining the diversity order obtained. However, the achieved gain comes at the expense of spectral efficiency, since a single symbol requires transmission over several time intervals. Furthermore, a repetition code does not effectively exploit all the DoFs available in a fading channel.

Instead, more sophisticated coding schemes, such as linear block codes or convolutional codes, along with interleaving can be employed to realize additional coding gain.[†] In this way, the encoded symbols can be efficiently dispersed over different coherence periods, thus taking full advantage of the inherent time diversity in time-selective channels. Nevertheless, as the fading rate (i.e., the B_d) of a fading channel decreases, for example, due to mobile terminals that slow down, it may not be possible to obtain time diversity without introducing unacceptable delays, since in this case the inevitable large T_c limits the performance of a given interleaver.

4.4.4.4 Polarization Diversity

Early measurements on long-distance transmission revealed that propagation characteristics of a wireless medium are not the same for differently polarized waves. In addition, multiple reflections between the Tx and Rx cause depolarization of the radio waves, thus dispersing some energy of the transmitted signal into other polarization directions.[‡] Due to that attribute, linearly polarized transmitted waves can come out at the Rx end with an additional nontrivial orthogonal component. These observations motivated many researchers to investigate the fading statistics associated with radio signals received by antennas of various polarization modes.

Among the first, Glaser and Faber [66] demonstrated that the vertical and horizontal polarized components of the same received signal undergo almost statistically independent fading while propagating through certain wireless

---

[*] This implies that $(L-1)T \geq T_c \Rightarrow L \geq (T_c/T) + 1$, where T is the signaling interval.

[†] Diversity order provided by channel coding is directly related to the minimum Hamming distance of the codes.

[‡] Breit (1927) maybe was the first to elaborate a theory on this phenomenon.

environments. This fact has been utilized to improve radio system performance through the use of *polarization diversity*, where two or more spatially separated uni-polarized antennas are replaced by a single antenna structure of almost co-located elements by employing multiple polarizations. The simplest polarization diversity systems utilize dual-polarized antennas with two orthogonal components, that is, copolarized and cross-polarized, rendering them very handy for use in cellular radio networks owing to the minimized antenna installation space and the fact that mobile terminals can experience reliable communication whatever the hand-set angle of tilt.

4.4.4.5 Pattern Diversity

Antenna *pattern diversity* comprises another form of exploiting the inherent DoFs of multipath propagation environment. The concept behind this approach lies in the relationship that holds between the received signal fluctuation and the direction of the main lobe of the Rx antenna pattern. The first experimental results validating this statement were reported by Bruce [67] and Bruce and Beck [68], who found remarkable reduction in fading by using a horizontal rhombic antenna with an extremely sharp directional pattern. The success of their attempt was based on the existence of stable angular separation among signal components following different paths, thus enabling a steerable antenna with sufficiently sharp directivity to accept only one of them, maybe the best after some calibration, at any time.

Friis et al. [69] clearly demonstrated the qualitative relation between angles of arrival and propagation delays of the received signal components; the greater the delay the greater the angle above the horizontal (or azimuth) plane. In [70], Friis and Feldman (1937) exploited this phenomenon with the aid of a special arrangement whereby individual wave groups arriving at different vertical angles were received separately and, after delay equalization, combined coherently to offer *directional diversity* or *directivity diversity*. They accomplished that with the multiple unit steerable antenna (MUSA) system. According to Gregory and Newsome (2010), the MUSA array was the last major technological development in the short-wave communication era, representing the ultimate receiving system [71]. It was probably the most complex radio Rx ever built and gave valuable service between the 1940s and 1960s. Later, Vogelman et al. (1959) extended the concept of a single beam steerable to a multibeam antenna, where signal components associated with individual beams were particularly combined to offer *angle diversity* or *angle-of-arrival diversity* [72]. However, the effectiveness of this system depends upon the antenna characteristics, which have to be properly chosen so that the angle between adjacent beams is always smaller than the *angular spread* of the channel and that low correlation among different beams is ensured.

Obviously, pattern diversity can also be provided via several antennas with diverse patterns spaced apart from each other. An interesting case arises when antennas with similar patterns are placed close to each other with respect to the coherence space defined in Section 4.4.4.1. Particularly, mutual coupling effects alter their antenna pattern, thus yielding again a quite effective pattern diversity scheme. In addition, *field component diversity*, first proposed by Gilbert [73], utilizes sophisticated energy density antenna arrangements to receive uncorrelated electric and magnetic field components of the transmitted signal so as to reduce fading in mobile radio. The conception of the energy density antenna as a means to mitigate signal fading in mobile radio channels is credited to Pierce, who communicated this idea to other researchers, such as Gilbert (1965) and Lee (1967).

4.4.4.6 Other Forms of Diversity

It has been mentioned earlier that the inherent diversity of frequency-selective fading channels can be exploited by spread spectrum techniques (e.g., DS-SS) with RAKE reception. However, the RAKE receiver is optimal for time-dispersive channels only, whereas for rapid temporal variations of the channel (i.e., fast fading), the resulting Doppler spread induces performance degradation due to inaccurate channel estimation. This fact motivated Sayeed and Aazhang [74] to suggest the dual version of the conventional RAKE receiver, called *Doppler RAKE receiver*, that is capable of resolving Doppler components at different frequency shifts. By adopting a joint time-frequency representation for the received signal, they first revealed the inherent diversity mechanism of a doubly-dispersive channel and further proved that it can be exploited by appropriate signal processing. This new receiver structure, which actually realizes B_d as another DoF, performs optimally in frequency-dispersive channels and offers *Doppler diversity* of order proportional to TB_d. In the more general case of a doubly-dispersive channel, Sayeed and Aazhang proposed again the *time-frequency RAKE receiver* that can offer in an optimal fashion joint multipath-Doppler diversity of order proportional to the product of the channel spread factor and the processing gain of the spread spectrum signaling technique [74].

Knopp and Humblet [75] studied the optimal power control scheme that maximizes information capacity in the uplink of single-cell multiuser communications corrupted by flat fading. They proved that the best strategy, that is scheduling at any time only the user with the largest SNR, provides *multiuser diversity* of order proportional to the number of users. Since then, the presence of multiple users on the same system has been conceived as a new DoF that was very effectively exploited by Sendonaris et al. [76] and a little later by Laneman et al. [77] through the introduction of the *user cooperation diversity* and *cooperative diversity* concepts, respectively. These diversity gains can be achieved via the cooperation of intra-cell users that share their antennas and other resources so as to create a virtual array in a distributed manner. For this reason, Laneman and Wornell (2000) adopted also the term *distributed spatial diversity* [78].

A related concept, suggested by Boyer et al. [79], is the *multihop diversity*, where the benefits of spatial diversity are achieved from the concurrent reception of signals that have been transmitted by multiple previous terminals along a single primary route. Particularly, two models of multihop diversity channels were defined in [79]; the *decoded relaying* and the *amplified relaying*. In the former, each intermediate terminal combines, decodes, and reencodes the received signals from all preceding terminals, whereas in the latter, the intermediate terminals simply combine and amplify them, before retransmission. In addition, Boutros and Viterbo [80] introduced the *signal space diversity* or *modulation diversity*, whereby significant coding gains can be realized over fading channels with the aid of specially designed multidimensional QAM constellations. The key point behind this technique lies in the employment of particular lattice rotation, IQ interleaving/deinterleaving and decoding procedures.

Finally, in a different context, Callaghan and Longstaff [81] proposed the *waveform diversity*; a technique employing multiple transmit waveforms for suppressing synthetic aperture radar (SAR) range ambiguities (i.e., ambiguous range target returns). The waveform diversity scheme embeds, let say, N_b bits of information per pulse by selecting the waveform on a pulse-to-pulse basis from a set of 2^{N_b} waveforms, where each waveform represents a distinct N_b-bit symbol. The Rx obtains the embedded information in each pulse by determining the radar waveform that has been transmitted [82]. It is interesting to note that, according to the IEEE Standard Radar Definitions (IEEE Std 686-2008), this term is defined as the adaptivity of the waveform to dynamically optimize radar performance, while exploiting other domains such as time, frequency, coding, antenna radiation pattern, and polarization.

4.4.4.7 High-Level List of Pure Diversity Techniques

In Table 4.2, a quite extensive list of pure diversity techniques proposed so far in the literature is given. With the term *pure* it is implied here that only the diversity techniques which utilize a single basic principle or concept are included. Therefore, in Table 4.2, other very popular diversity strategies, such as space-time coding, space-time-frequency coding, and so on, are not included.

Table 4.2   List of Pure Diversity Techniques in Chronological Order

Diversity	Details	Suggested by	Year, Reference
Spatial	RxD techniques	Several	1920s
	TxD techniques	Several	1990s
Time	Antiparasitic device	Verdan and Loiseau	1928, [54]
	Coding with interleaving	Several	1950s
Polarization	Rx apparatus with aerials	Eckersley	1929, [53]
	Dual-polarized antenna	Glaser and Faber, Jr.	1953, [66]
Antenna pattern	Steerable antenna	Bruce	1931, [67]
	Directional diversity	Friis and Feldman	1937, [70]
	Angle diversity	Vogelman et al.	1959, [72]
	Field component diversity	Gilbert	1965, [73]
Frequency	FDM	Beverage and Peterson	1931, [52]
	FH-SS	Several	1940s
	OFDM	Chang	1966, [49]
Multipath	RAKE receiver	Price and Green	1958, [65]
Multiuser	–	Knopp and Humblet	1995, [75]
Interference	–	Olofsson et al.	1995, [64]
Waveform	–	Callaghan and Longstaff	1997, [81]
User cooperation	–	Sendonaris et al.	1998, [76]
Signal space	a.k.a. modulation diversity	Boutros and Viterbo	1998, [80]
Doppler	–	Sayeed and Aazhang	1999, [74]
Distributed spatial	–	Laneman and Wornell	2000, [78]
Cooperative	–	Laneman et al.	2001, [77]
Multihop	–	Boyer et al.	2001, [79]

References

1. J. G. Proakis, *Digital Communications*, 4th ed. New York: McGraw-Hill, 2000.
2. T. S. Rappaport, *Wireless Communications: Principles and Practice*. Englewood Cliffs, NJ: Prentice Hall, 1996.
3. M. K. Simon and M.-S. Alouini, *Digital Communication over Fading Channels*, 2nd ed. Hoboken, NJ: John Wiley & Sons, 2005.
4. G. L. Stüber, *Principles of Mobile Communication*, 2nd ed. New York: Kluwer Academic Publishers, 2002.
5. D. Tse and P. Viswanath, *Fundamentals of Wireless Communication*. Cambridge, UK: Cambridge University Press, 2005.
6. B. Sklar, Fading channels, Chapter 4, In *Handbook of Antennas in Wireless Communications*, Godara, L.C., (Ed.). Boca Raton, FL: CRC Press, 2002.
7. A. F. Naguib and A. R. Calderbank, Diversity in wireless systems, Chapter 3, In *Space-Time Wireless Systems: From Array Processing to MIMO Communications*, Bölcskei, H., Gesbert, D., Papadias, C.B., and van der Veen, A.-J., (Eds.). Cambridge, UK: Cambridge University Press, 2006.
8. E. Dahlman, S. Parkvall, and J. Sköld, *4G LTE/LTE-Advanced for Mobile Broadband*. Oxford, UK: Academic Press, 2011.
9. E. Biglieri, J. Proakis, and S. Shamai, Fading channels: Information-theoretic and communications aspects, *IEEE Trans. Inf. Theory*, 44, 6, 2619–2692, 1998.
10. V. M. Kapinas, Optimization and performance evaluation of digital wireless communication systems with multiple transmit and receive antennas, PhD dissertation, Aristotle University of Thessaloniki, Thessaloniki, Greece, January 2014.
11. A. J. Coulson, A. G. Williamson, and R. G. Vaughan, A statistical basis for lognormal shadowing effects in multipath fading channels, *IEEE Trans. Commun.*, 46, 4, 494–502, 1998.
12. L. Rayleigh, On the resultant of a large number of vibrations of the same pitch and of arbitrary phase, *Phil. Mag.*, 10, 60, 73–78, 1880.
13. R. G. Gallager, *Circularly-Symmetric Gaussian Random Vectors, Appendix to Principles of Digital Communication*. Cambridge, UK: Cambridge University Press, 2008.
14. U. Charash, A study of multipath reception with unknown delays, PhD dissertation, University of California, Berkeley, CA, January 1974.
15. S. O. Rice, Statistical properties of a sine wave plus random noise, *Bell Sys. Tech. J.*, 27, 1, 109–157, 1948.
16. M. Nakagami, The *m*-distribution–A general formula of intensity distribution of rapid fading, In *Statistical Method of Radio Propagation*, Hoffman, W.C., (Ed.). Oxford, UK: Pergamon Press, 1960, pp. 3–36.
17. M. Abramowitz and I. A. Stegun, *Handbook of Mathematical Functions with Formulas, Graphs, and Mathematical Tables*, 9th ed. New York: Dover Publications, 1972.
18. R. S. Hoyt, Probability functions for the modulus and angle of the normal complex variate, *Bell Sys. Tech. J.*, 26, 2, 318–59, 1947.
19. I. S. Gradshteyn and I. M. Ryzhik, *Table of Integrals, Series, and Products*, 7th ed. San Diego, CA: Academic Press, 2007.
20. E. W. Stacy, A generalization of the gamma distribution, *Ann. Math. Statist.*, 33, 3, 1187–1192, 1962.
21. M. D. Yacoub, The $\alpha - \mu$ distribution: A physical fading model for the Stacy distribution, *IEEE Trans. Veh. Technol.*, 56, 1, 27–34, 2007.

22. V. A. Aalo, T. Piboongungon, and C.-D. Iskander, Bit-error rate of binary digital modulation schemes in generalized gamma fading channels, *IEEE Commun. Lett.*, 9, 2, 139–141, 2005.

23. P. S. Bithas, N. C. Sagias, and P. T. Mathiopoulos, GSC diversity receivers over generalized-gamma fading channels, *IEEE Commun. Lett.*, 11, 12, 964–966, 2007.

24. P. M. Shankar, Error rates in generalized shadowed fading channels, *Wirel. Pers. Commun.*, 28, 4, 233–238, 2004.

25. P. S. Bithas, N. C. Sagias, P. T. Mathiopoulos, G. K. Karagiannidis, and A. A. Rontogiannis, On the performance analysis of digital communications over generalized-*K* fading channels, *IEEE Commun. Lett.*, 10, 5, 353–355, 2006.

26. E. W. Marchant, Conditions affecting the variations in strength of wireless signals, *J. IEEE*, 53, 243, 329–340, 1915.

27. J. E. Taylor, Wireless telegraphy in relation to interferences and perturbations, *J. IEEE*, 47, 208, 119–140, 1911.

28. F. G. Loring et al., Discussion on wireless telegraphy in relation to interferences and perturbations, *J. IEEE*, 47, 208, 140–166, 1911.

29. R. H. Marriott, Radio range variation, *Proc. IRE*, 2, 1, 37–53, 1914.

30. L. W. Austin, Seasonal variation in the strength of radiotelegraphic signals, *Proc. IRE*, 3, 2, 103–106, 1915.

31. C.-P. Yeang, Characterizing radio channels: The science and technology of propagation and interference, 1900–1935, PhD dissertation, Program in Science, Technology and Society, M.I.T., Cambridge, MA, September 2004.

32. R. Bown and G. D. Gillett, Distribution of radio waves from broadcasting stations over city districts, *Proc. IRE*, 12, 4, 395–409, 1924.

33. R. Bown, D. K. Martin, and R. K. Potter, Some studies in radio broadcast transmission, *Proc. IRE*, 14, 1, 57–131, 1926.

34. L. De Forest, Recent developments in the work of the Federal Telegraph Company, *Proc. IRE*, 1, 1, 37–51, 1913.

35. C. R. Burrows, The history of radio wave propagation up to the end of World War I, *Proc. IRE*, 50, 5, 682–684, 1962.

36. A. H. Taylor and A. S. Blatterman, Variations in nocturnal transmission, *Proc. IRE*, 4, 2, 131–148, 1916.

37. G. W. Pickard, Short period variations in radio reception, *Proc. IRE*, 12, 2, 119–158, 1924.

38. J. H. Dellinger, L. E. Whittemore, and S. Kruse, A study of radio signal fading, In *Scientific Papers of the Bureau of Standards*, Vol. 19. Washington, DC: U.S. Government Printing Office, 1923, pp. 193–230.

39. G. K. Karagiannidis, *Telecommunication Systems*, (in Greek) 2nd ed. Thessaloniki, Greece: Tziolas Publications, 2011.

40. W. H. Tranter, K. S. Shanmugan, T. S. Rappaport, and K. L. Kosbar, *Principles of Communication Systems Simulation with Wireless Applications*. Upper Saddle River, NJ: Prentice Hall, 2004.

41. Y. Shmaliy, *Continuous-Time Systems*. Dordrecht, the Netherlands: Springer, 2007.

42. T. Kailath, *Linear Systems*. Englewood Cliffs, NJ: Prentice Hall, 1980.

43. T. Kailath, Sampling models for linear time-variant filters, M.I.T. Research Laboratory of Electronics, Cambridge, MA, Technical Report 352, May 1959.

44. S. Barbarossa and A. Scaglione, Time-varying fading channels, Chapter 4, In *Signal Processing Advances in Wireless and Mobile Communications, Volume II: Trends in Single- and Multi-User Systems*, Stoica, P., Giannakis, G., Hua, Y., and Tong, L., (Eds.). Upper Saddle River, NJ: Prentice Hall, 2000.

45. A. Papoulis, *Probability, Random Variables, and Stochastic Processes*, 3rd ed. New York: McGraw-Hill, 1991.

46. C. Oestges and B. Clerckx, *MIMO Wireless Communications: From Real-World Propagation to Space-Time Code Design*. San Diego, CA: Academic Press, 2007.

47. G. J. Foschini and M. J. Gans, On limits of wireless communications in a fading environment when using multiple antennas, *Wirel. Pers. Commun.*, 6, 3, 311–335, 1998.

48. G. D. Forney, Jr., Maximum-likelihood sequence estimation of digital sequences in the presence of intersymbol interference, *IEEE Trans. Inf. Theory*, 18, 3, 363–378, 1972.

49. R. W. Chang, Synthesis of band-limited orthogonal signals for multichannel data transmission, *Bell Syst. Tech. J.*, 45, 10, 1775–1796, 1966.

50. C. B. Dietrich, Jr., K. Dietze, J. R. Nealy, and W. L. Stutzman, Spatial, polarization, and pattern diversity for wireless handheld terminals, *IEEE Trans. Antennas Propag.*, 49, 9, 1271–1281, 2001.

51. Z. Wang and G. B. Giannakis, A simple and general parameterization quantifying performance in fading channels, *IEEE Trans. Commun.*, 51, 8, 1389–1398, 2003.

52. H. H. Beverage and H. O. Peterson, Diversity receiving system of RCA Communications, Inc., for radiotelegraphy, *Proc. IRE*, 19, 4, 531–561, 1931.

53. T. L. Eckersley, An investigation of short waves, *J. IEEE*, 67, 392, 992–1029, 1929.

54. C. Verdan and L. Loiseau, Signaling system applicable to telegraphy and telemechanical transmission, U.S. Patent 1 677 062, July 1928.

55. H. H. Harrison, Developments in machine telegraph systems and methods of operation, *J. IEEE*, 68, 407, 1369–1453, 1930.

56. F. Adcock, Improvement in means for determining the direction of a distant source of electromagnetic radiation, British Patent 130 490, August 1919.

57. R. L. Smith-Rose, A theoretical discussion of various possible aerial arrangements for rotating-beacon transmitters, *J. IEEE*, 66, 375, 270–279, 1928.

58. G. E. Hansell, Transmitter space diversity as applied to shipboard reception, *IRE Trans. Prof. Commun. Syst.*, 3, 1, 44–46, 1955.

59. I. Brodsky, How Reginald Fessenden put wireless on the right technological footing, In *Proceedings of IEEE Global Telecommunications Conference*, New Orleans, LO, November 2008, pp. 1–5.

60. J. S. Belrose, Reginald Aubrey Fessenden and the birth of wireless telephony, *IEEE Antennas Propag. Mag.*, 44, 2, 38–47, 2002.

61. J. S. Belrose, Fessenden and Marconi: Their differing technologies and transatlantic experiments during the first decade of this century, In *Proceedings of 1995 International Conference on 100 Years of Radio*, London, UK, September 1995, pp. 32–43.

62. J. E. Brittain and A. Reginald, Fessenden and the origins of radio [scanning the past], *Proc. IEEE*, 84, 12, 1852–1853, 1996.

63. J. E. Brittain and A. Reginald, Electrical engineering hall of fame: Reginald A. Fessenden, *Proc. IEEE*, 92, 11, 1866–1869, 2004.

64. H. Olofsson, J. Naslund, and J. Skold, Interference diversity gain in frequency hopping GSM, In *Proceedings on IEEE Vehicular Technology Conference*, Chicago, IL, July 1995.

65. R. Price and E. Green, A communication technique for multipath channels, *Proc. IRE*, 46, 3, 555–570, 1958.

66. J. L. Glaser and L. P. Faber, Jr., Evaluation of polarization diversity performance, *Proc. IRE*, 41, 12, 1774–1778, 1953.

67. E. Bruce, Developments in short-wave directive antennas, *Proc. IRE*, 19, 8, 1406–1433, 1931.

68. E. Bruce and A. C. Beck, Experiments with directivity steering for fading reduction, *Proc. IRE*, 23, 4, 357–371, 1935.

69. H. T. Friis, C. B. Feldman, and W. M. Sharpless, The determination of the direction of arrival of short radio waves, *Proc. IRE*, 22, 1, 47–78, 1934.

70. H. T. Friis and C. B. Feldman, A multiple unit steerable antenna for short-wave reception, *Proc. IRE*, 25, 7, 841–917, 1937.

71. D. Gregory and S. Newsome, Cooling radio station, Hoo Peninsula, Kent: An archaeological investigation of a short-wave receiving station, English Heritage Research Department, Portsmouth, UK, Technical Report 110-2010, 2010.

72. J. H. Vogelman, J. L. Ryerson, and M. H. Bickelhaupt, Tropospheric scatter system using angle diversity, *Proc. IRE*, 47, 5, 688–696, 1959.

73. E. N. Gilbert, Energy reception for mobile radio, *Bell Sys. Tech. J.*, 44, 8, 1779–1803, 1965.

74. A. M. Sayeed and B. Aazhang, Joint multipath-doppler diversity in mobile wireless communications, *IEEE Trans. Commun.*, 47, 1, 123–132, 1999.

75. R. Knopp and P. A. Humblet, Information capacity and power control in single-cell multiuser communications, In *Proceedings on IEEE International Conference on Communications*, Seattle, WA, June 1995.

76. A. Sendonaris, E. Erkip, and B. Aazhang, Increasing uplink capacity via user cooperation diversity, In *Proceedings on IEEE International Symposium on Information Theory*, Cambridge, MA, August 1998.

77. J. N. Laneman, G. W. Wornell, and D. N. C. Tse, An efficient protocol for realizing cooperative diversity in wireless networks, In *Proceedings on IEEE International Symposium on Information Theory*, Washington, DC, June 2001.

78. J. N. Laneman and G. W. Wornell, Exploiting distributed spatial diversity in wireless networks, In *Proceedings on Allerton Conference on Communication, Control, and Computing*, Urbana-Champagne, IL, October 2000.

79. J. Boyer, D. Falconer, and H. Yanikomeroglu, A theoretical characterization of the multihop wireless communications channel with diversity, In *Proceedings on IEEE Global Communications Conference*, San Antonio, TX, November 2001.

80. J. Boutros and E. Viterbo, Signal space diversity: A power- and bandwidth-efficient diversity technique for the Rayleigh fading channel, *IEEE Trans. Inf. Theory*, 44, 4, 1453–1467, 1998.

81. G. D. Callaghan and I. D. Longstaff, Wide-swath space-borne SAR and range ambiguity, In *Proceedings on Radar 97 (Conference Publication No 449)*, Edinburgh, UK, October 1997.

82. A. Hassanien, M. G. Amin, Y. D. Zhang, and F. Ahmad, Signaling strategies for dual-function radar communications: An overview, *IEEE Aerosp. Electron. Syst. Mag.*, 31, 10, 36–45, 2016.

83. The Electrical World and Engineer, vol. XL, no. 20, Nov. 15, 1902.

84. The Electrical World and Engineer, vol. XL, no. 23, Dec. 6, 1902.

Chapter 5

Multicarrier Modulation Schemes—Candidate Waveforms for 5G

Konstantinos Maliatsos and Athanasios G. Kanatas

Contents

5.1 Introduction

Wireless communications have been continuously evolving with the provision of new sets of applications and services to the mobile user. In order to achieve the demanding objectives and requirements imposed by the increasing use of mobile services and applications, wireless communication evolution is triggered by the need for (Andrews et al. 2014; NGMN Alliance 2015; Bockelmann et al. 2016):

- higher data rates that will allow users to download huge volumes of data through the radio channel
- efficient utilization and sharing of the finite radio resources among the rapidly increasing number of users of mobile services and applications
- reliable communication links that ensure adequate quality of service (QoS) to all users
- seamless connectivity regardless of location, mobility conditions, and density of connected users

Challenges in wireless communications are amplified due to the fact that they rely on a finite and scarce resource, the radio spectrum. Standardization of waveforms and access protocols, as well as licensed use of spectrum were necessary actions in order to facilitate the development of efficient and reliable mobile radio systems.

The road to 5th-generation (5G) communication systems has marked a change in the objectives and orientation of wireless evolution. 5G systems will incorporate a large variety of emerging services and applications that deviate from the conventional development steps of cellular networking.

5G marks a transition to machine-type communications (Militano et al. 2015). Device-to-device (D2D) communications and vehicular communications (Hobert et al. 2015; Uhlemann 2015) will enable connectivity of billions of intelligent machines, allowing remote control and monitoring, as well as unmanned operation and coordination. Internet of things (IoT) (Mavromoustakis et al. 2016) applications will provide a new range of capabilities and emerging technologies like the smart cities, the connected vehicle, the smart grid, and many more. Recently, Tactile Internet has also emerged as a 5G-enabled technology. Tactile Internet refers to extremely low-latency communications with high availability and reliability

(Simsek et al. 2016). Tactile Internet delivers a variety of public services ensuring remote sensing and real-time control of cyber-physical systems, for example, unmanned aerial/surface/ground vehicles, robots, automated machinery, and many more.

5G aims to higher data rates in order to satisfy the increasing demand for throughput from all users and applications using a finite set of radio resources. In order to maximize the utilization of spectrum resources, composite network structures are considered that differentiate from the conventional cellular network layout. Cooperative schemes, Massive MIMO (Larsson et al. 2014), cloud radio access network (C-RAN) (Wu et al. 2015), coordinated multipoint (Jungnickel et al. 2014), and full-duplex (Sabharwal et al. 2014) are some of the concepts adopted in the design of 5G networks. In addition, wireless regional area networks (WRAN) that bring high-speed broadband Internet access in remote, isolated areas are also a 5G objective.

The transition to unconventional network structures and machine-type communications also marks a transition from the conventional cell-centric model to distributed modes. Moreover, 5G has to deal with spectrum scarcity in order to maximize radio resource utilization and it should be able to operate over non-contiguous, fragmented spectrum. Expansion to millimeter-Wave (mm-Wave) frequencies is also considered, as means for high-rate connectivity used for back-hauling, fronthauling (De La Oliva et al. 2015), WRAN solutions, as well as in some cases for low-range radio access.

Therefore, 5G aims to the design of a radio network that: (1) maximizes spectral efficiency, (2) is highly tolerant and robust to interference, (3) is able to achieve accurate synchronization over a diverse set of radio and network conditions, and (4) is able to harmonically coexist with a large number of users (people or devices) with different requirements and objectives.

The research community is currently very active on designing, proposing, and evaluating waveform candidates for 5G systems (Banelli et al. 2014). The objective of this chapter is to investigate waveforms that will help achieve the 5G goals, that is, to propose modulation schemes with exceptional spectral efficiency, low out-of-band emissions (OOB), robustness against interference and synchronization errors and suitable for use in a fragmented radio spectrum.

Orthogonal frequency division multiplexing (OFDM) and OFDM-based transmission schemes have dominated current wireless communication standards including WLANs, LTE, and DVB. OFDM has become popular due to specific advantages:

- Robustness against frequency selective fading with the division of information symbols in parallel narrowband channels
- Efficient use of spectrum due to overlapping transmission of orthogonal parallel narrowband channels
- Low-complexity implementation with the use of Fast Fourier Transforms

- Low-complexity channel equalization compare with single-carrier solutions
- Robustness against intersymbol interference (ISI) with the use of the cyclic prefix
- Robustness against impulsive noise
- Simple integration of multiple-input multiple-output (MIMO) systems in the OFDM transmitter/receiver chain
- Ability to easily integrate adaptive modulation and coding techniques to efficiently exploit the radio channel
- Provision of direct extension to a multiplexing scheme with orthogonal frequency division multiple access (OFDMA) for resource sharing

For all these reasons, OFDM has given a great boost to the evolution of wireless communications. However, as requirements become more demanding, OFDM disadvantages become a limiting factor in the route toward 5G. More specifically:

- Due to the use of rectangular pulses in the time domain, OFDM presents powerful sidelobes, that is, high OOB emissions. This fact imposes the use of null guard subcarriers at the edges of the transmitted signal spectrum. As a result, spectrum efficiency is reduced and the use of OFDM in fragmented spectrum is not recommended.
- The use of cyclic prefix is beneficial in many aspects; however, it reduces spectral efficiency significantly.
- OFDM requires synchronization accuracy in time and frequency. Despite the fact, that the cyclic prefix can provide protection against timing synchronization errors, high accuracy in frequency alignment is difficult to achieve in a multiuser environment with distributed coordination. Errors in frequency synchronization can be extremely harmful in OFDM, since they lead to InterCarrier Interference (ICI).
- OFDM exhibits large peak-to-average power ratio (PAPR) and thus is not recommended for use in low-cost user equipment (UE) hardware, where poor amplifier performance and low Digital-to-Analog converter resolution may lead to unrecoverable signal distortion.

In this chapter, after a presentation of the OFDM modulation, a review of the most popular OFDM-alternative waveforms is given. For each modulation, there is a short description of the transmitter/receiver chains and their main advantages and disadvantages are denoted. More specifically, FBMC-OQAM, UFMC, and GFDM are analyzed.

5.2 Multicarrier Modulation for ISI Avoidance

The problem of ISI in wideband communications can be summarized in the following sentence: When the time dispersion caused by the radio channel is comparable with the period of the transmitted symbol T, then during reception each symbol suffers from interference by delayed versions of previous symbols.

Generally, in wideband transmissions $T_m \gg T$, where T_m is the maximum excess delay induced by the channel. Multicarrier modulations resolve this issue using a simple rationale. Instead of transmitting symbols with short pulses, they are divided in N parallel streams and transmitted using long narrowband pulses, so that $NT \gg T_m$. This method limits the effect of ISI. In addition, guard intervals (in time and/or frequency) or cyclic signal extensions can completely cancel the effect of the dispersive and selective radio channel. Moreover, the multicarrier modulation scheme should minimize possible interference among the N narrowband streams that are transmitted in parallel. The interference among the N subchannels in OFDM is called ICI.

If we omit details and particularities of each multicarrier scheme, multicarrier modulation can be roughly described as a pair of filter banks (transmitter and receiver) applied to the N parallel signal streams. A simplified, generic block diagram to describe the concept of multicarrier modulations in the analog time domain is presented in Figure 5.1.

Based on Figure 5.1, a multicarrier scheme is able to completely eliminate ISI, when the set of filter functions $\phi_n(t)$ $(n = 0...N-1)$ is orthogonal with their mT $(m \in \mathbb{Z})$ time-shifted versions. This filter configuration in the transmitter and receiver forms a pair of matched filters. In addition, the filters should be ideally orthogonal with the rest of the bank filters in order to avoid ICI. Ideally, the filter bank forms an orthogonal basis where:

$$\int_{-\infty}^{+\infty} \phi_k(t - nT)\phi_l^*(t - mT)dt = c\delta[k - l]\delta[n - m] \tag{5.1}$$

for $k,l = 0...N-1$, $m,n \in \mathbb{Z}$ and c is a constant. If $s_{n,k}$ is the information symbol transmitted by the kth subchannel at the nth time instance, then the transmitted signal can be expressed as

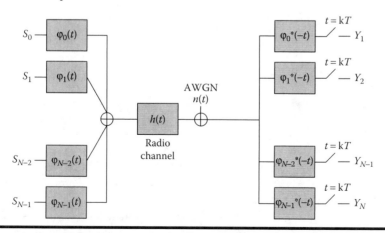

Figure 5.1   Generic multicarrier modulation transmission.

$$s(t) = \sum_{n=-\infty}^{+\infty} \sum_{k=0}^{N-1} s_{n,k} \phi_k(t - nT) \tag{5.2}$$

Assuming a flat radio channel with additive white Gaussian noise (AWGN) $n(t)$, the received signal is given by

$$y_{n,k} = \int_{-\infty}^{+\infty} (s(t) + n(t)) \phi_k^*(t - nT) dt = s_{n,k} + N_{n,k} \tag{5.3}$$

where $N_{n,k}$ is the noise component filtered by the receiver. Based on the aforementioned description, the functions $\phi_n(t)$ ideally constitute an orthonormal basis for the signal space.

The results in Equations 5.2 and 5.3 present a system without ISI and ICI, but assume flat radio channel. In case of a frequency selective channel, the transmitted pulse is filtered by:

$$p_k(t) = \phi_k(t) * h(t) \tag{5.4}$$

where $*$ indicates convolution and $h(t)$ indicates the impulse response of the radio channel. In this case, in order to achieve ISI/ICI-free orthogonal transmission, the selection of the filter responses $\phi_n(t)$ should form a set of orthonormal eigenfunctions defined by the channel autocorrelation function $r(t) = h(t) * h^*(-t)$ (Li and Stuber 2006). Moreover, it is desired that $\phi_n(t)$ has duration limited to the assumed symbol period T, that is, $\phi_n(t) \neq 0$ for $t \in [-T/2, T/2)$. In a practical system, it is not realistic to assume that the eigenfunctions, that is, the filters, will be computed for each given channel $h(t)$, especially since the radio channel is time variant.

Therefore, in realistic transmission schemes, the frequency-selective channel imposes limitations and requirements that cannot be exceeded without a cost.

Generally, the baseband discrete-time representation of a multicarrier scheme provides more insight in the operation of each modulations scheme.

5.3 Orthogonal Frequency Division Multiplexing

OFDM (Li and Stuber 2006) follows the general principles of multicarrier modulations. The available bandwidth B is divided in N subchannels, called subcarriers. No guard band is kept between subchannels. The available data are divided in blocks of N parallel streams. After applying the OFDM modulation, each stream constitutes the OFDM symbol with duration T.

The selected base for OFDM is the uniformly-modulated rectangular pulse over the available bandwidth. Assuming a baseband equivalent representation, the base functions are given by

$$\varphi_n(t) = u(t)e^{2\pi j f_n t}$$

$$\text{where } u(t) = \begin{cases} 1, & 0 \le t < T \\ 0, & \text{elsewhere} \end{cases} \tag{5.5}$$

$$f_n = n\frac{B}{N}, n = -\frac{N}{2}, ..., \frac{N}{2} - 1$$

For AWGN transmission in ideal conditions, OFDM achieves maximum efficiency as it maintains orthogonality in both time and frequency while the time-bandwidth product for each data symbol is unity, that is, $(B/N)T = 1$. As a next step, we focus on an individual OFDM symbol. If S_n is the data symbol for the nth subcarrier for the given OFDM symbol, then:

$$s(t) = \begin{cases} \displaystyle\sum_{n=-N/2}^{N/2-1} S_n e^{2\pi j f_n t}, 0 \le t < T \\ 0, \text{elsewhere} \end{cases} \tag{5.6}$$

It is more convenient to express Equation 5.6 in the baseband discrete-time representation assuming Nyquist sampling, that is, sampling period $T_s = 1/B = T/N$. Then, for $0 \le t < T$:

$$s(mT_s) = \sum_{n=-N/2}^{N/2-1} S_n e^{2\pi j f_n t} = \sum_{n=-N/2}^{N/2-1} S_n e^{2\pi j n m \frac{BT}{N^2}}$$

$$= \sum_{n=-N/2}^{N/2-1} S_n e^{2\pi j \frac{nm}{N}} = \sum_{n=0}^{N-1} S_n e^{2\pi j \frac{nm}{N}} \tag{5.7}$$

where the last step is true due to spectrum folding in the digital domain (assuming $S_{n+N/2} = S_n$ for $n < 0$). Given that analysis is performed in the Nyquist rate, the duration of the OFDM symbol in samples is also N. The final result also denotes a major advantage of OFDM modulation, as it can be implemented with a simple inverse discrete fourier transform (DFT). Moreover, the use of the fast Fourier transform (FFT) will reduce the necessary workload furthermore. This fact explains the statement that *in OFDM the data symbols are provided in the frequency domain and through the inverse DFT, the OFDM symbols are produced and transmitted in the time domain.*

It is clear that for an AWGN channel, the receiver design is simplified to an FFT:

$$X_n = \sum_{m=0}^{N-1} (s(mT_s) + w(mT_s))e^{-2\pi j \frac{nm}{N}}$$

$$= S_n + \sum_{m=0}^{N-1} w(mT_s)e^{-2\pi j \frac{nm}{N}} = S_n + W_n \tag{5.8}$$

5.3.1 Cyclic Extension and Frequency Selective Channels

Maximum spectral efficiency is achieved for nonselective radio channel. However, when a signal passes through the radio channel, it is extended in time. If $v+1$ is the length of the radio channel impulse response in samples, then the only way to cancel ISI is to use a guard time interval with v samples. As a matrix relationship, the convolution result for N output samples is expressed as

$$
\begin{bmatrix} y_{N-1} \\ y_{N-2} \\ \vdots \\ y_0 \end{bmatrix} = \begin{bmatrix} p_0 & p_1 & \cdots & p_v & 0 & \cdots & 0 \\ 0 & p_0 & \ddots & p_1 & p_v & \ddots & 0 \\ \vdots & \ddots & \ddots & \ddots & \ddots & \ddots & \vdots \\ 0 & \cdots & 0 & p_0 & p_1 & \cdots & p_v \end{bmatrix} \begin{bmatrix} s_{N-1} \\ \vdots \\ s_0 \\ s_{-1} \\ \vdots \\ s_{-v} \end{bmatrix} + \begin{bmatrix} w_{N-1} \\ w_{N-2} \\ \vdots \\ w_0 \end{bmatrix} \tag{5.9}
$$

where $p_m, m = 0,\ldots,v$ is the channel impulse response and $s_{-v},\ldots,s_{-1}$ are the guard interval samples. In OFDM, the guard interval is called cyclic prefix, that is, the OFDM symbol is cyclicly extended. Thus, the cyclic prefix is given by

$$
\begin{aligned} s_{-l} &= s_{N-l} \\ l &= 1,\ldots,v \end{aligned} \tag{5.10}
$$

The cyclic extension has a special, beneficial role for the OFDM functionality. Due to the cyclic prefix, the relationship in Equation 5.9 is transformed as

$$
\begin{bmatrix} y_{N-1} \\ y_{N-2} \\ \vdots \\ y_0 \end{bmatrix} = \begin{bmatrix} p_0 & p_1 & \cdots & p_v & 0 & \cdots & 0 \\ 0 & p_0 & \ddots & p_{v-1} & p_v & \ddots & 0 \\ 0 & \ddots & \ddots & \ddots & \ddots & \ddots & 0 \\ 0 & \cdots & 0 & p_0 & p_1 & \cdots & p_v \\ p_v & 0 & \cdots & 0 & p_0 & \cdots & p_v \\ \vdots & \ddots & \ddots & \ddots & \ddots & \ddots & \vdots \\ p_1 & \cdots & p_v & 0 & \cdots & 0 & p_0 \end{bmatrix} \begin{bmatrix} s_{N-1} \\ \vdots \\ s_1 \\ s_0 \end{bmatrix} + \begin{bmatrix} w_{N-1} \\ w_{N-2} \\ \vdots \\ w_0 \end{bmatrix}, \quad y = Px + n
$$

$$\tag{5.11}$$

From Equation 5.11 it is seen that, besides its operation as a guard band, the cyclic prefix transforms the linear convolution of the signal with the channel response, to circular convolution. The channel matrix **P** is now a $N \times N$ circulant matrix. Circulant matrices are special cases of Toeplitz matrix. Each row follows from the previous row with right-hand rotation by one element. The overflown matrix elements are not discarded but circularly reinserted at the beginning of the row.

If we now perform eigenvalue decomposition of square matrix $\mathbf{P}$, then:

$$\mathbf{P} = \mathbf{MUM}^*\tag{5.12}$$

As $\mathbf{P}$ is a normal matrix (i.e., $\mathbf{PP}^* = \mathbf{P}^*\mathbf{P}$), then $\mathbf{U} = \Lambda$ is diagonal and $\mathbf{M}$ is unitary (i.e., $\mathbf{M}^*\mathbf{M} = \mathbf{MM}^* = \mathbf{I}$). Thus, the first useful property of the circulant matrix is that eigenvalue decomposition also performs diagonalization simultaneously. The second valuable property is that, according to the spectral theorem, for all circulant matrices the matrix of eigenvectors $\mathbf{M}$ is given by the inverse DFT matrix, that is, $\mathbf{M} = \mathbf{F}^*$ where

$$\mathbf{F} = \begin{bmatrix} e^{-j\frac{2\pi}{N}(N-1)(N-1)} & \cdots & e^{-j\frac{2\pi}{N}2(N-1)} & e^{-j\frac{2\pi}{N}(N-1)} & 1 \\ e^{-j\frac{2\pi}{N}(N-1)(N-2)} & \cdots & e^{-j\frac{2\pi}{N}2(N-2)} & e^{-j\frac{2\pi}{N}(N-2)} & 1 \\ \vdots & \cdots & \vdots & \vdots & \vdots \\ e^{-j\frac{2\pi}{N}(N-1)} & \cdots & e^{-j\frac{2\pi}{N}2} & e^{-j\frac{2\pi}{N}} & 1 \\ 1 & \cdots & 1 & 1 & 1 \end{bmatrix}\tag{5.13}$$

The eigenvalues of the circulant channel matrix $\mathbf{P}$ that appear in the diagonal of Λ form the channel transfer function. The great benefit of the circulant channel is that eigenvector matrices are channel-independent. OFDM transmission/reception is described by a sequence of matrix relationships:

■ In the transmitter, the signal vector $\mathbf{S}$ is multiplied with $\mathbf{F}^*$, that is, $s = \mathbf{F}^*\mathbf{S}$ to produce the OFDM symbol.
■ The signal propagates through the channel and the received signal $\mathbf{y}$ is demodulated by multiplying it with the DFT matrix, that is, $\mathbf{Y} = \mathbf{Fy}$.
■ The diagonal shape of Λ ensures ICI-free transmission (the AWGN component is omitted). Overall:

$$\mathbf{Y} = \mathbf{Fy} = \mathbf{FF}^*\Lambda\mathbf{Fx} = \mathbf{FF}^*\Lambda\mathbf{FF}^*\mathbf{S} = \Lambda\mathbf{S}\tag{5.14}$$

Matrix multiplication with the conjugate matrix $\mathbf{F}$ is equivalent with the inverse DFT. Respectively, multiplication with $\mathbf{F}$ performs the DFT.

The great provided advantage is that the equalization procedure is greatly simplified. Based on Equation 5.14, it can be seen that equalization can be achieved with a single multiplication per subcarrier. For example, the zero-forcing equalizer is implemented using Λ^{-1}, which is also diagonal with inverted elements of Λ.

Let's assume that a series of L OFDM symbols is sequentially transmitted. The transmitted signal can be expressed using the following formula:

$$x(m) = \sum_{l=0}^{L-1} s_l(m - lT(1+\nu))$$

$$\text{where } s_l(m) = \begin{cases} \sum_{n=0}^{N-1} S_{l,n} e^{2\pi j \frac{nm}{N}}, & m = -\nu, -\nu+1, ..., 0, ..., N-1 \\ 0, & \text{elsewhere} \end{cases}$$

(5.15)

where $S_{l,n}$ expresses the frequency-domain symbol for OFDM symbol l in subcarrier n and $s_l(m)$ the time-domain samples for OFDM symbol l. It is noted that due to the periodicity of the DFT, for $m = 1, ..., \nu$, $s_l(-m) = s_l(N-m)$, in accordance with Equation 5.10. After a synchronization procedure, the receiver isolates samples $s_l(m)$ with $m = 0, ..., N-1$ and performs equalization using matrix Λ.

5.3.2 Out-of-Band Emissions in OFDM

OFDM modulation achieves maximum spectral efficiency for ideal AWGN transmission, where a cyclic prefix is not needed. At this point, we should mention the Balian–Low theorem from Gabor theory, that states that there is no function basis to simultaneously satisfy the following criteria:

1. All functions are mutually orthogonal in time and frequency.
2. They are well localized in time and frequency domain.
3. Maximal spectral efficiency is achieved.

Well-localized functions are the functions that provide compact support in a given time-frequency grid. OFDM in ideal AWGN transmission provides orthogonal functions and maximum efficiency; however, it is not well-localized in the time-frequency grid defined by T and B. On the contrary, OFDM spectrum is infinite since the Fourier transform of the rectangular pulse is a *sinc* function. This leads to the first significant disadvantage, the OFDM sidelobes, and high OOB radiation. In Figure 5.2, the spectrum shape for two adjacent subcarriers is presented.

The strong sidelobes cause important implications in OFDM operation:

▪ Strong OOB emissions cause adjacent channel interference. Respectively, the OFDM receiver gathers significant power from OOB interfering signals. In real-world conditions OFDM requires large spectral guard bands from adjacent emissions resulting in significant reduction of spectral efficiency.
▪ As OFDM subcarriers are not well-localized in frequency, OFDM is sensitive in frequency shifts and phase noise. Frequency shifts are caused from local oscillator deviations (in the transmitter or/and the receiver), or due to Doppler effects. These shifts will cause misalignment of the subcarriers resulting in loss

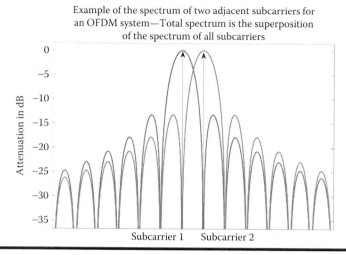

Figure 5.2   Sinc-shaped spectrum for two adjacent OFDM subcarriers.

of orthogonality. As a result, ICI appears. From Figure 5.2, it can be seen that a shift equal to 0.25 of the subcarrier distance will reduce the signal power by approximately 3 dB and will induce more than −10 dB ICI (using the signal power under ideal reception as reference).

5.3.3 Cyclic Prefix and Spectral Efficiency

In frequency selective channels, the cyclic prefix reduces spectral efficiency even more. The signal is extended beyond the period T defined by the Nyquist rate. If β is the cyclic prefix percentage in terms of OFDM symbol duration, then the time-bandwidth product is given by $(B/N)T(1+\beta)>1$ and the spectral efficiency is reduced. For highly frequency selective channels, the cyclic prefix size should increase accordingly. In existing standards, for example, LTE with extended prefix (Long-Term Evolution (LTE) standard 2008) or IEEE 802.11 (2007), the cyclic prefix is one-quarter of the OFDM symbol duration.

Moreover, the use of the cyclic prefix implies that $\beta/(1+\beta)$ of the transmitted power carries redundant information. Therefore, OFDM energy efficiency is also reduced. It should be noted that OFDM versions with the use of zero-prefix can also be considered in order to avoid energy expenditure with a small cost in the receiver design complexity (Muquet et al. 2000).

5.3.4 Synchronization in OFDM

Time synchronization is also crucial in OFDM (Maliatsos et al. 2008). Timing errors may cause severe degradation of SNR. However, OFDM is not considered sensitive in time synchronization errors due to the protection provided by the cyclic prefix. If ν is the cyclic prefix length and μ is the channel impulse response length with

$\mu < \nu$, then it can easily be proved that for timing errors $\varepsilon = -\nu + \mu, ..., -1$, the performance of the OFDM receiver does not degrade. As no ISI appears, the only effect is a phase shift that can be compensated by the equalizer without any modification.

For positive errors, or $\varepsilon < -\nu + \mu$, ISI appears in the reception procedure that will cause distortion. However, it should be noted that for $\varepsilon < -\nu + \mu$, or even in cases where $\mu \geq \nu$ (for relatively small values), performance degradation is performed smoothly, in contrast with the performance collapse that takes place in single carrier systems when the equalizer tolerance is exceeded. This is also an advantage for OFDM.

5.3.5 Peak to Average Power Ratio

Another OFDM disadvantage is the high values of peak-to-average power ratio (PAPR). OFDM time samples are produced as a sum of the frequency-domain symbols multiplied with the inverse DFT basis. Assuming that all available subcarriers are used, if we consider the data symbols as independent random variables and given the fact that the number of subcarriers is adequately high (≥ 32), the Central Limit theorem can be invoked. Therefore, the OFDM time-domain samples are independent Gaussian variables, that is, the OFDM waveform has the same statistical behavior with AWGN. Thus, the signal has a highly irregular behavior in the time domain and since the time samples are uncorrelated, peaks may appear suddenly at any time. As an example, for mean signal power 1, one out of every 1000 samples is expected to have power higher than 4.5. This means that the PAPR is expected to exceed 4.5 for and OFDM system with $N = 1024$.

High PAPR may introduce serious degradation in the system performance when the signal passes through nonlinear high power amplifiers (HPA) during transmission. This may cause spectral growth of the signal in the form of intermodulation and spurious emissions. In order to avoid these effects, the HPA should operate in its linear region with inefficient power conversion. Moreover, if there are specific applications, regulatory restrictions, or constraints imposed by the digital-to-analog converter (DAC) regarding the peak transmit power, then the mean signal power should also be reduced. In turn, this reduces the range of multicarrier transmission.

In order to avoid PAPR effects, sophisticated and expensive hardware components at the transceiver are required, or algorithms that can reduce PAPR should be applied. There are plenty of PAPR reduction techniques for OFDM, from simple clipping to transmit signal precoding (Han and Lee 2005). However, the application of PAPR reduction techniques may either degrade system performance under circumstances, or entail the need for more complicated transceiver designs.

5.3.6 Summary—Advantages and Disadvantages

OFDM advantages can be separated in multicarrier modulation-specific, that is, advantages that can be found in many multicarrier modulations and OFDM-specific (Li and Stuber 2006):

- Multicarrier modulation advantages
  - As a modulation scheme, it is able to integrate adaptive modulation and coding procedures in order to optimize data throughput.
  - It inherently supports MIMO transmission and it can use narrowband MIMO-related algorithms per narrowband subcarrier.
  - OFDM can also be used as a multiple access technique, that is, the OFDMA.
  - Frequency domain interleaving and scrambling can protect signals from burst errors, in case some subcarriers suffer from deep fading. In this sense, OFDM is robust in narrowband interference, as only a small number of subcarriers is affected.
- OFDM-specific advantages
  - OFDM achieves maximal spectral efficiency for ideal transmission over AWGN channel. Despite the fact that spectral efficiency is significantly reduced under real-world conditions, OFDM has been proved highly efficient in wideband channels compared to single carrier modulations.
  - OFDM is implemented with minimization of the computational workload in the transmitter and the receiver. This is caused by the fact that modulation/demodulation is performed with IFFT/FFT, the computationally efficient implementations of the inverse and direct DFT.
  - Equalization is performed with a simple multiplication per subcarrier by taking advantage the circular extension of the signal.
  - ISI is canceled with the use of the cyclic prefix. Even if the cyclic prefix does not fully protect the OFDM symbol due to extended channel impulse response, the receiver will not collapse, but its performance will degrade smoothly.
  - Time synchronization is crucial for OFDM; however, the cyclic prefix provides additional protection and thus the OFDM receiver can tolerate and manage a range of possible synchronization errors.

On the other hand, the OFDM disadvantages are summarized in the following:

- OFDM OOB emissions are quite high causing adjacent channel interference.
- On the other hand, unfiltered OFDM is vulnerable to interference from adjacent channels (Maliatsos et al. 2013). The high OFDM sidelobes reduce spectral efficiency, since spectral guard bands should be kept to avoid interfering effects.
- OFDM is sensitive in frequency shifts and there is need for elaborate compensation algorithms in the receiver (Pollet et al. 1995; Maliatsos et al. 2008).
- The use of the cyclic prefix is beneficial for OFDM in plenty of ways. However, the cyclic prefix reduces spectral efficiency as well as energy efficiency.
- OFDM has high PAPR and this may cause significant degradation in system performance.

The provided analysis shows that OFDM has some significant disadvantages that have to be mitigated in order to achieve the 5G objectives. However, even at this point, OFDM performs exceptionally well balancing the trade-off between performance and implementation complexity.

Given the increasing potential of the current transceivers, the research community searches for candidate waveforms that balance better among the tradeoffs defined by the Balian–Low theorem and the computational workload needed for the implementation of the systems.

5.4 FilterBank MultiCarrier Modulation with Offset QAM

OFDM without cyclic prefix achieves maximum spectral efficiency for AWGN channel under ideal conditions. According to Gabor theory and Balian–Low theorem (Mallat 1999; Feichtinger and Strohmer 2012), there is no pulse other than the rectangular (used in OFDM) that can achieve orthogonal transmission with maximum efficiency for symbol duration T with critical Nyquist sampling.

However, a simple trick can be used to bypass the Balian–Low constraints and to achieve maximal spectral efficiency with critical sampling using pulses that are well-localized in both time and frequency. This can be achieved using a time offset in the QAM constellation of the frequency-domain data symbols. FilterBank MultiCarrier modulation with Offset QAM (FBMC-OQAM) is based on the aforementioned technique (Saltzberg 1967; Hirosaki 1981).

Two keystones are defined in order to achieve orthogonality in pulse-shaped FBMC-OQAM. First, a time offset is inserted between the real and imaginary part of the QAM symbol. Second, the design is performed so that for two adjacent subchannels, when the imaginary part is delayed by the offset on the one subchannel, then the real part is delayed on the other. Let's assume that there are M subchannels (M even) and that $c_{m,n}$ is the n-th QAM symbol of the m-th subchannel. The M subchannels are defined by the set of carrier frequencies $f_m = m\Delta f_c$, $m = 0,1. M-1$ where Δf_c defines the spectral distance between adjacent carriers. In order to achieve maximum spectral efficiency $\Delta f_c = (B/M)$. A basis using functions $\Psi_{m,n}$ is defined as (Du and Signell 2007):

$$\Psi_{m,n}(t) = h(t - nT)e^{2\pi j f_m t} \tag{5.16}$$

where:
$h(t)$ is the baseband prototype pulse
T is the wideband symbol period

The modulated signal occurs as a linear combination of the bases multiplied by the data symbols. However, because of the offset insertion the FBMC-OQAM transmitted signal is expressed as

$$s(t) = \sum_{n=-\infty}^{\infty} \sum_{m=0}^{M/2-1} \begin{array}{l} (\Re(c_{2m,n})h(t-nT) + j\Im(c_{2m,n})h(t-nT-T/2))e^{2\pi j 2m\Delta f_c t} \\ + (j\Im(c_{2m+1,n})h(t-nT) + \Re(c_{2m+1,n})h(t-nT-T/2))e^{2\pi j(2m+1)\Delta f_c t} \end{array}$$

(5.17)

With proper re-assignment of $c_{m,n}$, the transmitted signal can be expressed as a function of real-valued symbols. Thus, if:

$$
\begin{array}{ll}
a_{2m,2n} = \Re(c_{2m,n}), & \phi_{2m,2n} = 0, \\[2mm]
a_{2m,2n+1} = \Im(c_{2m,n}), & \phi_{2m,2n+1} = \dfrac{\pi}{2}, \\[2mm]
a_{2m+1,2n} = \Im(c_{2m+1,n}), & \phi_{2m+1,2n} = \dfrac{\pi}{2}, \\[2mm]
a_{2m+1,2n+1} = \Re(c_{2m+1,n}) & \phi_{2m+1,2n+1} = 0,
\end{array}
$$

(5.18)

then, the transmitted signal is given by

$$s(t) = \sum_{m=0}^{M-1} \sum_{n=-\infty}^{\infty} a_{m,n} h(t-nT/2)e^{2\pi jm\Delta f_c t} e^{j\phi_{m,n}}$$

(5.19)

From Equation 5.19, it can be seen that the filter operates at twice the rate, but nevertheless, it requires the same number of multiplications as the complex convolution at the Nyquist rate. The oversampling bypasses the Balian–Low constraints with no cost in complexity workload. Conversion to real-valued symbols is done with the offset insertion in either the real or imaginary part of the original symbol, following the rule that a real pulse is immediately adjacent only with purely imaginary pulses in time and frequency.

In the FBMC-OQAM receiver, demodulation is performed by the following relationship:

$$\hat{a}_{m,n} = \Re\left\{ \int_{-\infty}^{\infty} h(t-nT/2)e^{-2\pi j\Delta f_c t} e^{-j\phi_{m,n}} s(t)dt \right\}$$

(5.20)

From Equation 5.20, it can be seen that only the projection of the real part of the signal onto the basis is taken into account. For ideal transmission, the symbols can be acquired with perfect reconstruction (orthogonality in time and frequency), if the pulse has the following property:

$$\Re\left\{ \int_{-\infty}^{\infty} h(t-nT_0/2)h(t-n'T_0/2)e^{2\pi j(m-m')\Delta f_c t} e^{j(\phi_{m',n'}-\phi_{m,n})} dt \right\} = \delta(m-m')\delta(n-n')$$

(5.21)

This relationship is the necessary condition to ensure orthogonality (in the presented analysis we assume that filter energy is normalized in unity to maintain the same average signal power).

5.4.1 FBMC-OQAM in the Discrete Time Domain

Critical time sampling is assumed, that is, $T_s = T/M$. The discrete baseband filter is $h(k)$, where k is the time variable/sample index. The filter is considered a causal, symmetrical, real-valued finite impulse response (FIR) filter in order to ensure linear phase response and exploit symmetry to reduce computational workload. The equivalent baseband discrete signal is given by

$$s(k) = \sum_{m=0}^{M-1} \sum_{n=-\infty}^{\infty} a_{m,n} h(k - nM/2) e^{\frac{2\pi jm\left(k - \frac{D-1}{2}\right)}{M}} e^{j\phi_{m,n}} \qquad (5.22)$$

where D is the number of filter coefficients. The reception is expressed as

$$\hat{a}_{m,n} = \Re\left\{ \sum_{k=-\infty}^{\infty} h(k - nM/2) e^{\frac{-2\pi j\left(k - \frac{D-1}{2}\right)m}{M}} e^{-j\phi_{m,n}} s(k) \right\} \qquad (5.23)$$

Orthogonality is achieved if:

$$\left\{ \sum_{k=-\infty}^{\infty} h(k - nM/2) h(k - n'M/2) e^{j(\phi_{m,n} - \phi_{m',n'})} e^{\frac{2\pi j(m-m')\left(k - \frac{D-1}{2}\right)}{M}} \right\} = \delta(m - m')\delta(n - n')$$

$$(5.24)$$

In FBMC-OQAM, transmission can be seen as the result of a synthesis filterbank with M subchannels operating in twice the critical sampling rate. In order to express transmission as the result of a synthesis filterbank, phase $\phi_{m,n}$ is expressed as (Siohan et al. 2002):

$$\phi_{m,n} = \frac{\pi}{2}(n + m) - \pi nm \qquad (5.25)$$

The definition in Equation 5.25 is different from Equation 5.18; however, the filtering result is not affected. The filterbank-modulator is defined as follows:

$$\text{Filters}: f_m(k) = h(k) e^{2\pi j \frac{m\left(k - \frac{D-1-M/2}{2}\right)}{M}}$$

$$m\text{-th Input}: x_m(n) = a_{m,n} e^{j\frac{\pi}{2}n} \qquad (5.26)$$

$$\text{Tx Signal}: s(k) = \sum_{m=0}^{M} \sum_{n=-\infty}^{\infty} x_m(n) f_m(k - nM/2)$$

If D is analyzed in two integers using the rule $D-1 = \alpha M/2 - \beta, (\alpha > 0, 0 \leq \beta < M/2)$, then reception is performed using the formula:

$$
\begin{aligned}
\hat{a}_{m,n-\alpha} &= \Re\left\{ e^{-j\frac{\pi}{2}(n-\alpha)} \sum_{k=-\infty}^{\infty} h(nM/2 - k - \beta)e^{j\frac{2\pi m\left(nM/2 - k + \frac{D-1-M/2}{2} - \frac{\alpha M}{2}\right)}{M}} s(k) \right\} \\
&= \Re\left\{ e^{-j\frac{\pi}{2}(n-\alpha)} \sum_{k=-\infty}^{\infty} h(nM/2 - k)e^{j\frac{2\pi m\left(nM/2 - k - \frac{D-1+M/2}{2}\right)}{M}} s(k-\beta) \right\}
\end{aligned}
\tag{5.27}
$$

This result is equivalent with the use of the following analysis filter bank at the receiver:

$$
\text{Filters} : g_m(k) = h(k)e^{2\pi j\frac{k - \frac{D-1+M/2}{2}}{M}} = f_m^*(D-1-k)
$$

$$
\text{Input} : u(k) = s(k-\beta)
\tag{5.28}
$$

$$
\text{Output:} \ \hat{a}_{m,n-\alpha} = \Re\left\{ y_m(n)e^{-j\frac{\pi}{2}(n-\alpha)} \right\}
$$

FBMC-OQAM implementation using filterbanks is presented in Figure 5.3 (Siohan et al. 2002). Filters are expressed using their z-transform.

Delay β is used to align phases between transmitter and receiver. The total filtering latency is α.

5.4.2 Polyphase Filters

The synthesis/analysis filters constitute a uniform DFT filterbank. The filterbank can be efficiently implemented using the polyphase decomposition that drastically reduces the computational workload (Harris 2004). The z-transform of the polyphase components of $h(k)$ is given by

$$
H(z) = \sum_{l=0}^{M-1} z^{-l} H_l(z^M)
$$

$$
H_l(z) = \sum_{n} h(l+nM)z^{-n}
\tag{5.29}
$$

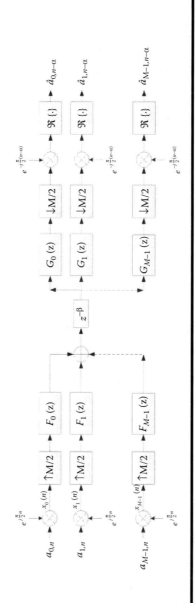

Figure 5.3 FBMC-OQAM transmission and reception filterbanks.

The synthesis and analysis filters can also be expressed as: ($W_M = e^{-j2\pi/M}$):

$$F_m(z) = \sum_{l=0}^{M-1} W_M^{-m\left(l-\frac{D-1-M/2}{2}\right)} z^{-l} H_l(z^M)$$

$$G_m(z) = \sum_{l=0}^{M-1} W_M^{-m\left(l-\frac{D-1+M/2}{2}\right)} z^{-l} H_l(z^M)$$

(5.30)

The term W_M indicates the existence of inverse and direct DFT in the synthesis and analysis banks, respectively. If we use the polyphase filters to assemble the uniform DFT bank, the FBMC-OQAM modulator and demodulator are implemented as depicted in Figure 5.4.

5.4.3 *Filter Specifications*

In order to ensure that orthogonality in time and frequency is achieved, the synthesis and analysis filterbanks should retain the perfect reconstruction (PR) property. According to (Vaidyanathan 1993), PR cannot be achieved for critically-sampled uniform DFT banks. Nevertheless, since the used filterbanks are not operating in the Nyquist rate, they have the same properties with cosine-modulated filterbanks (Nguyen and Koilpillai 1996) and, especially, the modified DFT (MDFT) filter banks (Karp and Fliege 1999). The necessary condition in order to obtain the PR feature for cosine-modulated and MDFT filterbanks is given by

$$H_k(z)\tilde{H}_k(z) + H_{k+M/2}(z)\tilde{H}_{k+M/2}(z) = \frac{2}{M}$$

(5.31)

where $H_k(z)$ is the polyphase filters and based on (Karp and Fliege 1999), $\tilde{H}_k(z)$ can be given by ($\beta \neq 0$):

$$\tilde{H}_k(z) = \begin{cases} \alpha \text{ odd} = z^{\frac{a-1}{2}} \begin{cases} H_{\frac{M}{2}-\beta-k}(z),\ k < \frac{M}{2}-\beta+1 \\ z^{-1}H_{M+\frac{M}{2}-\beta-k}(z),\ k \geq \frac{M}{2}-\beta+1 \end{cases} \\ \\ \alpha \text{ even} = z^{\frac{a-2}{2}} \begin{cases} H_{M-\beta-k}(z),\ k < M-\beta+1 \\ z^{-1}H_{2M-\beta-k}(z),\ k \geq M-\beta+1 \end{cases} \end{cases}$$

(5.32)

There is a plethora of algorithms that can be used to implement MDFT banks with the PR property (Koilpillai and Vaidyanathan 1992; Karp and Fliege 1996, 1999; Nguyen and Koilpillai 1996; Bolcskei and Hlawatsch 1998, Lu et al. 2004).

In Karp and Fliege (1999), an MDFT implementation is presented, where the oversampled uniform DFT bank of Figure 5.4 is divided in two critically sampled filterbanks. In Figures 5.5 and 5.6, we present a novel representation of the

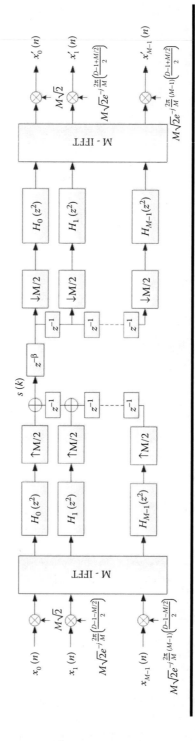

Figure 5.4 Implementation of FBMC-OQAM transmission/reception using polyphase filters.

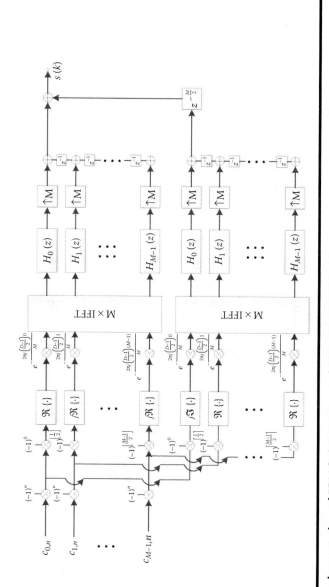

Figure 5.5 Implementation of FBMC-OQAM transmitter with MDFT bank.

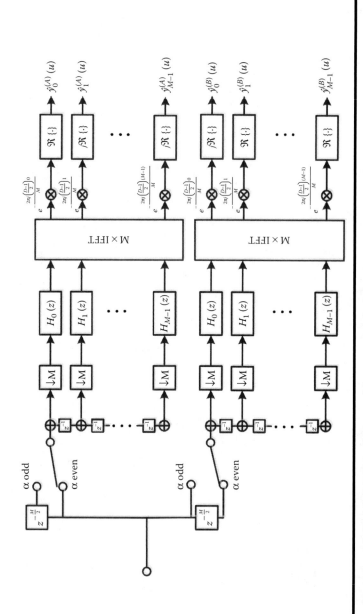

Figure 5.6   Structure of FBMC-OQAM receiver with MDFT analysis bank.

transmitter and the receiver, respectively, using causal filters and polyphase filters of Type 1 (Harris 2004), that follow the guidelines of (Karp and Fliege 1999).

The acquired symbols from the receiver of Figure 5.6 are obtained from the analysis bank outputs using the following formulas:

$$\alpha \text{ even, } \alpha' = \alpha/2:$$

$$\hat{a}_{m,2u-\alpha} = \Re\left(e^{-j\pi(u-\alpha')} e^{-j\frac{\pi m}{2}} y_A(u) \right) = (-1)^{u-\alpha'}(-1)^{\left\lfloor\frac{m}{2}\right\rfloor} \times \begin{cases} \Re(y_m^{(A)}(u)), \, m \text{ even} \\ \Im(y_m^{(A)}(u)), \, m \text{ odd} \end{cases}$$

$$\hat{a}_{m,2u-1-\alpha} = \Re\left(e^{-j\pi(u-\alpha'-1)} e^{-j\frac{\pi(m+1)}{2}} y_B(u) \right) = (-1)^{u-\alpha'-1}(-1)^{\left\lfloor\frac{m+1}{2}\right\rfloor} \times \begin{cases} \Im(y_m^{(B)}(u)), \, m \text{ even} \\ \Re(y_m^{(B)}(u)), \, m \text{ odd} \end{cases}$$

$$\alpha \text{ odd, } \alpha' = \lfloor \alpha/2 \rfloor:$$

$$\hat{a}_{m,2u-\alpha} = \Re\left(e^{-j\pi(u-\alpha')} e^{-j\frac{\pi(m-1)}{2}} y_A(u) \right) = (-1)^{u-\alpha'}(-1)^{\left\lfloor\frac{m-1}{2}\right\rfloor} \times \begin{cases} \Im(y_m^{(A)}(u)), \, m \text{ even} \\ \Re(y_m^{(A)}(u)), \, m \text{ odd} \end{cases}$$

$$\hat{a}_{m,2u-1-\alpha} = \Re\left(e^{-j\pi(u-\alpha'-1)} e^{-j\frac{\pi m}{2}} y_B(u) \right) = (-1)^{u-\alpha'-1}(-1)^{\left\lfloor\frac{m}{2}\right\rfloor} \times \begin{cases} \Re(y_m^{(B)}(u)), \, m \text{ even} \\ \Im(y_m^{(B)}(u)), \, m \text{ odd} \end{cases}$$

$$(5.33)$$

5.4.4 FBMC-OQAM in Real-World Conditions

The presented analysis proves that an FBMC-OQAM system can be orthogonal with the use of filters with PR property. Moreover, FBMC-OQAM also achieves maximum spectral efficiency. However, in a time-variant and frequency selective channel, the orthogonality is lost.

In real-world operation, filters without precise PR properties may perform better due to the radio channel effects, if they are well-localized in time and frequency. This means that an FBMC-OQAM filter should concentrate the maximum energy possible to the time-frequency point (m,n). Empirically, an FBMC-OQAM system is expected to operate properly, if:

$$B_D \ll \Delta f_c$$
$$\tau_{\max} \ll T$$
$$\hat{f} \ll B_C \qquad (5.34)$$
$$\hat{t} \ll T_C$$

where:

B_D is the Doppler Spread

T_C the channel coherence time

B_C the coherence bandwidth

$\tau_{\max}$ is the maximum excess delay

$$\hat{f}^2 = \frac{1}{E} \int_{-\infty}^{\infty} f^2 |H(f)|^2 \, df, \, \hat{t}^2 = \frac{1}{E} \int_{-\infty}^{\infty} t^2 |h(t)|^2 \, dt.$$

(E is the total filter energy and $H(f)$ is the filter Fourier transform.) Parameters $\hat{f}$ and $\hat{t}$ express the well-localization property of the filter.

In case one of the aforementioned conditions is not satisfied, then the issue of proper signal equalization becomes difficult. This is an important disadvantage of FBMC-OQAM compared to OFDM. With the use of the cyclic prefix and with low computational load, OFDM performs frequency-domain equalization (FEQ). FBMC-OQAM does not use cyclic prefix but it is quite tolerant in ISI and ICI due to the use of well-localized pulses. If the conditions of Equation 5.34 are satisfied, then FEQ can also be used in FBMC-OQAM with negligible distortion. Empirical tests have shown that well-localized filters provide FEQ support for $\tau_{max} \leq 0.1T$ and $B_D \leq 0.15\Delta f_c$. If the conditions of Equation 5.34 are not satisfied, then time domain equalization (TEQ) procedures should be applied for each individual subchannel (Nedic and Popovic 2002; Farhang-Boroujeny and Lin 2003; Lin et al. 2009). TEQ procedures require heavy computational workload. Therefore, FBMC-OQAM imposes specific constraints that depend on the radio channel coherence and dispersion measures, as well as the quality of the used filters.

5.4.5 Advantages and Disadvantages

FBMC-OQAM has the following advantages:

■ FBMC-OQAM has significantly lower sidelobes due to the use of well-localized filters. This means that the FBMC-OQAM waveform does not cause high adjacent channel interference. In addition, the FBMC-OQAM receivers have higher tolerance in adjacent channel interference. As an example, Figure 5.7 depicts the spectrum of the OFDM rectangular pulse versus an FBMC-OQAM pulse proposed in the context of the EU project PHYDYAS (Bellanger et al. 2010). The pulse impulse response is given by

$$h(t) = 1 + 2\sum_{k=1}^{K-1} H_k \cos\left(\frac{2\pi kt}{KT}\right) \tag{5.35}$$

with $K = 4$, $H_0 = 1$, $H_1 = 0.97196$, $H_2 = \sqrt{2}/2$, $H_3 = 0.235147$.

■ FBMC-OQAM is not using cyclic prefix. Hence, spectral and energy efficiency is not reduced. FBMC-OQAM is functional, if the used well-localized pulses satisfy the conditions of Equation 5.34.

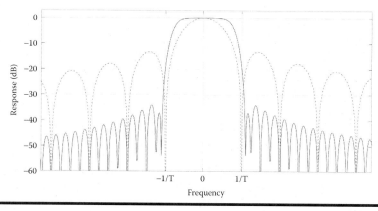

Figure 5.7   FBMC-OQAM pulse versus OFDM rectangular pulse.

■ FBMC-OQAM is not sensitive in frequency offsets due to the excellent frequency localization properties of the subchannel filters. In addition, FBMC-OQAM is significantly less sensitive than OFDM in higher Doppler spreads. Thus, FBMC-OQAM is robust and functional for high-user mobility and non-negligible frequency alignment errors.

The FBMC-OQAM disadvantages in comparison with OFDM are:

■ The implementation of the FBMC-OQAM transmitter and receiver requires much more computational workload. FBMC-OQAM contains DFT transforms in conjunction with a filtering operation. However, due to the fact that plenty of work has been done for improved FBMC-OQAM complexity (e.g., Dandach and Siohan 2011; Nadal et al. 2016, 2014) combined with the evolution in digital processor hardware, FBMC-OQAM is feasible and implementable in modern radio systems.

■ The FBMC-OQAM robustness in radio channel frequency-selectivity depends on the filter selection and the delay spread/coherence bandwidth of the channel. As no cyclic prefix exists, FBMC-OQAM allows ISI; however, due to the well-localization properties of the filters, ISI effects can be negligible. If ISI exceeds system tolerance, then computationally hard TEQ should be performed per subchannel.

■ The absence of cyclic prefix makes FBMC-OQAM sensitive in time synchronization errors. However, with the use of well-localized filters, FBMC-OQAM is able to remain functional for low values of timing offsets.

FBMC-OQAM has been in the center of activities for several EU research projects (FP7 Project EMPhAtiC 2015; FP7 Project METIS 2015; FP7 Project PHYDYAS 2010). This fact, along with its significant advantages, proves that FBMC-OQAM is a strong candidate for the role of the 5G waveform.

5.5 Universal Filtered Multicarrier

Universal Filtered Multicarrier (UFMC) (Vakilian et al. 2013) or Universal Filtered OFDM (UF-OFDM) (Wild et al. 2014) focuses on the drastic reduction of OFDM sidelobes. The main concept behind UFMC is to divide an OFDM waveform into components and filter each component individually.

Let us assume that the available bandwidth B contains N subcarriers. The overall bandwidth B is divided into M subbands. Each subband contains N_m subcarriers with $m = 0,...,M-1$. Thus, $N = \sum_{m=0}^{M-1} N_m$. Subcarrier groups are allowed to carry a different number of subcarriers; however, in the most common UFMC implementation, each group carries the same number of subcarriers, that is, $N_M = N_m = N/M$ for all m. This simplification allows the use of a critically sampled uniform DFT filterbanks that will drastically reduce the filtering computational workload with the use of polyphase decomposition. On the other hand, a critically sampled uniform DFT bank excludes the use of filters with the PR property. In the following analysis, we will assume that the subcarrier groups are of equal size.

If $\mathbf{x}$ is the vector containing the data symbols, then the subvectors $\mathbf{x}_m$ that contain the data symbols per subband are defined as

$$\mathbf{x} = [\mathbf{x}_0^T \mid \mathbf{x}_1^T \mid \quad ... \quad \mathbf{x}_{M-1}^T]^T \tag{5.36}$$

In Figure 5.8, the main procedures of the UFMC transmitter are presented.

For each subband, an N-point inverse DFT is performed separately. Each subvector $\mathbf{x}_m$ is placed in the same subcarrier indices as it was in the original vector $\mathbf{x}$. Subcarriers outside the given subband are zero-padded. If $\mathbf{0}_{N_m}$ is a $(N_m \times 1)$ vector of zeros, then the input vector for the inverse DFT of the lth subband is given by

$$\mathbf{x}_l^{iDFT} = [\mathbf{0}_0^T \quad ... \quad \mathbf{0}_{l-1}^T \quad \mathbf{x}_l^T \quad \mathbf{0}_{l+1}^T \quad ... \quad \mathbf{0}_{M-1}^T]^T \tag{5.37}$$

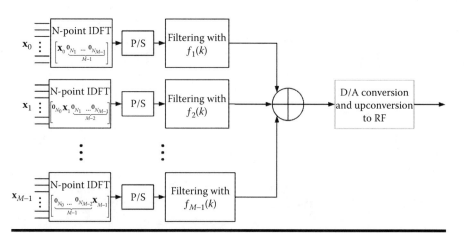

Figure 5.8 Block diagram of the UFMC transmitter operations.

Alternatively, the output of the inverse DFT of the *l*th subband is given by

$$z_k^{(l)} = \sum_{n=lN_M}^{(l+1)N_M-1} x_n e^{-2\pi j\frac{nk}{N}} \tag{5.38}$$

From Equation 5.38, it is clear, that despite the fact that M DFT operations are performed, the computational workload is not increased, as the signals are zero-padded. In fact, summation of the result of Equation 5.38 for all subbands will provide the OFDM equivalent signal.

At this point, data from all subbands have been transformed in the time domain (k is the sample index in the time domain). A parallel-to-serial block follows and each signal is filtered with an FIR filter with impulse response $f_m(k)$ $(m = 0,...,M-1)$. Each filter is designed as passband for the given subband. Assuming that all filters have the same number of coefficients L, then the output of the *l*th filter is given by

$$y_k^{(l)} = \sum_{i=0}^{L-1} f_l(i)z_{k-i}^{(l)} = \sum_{i=0}^{L-1} \sum_{n=lN_M}^{(l+1)N_M-1} f_l(i)x_n e^{-2\pi j\frac{n(k-i)}{N}} \tag{5.39}$$

The filtering result has $N + L - 1$ samples since linear convolution between $z_k^{(l)}$ and $f_l(k)$ is performed.

The filter is introduced in order to restrict the OOB emissions of the rectangular pulse in OFDM modulation. This is presented in the example of Figure 5.9, where the OFDM power spectral density is compared with the UFMC power spectral density ($N = 1024$, $M = 16$, $L = 50$).

In its original form, UFMC does not use cyclic prefix, although versions that consider the use of a cyclic extension have emerged (Zhang et al. 2016). UFMC uses an additional symbol duration for the $L - 1$ extra samples that occur due to

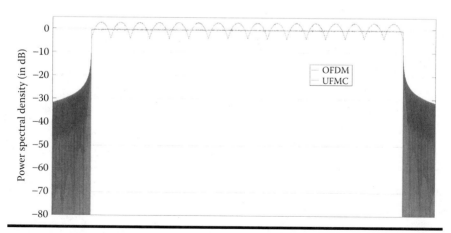

Figure 5.9 Sidelobe suppression in UFMC.

filtering. Generally, it is assumed that the filter length L will be in the order of the equivalent OFDM cyclic prefix. After filtering, all outputs are added to produce the UFMC waveform $y_k = \sum_{m=0}^{M-1} y_k^{(m)}$, with $k = 0,...,N + L - 1$. The total symbol duration is given by $T(1 + L/N)$.

Let us now assume that the signal is filtered by the radio channel with impulse response $p(k)$ of length U. Then, the received signal is given by

$$r_k = \sum_{u=0}^{U-1} p(u) y_{k-u} = \sum_{u=0}^{U-1} \sum_{m=0}^{M-1} p(u) y_{k-u}^{(m)} \qquad (5.40)$$

In Schaich and Wild (2014) and (Deliverable D3.2 5G Waveform Candidate Selection 2015), a windowing option is proposed. It is assumed that the incoming signal is weighted in the time domain with a window $w(k)$. As an example a raised-cosine window is proposed:

$$w(k) = \begin{cases} \dfrac{1}{2}\left(1 - \cos\left(\dfrac{k}{L/2-1}\pi\right)\right), & k = 0, \dots L/2 - 1 \\[2ex] 1, & k = L/2,...N + L/2 \\[2ex] \dfrac{1}{2}\left(1 + \cos\left(\dfrac{k-N-L/2+1}{L/2-1}\pi\right)\right), & k = L/2 + N + 1,...,N + L - 1 \end{cases} \qquad (5.41)$$

The window is introduced to relax timing synchronization requirements. Therefore, if we assume that a synchronization error exists, then the window will reduce ISI coming from the adjacent UFMC symbols. Nevertheless, the application of the window will also cause signal distortion. Windowing in the time domain is equivalent to circular convolution in the frequency domain. The effect of convolution will be to increase ICI, especially among subcarriers of the same subband. Hence, when strict timing synchronization requirements are considered, windowing should be avoided. The receiving operations are presented in Figure 5.10.

After windowing, the receiver has to process $N + L - 1$ samples. From Equation 5.40, it is seen that due to the radio channel each UFMC symbol extends to $N + L + U - 2$ samples. This extension in the absence of the cyclic prefix will cause ISI. Therefore, in UFMC ISI cannot be completely mitigated as in OFDM. The receiver either isolates $N + L - 1$ samples, or it uses the windowing procedure to extract them. Similarly to FBMC-OQAM, if the conditions of Equation 5.34 apply, then ISI will be negligible and FEQ can be used.

A $2N$-DFT operation is performed to the receiver as the UFMC symbol has been extended to $N + L - 1$ samples. The input of the DFT is zero-padded accordingly:

$$R_n = \frac{1}{\sqrt{N}} \sum_{k=0}^{N+L-2} r_k e^{-2\pi j \frac{kn}{2N}}, k = 0,1,..,2N - 1 \qquad (5.42)$$

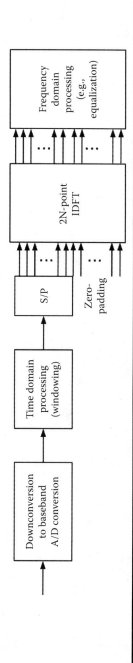

Figure 5.10   Block diagram of the UFMC receiver operations.

If we assume negligible ISI and ICI, then the received signal in the frequency domain can be expressed as

$$R_n = P(n) \sum_{m=0}^{M-1} F_m(n) \bar{X}_n + W(n) \tag{5.43}$$

where:

$P(n)$ and $F_m(n)$ is the discrete frequency transfer function of the radio channel m-th subband filter, respectively

$W(n)$ is AWGN

$\bar{X}_n$ are the $2\times$ frequency-oversampled frequency domain symbols. More specifically, if X_n are the transmitted frequency domain symbols then:

$$\bar{X}_n = \sum_{k=0}^{N-1} \left(\sum_{u=0}^{N-1} X_n e^{2\pi j \frac{ku}{N}} \right) e^{-2\pi j \frac{kn}{2N}} \tag{5.44}$$

It can be easily proved that $\bar{X}_{2n} = X_n$, which means that only the even frequency domain samples of $\bar{X}_n$ and consequently of R_n contain useful information. Odd samples contain interfering signals and they are dropped.

If the subband filters are able to isolate each subband successfully, then the result in Equation 5.43 can simplified even more:

$$R_n^{(m)} = P(n) F_m(n) X_n^{(m)} + W(n) \tag{5.45}$$

where $X_n^{(m)}$ are the symbols on the subcarriers of the mth subband and $R_n^{(m)}$ are the frequency domain samples for the mth subband. Subcarriers from other subbands are omitted. It is reminded that odd samples from Equation 5.43 have been dropped. FEQ can now be applied to acquire estimates of the transmitted symbols.

Problems in UFMC operation may arise if:

■ The filters fail to isolate each subchannel. Filtering breaks the orthogonality among subcarriers of different subbands. Poor filter design will cause ICI, especially for the edge subcarriers of each subband.

■ ISI and ICI will arise for highly selective and dispersive channels. Filter ramp-up and ramp-down will provide soft protection against highly frequency selective channels, because ISI will be attenuated. UFMC is expected to operate well, if the channel delay spread is less than a percentage of the filter length L.

Generally, UFMC reception for the mth subband can be expressed as

$$R_n^{(m)} = P(n) F_m(n) X_n^{(m)} - \beta X_n^{(m)} + I_{\text{ISI}} + I_{\text{ICI}} + W(n) \tag{5.46}$$

where:
I_{ISI} is the ISI interference
I_{ICI} is the ICI interference
β is a reduction factor due to lack of orthogonality

5.5.1 Advantages and Disadvantages

UFMC has the following advantages:

- Due to low OOB emissions, UFMC provides higher spectral efficiency than OFDM. In addition, UFMC seems more appropriate for use in white spaces and cognitive radios.
- Since no cyclic prefix is used, UFMC is more energy efficient than OFDM.
- Filtering in UFMC is performed in groups of subcarriers, unlike FBMC-OQAM where filtering is performed per subcarrier. This means that the filter length in UFMC is much shorter compared with FBMC-OQAM filters. As a result, UFMC is computationally more attractive and it is proved more suitable for low-latency, short-burst communications than FBMC-OQAM.
- UFMC can be seen as a linear combination of filtered OFDM signals. Thus, the majority of algorithms (e.g., channel estimation, synchronization) used in the well-documented OFDM can be directly applied in UFMC.
- UFMC is more robust than OFDM in frequency offsets and Doppler spreads due to filtering.

On the other hand, UFMC has the following disadvantages.

- It is inherently nonorthogonal. This means that under any circumstances interference between subcarriers will exist. Therefore, a noise floor appears that (depending on the quality of used filters) may impose constraints in the system performance.
- UFMC may not use cyclic prefix; however, the signal is extended by $L-1$ samples. This leads to spectral efficiency reduction compared with the performance of FBMC-OQAM.
- Similarly to FBMC-OQAM, UFMC provides soft protection from ISI. For highly selective channels, ISI levels will increase and FEQ may not be suitable for data symbol acquisition.
- UFMC is more sensitive in frequency offsets than FBMC-OQAM. The UFMC filter is not able to protect subcarriers from the same group when frequency alignment is lost.
- Since no cyclic prefix is used, UFMC is more sensitive in timing errors than OFDM. Windowing may be used to suppress their effects; however, ICI will be increased.

■ UFMC depends highly on internal tradeoffs. For example, high L will provide good subband isolation but it will decrease spectral efficiency. Windowing will relax timing requirements but will cause ICI, and so on.

UFMC was extensively studied from the EU project 5Gnow (5th-generation non-orthogonal waveforms for asynchronous signaling) and it was found more suitable for short-burst communications than other multicarrier modulation candidates.

5.6 Generalized Frequency Division Multiplexing

Generalized frequency division multiplexing (GFDM) was introduced in Fettweis et al. (2009). GFDM is a generalization of OFDM with use of nonrectangular pulses. Its implementation relies on a conventional digital filter-bank. It was initially proposed as a modulation scheme targeting on the design of cognitive radio waveforms able to occupy white spaces of fragmented licensed spectrum. A complete analysis of GFDM is presented in Michailow et al. (2014).

A basic differentiation of GFDM from OFDM and UFMC is the fact that it is a block modulation scheme. GFDM considers a time-frequency grid where multiple symbols are sequentially transmitted per subcarrier in time slots. If we assume that the block contains N data symbols, then these symbols are organized in a grid of K subcarriers and M time slots each, that is, $N = KM$. OFDM can be seen as a special case of GFDM where $M = 1$ (one time symbol per subcarrier) and the rectangular pulse is used.

5.6.1 The GFDM Transmitter

The data symbols form a $K \times M$ matrix $\boldsymbol{D}$, which is the GFDM block. The element $d_{k,m}$ of the matrix is the transmitted symbol at the mth time slot from the kth subcarrier. A prototype baseband filter $g(n)$ is defined. Each data symbol is transmitted using a time and frequency shifted version of $g(n)$ given by

$$g_{k,m}(n) = g((n - mK) \bmod N)e^{2\pi j \frac{kn}{K}} \tag{5.47}$$

It is noted that in the time domain circular shift is performed. The application of a circular pulse allows GFDM to embed the cyclic prefix in its operation and use FEQ due to the duality of circular convolution (in the time domain) with multiplication (in the frequency domain). The inverse DFT exponents modulate the filter in a given subcarrier. As GFDM considers a critically sampled system, the time index for the GFDM block is $n = 0, \ldots, N - 1$ (Figure 5.11).

The GFDM transmitted block is given by

$$x(n) = \sum_{k=0}^{K-1} \sum_{m=0}^{M-1} g_{k,m}(n) d_{k,n} \tag{5.48}$$

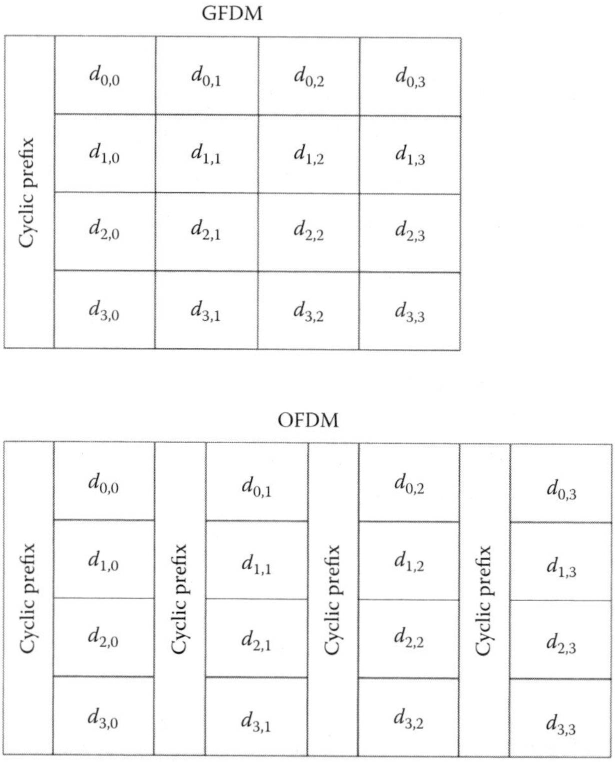

GFDM

Cyclic prefix				
	$d_{0,0}$	$d_{0,1}$	$d_{0,2}$	$d_{0,3}$
	$d_{1,0}$	$d_{1,1}$	$d_{1,2}$	$d_{1,3}$
	$d_{2,0}$	$d_{2,1}$	$d_{2,2}$	$d_{2,3}$
	$d_{3,0}$	$d_{3,1}$	$d_{3,2}$	$d_{3,3}$

OFDM

Cyclic prefix		Cyclic prefix		Cyclic prefix		Cyclic prefix	
	$d_{0,0}$		$d_{0,1}$		$d_{0,2}$		$d_{0,3}$
	$d_{1,0}$		$d_{1,1}$		$d_{1,2}$		$d_{1,3}$
	$d_{2,0}$		$d_{2,1}$		$d_{2,2}$		$d_{2,3}$
	$d_{3,0}$		$d_{3,1}$		$d_{3,2}$		$d_{3,3}$

Figure 5.11   GFDM blocks versus OFDM symbols ($K = 4$, $M = 4$).

In Gasper et al. (2013), a convenient GFDM transmitter model is proposed, where the transmitted vector **x** that contains $x(n)$ is expressed as

$$\mathbf{x} = \boldsymbol{A}\boldsymbol{d} \qquad (5.49)$$

where $\boldsymbol{d}$ is a vector that contains all elements of $\boldsymbol{D}$ row-wise, that is, $[\boldsymbol{d}]_0 = [\boldsymbol{D}]_{0,0}, [\boldsymbol{d}]_1 = [\boldsymbol{D}]_{0,1}, ..., [\boldsymbol{d}]_{K-1} = [\boldsymbol{D}]_{0,K-1}, [\boldsymbol{d}]_K = [\boldsymbol{D}]_{1,0} ....$ and so on. Moreover, if we collect all pulse samples $g_{k,m}(n)$ into a vector $\mathbf{g}_{k,m}$, then $\boldsymbol{A}$ is a $N \times N$ matrix given by

$$A = \begin{bmatrix} \boldsymbol{g}_{0,0} & \cdots & \boldsymbol{g}_{K-1,0} & \cdots & \boldsymbol{g}_{0,1} & \cdots & \boldsymbol{g}_{K-1,1} & \cdots & \boldsymbol{g}_{0,M-1} & \cdots & \boldsymbol{g}_{K-1,M-1} \end{bmatrix} \qquad (5.50)$$

5.6.2 Cyclic Prefix and OOB Radiation Control

As mentioned before, GFDM uses cyclic prefix to avoid ISI. Once again, the cost of cyclic prefix is spectrum and energy efficiency. On the other hand, the cyclic prefix transforms the linear convolution with the radio channel to circular enabling FEQ.

The use of the cyclic prefix and circular filtering introduces discontinuities, as the pulses at the edges of the block do not ramp-up/ramp-down. This leads to increased sidelobes and OOB emissions, despite the use of smoother pulses. Even so, GFDM outperforms OFDM in terms of OOB radiation. However, the use of windowed cyclic prefix is considered in order to reduce its sidelobes furthermore. The technique is known as pinching the block boundary (Bala et al. 2013).

When pinching the block boundary, we exploit the redundancy introduced by the cyclic prefix to apply a time domain window to the transmitted data. Each GFDM block in the time domain is multiplied by a window function $w(n)$, that provides a smooth fade-in and fade-out for the block.

In order to understand the technique, an example is presented (Gaspar 2016). Let us assume the following window function $w(n)$:

$$
w(n) = \begin{cases}
w_{up}(n), -N_W - N_{CP} \leq n < -N_{CP} \\
1, -N_{CP} \leq n < N - 1 \\
w_{down}(n), N \leq n < N + N_W \\
0
\end{cases}
\tag{5.51}
$$

where:

N_{CP} is the cyclic prefix length

N_W is the length of the cyclic extension expended with ramp up $w_{up}(n)$ and ramp down $w_{down}(n)$ smoothing functions

The initial block is multiplied with the window:

$$
\tilde{x}(n) = w(n)x(n)
\tag{5.52}
$$

In the receiver side, after the signal acquisition, the N_W front samples are extracted and added at the tail part of the received block. The procedure reestablishes circular periodicity, in a similar manner with zero-prefix OFDM (Van Waterschoot et al. 2010). The pinching the block boundary procedure is presented in Figure 5.12.

Another method to reduce OOB radiation is the insertion of guard symbols (Deliverable D3.2 5G Waveform Candidate Selection 2015), (Matthé et al. 2014). Assuming cyclic prefix length of $rK, r \in \mathbb{N}$, the signal will remain constant at the block boundaries, if symbols carried at time slots 0 and $M - r$ remain at a fixed value. Therefore, by setting the guard symbols to zero, the GFDM OOB emissions are strongly attenuated. The guard symbols introduce reduction in spectral efficiency by $(M - 2)/M$. Nevertheless, for M adequately high, the spectral efficiency loss is negligible.

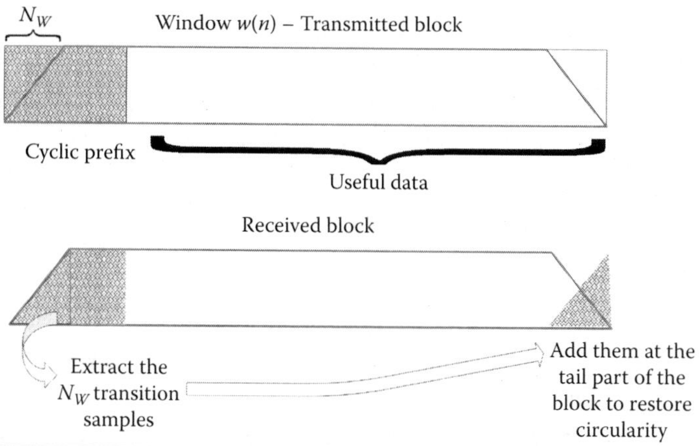

Figure 5.12   Pinching the block boundary.

5.6.3 The GFDM Receiver

The receiver obtains time and frequency synchronization. Then, following the OFDM demodulation steps, the cyclic prefix is removed. In addition, the cyclic prefix has transformed the linear convolution of the signal with the channel to circular. Assuming that matrix **H** is a circular convolution matrix that contains the impulse response of the radio channel, the received block is given by

$$\mathbf{y} = \mathbf{HAd} + \mathbf{w} \tag{5.53}$$

The OFDM-FEQ can now be applied via the DFT eigenvalue decomposition. Thus, after equalization the signal is given by (zero-forcing equalization is assumed):

$$\mathbf{z} = \mathbf{H}^{-1}\mathbf{HAd} + \mathbf{H}^{-1}\mathbf{w} \tag{5.54}$$

The symbol estimates can now be extracted with the following matrix multiplication:

$$\hat{\mathbf{d}} = \mathbf{Bz} \tag{5.55}$$

B is a $N \times N$ matrix that cancels the effects of the transmitting filter. Possible selections for **B** are:

- The matched filter, where $\mathbf{B} = \mathbf{A}^H$. The matched filter maximizes SNR; however, if the selected filters are nonorthogonal, then ICI between subcarriers will be introduced.
- Zero-forcing, where $\mathbf{B} = \mathbf{A}^{-1}$. Zero-forcing does not cause ICI; however, it enhances noise and it may cause severe SNR degradation. It is noted that in some cases, **A** is not invertible.

■ The linear minimum mean square error (MMSE) receiver, where $\mathbf{B} = (\mathbf{R}_w + \mathbf{A}^H \mathbf{A})^{-1} \mathbf{A}^H$. $\mathbf{R}_w$ is the noise autocorrelation matrix that is assumed known. For AWGN, $\mathbf{R}_w = \sigma_w^2 \mathbf{I}_N$ where σ_w^2 is the noise power. The MMSE receiver makes a compromise between ISI and noise enhancement.

5.6.4 Advantages and Disadvantages

The GFDM waveform has the following advantages:

■ With the use of proper techniques, GFDM provides significant gain in the reduction of OOB emissions compared with OFDM.

■ GFDM is more robust than UFMC and OFDM in frequency offsets and Doppler spread.

■ GFDM uses cyclic prefix and it is fully protected from ISI inflicted by the radio channel, if the channel impulse response length is less than the cyclic prefix.

■ As the cyclic prefix exists, GFDM can take advantage of OFDM techniques and algorithms that utilize the cyclic prefix.

■ OFDM is a special case of GFDM. Therefore, a simple way to design efficient algorithms is to attempt to generalize their results for GFDM.

■ For non-orthogonal pulses, there are receiver designs that can help balance between ICI-ISI and noise enhancement.

On the other hand GFDM has the following disadvantages:

■ GFDM without the use of guard symbols or pinching is clearly outperformed by UFMC and FBMC-OQAM in OOB radiation. The use of the aforementioned techniques will reduce OOB emissions; however, there is significant cost in spectral efficiency.

■ GFDM uses cyclic prefix and therefore it has the corresponding cost in spectral and energy efficiency.

■ Given the fact that each subcarrier is individually filtered, the filter length is generally larger than the corresponding length in UFMC, which leads to increased computational workload. Moreover, as GFDM benefits from large M values, it does not perform well for low-latency and short-burst communications.

■ Orthogonality (in time and frequency) is not given for GFDM. It is noted that orthogonality in the critical rate cannot be achieved through computational efficient structures (e.g., uniform DFT polyphase banks). Thus, GFDM has to balance in the trade-off between increased complexity and ICI-ISI tolerance.

5.7 Conclusion

In this chapter, an extensive presentation of the most popular multicarrier modulation techniques (OFDM, FBMC-OQAM, UFMC, and GFDM) was provided. Each modulation has various advantages and disadvantages that depend on the way they manage the tradeoffs defined by the Balian–Low theorem. Besides the presented conventional approaches, a large number of modifications is found in the literature, that attempt to balance efficiently among the defined tradeoffs. From windowed OFDM (Achaichia et al. 2011), unique word OFDM (UW-OFDM [Hofbauer et al. 2010]) and WCP-COQAM (Lin and Siohan 2014) to TS-OQAM-GFDM (Gasper et al. 2015) and single-carrier extensions (SC-FDMA [Long-Term Evolution (LTE) standard 2008]), research activity toward the definition of the 5G waveform is blossoming. Will the road to the 5G waveform be *evolutionary* with continuation of OFDM dominance due to its implementation advantages, or *revolutionary* with the definition of an OFDM successor that balances efficiently among the challenges?

References

5th Generation non-orthogonal waveforms for asynchronous signalling. http://www.5gnow. eu/.FP7 project.

Andrews J. G., Buzzi, S., Choi, W., Hanly, S. V., Lozano, A., Soong, A. C. K., and Zhang, J. C. What will 5G be? *IEEE Journal on Selected Areas in Communications*, 32(6):1065–1082, 2014.

Achaichia, P., Le Bot, M., and Siohan, P. Windowed OFDM versus OFDM/OQAM: A transmission capacity comparison in the HomePlug AV context. In *Power Line Communications and Its Applications (ISPLC), 2011 IEEE International Symposium on*, pp. 405–410. IEEE, 2011.

Bala, E., Li, J., and Yang, R. Shaping spectral leakage: A novel low-complexity transceiver architecture for cognitive radio. *IEEE Vehicular Technology Magazine*, 8(3):38–46, 2013.

Banelli, P., Buzzi, S., Colavolpe, G., Modenini, A., Rusek, F., and Ugolini, A. Modulation formats and waveforms for 5G networks: Who will be the heir of OFDM?: An overview of alternative modulation schemes for improved spectral efficiency. *IEEE Signal Processing Magazine*, 31(6):80–93, 2014.

Bellanger, M., Le Ruyet, D., Roviras, D., Terré, M., Nossek, J., Baltar, L., Bai, Q., Waldhauser, D., Renfors, M., Ihalainen T. et al. FBMC physical layer: A primer. *PHYDYAS*, 2010.

Bockelmann, C., Pratas, N., Nikopour, H., Au, K., Svensson, T., Stefanovic, C., Popovski, P., and Dekorsy, A. Massive machine-type communications in 5G: Physical and MAC-layer solutions. *IEEE Communications Magazine*, 54(9):59–65, 2016.

Bolcskei, H. and Hlawatsch, F. Oversampled cosine modulated filter banks with perfect reconstruction. *IEEE Transactions on Circuits and Systems II: Analog and Digital Signal Processing*, 45(8):1057–1071, 1998.

Deliverable D3.2 5G Waveform Candidate Selection. FP7 Project - 5th Generation non-orthogonal waveforms for asynchronous signalling. http://www.5gnow.eu/, 2015.

Dandach, Y. and Siohan. P. FBMC/OQAM modulators with half complexity. In *2011 IEEE Global Telecommunications Conference–GLOBECOM 2011*, pp. 1–5, 2011.

De La Oliva, A., Pérez, X. C., Azcorra, A., Di Giglio, A., Cavaliere, F., Tiegelbekkers, D., Lessmann, J., Haustein, T., Mourad, A., and Iovanna, P. Xhaul: Toward an integrated fronthaul/backhaul architecture in 5G networks. *IEEE Wireless Communications*, 22(5):32–40, 2015.

Du, J. and Signell, S. Classic OFDM systems and pulse shaping OFDM/OQAM systems. Technical Report 07:01, KTH, Electronic, Computer and Software Systems, ECS, 2007.

Farhang-Boroujeny, B. and Lin, L. Analysis of post-combiner equalizers in cosine-modulated filterbank-based transmultiplexer systems. *IEEE Transactions on Signal Processing*, 51(12):3249–3262, 2003.

Feichtinger, H. G. and Strohmer, T. *Gabor Analysis and Algorithms: Theory and Applications*. New York: Springer Science & Business Media, 2012.

Fettweis, G., Krondorf, M., and Bittner, S. GFDM-generalized frequency division multiplexing. In *Vehicular Technology Conference, 2009. VTC Spring 2009. IEEE 69th*, pp. 1–4. IEEE, 2009.

FP7 Project EMPhAtiC. Enhanced multicarrier techniques for professional ad-hoc and cell-based communications. http://www.ict-emphatic.eu/, 2015.

FP7 Project METIS. Mobile and wireless communications enablers for twenty-twenty (2020) information society. https://www.metis2020.com/, 2015.

FP7 Project PHYDYAS. Physical layer for dynamic spectrum access and cognitive radio. http://www.ict-phydyas.org/, 2010.

Gaspar, I. S. Waveform advancements and synchronization techniques for generalized frequency division multiplexing. PhD thesis, Technishe Universitat Dresden, 2016.

Gaspar, I., Matthé, M., Michailow, N., Mendes, L. L., Zhang, D., and Fettweis, G. Frequency-shift offset-QAM for GFDM. *IEEE Communications Letters*, 19(8):1454–1457, 2015.

Gaspar, I., Michailow, N., Navarro, A., Ohlmer, E., Krone, S., and Fettweis, G. Low complexity GFDM receiver based on sparse frequency domain processing. In *Vehicular Technology Conference (VTC Spring), 2013 IEEE 77th*, pp. 1–6. IEEE, 2013.

Han, S. H. and Lee, J. H. An overview of peak-to-average power ratio reduction techniques for multicarrier transmission. *IEEE Wireless Communications*, 12(2):56–65, 2005.

Harris, F. *Multirate Signal Processing for Communication Systems*. Upper Saddle River, NJ: Prentice Hall, 2004.

Hobert, L., Festag, A., Llatser, I., Altomare, L., Visintainer, F., and Kovacs, A. Enhancements of V2X communication in support of cooperative autonomous driving. *IEEE Communications Magazine*, 53(12):64–70, 2015.

Hofbauer, C., Huemer, M., and Huber, J. B. Coded OFDM by unique word prefix. In *Communication Systems (ICCS), 2010 IEEE International Conference*, pp. 426–430. IEEE, 2010.

Hirosaki, B. An orthogonally multiplexed QAM system using the discrete Fourier transform. *IEEE Transactions on Communication*, 29(7):982–989, 1981.

IEEE Std 802.11-2007. IEEE standard for information technology—Telecommunications and information exchange between systems—Local and metropolitan area networks—Specific requirements—Part 11: Wireless LAN medium access control (MAC) and physical layer (PHY) specifications, June 2007.

Jungnickel, V., Manolakis, K., Zirwas, W., Panzner, B., Braun, V., Lossow, M., Sternad, M., Apelfrojd, R., and Svensson, T. The role of small cells, coordinated multipoint, and massive MIMO in 5G. *IEEE Communications Magazine*, 52(5):44–51, 2014.

Karp, T. and Fliege, N. J. Computationally efficient realization of MDFT filter banks. In *European Signal Processing Conference, 1996. EUSIPCO 1996. 8th*, pp. 1–4. IEEE, 1996.

Karp, T. and Fliege, N. J. Modified DFT filter banks with perfect reconstruction. *Circuits and Systems II: Analog and Digital Signal Processing, IEEE Transactions on*, 46(11):1404–1414, 1999.

Koilpillai, R. D. and Vaidyanathan, P. P. Cosine-modulated FIR filter banks satisfying perfect reconstruction. *IEEE Transactions on Signal Processing*, 40(4):770–783, 1992.

Larsson, E. G., Edfors, O., Tufvesson, F., and Marzetta, T. L. Massive MIMO for next generation wireless systems. *IEEE Communications Magazine*, 52(2):186–195, 2014.

Li, Y. G. and Stuber, G. *Orthogonal Frequency Division Multiplexing for Wireless Communications*. New York: Springer Science & Business Media, 2006.

Lin, H. and Siohan, P. Multi-carrier modulation analysis and WCP-COQAM proposal. *EURASIP Journal on Advances in Signal Processing*, 2014(1):79, 2014.

Lin, H., Siohan, P., Tanguy, P., and Javaudin, J. P. An analysis of the EIC method for OFDM/OQAM systems. *Journal of Communications*, 4(1), 2009.

Lu, W. S., Saramaki, T., and Bregovic, R. Design of practically perfect-reconstruction cosine-modulated filter banks: A second-order cone programming approach. *IEEE Transactions on Circuits and Systems I: Regular Papers*, 51(3):552–563, 2004.

Long Term Evolution (LTE) standard. 3GPP TS 36.211 Evolved universal terrestrial radio access (E-UTRA) physical channels and modulation (release 8), December 2008.

Maliatsos, K., Adamis, A., and Constantinou, P. SNR degradation due to timing and frequency synchronization errors for OFDMA systems with subband carrier allocation. In *2008 14th European Wireless Conference*, pp. 1–6, IEEE, 2008.

Maliatsos, K., Adamis, A., and Kanatas, A. G. Interference versus filtering distortion trade-offs in OFDM-based cognitive radios. *Transactions on Emerging Telecommunications Technologies*, 24(7–8):692–708, 2013.

Mallat, S. *A Wavelet Tour of Signal Processing*. San Diego, CA: Academic Press, 1999.

Matthé, M., Michailow, N., Gaspar, I., and Fettweis, G. Influence of pulse shaping on bit error rate performance and out of band radiation of generalized frequency division multiplexing. In *Communications Workshops (ICC), 2014 IEEE International Conference on*, pp. 43–48. IEEE, 2014.

Mavromoustakis, C. X., Mastorakis, G., and Batalla, J. M. Internet of things (IoT) in 5G mobile technologies. *Modeling and Optimization in Science and Technologies*, 56:93, 2016.

Militano, L., Araniti, G., Condoluci, M., Farris, I., and Iera, A. Device-to-device communications for 5G internet of things. *IOT, EAI*, 2015.

Muquet, B., de Courville, M., Dunamel, P., and Giannakis, G. OFDM with trailing zeros versus OFDM with cyclic prefix: Links, comparisons and application to the HIPERLAN/2 system. In *2000 IEEE International Conference on Communications. ICC 2000*, vol. 2, pp. 1049–1053. IEEE, 2000.

Michailow, N., Matthé, M., Gaspar, I. S., Caldevilla, A. N., Mendes, L. L., Festag, A., and Fettweis, G. Generalized frequency division multiplexing for 5th generation cellular networks. *IEEE Transactions on Communications*, 62(9):3045–3061, 2014.

Nadal, J., Nour, C. A., and Baghdadi, A. Low-complexity pipelined architecture for FBMC/OQAM transmitter. *IEEE Transactions on Circuits and Systems II: Express Briefs*, 63(1):19–23, 2016.

Nadal, J., Nour, C. A., Baghdadi, A., and Lin, H. Hardware prototyping of FBMC/OQAM baseband for 5G mobile communication. In *Proceedings of the 2014 25th IEEE International Symposium on Rapid System Prototyping*, New Delhi, India, vol. 1617, pp. 7277, 2014.

Nedic, S. and Popovic, N. Per-bin DFE for advanced OQAM-based multi-carrier wireless data transmission systems. In *Broadband Communications, 2002. Access, Transmission, Networking. 2002 International Zurich Seminar on*, pp. 38–1–38–6, 2002.

NGMN Alliance. 5G white paper. Next generation mobile networks, white paper, 2015.

Nguyen, T. Q. and Koilpillai, R. D. The theory and design of arbitrary-length cosine-modulated filter banks and wavelets, satisfying perfect reconstruction. *IEEE Transactions on Signal Processing*, 44(3):473–483, 1996.

Pollet, T., Van Bladel, M., and Moeneclaey, M. BER sensitivity of OFDM systems to carrier frequency offset and Wiener phase noise. *IEEE Transactions on Communications*, 43(234):191–193, 1995.

Saltzberg, B. Performance of an efficient parallel data transmission system. *IEEE Transactions on Communication Technology*, 15(6):805–811, 1967.

Sabharwal, A., Schniter, P., Guo, D., Bliss, D. W., Rangarajan, S., and Wichman, R. In-band full-duplex wireless: Challenges and opportunities. *IEEE Journal on Selected Areas in Communications*, 32(9):1637–1652, 2014.

Schaich, F. and Wild, T. Relaxed synchronization support of universal filtered multi-carrier including autonomous timing advance. In *Wireless Communications Systems (ISWCS), 2014 11th International Symposium on*, pp. 203–208. IEEE, 2014.

Siohan, P., Siclet, C., and Lacaille, N. Analysis and design of OFDM/OQAM systems based on filterbank theory. *IEEE Transactions on Signal Processing*, 50(5):1170–1183, 2002.

Simsek, M., Aijaz, A., Dohler, M., Sachs, J., and Fettweis, G. 5G-enabled tactile internet. *IEEE Journal on Selected Areas in Communications*, 34(3):460–473, 2016.

Uhlemann, E. Introducing connected vehicles [connected vehicles]. *IEEE Vehicular Technology Magazine*, 10(1):23–31, 2015.

Vaidyanathan, P. P. *Multirate Systems and Filter Banks*. Englewood Cliffs, NJ: Prentice Hall, 1993.

Vakilian, V., Wild, T., Schaich, F., ten Brink, S., and Frigon, J. F. Universal-filtered multi-carrier technique for wireless systems beyond LTE. In *GLOBECOM-2013*, Atlanta, GA, December 2013.

Van Waterschoot, T., Le Nir, V., Duplicy, J., and Moonen, M. Analytical expressions for the power spectral density of CP-OFDM and ZP-OFDM signals. *IEEE Signal Processing Letters*, 17(4):371–374, 2010.

Wild, T., Schaich, F., and Chen. Y. 5G air interface design based on universal filtered (UF-) OFDM. In *19th International Conference on Digital Signal Processing (DSP-2014)*, Hong Kong, August 2014.

Wu, J., Zhang, Z., Hong, Y., and Wen, Y. Cloud radio access network (C-RAN): A primer. *IEEE Network*, 29(1):35–41, 2015.

Zhang, L., Xiao, P., and Quddus, A. Cyclic prefix-based universal filtered multicarrier system and performance analysis. *IEEE Signal Processing Letters*, 23(9):1197–1201, 2016.

Chapter 6

RF Planning for Next-Generation Systems

George V. Tsoulos, Georgia E. Athanasiadou,
Dimitra Zarbouti, and Ioannis Valavanis

Contents

6.1 Introduction

The process of designing a radio network structure and determining network elements subject to different constraints is generally referred to as radio network planning and can be associated with network dimensioning and detailed planning.

Moreover, radio network planning optimization is the task that leads to a network configuration that achieves the best possible performance. Its development through time is mapped to the development of the radio access technologies and their corresponding limitations and capabilities. The first analogue radio networks had low capacity requirements and hence, radio network planning was entirely based on wide area coverage. As a consequence, omnidirectional antennas were used at base stations (BSs) positioned as high as possible (macrocells) so that the site density is kept low. For this reason, simple empirical pathloss propagation models such as Okumura–Hata and COST231 [1,2] were used for macrocellular coverage radio network planning. These were the early days of the over-simplified planning.

Together with the evolution of 2G digital systems, the site density increased due to higher capacity requirements, which also forced the transition of the cellular networks to cell splitting or sectorization. Strict requirements for increased spectral efficiency led to the adoption of several interference control mechanisms such as antenna tilting [3] and smart antennas [4].

Initially, the planning task involved only two issues: the number of BSs and their locations, also known as the radio BS location problem. In terms of the optimization process, which usually includes cost optimization, it is one of the classical problems in the field of mathematical constrained optimization, along with constraints imposed by the radio access technology. Note that even the single BS location problem may be *NP-hard* in continuous space [5].

The radio BS location problem [6–8], the frequency assignment problem (minimum cost-frequency assignment subject to assignment and interference constraints [9,10]), and the topological network design (minimum cost network topology able to connect the candidate BSs to a fixed telephone network [11]), are the three most widely studied problems in the context of radio network planning and optimization.

Although for 2G cellular systems, the network planning process could be divided into two generally unrelated processes, namely the coverage and capacity planning stages, this is not the case for 3G, 4G, and 5G systems, where capacity and coverage planning are highly interrelated and typically should be simultaneously treated. In 2012, ~10% of the population was covered by 4G LTE systems, while projections show that this percentage will reach ~60% by 2018 [12] and by 2020, 50% of all broadband mobile subscriptions (~3.8 billions) will be LTE subscribers [13]. The plethora of data services (web browsing, social networking, audio, video, software downloading, etc.) offered even by current 4G LTE deployments, when combined with capacity "hungry" future applications (either in terms of required data rate, or in terms of numbers of supported users-devices, or both), lead to the need of an accurate forecast and geographical mapping of user throughput requirements, which is rather difficult to achieve. In addition, these requirements lead to mixed cell-structure deployments that comprise both high (macro) and low (relay, micro, femto) network nodes [14,15], as well as heterogeneous wireless system deployments.

However, small cell operation and deployment density is a challenging task that directly affects the overall interference in the network, and hence, they have a major impact on the allocation and configuration of macrocells, especially when considered in the context of mixed or hierarchical cell structures. Furthermore, mixed cell structures entail major changes for the cell planning paradigm, especially when seen in conjunction with energy efficiency and total cost.

The overall cost and complexity of a network directly relates to the number of BSs required to achieve the objectives of the provided services. Positioning BSs is not an easy task, while the optimal network deployment requires solving a complex optimization problem of combinatorial nature.

Although BS positioning dates back to the deployment of the first cellular wireless systems, the requirement for geographically inhomogeneous capacity provision, nowadays, has led to the adoption of hierarchical cell structures together with relays, where the cost significantly varies for the different deployments. One of the key characteristics of these systems with regards to the BS location issue is that the optimum solution can range from a small group of macrocells to a large set of microcells.

Overall network planning and optimization is an important research area with a considerable amount of published work ranging from BS location and coverage planning to antenna parameter configuration and cell load balancing [16–18]. Moreover, green radio network planning via optimization of BS antenna patterns and transmitted powers in heterogeneous 4G networks with geographically inhomogeneous throughput requirements is not an easy problem to tackle, and hence, frequently omitted. Moreover, the initial phase of network dimensioning and planning often fails to meet growing throughput demands, and hence, network operation becomes problematic, especially around hotspots (HSs).

In LTE systems where interference strongly interrelates capacity and coverage, network planning becomes even more complicated given this new constraint. In order to deal with such a complicated planning task, these criteria are most frequently targeted within an optimization framework. Related research targets, among other things, the optimal configuration of BS locations, antenna parameters, relaying, energy consumption, and cell load balancing [19–22]. Optimization of antenna-related parameters (e.g., TX power, antenna tilting, and azimuth orientations of sectors) in LTE systems appeared recently in [23–25]. Nature-inspired optimization methods, in general, and evolutionary algorithms, more specifically, have recently gained interest in this field [20,23].

The rest of the chapter is organized as follows:

First, Section 6.2 describes the BS positioning problem in the context of mixed cell structures, Section 6.3 presents the simulation methodology, which includes the propagation, system and cost modules and Section 6.4 the BS location optimization algorithm.

In the context of this chapter, a rigorous generic LTE system model is employed that works for a general network topology and explicitly accounts for mixed cell structures and relay nodes (RNs) as well as nonuniform traffic demand (HSs) [26,27]. Due to the modular approach that the simulation tool is developed, it provides the flexibility to consider more complex scenarios, for example, in the context of 5G systems with another underlying system model. In addition, a partially combinatorial optimization algorithm is used that exploits specific characteristics of the LTE system in order to automatically identify an optimum solution that meets the operator's requirements in terms of coverage and capacity in a given area, with the minimum number of BSs and hence, cost.

Following the BS location optimization analysis, Section 6.5 formulates the planning optimization problem with a multiobjective approach, and then describes in detail the green planning optimization algorithm.

Given the optimal BS locations, a well-established, multiobjective evolutionary algorithm (nonsorting dominated genetic algorithm-II, NSGA-II) is then used in order to identify in a stochastic manner the optimum solution for macro and micro BSs planning in terms of transmitted power settings, panning, and fanning, that is, altering the sector azimuth orientation and beamwidth, respectively. The algorithm can be used to provide an optimal LTE network configuration, that minimizes the total transmitted power in the test area given the coverage and capacity criteria. In order to test the presented methodology, scenarios of varying capacity requirements are employed, featuring a number of control nodes (CNs) and HSs. Given the emerging capacity problems caused most frequently by HSs, the optimization algorithm is applied to reconfigure the BS radiation patterns (it optimizes the BS antenna patterns with respect to their pointing direction, 3 dB beamwidth and transmitted power) and thus satisfy the set criteria using the most energy-efficient network setup with varying area capacity requirements.

Section 6.6 presents results both for location optimization and green planning. Results shown in Section 6.6.1 demonstrate the flexibility of the proposed BS location optimization algorithm to achieve optimization for the planning triplet (coverage, capacity, cost), as well as a trend for solutions with very small cells if increased capacity is required.

Results in Section 6.6.2 show that the reconfigured network setup featured neatly adjusted radiation patterns, increased capacity capability, reduced network cost and energy consumption, and, most importantly, improved public safety with regards to reduced power emissions. Hence, the overall presented analysis achieves optimization for the quadruplet (coverage, capacity, cost, TX power).

6.2 The Cellular Layout

Naturally, the initial set of candidate BS locations is user supplied since they are usually locations where the operator has obtained planning permissions. It must be clarified here that the initial set of sites determines to a great extent the final solution as no additional locations are identified or added during the optimization process and the user selected positions are not modified. Nevertheless, this is the

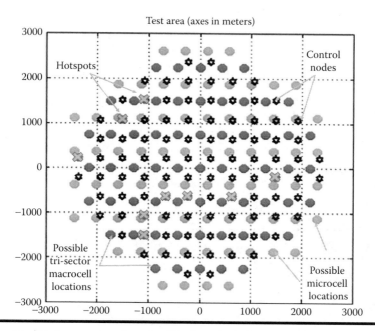

Figure 6.1 **Test area showing possible locations for sectorized macrocell BSs, micro BSs, control nodes, and hotspots.**

most meaningful choice because real-world radio network planning and site positioning are limited mainly from practical issues. A number of user supplied points or "control nodes" (CNs) are used to represent the capacity and coverage requirements in the operational area [26–28].

Figure 6.1 shows a typical scenario employed to test the optimization algorithm. The BS and the CN positions are almost uniformly distributed in a circular area of ~2.5 km² radius, that is, the total urban area under study is ~20 Km². Both uniform and nonuniform service requirements can be accommodated by choosing the percentage of CNs with specific data rate requirements.

Figure 6.2 shows that each macro BS employs three sectors pointing toward 30°, 150°, and 270° in azimuth, while the radiation pattern from [30] is used for each sector. Between the different sectors, as well as in the directions of maximum power, each macrocell can also have omnidirectional RNs. If a CN cannot be connected directly to a BS, then an RN is activated and the CN connects to a BS via the RN (mainly due to higher SINR and hence, lower bandwidth requirements, as it will be explained later). On the other hand, micro BSs are omnidirectional and they are not combined with RNs. The RNs are placed 800 m away from the BS, there are 96 CNs, and the candidate nodes are 74 microcells and 61 macrocells with 366 RNs in the test area.

Relay node (RN) positions (2 per sector)

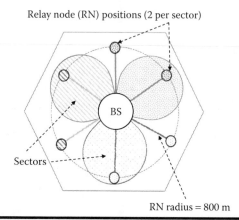

RN radius = 800 m

Figure 6.2 The sector orientation and the position of possible RNs for each tri-sector macrocell.

6.3 The Simulation Methodology

The simulation tool that was developed in order to solve the BS location optimization problem consists of four main modules.

6.3.1 The Propagation Module

It employs the WINNER model for macro, micro, and RNs scenarios with the LOS distance set to 500 m [31]. The propagation parameters employed in the simulations are summarized in Table 6.1. It must be emphasized here that more recent versions of the proposed tool involve deterministic propagation models with geographical databases, in order to tackle the problem of the site specific nature of radiowave propagation characteristics in smaller cell operational environments.

Table 6.1 Propagation Models

Macro BS Propagation model	WINNER Path Loss Models: Scenario *C2 NLOS* with $h_{Tx} = 25$ m
Micro BS Propagation model	WINNER Path Loss Models: Scenario *B1 NLOS* with $h_{Tx} = 10$ m, $d_{LOS} = 500$ m
Relay Node Propagation model	WINNER Path Loss Models: Scenario *B1 NLOS* with $h_{Tx} = 5$ m, $d_{LOS} = 500$ m

6.3.2 The System Module

We consider the downlink of an LTE heterogeneous network, that is, a mixed architecture with macrocells and microcells. Each microcell site comprises one BS, while each macrocell three sectors with one BS each. Assume that B_m is the set of all micro BSs, B_M the set of all macro BSs and B is their union representing all the BSs available in the network area. The micro BSs employ omnidirectional antennas, that is, the normalized antenna gain is $A_b = 1, \forall b \in B_m$, while the macro BSs are equipped with directional $120°$ sector antennas and their normalized antenna pattern is produced with $A_b = -\min(12(\phi/\phi_{3dB})^2 - 25\,\mathrm{dB})\forall b \in B_m$ [32]. The default value for the combined (elevation and azimuth) maximum antenna gain for all macro BSs is set to $A_{\max}^{\mathrm{Macro}} = 18$ dBi when $\phi_{3dB} = 70°$ and $\theta_{3dB} = 10°$. The default pointing directions for the sectors are $\{\phi_0^{b,1}, \phi_0^{b,2}, \phi_0^{b,3}\} = \{30°, 150°, 270°\}, \forall b \in B_m$ and the 3 dB beamwidth is set to $\phi_{3dB}^{b,s} = 70°, \forall b \in B_m, s \in \{1,2,3\}$. More details on the system parameters used in simulations can be found in Table 6.2. It should be pointed out that for the fanning process, the maximum gain of each sector $(A_{\max}^{\mathrm{Macro}})$ changes according to the ϕ_{3dB}. For instance, when the antenna pattern narrows down to $\phi_{3dB} = 40°$ the maximum gain is ~20.5 dB.

Each BS reuses the entire system bandwidth. The coverage and throughput requirements in the network area are defined by the CNs and HSs. Let us

Table 6.2 Simulation Parameters

Macro BS Tx power P_{tx}^{Macro}	46 dBm default or reconfigurable for planning purposes
Macro BS Antenna Gain (Sector) $A_{\max}^{\mathrm{Macro}}$	18 dBi
Micro BS Tx power P_{tx}^{Micro}	33 dBm default or reconfigurable for planning purposes
Micro BS Antenna Gain (Omnidirectional) $A_{\max}^{\mathrm{Micro}}$	4 dBi
Relay Node Tx power	30 dBm
Relay Node Antenna Gain (Omnidirectional)	5 dBi
User antenna pattern (Omnidirectional)	0 dBi
Total cable losses	2 dB
Central frequency	2.12 GHz
Bandwidth	20 MHz
Noise	−174 dBm/Hz

consider K CNs (or HSs) distributed in the network area, where the SINR at the kth CN (or HS) is

$$\text{SINR}_k = \frac{P_{tx}^{b_k} A_{b_k}^{\max} A_{b_k}^{\phi \rightarrow k} G_{b_k,k}^{ch}}{\displaystyle\sum_{b \in \{B \setminus b_k\}} P_{tx}^{b} A_{b}^{\max} A_{b_k}^{\phi \rightarrow k} G_{b_k,k}^{ch} + P_N} \tag{6.1}$$

where:

b_k is the serving BS for the kth CN (or HS)

$P_{tx}^{b_k}$ and $A_{b_k}^{\max}$ are the transmitted power and maximum antenna gain for the b_k BS, respectively

$A_{b_k}^{\phi \rightarrow k}$ is the normalized antenna gain of the b_k BS toward the kth CN

$G_{b_k,k}^{ch}$ is the channel gain between them. Similarly, for the interference calculation

$A_{b}^{\phi \rightarrow k}$ is the normalized antenna gain of the bth interfering BS toward the served CN

P_N is the noise power at the receiver

Each CN is associated with a target throughput requirement. But the throughput requirements of the kth CN (or HS), r_k, corresponds to a specific number of resource blocks (RBs), n_k, that the serving BS, b_k, has to allocate to this link. Given that each BS has in total 100 RBs to allocate (20 MHz LTE system), the optimization process proposed in this chapter takes into account that there is a limitation for the number of CNs that each BS can serve. This limitation is considered for both optimization problems formulated in Sections 6.4 and 6.5.

The number of required RBs per CN (or HS), n_k, depends on the channel quality indicator (CQI) and rank index (RI) feedback [33] of the CN (or HS) along with its specific rate requirement (r_k). In the context of this work, [34] is used in order to map the calculated SINR of Equation 6.1 to a CQI value, and Table 7.2.3-1 of [33] in order to map the CQI value to a specific modulation order (Q_m) and CR.

More specifically:

1. The SINR calculated at the CN is mapped to a single CQI value using [34].
2. The CQI is mapped to a proper modulation level (Q_m) and CR. This mapping is derived according to Table 7.2.3-1 of [33].
3. Equation 6.2 estimates the number of RBs, n_k, required by the kth CN.

$$n_k = \left\lceil \frac{r_k t_{sf} - \text{CRC}}{\text{RI} \cdot n_{\text{RE}} \cdot Q_m \cdot \text{CR}} \right\rceil \tag{6.2}$$

where t_{sf} is the LTE subframe duration, that is, 1 ms, CRC is set to 24-bits, which is the standard value for LTE [35], and n_{RE} is the number of data resource elements (REs) per RB set to 126 (one antenna port and the control format indicator set to 3). Clearly, the better the SINR at the CN side, the higher the modulation level and CR that the system can support, leading to fewer RBs.

When the CN (or HS) is connected to an RN, Equation 6.1 is again employed but now b_k is the macro BS serving the RN at backhaul. The RNs are considered to be out-band, that is, the backhaul link uses different bandwidth resources from the access link. The RBs needed by the users connected to an RN are subtracted from the pool of RBs of the corresponding BS to prevent intracell downlink interference. It should be noted that for the purposes of this work a capacity calculation module was developed that uses the SINR, as calculated in Equation 6.1, along with the rate requirements of each CN/HS, and returns either the number of necessary RBs as in Equation 6.2, or a NaN value if the SINR is too low to support any throughput [26].

6.3.3 The Cost Module

A set of BSs and RNs is considered as a possible solution when it can provide the bandwidth resources according to the SINR and data rate request, as well as coverage to all the control nodes. For each solution, the deployment cost is calculated. Given that macrocells are much more expensive to deploy than microcells and RNs, appropriate corresponding weights (×20, ×2, and ×1) are applied, respectively. These weights measure the relative total cost (capital and operational expenditures) for each antenna type (see Chapter 8 in [36]).

$$\text{Weighted\_Cost} = (\text{Number of activated macrocells}) \times 20$$

$$+ (\text{Number of activated RNs}) \times 1 \qquad (6.3)$$

$$+ (\text{Number of activated microcells}) \times 2$$

Note that the cost of a macrocell is considered to be the same irrespective of the number of active sectors. Also, the user can change the relative cost of each BS, based on additional criteria (e.g., if a possible site is an existing 3G site it costs less than a BS in a completely new site). Among the produced solutions, the one with the minimum weighted cost is preferred.

6.4 The BS Location Optimization Algorithm

The proposed BS location optimization algorithm follows a partially combinatorial approach. Similar approaches have been used for automatically locating BSs for 2G systems with coverage and traffic criteria [28,36]. The combination approach fits perfectly with the idea of preselected antenna positions. The idea is that if all possible combinations of BS positions are examined, the optimum combination for this set of BSs will be found. However, the number of combinations dramatically increases as the number of possible locations is increased. To avoid this problem the algorithm segments the total number of possible locations into smaller groups, which are randomly selected, and for each smaller group a search for solution is performed. This procedure of searching for solutions into smaller random groups of candidate BSs is performed for a large number of iterations (user defined).

A rather peculiar thing with heterogeneous systems is that the optimum solution can be anything between a small group of macrocells and a large number of microcells. As a consequence, the algorithm cannot be based on continuously examining exhaustively smaller groups of BSs in order to find a better solution since, for example, in some cases a cheaper solution can be produced if an expensive macrocell is replaced by a (small) number of microcells. As a result, the algorithm must continuously examine relatively large groups of candidate BSs. This, however, means prohibited large number of possible combinations (up to the order of several millions). In order to keep the number of examined combinations manageable, every time the number of combinations reaches a predefined threshold, a BS is removed from the group of candidates, as illustrated in the Pseudocode of Figure 6.3. This is possibly the main advancement with respect to the initial approach presented in [26] where the algorithm searches for the solution with the minimum number of BSs, and from them, the set with the smaller cost is chosen.

As also shown in Figure 6.3 (in *Check_for_a_Solution*), the algorithm examines a combination of possible BSs, only if its cost is equal or smaller than the cost of the existing solutions. Then it checks if all CNs are covered, that is, if the received power is above the threshold of −95 dBm, as calculated by the LTE link budget. Although this is a relatively high threshold that corresponds to a link with a large bandwidth allocation, as coverage at the control nodes is essential, it is reasonable to keep the power threshold at this high level. If any CNs are not covered, then the proposed solution is invalid, and the algorithm proceeds to the next combination of possible BSs.

If the solution passes the coverage test, the SINR and the required RBs for all possible links are calculated. The CNs covered by only one BS are the first to be allocated to a BS (microcell or macrocell sector). Then, starting from the CNs, which require the highest number of RBs, each CN is allocated to the BS with the best link (i.e., the highest SINR). If this BS has not enough RBs available, the next BS is considered, until there are no BSs left. If a CN cannot be served by any of the BSs, the solution is invalid. If all CNs are allocated to BSs, the combination is stored as one possible solution and the algorithm continues the search (for other solutions with equal or smaller total cost) until a predefined number of trials that guarantees convergence is reached (e.g., 2000).

Input data

Calculate/load received power from each transmitting unit (macrocell, microcell or RN) to each user CN

for i=1:NUM_COUNTS

    Put BSs (macrocells & microcells) in random order

    Divide BS in subgroups

    for each subgroup

        for i=1:size_of_subgroup

            while 'the number of all possible combinations of the subgroup' > threshold

                remove the last BS of the subgroup

            end

            for all possible combinations of the remaining BSs

                Check_for_a_Solution

            end

        end

    end

    Save the solution with the minimum cost

end

Function ***Check_for_a_Solution***

if the cost of the solution under investigation is equal or smaller than the cost of the solutions found so far

and if all the CNs are covered by a transmitting unit

    Calculate the SINR of each possible link and

    Estimate the number of RBs needed for each link

    Allocate CNs to BSs (macrocells & microcells) starting from the most demanding links and the CNs which can be served by only one transmitting unit.

    if a CN cannot be served because the corresponding BSs have allocated all their RBs, the solution is invalid

    else calculate the cost of the solution

    end

end

Figure 6.3 Pseudocode of the optimization algorithm.

6.5 Transmit Power Minimization for Green Planning

The goal here is to exploit the reconfigurability of the BS antenna radiation patterns in order to minimize the transmitted power in the entire network area, with the coverage and capacity constraints set by CNs and HSs. For this reason, a network planning algorithm is proposed that performs adaptive panning and fanning (see Figure 6.4), while adjusting the transmitted power of each sector, as well as microcell, in order to *minimize the total transmitted power* of the network. The final solution of the algorithm comprises the "greener" power allocation to the network BSs that can also meet capacity and coverage criteria. Equations 6.4 through 6.7 formulate the problem.

$$\{\mathbf{\Phi}_{0,opt}, \mathbf{\Phi}_{3\,dB,opt}, \mathbf{P}_{tx,opt}\} \leftarrow \underset{\substack{\phi_0^{b,s} \in \mathbf{D}_0, \phi_{3dB}^{b,s} \in \mathbf{D}_{3dB}, \\ P_{tx}^b \in \{\mathbf{P}_{tx}^b\}}}{\arg\min} \left(\sum_{b \in B} P_{tx}^b \right) \tag{6.4}$$

Subject to:

$$P_{rx}^k \geq -95 \text{ dBm}, \forall k \in K \tag{6.5}$$

$$\sum_{k \in \{K_b\}} n_k \leq 100, \forall b \in B \tag{6.6}$$

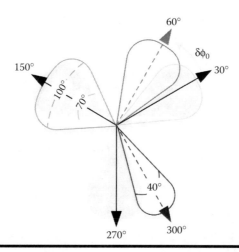

Figure 6.4 Panning/fanning operations: Sector initially pointing at 30° now points at 60° (panning). Sector pointing at 150° changes its 3 dB beamwidth to 100° (fanning). Sector initially pointing at 270° now points at 300° and changes its 3 dB beamwidth to 40° (panning+fanning).

$$r_k \le \frac{n_k(\text{RI} \cdot n_{\text{RE}} \cdot Q_m \cdot \text{CR})}{t_{\text{sf}}}, \forall k \in K \tag{6.7}$$

where:

$\mathbf{\Phi}_{0,opt}$ and $\mathbf{\Phi}_{3\,\text{dB},opt}$ are $1 \times |B_M|$ vectors of optimum pointing directions and beamwidths for all the sectors

$\mathbf{P}_{tx,opt}$ is $1 \times |B|$ vector of optimum transmit powers for all the BSs in the network

$\mathbf{D}_0$, $\mathbf{D}_{3\text{dB}}$, $\mathbf{P}_{tx}^b$ are the sets of allowed values for the sector pointing directions, their 3 dB beamwidth and the power levels, respectively, (see also Equations 6.8 through 6.10 in this section)

Finally, in Equation 6.6 K_b is the set of CNs that are served by the bth BS. The minimum received power constraint in Equation 6.5 is the coverage requirement for the kth network CN/HS, calculated via a typical LTE link budget. The constraints in Equations 6.6 and 6.7 set the LTE limit for the maximum number of RBs per BS and the CN throughput requirements, respectively.

The proposed solution is based on genetic algorithms (GAs), which are popular stochastic optimization methods, as they can deal with problems with a vast and complex search space. The standard GA or its advanced versions (steady state GA, distributed GA, and NSGA-II) have been used for the optimum BS location [20] or cell planning [37] problems. The stochastic methodology used here for LTE planning applies the NSGA-II, a standard for evolutionary multiobjective optimization that has been shown to outperform other alternatives [38]. NSGA-II belongs to the group of multiobjective approaches that aim to converge to a pareto front, that is, a certain percentage of GA population that is optimal given the different optimization objectives. LTE planning is a multiobjective problem since coverage and capacity must be satisfied, while the total transmitted power in the whole network is minimized.

More specifically, the multiobjective GA tries to find the optimum pointing direction$(\phi_{0,opt}^{b,s} \forall b \in B_M, s \in \{1, 2, 3\})$ and 3 dB beamwidth $(\phi_{3\text{dB},opt}^{b,s} \forall b \in B_M, s \in \{1, 2, 3\})$ of all macro BSs along with the optimum transmit power of all BSs $(P_{tx,opt}^b \ \forall b \in B)$ by searching into a set of allowed predefined values. The search areas for $\phi_{0,opt}^{b,s}$ and $\phi_{3\text{dB},opt}^{b,s}$ are given by Equations 6.8 and 6.9, respectively, and for $P_{tx,opt}^b$ by Equation 6.10.

$$\mathbf{D}_0 = \{\phi_0^s - 3\delta\phi_0, \phi_0^s - 2\delta\phi_0, \dots, \phi_0^s + 3\delta\phi\}, s \in \{1, 2, 3\}, \delta\phi_0 = 10° \tag{6.8}$$

$$\mathbf{D}_{3\text{dB}} = \{\phi_{3\text{dB}} - 2\delta\phi_{3\text{dB}}, \phi_{3\text{dB}}, \ \delta\phi_{3\text{dB}}, \dots, \phi_{3\text{dB}} + 2\delta\phi_{3\text{dB}}\}, \delta\phi_{3\text{dB}} = 15° \tag{6.9}$$

$$\mathbf{P}_{tx}^b = \begin{cases} \{-\text{Inf}, 11, 22, 33\}\,\text{dBM}, & \textit{if } b \in B_m \\ \{-\text{Inf}, 22, 26, 30, 34, 38, 42, 46\}\,\text{dBM}, & \textit{if } b \in B_M \end{cases} \tag{6.10}$$

Note that the minimum allowed value for the transmitted power is –Inf. This way we can consider the scenario where one BS (macro or micro) can be switched off.

6.5.1 The Algorithm: minPower

GA applies biologically inspired operators on a randomly chosen initial population of candidate solutions, in order to minimize a cost function. After GA *initialization*, the operators of *selection*, cost function *evaluation,* and *crossover* or *mutation* are iteratively applied in order to produce new offsprings. This continues until GA convergences, and in the case of NSGA-II this corresponds to an optimal pareto front. All operators applied as well as the multiobjective function used here are described next.

> *Initialization*: An initial population of N_{pop} random chromosomes is created. Each chromosome represents a candidate solution for the tasks of panning, fanning, and transmitted power setting for the macro BSs, and transmitted power setting for the micro BSs. Given that $|B_M|$ macro BSs (sectors) and $|B_m|$ micro BSs exist in the area setup, each chromosome is a mask of $N = 3 \times |B_M| + |B_m|$ integers: 3 settings to optimize (orientation, beamwidth, and power) for each of the $|B_M|$ sectors and 1 setting (power) for each micro. Each integer corresponds to a possible value either for orientation, beamwidth, or power (e.g., for a macro sector with a default orientation 30°, [−3 −2 −1 0 1 2 3] maps to [0° 10° 20° 30° 40° 50° 60°], see Equation 6.9).
>
> *Evaluation*: The constraints of Equations 6.5 through 6.7 are translated to certain objectives in the GA method, and considered within the multiobjective function F to be minimized Equation 6.11. Each of the chromosomes in the current (initial or next) population is evaluated as follows:
>
> 1. The selected settings for macro BS panning/fanning and transmitted power, as implied by the chromosome are identified.
> 2. Each CN/HS is assigned serially to all active BSs ($P_{tx} > $ –Inf) that can cover it ($P_{rx} \geq$ –95 dBm) and all possible SINR values are calculated. Each CN/HS is then assigned to the BS antenna providing the best SINR.
> 3. Given the allocation of CNs/HSs to BS antennas, the number of RBs used per CN/HS are calculated in the capacity module, as described earlier. The total RBs used by each BS are also calculated.
> 4. The F function evaluates the solution encoded in the chromosome: it measures the total transmitted power in the network (Equation 6.11, F(1)). Two other mechanisms embedded in the calculation of the F function (Equation 6.11, F(2), F(3)) evaluate if the finally proposed subset of antennas can serve all CNs/HSs in terms of coverage and required throughput. Thus, F(2) considers the number of CNs/HSs that have been assigned a NaN value as number of RBs allocated to them. This can be

considered both as coverage and throughput. Finally, no more than the maximum 100 RBs can be used per BS (capacity criterion). The extra number of RBs (>100) needed per BS is summarized for all BSs and is the F(3) value.

$$F(1) = \sum_{b \in B} P_{tx}^b$$

$$F(2) = |K_a|, \; K_a = \{k \mid n_k \leftrightarrow \text{NaN}\} \tag{6.11}$$

$$F(3) = \sum_{b \in B} \left(\sum_{k \in K_b} (n_k) - 100 \right)$$

Selection: The selection operator selects, through a tournament selection scheme [38], the chromosomes to participate next in the operators of crossover and mutation. This is done after a nondominating sorting of chromosomes in the current population based on the F values (a chromosome is said to dominate another if its objective functions are no worse than the others and at least one of its objective functions is better than the others). The nondominated sorting produces the pareto front (a predefined fraction of population is applied), while the rest of the population is sorted in a dominated manner [38].

Crossover/Mutation: Given the selection of chromosomes to participate in crossover and mutation operators, these two chromosome variation operators are applied. The application of these variation parameters will produce the next offspring, a new population of candidate solutions that has the same size as the initial population. The crossover operator mates random pairs of the selected chromosomes (binary crossover), while integers within a chromosome can be mutated (change to alternative value).

The whole procedure (population *evaluation* and *selection*, *crossovers/mutations*) is repeated until the stopping criterion of nonimproving the pareto front is reached (average pareto spread and distance criteria are used [38]). The algorithm outputs a series of solutions corresponding to the pareto front that minimize the F function. If the coverage and capacity criteria are met, the best solution will feature F(2) = F(3) = 0 and a total transmitted power calculated as F(1) value. In some cases, solutions satisfying the criteria but corresponding to different designs of the same cost values could be outputted. The user could then consider any of those solutions. In case that coverage and capacity criteria cannot be met (F(2) > 0 and/or F(3) > 0), the pareto front will include the best possible solutions, each one outperforming the other for at least one of the three objectives.

6.6 Radio Network Planning Results

6.6.1 Base Station Location Optimization

First, the performance of the optimization algorithm is studied as the capacity requirements increase. All the CNs (including those depicted as HSs) require the same data rate. Figure 6.5a shows that with a low-capacity requirement of

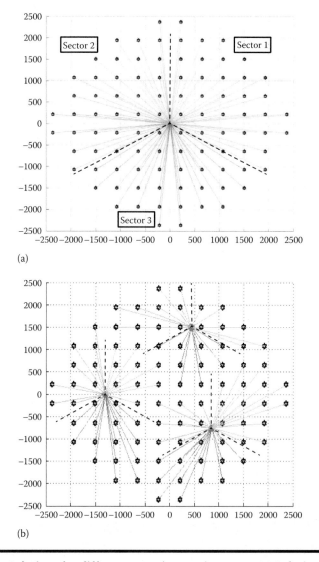

(a)

(b)

Figure 6.5 Solutions for different capacity requirements. (a) Solution for capacity requirement 0.5 Mbps/CN (umbrella cell scenario), (b) solution for capacity requirement 1 Mbps/CN. **(Continued)**

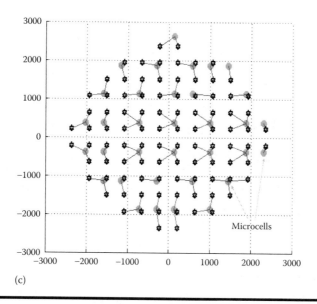

(c)

Figure 6.5 (Continued) Solutions for different capacity requirements. (c) solution for capacity requirement 6 Mbps/CN.

0.5 Mbps/CN (2.4 Mbps/Km²) one macrocell (3 sectors/BSs) is capable of covering the whole area. Hence, when capacity is not an issue, coverage can be provided by just a single (or a few for larger areas) macrocell. Due to the symmetry of the problem, the algorithm picked the macrocell in the center of the set-up and each CN connects to the sector that illuminates its position. This may very well be the scenario for an umbrella cell in a hierarchical cell structure, where the large cell supports all the problematic situations, for example, fast handovers, sudden power loss due to the street corner effect, coverage or capacity "holes," and so on.

However, in order to support higher data rates more BSs are needed as shown in Figure 6.5b, where in order to provide 1 Mbps/CN (4.8 Mbps/Km²), the solution consists of three macrocells. As the data rate increase further, the microcells prevail as shown with Figure 6.5c. As the cost of each microcell is just 10% of the cost of a macrocell, the whole area is covered with 31 microcells, and at the same time there are enough resources (RBs) to satisfy the need for higher data rates. Moreover, the lower transmitted power of microcells does not cause high levels of interference between the BSs, as well as contributes toward a more "green" network planning approach.

Hence, for 28.8 Mbps/Km² or 6 Mbps/CN, the solution consists of 34 microcells (Figure 6.5c). If we force the algorithm not to choose microcells, as many as 8 macrocells (24 BSs) with 30 RNs are needed, as shown in Figure 6.6a. This solution is far more expensive as Figure 6.6b illustrates, where the cost of the different solutions is shown (relatively to the more expensive solution) as the capacity requirement varies from 0.5 to 6 Mbps per CN (i.e., from 48 to 576 Mbps in the

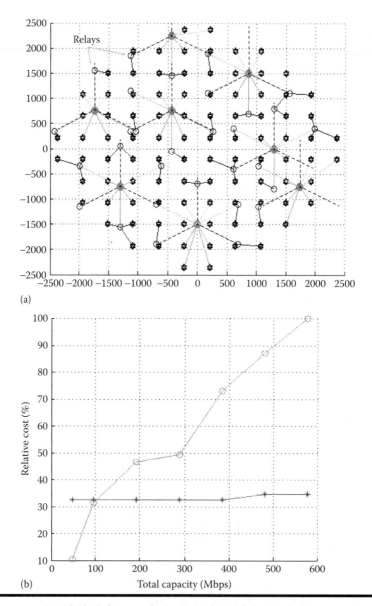

(a)

(b)

Total capacity (Mbps)

Figure 6.6 (a) Solution for 6 Mbps/CN only with macrocells and RNs (black circles). (b) Relative cost versus area capacity when only macrocells and RNs (circles), and only microcells (stars) are considered.

whole area). It can be clearly seen that in the optimization process the cost significantly increases when macrocells and RNs are exclusively considered.

For the same aforementioned scenarios, when microcells are also included, the algorithm soon turns to solutions with microcells only, due to the considerably smaller cost. In this case, as the capacity requirement increases from 0.5 to 6 Mbps per CN, the number of microcells remains practically unchanged (31–34 microcells are used) as the coverage requirement for the whole area must be fulfilled from the beginning, irrespective of the capacity requirements. As expected, microcells are not as good for coverage limited problems (i.e., up to 96 Mbps in the 20 Km² area), but they seem to be a clear winner for higher values of data rates. Indeed, microcells can provide ~600 Mbps to the whole area with just 32% of the cost of macrocells.

Figure 6.7 illustrates the solution for a case with inhomogeneous capacity requirements where 10% of the CNs need tenfold data rate (HSs), that is, 88.5 Mbps in total for the whole area. In the proposed solution, apart from three macrocells, a microcell is placed close to HSs, while a couple of RNs are also used to serve nearby HSs or other CNs with better SINR (less resources) so that their sector can allocate the necessary RBs to distant HSs. Nevertheless, if the total capacity requirement further increases, the algorithm produces solutions using only microcells, as in the case of the homogeneous scenarios. This is not only due to the significantly lower cost of small cells, but also due to smaller interference between BSs, and considerably more planning flexibility.

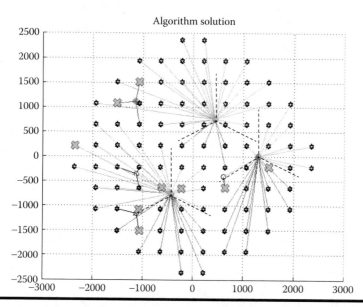

Figure 6.7 Solution for inhomogeneous capacity requirements: 0.5/5 Mbps/CN for 90/10% of the CN, respectively (*HSs* – crosses).

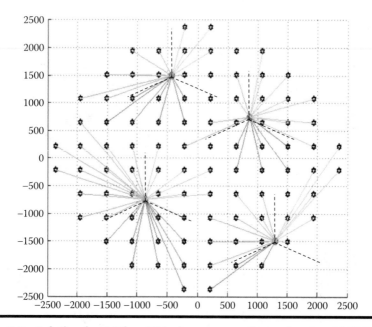

Figure 6.8   Solution for 6 Mbps/CN when macrocells employ 4 × 4 MIMO.

Finally, Figure 6.8 shows results for the high-capacity scenario (6 Mbps/CN) when the macro BSs employ 4 × 4 MIMO. The algorithm has produced a solution with half of the BSs and no RNs, compared to Figure 6.6a. Although the cost of the 4 × 4 MIMO BS is obviously higher than the conventional BS, it is clear that MIMO deployment will reduce the overall cost of the network due to the reduction of the required BSs.

6.6.2 Green Planning Results

Based on the BS location optimization algorithm in the 2.5 km × 2.5 km (6.25 km²) test area that is considered in this section, four candidate macro cell sites (12 BSs) and two candidate micro cell sites (2 BSs) are available in order to satisfy the throughput requirements of 64 CNs and 9 HSs, representing users or clusters of users (Figure 6.9). Various settings of data throughput were tested for the same setup, corresponding to (0.4, 0.6, 0.8, 1, 2) Mbps/CN, ×10 for HSs, respectively, and summing up a capacity requirement equal to approximately (9.5, 14, 18.5, 23, 46.5) Mbps/km². The minPower planning algorithm (N_{pop} = 5000) was applied in order to identify the network configuration with the minimum total transmitted power that also satisfies the coverage and capacity constraints.

For the various capacity requirements, the multiobjective GA converged to the optimal solution within 150 generations (up to 7 hours execution on a 12-CPU machine [2.3 GHz/CPU, 24 GB memory]). Figure 6.10 presents the BSs selected

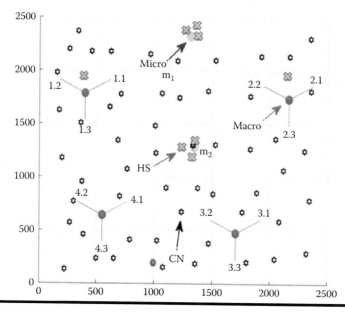

Figure 6.9 **Test area (2.5 km × 2.5 km): Following the BS location optimization algorithm, four macro sites (12 BSs) and two micro BSs (m_1, m_2) are available to satisfy the capacity requirements.**

to serve CNs/HSs and their transmitted power for the various capacity requirements. Results show that when capacity increases, the number of BSs and/or BS transmitted powers also increase (only for the two low capacity requirements, that is, 9.5 and 14 Mbps/km², the algorithm resulted to the same network configuration). It is shown that 9 macro BSs (out of 12) and both the micro BSs (with maximum power) are necessary to cover the capacity requirement of 46.5 Mbps/km². This is the maximum throughput that the setup of available macro/micro sites and CNs/HSs in Figure 6.9 permits. Results of the minPower algorithm not presented here show that more RBs (F(2)>0) would have to be allocated for capacity demands higher than 2 Mbps/CN (×10 for HSs) due to interference, even for the optimal configurations (this is possible if more bandwidth is allocated, e.g., in the context of a 5G system). For instance, the setting of 2.3 Mbps/CN (53.5 Mbps/km²) would need 10 more RBs (F(2) = 10) for the best solution that exploits all the available BSs (macro and micro).

We choose here to present in detail the solution provided by the minPower algorithm for the throughput setting 0.8 Mbps/CN (×10 for HSs), corresponding to 18.5 Mbps/km², as it provides some useful insights. For comparison reasons, we have applied an additional stochastic algorithm, previously proposed in [27], which selects the setup of minimum cost on the basis of relative cost for sites (×20 for macrocell, ×2 for microcell, [27], Chapter 8 in [36]). This algorithm, named here minCost, is applied on a macro/microcell site basis, that is, it cannot select single

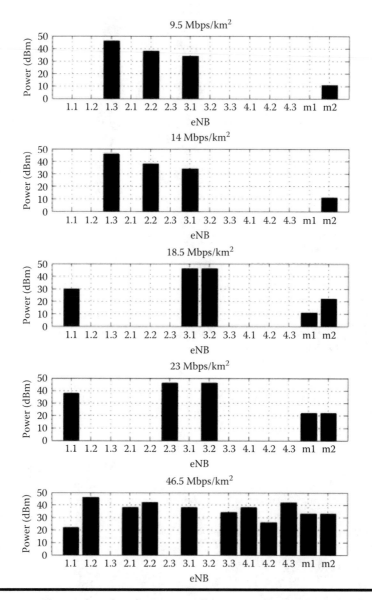

Figure 6.10 **BSs employed and corresponding transmitted powers for the solutions given by the minPower algorithm for the different capacity requirements.**

macro BSs, and uses the default values for macro BS sector orientation (30°, 150°, 270°) 3 dB beamwidth (70°), and power values (46 dBm for macro BSs, 33 dBm for micros). For the capacity requirement of 18.5 Mbps/km², the solution provided by minCost employed three macro sites (9 BSs), both the micro BSs and the total relative cost was 64 (Figure 6.11).

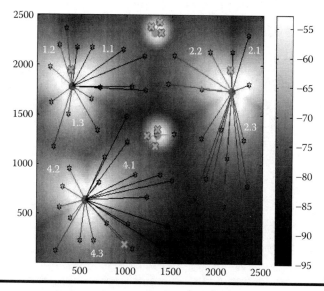

Figure 6.11   Network setup solution (minCost algorithm, 18.5 Mbps/km²).

The solution provided by the minPower algorithm employs only three single macro BSs (belonging to two separate macro sites) and the two available micros, a solution with a total relative cost equal to 44 (see Figure 6.12 for the solution, and Table 6.3 for details on the optimum antenna pattern and RBs used). The solution provided by the minPower algorithm transmits far less total power in the whole area than the one provided by the minCost algorithm (22%), and has less total relative cost as well (69%).

Furthermore, the regions colored white in Figures 6.11 and 6.12 are the areas where the received power exceeds the "strict" safety threshold $P_{rx,max}^{bio} = -52.7$ dBm* as proposed in [39]. It can be observed that the minPower algorithm provides a solution that features a smaller area exceeding the safety threshold compared to the one provided by the minCost algorithm.

With regards to the solution of the minPower algorithm in Figure 6.12, it is observed that the HSs in the center of the area led to the use of both micros, however, with a reduced power. All CNs in the upper left corner are served by Sector 1.1, which had to simultaneously broaden its beamwidth ($\phi_{3dB,opt}^{1.1} = 100°$) and steer its pointing direction ($\phi_{0,opt}^{1.1} = 50°$) away from the micro m_1. In addition, due to the fact that Sector 1.1 steered its lobe 20° to the left and the micro m_1 reduced its power, the CNs in the upper-right section of the area were able to be served by the distant Sector 3.1. Even though this sector is transmitting with full power (46 dBm), it still had to broaden its beamwidth to cover all the

---

* $P_{rx,max}^{bio} = P_D \cdot (\lambda^2 G_{rx}/4\pi)$, where $P_D = 0.3$ nW/cm² is the power spectral density given in , $G_{rx} = 1$, and $\lambda = 15$ cm for the central frequency.

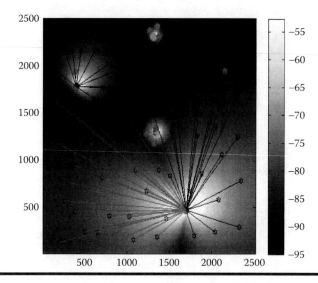

Figure 6.12   Network setup solution (minPower algorithm, 18.5 Mbps/km².).

Table 6.3   Network Configuration and Analysis for the Solution of the minPower Algorithm (18.5 Mbps/km²)

Sector	#RBs	Orientation	3 dB Beamwidth	Tx power (dBm)
1.1	59	50	100	30
3.1	96	30	100	46
3.2	90	180	55	46
m1	81	–	–	11
m2	65	–	–	22

CNs. Although m_2 serves fewer HSs than m_1, it transmits 11 dB more power. This is due to interference from Sector 3.1, which ends up serving CNs that are located closer to m_2. Finally, Sector 3.2 increased its directivity, but at the same time steered its beam 30° to the left, reducing the interference levels toward m_2 and Sector 1.1.

It has to be noted here that the initial setup with the default antenna pattern values tops up to 18.5 Mbps/km², utilizing 3 out of 4 macro sites (Figure 6.11), while the remaining macro resources cannot contribute anything more due to interference. However, the minPower algorithm exploits more degrees of freedom and finds solutions to the interference problem by neatly adjusting the antenna

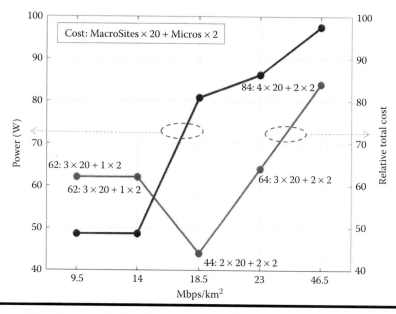

Figure 6.13 **Total transmitted power (W) in the test area and total relative cost as a function of the area capacity requirement (Mbps/km²).**

patterns of the separate BSs. As a result it can provide setups with capacity up to 46.5 Mbps/km² (Figure 6.10).

Figure 6.13 compares the total transmitted power (W) and the total relative cost for the five capacity scenarios (Mbps/km²) tested herein. As expected, the total transmitted power in the network increases with the throughput requirements, but this is not the case for the relative total cost. As we can see from Figure 6.13, the total cost of the network configuration for 18.5 Mbps/km² is less than that of 14 or even 9.5 Mbps/km². This is due to the nature of the minPower algorithm. Specifically, the transition from the 2nd to the 3rd scenario requires to switch off 1 macro site and to switch on 1 micro site. But since a micro site costs 10 times less than a macro site, the total relative cost of the network drops. On the other hand, two of the BSs in this scenario need to transmit in full power (46 dBm, see Table 6.3), raising the total transmit power to ~80 W. Hence, joint optimization of both power and cost, emerges as another option that will be addressed in future work.

Figure 6.14 shows the energy consumption rate (ECR) (power/throughput, Watt/Mbps) for the different capacity requirements. The results show that the ECR improves for increasing capacity, which proves that there is room for considerable improvement that the minPower algorithm can offer to the network, if considered in the planning and optimization process.

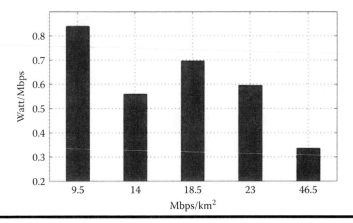

Figure 6.14 The network energy consumption rate for different capacity requirements.

References

1. Okumura, Y., Ohmori, E., Kawano, T., and Fukuda, K. Field strength and its variability in the VHF and UHF land mobile service. *Review Electronic Communication Laboratories*, 16(9/10): 825–873, 1968.
2. Hata, M. Empirical formula for propagation loss in land mobile radio services. *IEEE Transactions on Vehicular Technology*, VT-29(3): 317–325, 1980.
3. Lee, W. C. Y. *Mobile Cellular Telecommunications System*, New York: McGraw-Hill, pp. 194–199, 1990.
4. Tsoulos, G. V. Smart antennas for mobile communication systems: Benefits and challenges. *IEEE Electronics and Communication Engineering Journal*, 11(2): 84–94, 1999.
5. Garey, M. R. and Johnson, D. S. *Computers and Intractability: A Guide to the Theory of NP-Completeness*. New York: Freeman, 1979.
6. Amaldi, E., Capone, A., and Malucelli, F. Optimizing base station siting in UMTS networks. In *Proceedings of the 53rd IEEE Vehicular Technology Conference (VTC2001-Spring)*, May 2001.
7. Kocsis, I., Farkas, I. L., and Nagy, L. 3G base station positioning using simulated annealing. In *Proceedings of the 13th IEEE International Symposium on Personal, Indoor and Mobile Radio Communications (PIMRC 2002)*, pages 330–334, 2002.
8. Mathar, R. and Niessen, T. Optimum positioning of base stations for cellular radio networks. *Wireless Networks*, 6(6): 421–428, 2000.
9. Hale, W. K. Frequency assignment: Theory and applications. In *Proceedings of the IEEE*, vol. 68, pp. 1497–1514, 1980.
10. Naghshineh, M. and Katzela, I. Channel assignment schemes for cellular mobile telecommunication systems: A comprehensive survey. *IEEE Personal Communications*, 3: 10–31, 1996.
11. Mazzini, F. F., Mateus, G. R., and Smith, J. M. Lagrangean based methods for solving large-scale cellular network design problems. *Wireless Networks*, 9: 659–672, 2003.

12. Ericsson Mobility Report, June 2013, Available: http://www.ericsson.com/res/docs/2013/ericsson-mobility-report-june-2013.pdf.
13. Ericsson Mobility Report, June 2015, Available: http://www.ericsson.com/res/docs/2015/ericsson-mobility-report-june-2015.pdf.
14. Astely, D., Dahlman, E., Fodor, G., Parkvall, S., and Sachs, J. LTE release 12 and beyond. *Communications Magazine*, 51(7): 154–160, 2013.
15. Dahlman, E., Parkvall, S., and Sköld, J. *4G LTE/LTE-Advanced for Mobile Broadband*, Oxford, UK: Academic Press, 2011.
16. Hurley, S. Planning effective cellular mobile radio networks. *IEEE Transactions on Vehicular Technology*, 51(2): 243–253, 2002.
17. Amaldi, E., Capone, A., and Malucelli, F. Radio planning and coverage optimization of 3G cellular networks. *Wireless Networks*, 14: 435–447, 2008.
18. Mathar, R. M. and Niessen, T. Optimum positioning of base stations for cellular radio networks. *Wireless Networks*, 6: 421–428, 2000.
19. Amaldi, E., Capone, A., and Malucelli, F. Radio planning and coverage optimization of 3G cellular networks. *Wireless Networks*, 14(4): 435–447, 2008.
20. Lakshminarasimman, N., Baskar, S., Alphones, A., and Willjuice Iruthayarajan, M. Evolutionary multiobjective optimization of cellular base station locations using modified NSGA-II. *Wireless Networks*, 17(3): 597–609, 2011.
21. Gonzalez-Brevis, P., Gondzio, J., Fan, Y., Poor, H., Thompson, J., Krikidis, I., and Chung, P. J. Base station location optimization for minimal energy consumption in wireless networks. In *Vehicular Technology Conference (VTC Spring), 2011 IEEE 73rd*, May 2011, pp. 1–5.
22. Liu, Y., Tao, M., Li, B., and Shen, H. Optimization framework and graph-based approach for relay-assisted bidirectional OFDMA cellular networks. *IEEE Transactions on Wireless Communications*, 9(11): 3490–3500, 2010.
23. Awada, A., Wegmann, B., Viering, I., and Klein, A. Optimizing the radio network parameters of the long term evolution system using Taguchi's method. *IEEE Transactions on Vehicular Technology*, 60(8): 3825–3839, 2011.
24. Jaloun, M., Guennoun, Z., and Elasri, A. Use of genetic algorithm in the optimisation of the LTE deployment. *International Journal of Wireless & Mobile Networks (IJWMN)*, 3(3), 2011.
25. Kasem, F., Haskou, A., and Dawy, Z. On antenna parameters self optimization in LTE cellular networks. In *2013 Third International Conference on Communications and Information Technology (ICCIT)*, June 2013, pp. 44–48.
26. Athanasiadou, G., Zarbouti, D., and Tsoulos, G. Automatic location of base-stations for optimum coverage and capacity planning of LTE systems. In *2014 8th European Conference on Antennas and Propagation (EuCAP)*, 2014, pp. 2077–2081.
27. Valavanis, I. K., Athanasiadou, G., Zarbouti, D., and Tsoulos, G. V. Basestation location optimization for LTE systems with genetic algorithms. In *Proceedings of 20th European Wireless Conference (European Wireless 2014)*, May 2014, pp. 1–6.
28. Molina, A., Athanasiadou, G. E., and Nix, A. R. The automatic location of base-stations for optimised cellular coverage: A new combinatorial approach. *IEEE Vehicular Technology Conference*, 1999, pp. 606–610.
29. Athanasiadou, G. E., Tsoulos, G. V., and Zarbouti, D. A combinatorial algorithm for base-station location optimization for LTE mixed-cell MIMO wireless systems. *2015 9th European Conference on Antennas and Propagation (EuCAP)*, April 12–17, 2015.

30. Spatial channel model for multiple input multiple output (MIMO) simulations, 3GPP TR 25.996 v.6.1.0, 2003.
31. WINNER II WP1: Channel models. Deliverable D1.1.2, 30/11/2007.
32. Evolved Universal Terrestrial Radio Access (E-UTRA); Further advancements for E-UTRA physical layer aspects, 3GPP TR 36.814 Std., Rev. V9.0.0, 03 2010.
33. Evolved Universal Terrestrial Radio Access (E-UTRA); Physical layer procedures, 3GPP Std. TS 36.213, Rev. v10.1.0, 04 2011.
34. Ikuno, J. C., Wrulich, M., and Rupp, M. System level simulation of LTE networks. IEEE VTC 2010-Spring, 2010 IEEE 71st, pp. 1–5, May 16–19, 2010.
35. Evolved Universal Terrestrial Radio Access (E-UTRA); Multiplexing and channel coding, 3GPP Std. TS 36.212, Rev. v12.4.1, 04 2015.
36. Döttling, M., Mohr, W., and Osseiran, A. *Radio Technologies and Concepts for IMT-Advanced*, Hoboken, NJ: John Wiley & Sons, 2009.
37. Luna, F., Durillo, J. J., Nebro, A. J., and Alba, E. Evolutionary algorithms for solving the automatic cell planning problem: A survey. *Engineering Optimization*, 42(7): 671–690, 2010.
38. Deb, K., Pratap, A., Agarwal, S., and Meyarivan, T. A fast and elitist multiobjective genetic algorithm: NSGA-II. *IEEE Transactions on Evolutionary Computation*, 6(2): 182–197, 2002.
39. BioInitiative 2012. A rationale for biologically-based exposure standards for low-intensity electromagnetic radiation. Available: http://www.bioinitiative.org/.

Chapter 7

MIMO Techniques for 5G Systems

Athanasios G. Kanatas and Konstantinos Maliatsos

Contents

7.1 Introduction

One of the core integral parts of 5G systems will be based on multiantenna systems and the corresponding multiple-input multiple-output (MIMO) architectures. MIMO technology offers an unprecedented performance enhancement to wireless systems. The prospective characteristics of the next-generation systems include cooperation among base stations (BSs), called *coordinated multipoint* (CoMP), 3D beamforming capabilities offered by 2D antenna arrays, also called *full-dimension MIMO* (FD-MIMO), large-scale antenna systems, also called *massive-MIMO* systems, and millimeter wave communications combined with MIMO features, called *mm-Wave-MIMO* systems.

The inclusion of MIMO capabilities in wireless systems was evidenced around 2006 in WiFi and then in 3G cellular networks (Andrews et al. 2014). IEEE 802.11n, IEEE 802.11ac, LTE, and WiMAX are the systems where the MIMO techniques were consolidated. Especially in LTE, several MIMO variations have been proposed from the early releases of 3GPP standardization activities up to the current release 13, encompassing all the latest advancements in wireless communications theory.

The use of multiple antennas at one end of the link was known for many years to offer the advantages of diversity, as well as beamforming capabilities. By exploiting the spatial domain and putting many antenna elements at either the transmitter (multiple-input single-output, MISO) or the receiver (single-input multiple-output, SIMO) one can get *spatial diversity gain*, that is, improved error performance in fading channels compared to single-input single-output (SISO) systems. With the beamforming operation, the power of the desired signal is enhanced whereas, unwanted interference signals are suppressed, thus providing increased signal-to-interference-plus-noise ratio (SINR). The introduction of multiple antennas at both link ends, either concentrated or distributed, leads to the MIMO systems and may offer extra gains, the most important of which is the *spatial multiplexing gain*. Therefore, MIMO techniques increase system performance since they offer improved reliability, by applying space-time coding techniques to combat fading and/or increased throughput by applying spatial multiplexing techniques to exploit multipath components in rich scattering environments. Depending on the knowledge of channel state information (CSI) available at the receiver (CSIR) and/or the transmitter (CSIT), various communication techniques have been proposed and different performance gains are achieved.

Although the benefits of MIMO techniques were well-known for decades (Sibille et al. 2010), the first attempts to describe the gains in practical systems were recorded twenty years ago in the works of Paulraj and Kailath (1994), and Foschini and Gans (Foschini 1996; Foschini and Gans 1998). On the same period, Alamouti (1998), Tarokh et al. (1998) published landmark papers on space-time coding techniques. In 1999, Telatar for the first time set the framework for the investigation of the capacity offered in MIMO channels as well as its connection to the eigenvalues' distribution of random Wishart matrices (Telatar 1999). All the promised gains were conditioned on the so called *favorable propagation conditions*, that is, under the assumption of perfectly uncorrelated Rayleigh independent and identically distributed (i.i.d.) channels.

The use of multiple antennas at the BS or access point (AP) and the user equipment (UE) was also called point-to-point MIMO or single-user MIMO (SU-MIMO). Full CSIT is not feasible for practical SU-MIMO systems and, therefore, the communication techniques assume CSIR only, as shown in Figure 7.1a. Although for a single user this scenario offers a gain equal to the number of the equivalent parallel data streams that are multiplexed, the gain is constrained to favorable propagation conditions, which unfortunately is not true in realistic environments, especially at cell edges. Moreover, the multiplexing gain is limited by the number of antennas the UE is equipped with. A different approach to exploit the spatial degrees of freedom (DoF) is the MU-MIMO systems, (Caire and Shamai 2003; Vishwanath et al. 2003; Viswanath and Tse 2003). In MU-MIMO, the channel is spatially shared by the users, that is, MU-MIMO is a spatial multiple access technique where the BS is equipped with M antennas and the K users may have one or more antennas. The former is depicted in Figure 7.1b. The multiple antennas at the BS offer not only diversity gain but also the required DoF to spatially separate the users. The multiplexing gain equals the sum of the number of parallel data streams and it is limited by the number of BS antennas. The great advantage of MU-MIMO systems comes with the strong requirement for full CSI knowledge at the transmitter, since the precoding technique should cope with the

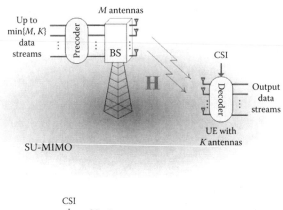

(a)

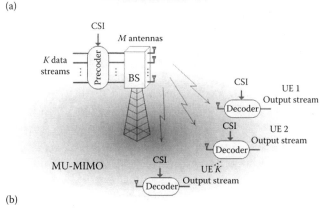

(b)

Figure 7.1 MIMO systems: (a) SU-MIMO system and (b) MU-MIMO with *K* single-antenna users.

multiuser interference. This implies the necessity for a feedback mechanism from the UE to the BS. In practical systems a major issue is the trade-off between feedback accuracy and uplink resources consumption. Moreover, the optimum transmit strategy in the downlink involves a theoretical interference pre-cancellation technique known as dirty paper coding (DPC) combined with user scheduling and power loading algorithm. Therefore, the complexity is increased and sub-optimum techniques have been proposed. In Section 7.3 several techniques for the broadcast channel (BC), that is, the downlink channel, and the multiple access channel (MAC), that is, the uplink channel, are discussed.

The term coordinated multipoint (CoMP) has been used widely for schemes where several transmission points cooperate to jointly design the Tx/Rx structure or to optimize the resource allocation from a global rather than a local perspective (Lee et al. 2012; Cui et al. 2014). A CoMP system can be considered as a large-scale MIMO system with several distributed antenna groups. Coordinated transmission from multiple BSs to their UE is very effective in mitigating the interference in multicell systems. There are several variations proposed in the literature for the implementation of CoMP techniques, which attempt to overcome the limitations imposed by synchronization issues, the large information exchange among BSs and the uncoordinated interference in dense cellular networks. Section 7.4 deals with these techniques.

Massive-MIMO is one of the key technologies that will manifest the arrival of 5G era. It is a direct extension of MU-MIMO systems where hundreds of antennas are placed at the BS, simultaneously serving many tens of terminals in the same time-frequency resource (Marzetta 2010). This increase in BS antennas offers a tremendous advantage in the propagation conditions of the links, called *favorable propagation*; the vector channels from the BS antenna array to the various UE are asymptotically orthogonal as the number of antennas at the BS, M, grows to infinity. The direct implication of this feature is that linear precoding/decoding techniques can be used, offering reduced complexity in hardware and signal processing techniques. Since massive-MIMO is a large scale MU-MIMO system, it relies on spatial multiplexing, that is, on the BS having adequate channel knowledge, on both the UL and the DL. This, as explained in detail in Section 7.5, implies that a time-division-duplex (TDD) technique is favorable. The large number of antennas allows for focusing energy into very small regions of space to bring huge improvements in throughput and radiated energy efficiency. There is an increasing interest in the wireless community for massive-MIMO systems and there are already pilot systems that bespoke the advantages (Rusek et al. 2013; Vieira et al. 2014). Nevertheless, there are still few issues to be addressed before the BSs are swamped with antennas (Larsson et al. 2014). The most known is the pilot contamination problem in realistic multicell networks and the channel reciprocity assumed in TDD operation, which are discussed in Section 7.5 (Figure 7.2).

Full-dimension MIMO (FD-MIMO) utilizes a large number of antennas in a 2D array configuration to implement spatially separated links to a large number of users (Kim et al. 2014; Razavizadeh et al. 2014). It constitutes an intermediate phase toward the massive-MIMO systems and a special study item has been setup and concluded by 3GPP recently for inclusion into Release 13 (3GPP). So far 3GPP activities have mostly considered antenna arrays that exploit the azimuth dimension. 3GPP with this study investigates how two-dimensional antenna arrays can further improve the LTE spectral efficiency by also exploiting the vertical dimension for beamforming and MIMO operations. The key features of FD-MIMO, the architectures, CSI feedback, and 3D channel modeling issues are discussed in Section 7.6.

In the absence of free available spectrum in the sub-6 GHz bands, the researchers have focused in the millimeter wave frequencies (30–300 GHz) where there is a plenty of underutilized bandwidth for cellular access technologies (Rappaport et al. 2013b; Roh et al. 2014; Hur et al. 2016). The significantly expanded channel bandwidth will allow for increased data rates and reduced latency, whereas, large

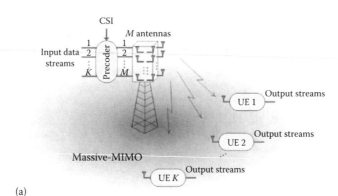

(a)

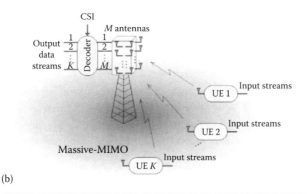

(b)

Figure 7.2 Massive-MIMO links: (a) downlink and (b) uplink.

antenna arrays of very high gain can be placed in really small dimensions and can be used to keep the antenna aperture constant, thus eliminating the frequency dependent pathloss and/or the increased noise bandwidth. Moreover, the mm-Wave technology is expected to help realize massive-MIMO systems. The introduction of mm-Wave technology in 5G systems affects not only the hardware and the corresponding components but also the channel modeling approach and the signal processing techniques (Heath et al. 2016). All the challenges related to mm-Wave MIMO systems are discussed in Section 7.7.

Notation: In the following paragraphs we use boldface to denote matrices and vectors and $\mathbb{E}[\cdot]$ for expectation. $|\mathbf{S}|$ denotes the determinant of a square matrix $\mathbf{S}$. For a matrix $\mathbf{H}$ we indicate its jth row, ith column, and (j,i)th element by $\mathbf{h}^j$, $\mathbf{h}_i$, and h_{ji} or, equivalently, by $[\mathbf{H}]_{j,i}$, respectively. $Tr(\mathbf{H})$ denotes the trace of the matrix. For symmetric matrices the notation $\mathbf{Q} \geq 0$ implies that $\mathbf{Q}$ is positive semidefinite.

7.2 Point-to-Point MIMO

Assume a point-to-point MIMO link with N_t transmit and N_r receive antennas, as shown in Figure 7.3. The typical system model for a narrowband nonfrequency selective channel is given by

$$y = \mathbf{H}\mathbf{x} + \mathbf{n} = \sum_{i=1}^{N_t} \mathbf{h}_i x_i + \mathbf{n} = \mathbf{h}_k x_k + \underbrace{\sum_{i \neq k} \mathbf{h}_i x_i}_{\text{Interference}} + \mathbf{n} \tag{7.1}$$

where:

$\mathbf{H} \in \mathbb{C}^{N_r \times N_t}$ represents the flat fading channel matrix

$\mathbf{n}$ is the $N_r \times 1$ i.i.d., zero mean, additive white Gaussian noise vector with covariance $\mathbb{E}[\mathbf{n}\mathbf{n}^H] = \sigma_n^2 \mathbf{I}_{N_r}$

The transmit covariance matrix of the input signal is $\mathbf{Q}_x = \mathbb{E}[\mathbf{x}\mathbf{x}^H]$ with an average power constraint P, that is, $Tr[\mathbf{Q}_x] = P$. The element h_{ji} of the channel matrix denotes the complex channel gain for the symbol transmission from the ith transmit antenna to the jth receive antenna. The typical assumption for the channel matrix elements is that they are i.i.d. complex Gaussian random variables with zero mean and unit variance. The precoder maps the input data streams to the transmit antennas, whereas the postcoder processes the signals from the receive antennas and estimates the input data streams. The signal given in Equation 7.1 is a composite signal and the detection of a stream suffers from interstream interference. This model presents a vector Gaussian channel and the main question in spatial multiplexing MIMO systems is how to transmit and receive with multiple antennas and achieve an equivalent system of multiple parallel independent subchannels. If the columns of matrix $\mathbf{H}$ were orthogonal, then one could multiply the received signal with the complex conjugate of the column of $\mathbf{H}$ that corresponds to the stream to be detected and all the remainder streams would be zeroed. Nevertheless, the channel is random and the columns will not in general be orthogonal. Therefore, we need an equivalent technique to decompose the vector channel to independent

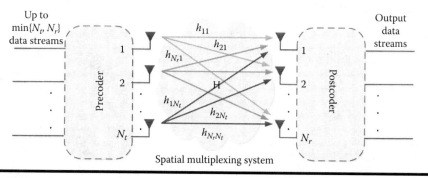

Figure 7.3 MIMO transceiver structure.

parallel scalar subchannels. This is achieved with the singular value decomposition (SVD) of the channel matrix. The basic assumption is that the channel is perfectly known at both the transmitter and the receiver. The SVD is used to decompose the channel matrix and write it in terms of a product of three matrices, that is, $\mathbf{H} = \mathbf{U}\mathbf{\Sigma}\mathbf{V}^H$, where $\mathbf{U}$ and $\mathbf{V}$ are unitary, that is, $\mathbf{U}^H \cdot \mathbf{U} = \mathbf{I}_{N_r}$ and $\mathbf{V}^H \cdot \mathbf{V} = \mathbf{I}_{N_t}$, and matrix $\mathbf{\Sigma}$ is a diagonal matrix with the singular values of $\mathbf{H}$. Matrix $\mathbf{U}$ contains the left and matrix $\mathbf{V}$ the right singular vectors of $\mathbf{H}$, or the eigenvectors of $\mathbf{H}^H\mathbf{H}$. Now define the following vectors, $\tilde{\mathbf{x}} = \mathbf{V}^H\mathbf{x}$, $\tilde{\mathbf{y}} = \mathbf{U}^H\mathbf{y}$, and $\tilde{\mathbf{n}} = \mathbf{U}^H\mathbf{n}$. Substituting in MIMO system model (Gore et al. 2003),

$$\tilde{\mathbf{y}} = \mathbf{U}^H\mathbf{y} = \mathbf{U}^H\mathbf{H}\mathbf{x} + \mathbf{U}^H\mathbf{n} = \mathbf{U}^H\mathbf{U}\mathbf{\Sigma}\mathbf{V}^H\mathbf{x} + \tilde{\mathbf{n}} = \mathbf{\Sigma}\tilde{\mathbf{x}} + \tilde{\mathbf{n}} \tag{7.2}$$

Matrix $\mathbf{\Sigma}$ is diagonal and therefore each output signal contains only one input signal. There are R_H SISO channels each of which consists of a scaled version of the transmitted signal plus noise. R_H is the rank of matrix $\mathbf{H}$. As shown in Figure 7.4, one can transmit R_H independent data streams toward the directions of the eigenvectors of matrix $\mathbf{H}^H\mathbf{H}$. This is why we call it *eigenbeamforming*.

The requirement of channel knowledge at the transmitter (CSIT) in eigenbeamforming is a quite strict one. When CSIR only is assumed, then more practical architectures may be utilized. A well known family of spatial multiplexing techniques is the layered space time (Foschini 1996). In horizontal encoding Bell Labs Layered Space Time (H-BLAST) architecture (Foschini et al. 2003), the information bit stream is demultiplexed to N_t substreams and each substream is coded, modulated and interleaved, and then transmitted by an antenna. The relationship of substreams and antennas remains fixed in time. The maximum achievable diversity gain is N_r. In vertical encoding BLAST architecture (V-BLAST) (Wolniansky et al. 1998), the bit stream is initially coded, modulated and interleaved, then is demultiplexed to N_t substreams and each substream is transmitted by an antenna. As each info bit transmitted by N_t antennas and received by N_r antennas, the diversity gain is at maximum N_tN_r. This implies greater complexity. In diagonal BLAST (D-BLAST) the architecture is a combination of the two techniques. The bit stream is horizontally encoded and the substreams enter a block that performs stream rotation in a round-robin fashion before entering the antennas. Thus, the substreams from the conventional encoders are distributed over space and the big advantage is the improved spatial diversity. Indeed, D-BLAST can achieve diversity gain greater than N_r and up to N_tN_r. The complexity of D-BLAST is the main problem compared to H-BLAST and V-BLAST.

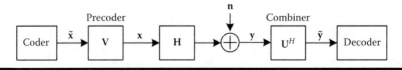

Figure 7.4   SVD-based MIMO transceiver structure.

7.2.1 Linear Receivers for Spatial Multiplexing

In spatial multiplexing MIMO systems the optimum receiver would be the maximum likelihood (ML) receiver. The main issue is that we cannot decompose the problem in scalar problems, since the columns of **H** are not orthogonal and we have to perform joint detection. The complexity of ML decoding exponentially increases with the number of receive antennas. Therefore, suboptimum in terms of diversity order, techniques with decreased complexity have been developed.

7.2.1.1 Zero-Forcing Receiver

The *zero forcing* (ZF) receiver, called also *decorrelator* or *interference nulling receiver*, attempts to cancel out the interstream interference using the Moore-Penrose pseudo-inverse of the channel matrix **H**. This matrix is either the *left inverse* of matrix **H**, that is, $\mathbf{H}^+\mathbf{H} = \mathbf{I}_{N_r}$ when $N_t \geq N_r$ and it holds that $\mathbf{H}^+ = (\mathbf{H}^H\mathbf{H})^{-1}\mathbf{H}^H$, or the *right inverse* of **H**, that is, $\mathbf{H}\mathbf{H}^+ = \mathbf{I}_{N_t}$ when $N_r \geq N_t$ and it holds that $\mathbf{H}^+ = \mathbf{H}^H(\mathbf{H}\mathbf{H}^H)^{-1}$. By premultiplying the received vector with the pseudo-inverse we get

$$\hat{\mathbf{x}} = \mathbf{H}^+\mathbf{y} = \mathbf{H}^+\mathbf{H}\mathbf{x} + \mathbf{H}^+\mathbf{n} = \mathbf{x} + \mathbf{H}^+\mathbf{n} \tag{7.3}$$

The interference is forced to zero and we can detect each stream separately with an ML detector. The difference of ZF and Maximal Ratio (Combining or Transmission) is that in maximal ratio techniques we premultiply **H** with $\mathbf{H}^H$ and not $\mathbf{H}^+$. Therefore, if the columns of **H** are orthogonal, the result is the same. But in general this is not true. The maximal ratio is also called *matched filtering* (MF). The main issue with the ZF receiver is the noise amplification. The power of the noise $\mathbf{n}_{ZF} = \mathbf{H}^+\mathbf{n}$ is now

$$\mathbb{E}\left[\|\mathbf{n}_{ZF}\|^2\right] = \mathbb{E}\left[\|\mathbf{V}\Sigma^{-1}\mathbf{U}^H\mathbf{n}\|^2\right] = \sigma_n^2 Tr\left(\Sigma^{-2}\right) = \sigma_n^2\sum_{i=1}^{N_t}\frac{1}{\lambda_i} \tag{7.4}$$

7.2.1.2 Minimum Mean Squared Error Receiver

Minimum mean square error (MMSE) is an alternative receiver that exploits the knowledge of noise variance to maximize the SINR. The detection is performed using the matrix

$$\mathbf{W}_{MMSE} = \left(\mathbf{H}^H\mathbf{H} + \sigma_n^2\mathbf{I}_{N_t}\right)^{-1}\mathbf{H}^H \tag{7.5}$$

It is proved that compared to ZF, the MMSE receiver avoids noise amplification but suffers from remaining interference. Overall, the performance of MMSE is better than that of ZF.

7.2.1.3 Ordered Successive Interference Cancellation

A method to improve the performance of linear detection techniques is the ordered successive interference cancellation (OSIC). It is a series of linear receivers, for example, ZF or MMSE, each of which detects one of the parallel streams, where the detected streams are successively subtracted by the received signal. Hence, at each step of the method we subtract from the received signal the detected stream and we send for subsequent detection the signal with the reduced interference.

7.2.2 Capacity

MIMO systems are classified according to channel knowledge at the transmitter into either *closed loop*, when the channel is known at both the Rx and the Tx, that is, full CSI is available, or *open loop*, when the channel is known only at the Rx, that is, no CSIT is available.

7.2.2.1 Closed-Loop Systems

Using the decomposition of the MIMO channel into parallel SISO subchannels one may express the capacity as the sum of the corresponding subchannels

$$C = \max_{P_i, \sum P_i \leq P} \sum_{i=1}^{R_H} \log_2\left(1 + \frac{\lambda_i P_i}{\sigma_n^2}\right) = \max_{P_i, \sum P_i \leq P} \log_2 \prod_{i=1}^{R_H}\left(1 + \frac{\lambda_i P_i}{\sigma_n^2}\right) \tag{7.6}$$

where:
P_i is the power allocated to each subchannel
P is the total transmitted power
λ_i are the eigenvalues of $\mathbf{H}^H\mathbf{H}$

Increasing the rank R_H of matrix $\mathbf{H}$, increases the capacity. This gain in capacity is called *spatial multiplexing gain*. It is proved that among the channels with equal power gain the one with equal eigenvalues preserves higher capacity. For high SNR values the optimum power allocation is to set equal power to each subchannel, and $C \approx R_H \log_2(P/\sigma_n^2)$. For low SNR values the optimum power allocation strategy is to set all power to the subchannel with the largest eigenvalue, and $C \approx (P/\sigma_n^2)\max(\lambda_i)\log_2 e$. Hence, in low SNR regime, the channel rank is not important. What really matters is the total power that is transferred from the Tx to the Rx. For intermediate SNR values the optimum power allocation is provided by the waterfilling algorithm (Gore et al. 2003).

7.2.2.2 Open-Loop Systems

According to Foschini and Gans (1998) and Telatar (1999) the capacity achieved is given by

$$
C = \begin{cases} \max\limits_{\mathbf{Q}_x : Tr[\mathbf{Q}_x] \leq P} \log_2\left[\det\left(\mathbf{I}_{N_r} + \dfrac{\mathbf{HQ}_x\mathbf{H}^H}{\sigma_n^2} \right) \right], & N_r < N_t \\[3mm] \max\limits_{\mathbf{Q}_x : Tr[\mathbf{Q}_x] \leq P} \log_2\left[\det\left(\mathbf{I}_{N_t} + \dfrac{\mathbf{H}^H\mathbf{Q}_x\mathbf{H}}{\sigma_n^2} \right) \right], & N_r > N_t \end{cases} \tag{7.7}
$$

If we assume equal power allocation to each transmit antenna then

$$
C = \begin{cases} \log_2\left[\det\left(\mathbf{I}_{N_r} + \dfrac{\rho}{N_t}\mathbf{HH}^H \right) \right], & N_r < N_t \\[3mm] \log_2\left[\det\left(\mathbf{I}_{N_t} + \dfrac{\rho}{N_t}\mathbf{H}^H\mathbf{H} \right) \right], & N_r > N_t \end{cases} \tag{7.8}
$$

where ρ is the SNR per receive antenna from all transmitters. It is proved that the bounds on the achievable rate are given by the following equation

$$
\log_2(1 + \rho N_r) \leq C \leq \min(N_t, N_r) \cdot \log_2\left(1 + \rho \frac{N_r}{\min(N_t, N_r)} \right) \tag{7.9}
$$

The lower limit is applicable when all but one of the singular values are equal to zero, and the upper limit when all of the $\min(N_t, N_r)$ singular values are equal. As for $x \to 0$ it holds that $\log_2(1 + x) \approx x\log_2 e$, the lower bound for low SNR values is given by: $C_{\rho \to 0} \approx \rho N_r \log_2 e = \rho(N_r/\ln 2)$. An excess number of Tx or Rx antennas, combined with asymptotic orthogonality of the propagation vectors, leads to upper bound. If we let N_t grow large while keeping N_r constant, the upper bound is given by: $C_{N_t \gg N_r} \approx N_r \cdot \log_2(1 + \rho)$. If we let N_r grow large while keeping N_t constant, the upper bound is given by: $C_{N_r \gg N_t} = N_t \cdot \log_2[1 + \rho(N_r/N_t)]$.

7.2.3 Space-Time Coding for MIMO Systems

In 1998, two important publications set the principles for the space-time coding techniques. The first was the article by Alamouti (1998), where a simple but efficient transmit diversity technique for two transmit antennas was proposed. The second one was the article by Tarokh et al. (1998), where the introduction of redundancy in space and time imposed correlation among the signals transmitted by different antennas in different time periods, providing improved reliability in a faded link. The space-time codes may be categorized in space-time block codes (STBCs) and space-time trellis codes (STTCs). The STBCs are further divided in orthogonal STBCs and non-orthogonal or quasi-orthogonal STBCs. There have been developed diverse criteria for the design of space-time codes depending on the gain we wish to maximize, that is, diversity or coding gain. At high SNR, the average error probability of an uncoded system is approximated by the equation $p_e \approx \left(G_c \cdot \overline{SNR}\right)^{-G_d}$, where G_c is the coding gain and G_d is the diversity gain or diversity order. Quasi-orthogonal STBCs can achieve full-rate coding for configurations other than $N_t = 2$. This advantage is achieved at the expense of diversity gain, that is, $G_d < N_t N_r$. Orthogonal STBCs and STTCs both achieve full diversity $G_d = N_t N_r$. Moreover, STTCs achieve also coding gain but at the expense of increased decoding complexity. In a period of T symbols the input-output equation is

$$\mathbf{Y} = \mathbf{HX} + \mathbf{N} \tag{7.10}$$

where $\mathbf{X}$ is the $N_t \times T$ codeword. The decoding is based on the assumption that the channel is known at the receiver, and the ML criterion is used, that is, the detection of the codeword is performed by

$$\hat{\mathbf{X}} = \arg\min_{\mathbf{X}} \left\| \mathbf{Y} - \mathbf{HX} \right\|_F^2 = \arg\min_{\mathbf{X}} \sum_{k=1}^{T} \left\| \mathbf{y}[k] - \mathbf{Hx}[k] \right\|_F^2 \tag{7.11}$$

where the minimization is over all possible codewords. The pairwise error probability (PEP) is the probability to detect the codeword $\mathbf{X}_B$ while the transmitted one was the $\mathbf{X}_A$. In Rayleigh fading and for high SNR values it is proved that

$$P[\mathbf{X}_A \rightarrow \mathbf{X}_B] \le \left(\prod_{i=1}^{R} \frac{1}{SNR\left(\lambda_i / 4N_t\right)} \right)^{N_r} = \left(\prod_{i=1}^{R} \lambda_i \right)^{-N_r} \left(\frac{SNR}{4N_t} \right)^{-RN_r} \tag{7.12}$$

where λ_i are the eigenvalues and R the rank of matrix $\left(\mathbf{X}_A - \mathbf{X}_B\right)\left(\mathbf{X}_A - \mathbf{X}_B\right)^H$. In order to minimize the PEP for each possible codeword pair $\mathbf{X}_p$, and $\mathbf{X}_q$, we should maximize the rank and the product of eigenvalues. These two criteria are the main ones used for STC design and the former is called the *rank criterion*, whereas the latter, the *determinant criterion*.

7.3 Multiuser MIMO

Multiuser MIMO (MU-MIMO) schemes have been proposed as an efficient way to deal with the shortcomings of practical SU-MIMO systems. In realistic networks, SU-MIMO systems are vulnerable in propagation conditions where either the SNIR is low, or due to the limited multipath richness and/or the presence of a strong line-of-sight (LOS) component the spatial correlation among antenna elements is high, or the number of antennas used at the mobile terminal (UE) is really small. All these factors affect the achievable diversity and spatial multiplexing gains. MU-MIMO antenna combining techniques in conjunction with resource allocation protocols among users provide unique benefits. The first is the robustness with respect to multipath richness, providing the capability to achieve full multiplexing gain regardless of the rank of each individual user. Furthermore, a compact antenna spacing at the BS is possible. The second benefit is the great opportunity to use single antenna terminals and still maintain the same diversity and multiplexing gains. To achieve these gains, however, the CSI of the users should be fed back to the BS. In a practical system this implies the design of a trade-off mechanism between feedback accuracy and uplink resources consumption, otherwise the practical settings are restricted to TDD or settings with low mobility, in order to extend in time the channel stationarity conditions.

7.3.1 Signal Models, Capacity Regions, and Sum-Rate Capacity

Assume an array of M antennas at the BS that serves simultaneously K autonomous users, with N_k antennas each. The maximum transmit power per user is P_k whereas, the total average transmit power is P. The users are typically separated by many wavelengths. The users do not collaborate to transmit or receive data. The MU-MIMO techniques imply the use of spatial sharing of the channel by the users, that is, it is a *spatial multiple access* scheme. The MU-MIMO multiple access channel (MAC), a terminology used in information theory for the cellular uplink channel, is a direct generalization of SU-MIMO techniques to multiuser case, whereas the downlink or broadcast channel (BC) is definitely more challenging.

7.3.1.1 MU-MIMO MAC

In the uplink, a discrete time Gaussian MAC is defined by the input-output relation:

$$\mathbf{y} = \sum_{k=1}^{K} \mathbf{H}_k^T \mathbf{x}_k + \mathbf{n} \tag{7.13}$$

where $\mathbf{x}_k$ is the $N_k \times 1$ kth user signal vector with covariance matrix $\mathbf{Q}_k = \mathbb{E}[\mathbf{x}_k \mathbf{x}_k^H]$ and each user is subject to an individual power constraint $Tr[\mathbf{Q}_k] \le P_k$. $\mathbf{H}_k \in \mathbb{C}^{N_k \times M}$ represents the flat fading channel matrix for the kth user, where the transpose operator is used for consistence with the BC notation (Gesbert et al. 2007). $\mathbf{n}$ is the $M \times 1$ i.i.d., zero mean, additive white Gaussian noise vector at the BS with covariance $\mathbb{E}[\mathbf{n}\mathbf{n}^H] = \sigma^2 \mathbf{I}_M$. It is assumed a perfect and instantaneous CSIR.

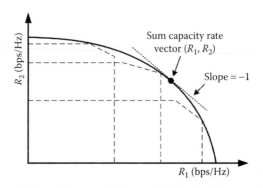

Figure 7.5   MU-MIMO MAC capacity region for $N_k > 1$.

The MAC capacity region is defined to be the convex closure of the set of achievable rate pairs for which messages may be conveyed with vanishingly small probability of error. For a K-user MU-MIMO channel, the performance limit is the K-dimensional capacity region consisting of all the vectors of rates $\mathbf{R} := (R_1,...,R_K)$ that can be achieved simultaneously by the K users. The MAC capacity region is a function of the users' channels $\mathbf{H}_k$ and the covariance matrices $\mathbf{Q}_k$. It is convenient to now consider the two-user case and depict a capacity region, as shown in Figure 7.5. This region is the union of all pentagons, each one of which corresponds to a different set of transmit covariance matrices (Goldsmith et al. 2003). The capacity of MAC is given then by

$$C^{\mathrm{MAC}}\left(\mathbf{H}_1,...,\mathbf{H}_k,\frac{P_1}{\sigma^2},...,\frac{P_k}{\sigma^2}\right) =$$

$$\bigcup_{\substack{\mathbf{Q}_k \geq 0 \\ Tr(\mathbf{Q}_k)\leq P_k,\forall k}} \left\{ \begin{array}{l} (R_1,...,R_K): \\ \sum_{k\in S} R_k \leq \log_2 \det\left(\mathbf{I}_M + \frac{1}{\sigma^2}\sum_{k\in S}\mathbf{H}_k\mathbf{Q}_k\mathbf{H}_k^H\right), \forall S \subseteq \{1,...,K\} \end{array} \right\} \tag{7.14}$$

A useful scalar metric that indicates a single *optimum* rate vector belonging to the capacity region is the *sum-rate capacity* defined as

$$\max_{\mathbf{R}\in C^{\mathrm{MAC}}} \sum_{k=1}^{K} R_k \tag{7.15}$$

The sum-rate is maximized by the rate vector corresponding to the point on the capacity region boundary at which the tangent line has slope -1. An efficient numerical technique to find the sum-rate maximizing covariance matrices is called *iterative waterfilling* (Yu et al. 2001).

In MAC each user's signal is received by the BS in the presence of interference from the other users. To cope with inter-user interference in an optimal way, the BS receiver uses multiuser detection to jointly detect the spatially multiplexed signals.

The receiver is based on an MMSE decoder followed by a successive interference cancellation (SIC). The complexity of this receiver is increased due to the absence of cooperation among the transmitting antennas.

7.3.1.2 MU-MIMO BC

In the downlink, the discrete time signal model of the BC for the kth user is given by

$$\mathbf{y}_k = \mathbf{H}_k \mathbf{x} + \mathbf{n}_k \tag{7.16}$$

where:

$\mathbf{H}_k \in \mathbb{C}^{N_k \times M}$ represents the flat fading channel matrix for the kth user

$\mathbf{n}_k$ is the $N_k \times 1$ i.i.d., zero mean, additive white Gaussian noise vector at the kth user

The transmit covariance matrix of the input signal is $\mathbf{Q}_x = \mathbb{E}[\mathbf{x}\mathbf{x}^H]$. The BS is subject to an average power constraint P, that is, $Tr[\mathbf{Q}_x] \leq P$. It has been proved, initially by Caire and Shamai (2003) for single-antenna users and then by Yu and Cioffi (2001) for multiantenna users, that the optimum transmit strategy for the MU-MIMO BC involves a theoretical interference precancellation technique known as *dirty paper coding* (DPC) combined with an implicit user scheduling and power loading algorithm. DPC is a joint encoding and modulation technique that allows known interference to be *presubtracted* at the transmitter without increasing the transmit power. The DPC encoder output provides data symbols for each of the K users which are precoded and transmitted simultaneously. DPC uses knowledge of the user's channels and data streams to perform coding in an ordered fashion among the users, and thereby to remove interference at the transmitter. Indeed, the transmitter first picks a codeword for receiver 1 (i.e., $\mathbf{x}_1$), then chooses a codeword for receiver 2 (i.e., $\mathbf{x}_2$), with full knowledge of the codeword intended for receiver 1 and therefore, the codeword of user 1 can be presubtracted such that receiver 2 is not interfered by the codeword for receiver 1. Similarly, the codeword for receiver 3 is chosen such that receiver 3 does not see the signals intended for receivers 1 and 2 as interference. This process continues for all users and provides a quite complicated capacity region. Receiver 1 sees the signals for all other users as interference, receiver 2 sees the signals for users 3 through K as interference and so on. Therefore, the ordering of the users is very important and each permutation π, in encoding operation, provides a different rate region given by

$$\mathbf{R}_\pi = \left\{ \left(R_{\pi(1)}, \ldots, R_{\pi(K)} \right) : R_{\pi(k)} = \log_2 \frac{\det\left[\mathbf{I}_M + \dfrac{1}{\sigma^2} \mathbf{H}_{\pi(k)} \left(\displaystyle\sum_{j=k}^{K} \mathbf{Q}_{\pi(j)} \right) \mathbf{H}_{\pi(k)}^H \right]}{\det\left[\mathbf{I}_M + \dfrac{1}{\sigma^2} \mathbf{H}_{\pi(k)} \left(\displaystyle\sum_{j=k+1}^{K} \mathbf{Q}_{\pi(j)} \right) \mathbf{H}_{\pi(k)}^H \right]} \right\} \tag{7.17}$$

where $\pi(k)$ denotes the index of the kth encoded user. The DPC region is defined as the convex hull of the union of all such rates vectors over all positive semidefinite covariance matrices $\mathbf{Q}_1,...,\mathbf{Q}_K$ such that $Tr(\mathbf{Q}_1+...+\mathbf{Q}_K)=Tr(\mathbf{Q}_x)\le P$ and over all permutations $[\pi(1),...,\pi(K)]$

$$C^{\text{DPC}}\left(\mathbf{H}_1,...\mathbf{H}_K,\frac{P}{\sigma^2}\right)\triangleq Co\left(\bigcup_{\pi,\mathbf{Q}_i}\mathbf{R}_\pi\right) \tag{7.18}$$

where:
   Co denotes the convex hull operation
   $\mathbf{R}_\pi$ the rate region for a given permutation π

The transmitted signal is $\mathbf{x}=\mathbf{x}_1+...+\mathbf{x}_K$ and the input covariance matrices are $\mathbf{Q}_i=\mathbb{E}[\mathbf{x}_i\mathbf{x}_i^H]$. The DPC yields uncorrelated signals $\mathbf{x}_1,...,\mathbf{x}_K$, which implies that $\mathbf{Q}_x=\mathbf{Q}_i+...+\mathbf{Q}_K$. The equations that provide the rates are neither a concave nor a convex function of the covariance matrices. Thus, the numerical solution is of great complexity since one should search the entire space of covariance matrices which meet the power constraint.

7.3.1.3 Duality and Scaling Laws

In Vishwanath et al. (2003) the authors prove a very useful theorem stating that the DPC region of MIMO BC with power constraint P is equal to the capacity region of the dual MIMO MAC with sum power constraint P

$$C^{\text{DPC}}\left(\mathbf{H}_1,...\mathbf{H}_K,\frac{P}{\sigma^2}\right)=\bigcup_{\sum_{k=1}^{K}P_k=P}C^{\text{MAC}}\left(\mathbf{H}_1,...,\mathbf{H}_k,\frac{P_1}{\sigma^2},...,\frac{P_k}{\sigma^2}\right) \tag{7.19}$$

Moreover, the authors in the same paper provide an explicit set of transformations to find covariance matrices in the BC/MAC that achieve the same rates. Therefore, one may numerically solve the convex problem of MAC, find the optimal MAC covariance matrices and then transform them to the corresponding optimal BC covariance matrices.

7.3.1.4 MAC Scaling Laws

At very low SNRs, the links are noise limited. Assume that all users transmit same power equal to P/K, and all users are equipped with the same number of antennas N. The limit for the sum rate capacity is given by Venkatesan et al. (2012):

$$\lim_{P/\sigma^2\to 0}C^{\text{MAC}}\left(\mathbf{H}_1,...,\mathbf{H}_k,\frac{P}{\sigma^2}\right)\approx\sum_{k=1}^{K}\frac{P}{\sigma^2 K}\lambda_{\max}^2\log_2 e \tag{7.20}$$

where λ_{max}^2 is the maximum eigenvalue of $\mathbf{H}_k$. Each user transmits with full power on its dominant eigenmode in order to maximize its own rate.

At very high SNRs the sum rate capacity is given by

$$\lim_{P/\sigma^2 \to \infty} C^{\mathrm{MAC}}\left(\mathbf{H}_1,\ldots,\mathbf{H}_k,\frac{P}{\sigma^2}\right) = \min\left(KN,M\right)\log_2\frac{P}{\sigma^2} \tag{7.21}$$

Thus, MAC scales with the minimum of the total number of transmit or receive antennas.

7.3.1.5 BC Scaling Laws

At very low SNRs,

$$\lim_{P/\sigma^2 \to 0} C^{\mathrm{DPC}}\left(\mathbf{H}_1,\ldots\mathbf{H}_K,\frac{P}{\sigma^2}\right) = \frac{P}{\sigma^2}\max_k\left[\lambda_{max}^2\left(\mathbf{H}_k\right)\right]\log_2 e \tag{7.22}$$

At very high SNRs, the sum-capacity of the BC is the same as the MAC's, given in Equation 7.21, if the BC power P is the same as the total power of the MAC users. The multiplexing gain, $\min(KN,M)$, achieved over the BC requires CSIT but does not require coordination between the users. This is an interesting result since, with full CSI at the transmitter and under the reasonable assumption that $K > M$, the sum rate capacity increases linearly in the number of transmit antennas at the BS. The number of simultaneously served users is arbitrary and the selection of the users to be served is done by the scheduler. Moreover, when $K > M$, the number of receive antennas per user N does not affect the scaling law. Thus, small and cheap single antenna terminals can be used and unlike in the SU-MIMO, the spatial multiplexing is maintained. The use of multiple antennas at the UE provides either diversity gain or an option for extra streams or interference cancellation. In addition, in MU-MIMO systems where the users are many wavelengths apart, the multiplexing gains are retained even in propagation conditions where the corresponding SU-MIMO systems would fail, that is, in cases with spatial correlation among antenna elements or in the existence of strong LOS component. If no CSIT is available at the BS, the capacity obtained is the same with that in point-to-point SU-MIMO systems

$$\lim_{P/\sigma^2 \to \infty} C^{\mathrm{DPC}}\left(\mathbf{H}_1,\ldots\mathbf{H}_K,\frac{P}{\sigma^2}\right) = \min\left(N,M\right)\log_2\frac{P}{\sigma^2} \tag{7.23}$$

For a large number of users, that is, $K \to \infty$, for fixed M and power P, and for full CSIT, the BC capacity is given by Hassibi and Sharif (2007)

$$\lim_{K \to \infty} \frac{C^{\mathrm{DPC}}\left(\mathbf{H}_1,\ldots\mathbf{H}_K,P/\sigma^2\right)}{\log_2\log_2\left(KN\right)} = M \tag{7.24}$$

This result implies that the multiplexing gain is M and it is obtained by the BS sending data to M carefully selected users out of K. The selection is performed by the scheduler based for example, on the criterion of sum rate maximization. Another interesting fact is that the number of receive antennas N plays very little role in the BC sum rate capacity. The scaling with respect to the number of users is doubly-logarithmic due to the tail of the Rayleigh distribution assumed in the model. If no channel knowledge is assumed at the BS, then the scaling with the number of users is

$$\lim_{K \to \infty} \frac{C^{\mathrm{DPC}}\left(\mathbf{H}_1, \ldots \mathbf{H}_K, P/\sigma^2\right)}{\log_2 \log_2 (K)} = 0 \qquad (7.25)$$

This results indicates that with no CSIT, the BS cannot exploit the users fading channels and their corresponding directions and may not be able to multiplex signals for multiple users, that is, to form spatial beams. Then the optimal scheme is to transmit to one user at a time.

7.3.2 Practical Issues and 3GPP Releases

Although DPC is the optimal encoding for MU-MIMO BC, it is a nonlinear technique with increased complexity. The precoding strategies that have been proposed for implementation in realistic systems include linear MMSE or ZF techniques and nonlinear approaches for example, vector perturbation (Hochwald et al. 2005) and Tomlinson-Harashima precoding (THP) (Zamir et al. 2002). The latter preserve additional transmit signal processing complexity in exchange for improved error rate performance.

As previously explained, to properly serve the spatially multiplexed users the MU-MIMO systems require CSI at the transmitter. CSIT, while not essential in SU-MIMO, it is of critical importance to MU-MIMO downlink precoding techniques. The strong need for CSIT feedback places a significant burden on uplink capacity, which is amplified when wideband waveforms are used, for example, OFDM, or when the user terminals preserve high mobility. An accurate CSI feedback would require an infinite number of bits and the reduced accuracy makes the system interference-limited and leads to decreased multiplexing gain. Realistic feedback strategies in the literature include vector quantization, dimension reduction, adaptive feedback, statistical feedback, and opportunistic SDMA.

Another challenge in MU-MIMO systems is the complexity of the scheduling procedure associated with the selection of a group of users that will be served simultaneously. Optimal scheduling involves exhaustive search. Therefore, other scheduling schemes include max-rate techniques, greedy user selection, and random user selection (Gesbert et al. 2007).

MU-MIMO techniques have been adopted in the LTE from the early releases (Lim et al. 2013). In LTE downlink the precoding is termed as *codebook-based* or *non-codebook-based* depending on whether it is based on cell-specific reference signals or UE-specific reference signals. Rel. 8 supported a primitive form of MU-MIMO based on codebook-based precoding. Each UE measures its spatial channel using the *common reference signal* (CRS) broadcasted from the BS, also known as eNB in LTE, selects the rank-1 precoder that best represents the channel from a predefined codebook set, and then feeds back the index of the codebook, called *precoding matrix indicator* (PMI), and the *channel quality information* (CQI) to the eNB via an uplink channel. The eNB based on the knowledge of CQI and PMI from different UE, selects those UE that experience a minimum level of inter-stream interference. In Rel. 8 MU-MIMO there is no UE-specific reference signal and the UE relies on CRS for both operations of channel estimation and demodulation of data. In Rel. 9 rank-2 transmissions, called dual layer beamforming, are supported using one more reference signal, the *demodulation RS* (DM-RS). Unlike the CRS, which is common to all users in a cell and hence not precoded, the DM-RS is UE-specific and precoded along with accompanying data. Moreover, in Rel. 9 the eNB may schedule two UE simultaneously with two layers for each UE, that is, four-stream supports multiplexing. In Rel. 10 there are three different reference signals in the downlink; CRS, DM-RS and channel state information (CSI)-RS. CRS is defined for only up to four transmit antennas to limit pilot over-head. DM-RS is used for demodulation of up to rank-8 and CSI-RS is defined for up to eight transmit antennas but with lower overhead compared to CRS. CSI-RS is transmitted in a fraction of subframes, whereas CRS is transmitted in every sub-frame. Therefore, a great flexibility and a trade-off between accuracy and signaling overhead is supported. In Rel. 10 no more than four UE are coscheduled, no more than two layers are allocated per UE, and no more than four layers are transmitted in total. The enhancements in MIMO technologies supported in later releases of 3GPP will be discussed in Sections 7.5 and 7.6.

7.4 Coordinated Multipoint

Coordinated multipoint (CoMP) transmission and reception belong to the category of advanced interference mitigation techniques proposed to improve not only the cell edge user throughput but also the average system throughput. Release 11 of 3GPP has included coordinated transmission in the downlink and coordinated reception in the uplink to manage inter-cell interference (ICI) and improve cell-edge coverage. In the former, the signals from multiple transmission points (TPs) are coordinated in a way to improve the received signal strength of the desired signal at a UE and/or to reduce the cochannel interference. In the latter, the uplink signal from a UE is received by a number of coordinated receiving points in order to improve the reliability and reduce the interference. These techniques may be applied to different deployment scenarios and by diverse cells, that is, macro, micro, pico, or even remote radio heads (RRH). The achieved performance of CoMP techniques depends mainly on the downlink CSI available at the BSs and this is obtained using uplink feedback from the UE to the BSs. Therefore, the feedback quality directly affects the performance.

The downlink coordinated transmission techniques may be categorized into coordinated scheduling and coordinated beamforming (CS/CB), joint transmission (JT), and dynamic point selection (DPS), or transmit point selection (TPS). In CS/CB, multiple coordinated transmission points (TPs) share the same CSI to coordinate scheduling/beamforming decisions, while data for a UE is available at and transmitted by a single TP. The coordinated scheduling and beamforming operations across multiple TPs are aligned to reduce the interference. In contrast, in JT scheme, data is simultaneously transmitted from multiple TPs to a single UE. Therefore, data is available simultaneously to all TPs. There are two JT schemes, the coherent and the noncoherent transmission. In the former, the signals from multiple TPs are jointly precoded to achieve coherent combining at the UE. Thus, the network has information related to the joint channel from all coordinated TPs to perform joint MIMO precoding. This requires very accurate synchronization between TPs. In the noncoherent JT, the network does not have information on the relationship of channels from the multiple TPs and the UE receives multiple transmissions individually precoded by each TP. The gain achieved is due to the increased power received by the UE. In a network with increased load, multiuser JT may be necessary in order to simultaneously transmit to multiple UE from the same set of TPs, leading to a combination of CoMP and MU-MIMO (Lee et al. 2012). In the TPS technique, the TP serving a UE may be changed at the subframe level according to the resources availability and the CSI. Again, data is available to all coordinated TPs.

The uplink coordinated reception techniques fall into two categories. The first is the dynamic selection of the best reception point according to the short-term channel quality and the second is the joint reception by many coordinated reception points. A major issue to be performed is the decoupling of the points that transmit downlink control signals from those that receive uplink data.

It is clear, that the coherent JT, more than any other technique, is sensitive to the accuracy of the feedback. Imperfect and/or outdated CSI directly affect the performance of JT. Moreover, it is necessary to keep tight synchronization and to share the CSI and the data for multiple UE among the coordinated TPs. These requirements make it hard to achieve the theoretical gains of JT in practical networks. It seems that the coordinated beamforming technique that comes with less stringent synchronization and coordination requirements, can achieve a satisfactory performance (Alexandropoulos et al. 2016).

7.5 Massive MIMO

The term *massive MIMO* is so appositely selected to depict not only the exceptionally large number of antennas at the BS but also the very serious implications to the cellular systems. Thomas L. Marzetta was the first to indicate the challenges and opportunities of scaling up MIMO to achieve huge improvements in throughput and energy efficiency (Marzetta 2010). The idea of massive MIMO builds upon the advantages offered by the MU-MIMO systems with the extra capability of being scalable technology. The very large number of service antennas relative to the number of active users in a cell, offer many benefits that include the great spatial focusing of energy into very small regions of space allowing for selectivity in transmitting and receiving data streams to/from the users, which in turn leads to greater throughput, increased energy efficiency, and effective power control. The main weapon in achieving all these improvements is the asymptotic orthogonality of channel vectors associated with distinct users, which in addition allows for greater simplicity in signal processing using linear precoding/decoding techniques. Other benefits are the use of inexpensive hardware, the reduced latency due to the avoidance of deep fades, the simplification of MAC layer protocol, and the robustness against intentional interference (Rusek et al. 2013). Nevertheless, there are also some issues to be addressed before massive MIMO becomes the 5G technology leader; these include the pilot contamination issue, the challenge of making many low-cost and low-precision components that work effectively together, the access of newly joined users, the development of novel cellular deployment scenarios, the power consumption of the BS RF front-end, and the overhead required for CSI acquisition.

The antenna arrays of massive MIMO systems contain hundreds of small and independently-controlled active antenna elements, fed via optical or electric circuits. Each antenna unit uses extremely low power, in the order of milliwatts. The deployment of the antenna array includes colocated and distributed scenarios. It is common in the respective literature to denote as M the number of BS antennas and as K the number of served users. Both are unconventionally large, but differ by a factor of two, four, or even an order of magnitude (Björnson et al. 2015). This is the reason massive MIMO systems offer huge spatial degrees-of-freedom and achieve strong signal gains, resilience to imperfect channel knowledge and low inter-user interference.

Massive MIMO is a large scale MU-MIMO and therefore, relies on spatial multiplexing, that is, on the BS having good enough channel knowledge, on both the UL and the DL. What is needed is a reliable transfer function information for the channels among all elements of the BS array and the users. For an FDD system, on the UL the K terminals send K orthogonal pilots, that is, known orthogonal training signals, based on which the BS estimates the CIR to each of the terminals. On the DL the BS sends M orthogonal pilot waveforms, based on which the terminals estimate the CIR, quantize the estimates, and feed them back to BS. This is a quite demanding operation and seems not feasible with hundreds of antennas. Moreover, the M estimates of the pilots are sent over UL slot, extending the portion of the slot dedicated to pilot information and decreasing the data portion. The case is even more difficult in high mobility scenarios where the fast time-varying channel calls for a frequent enough CSI update. The optimal DL pilots should be mutually orthogonal between the antennas, which means that the amount of time-frequency resources needed for DL pilots scales with M. The number of CIRs each terminal must estimate is proportional to the number of antennas and thus, the UL resources needed for the feedback also scale with M. The obvious solution is the TDD mode of operation (Marzetta 2015), based on the assumption that the UL and DL channels are reciprocal. Nevertheless, there are few proposals for the application of massive MIMO techniques in FDD systems (Chen and Lau 2014; Choi et al. 2014). It should be pointed out that the channel reciprocity in TDD systems breaks due to RF hardware asymmetry between UL and DL (Luo and Wang 2015; Luo 2016; Jiang et al. 2016) and calibration of the BS array is required (Vieira et al. 2014; Liu et al. 2015).

7.5.1 Pilot Contamination

Consider an OFDM-based MU-MIMO system with L cells and let g_{nmjkl} denote the channel coefficient between the mth BS antenna ($m = 1,\ldots,M$) in the jth cell ($j = 1,\ldots,L$), and the kth user terminal ($k = 1,\ldots,K$) in the lth cell ($l = 1,\ldots,L$) in the nth subcarrier ($n = 1,\ldots,N_{FFT}$), where N_{FFT} is the number of subcarriers. This coefficient equals a complex small-scale fading factor (h_{nmjkl}) times an amplitude factor (β_{jkl}) that accounts for geometric attenuation and shadow fading,

$$g_{nmjkl} = h_{nmjkl} \cdot \beta_{jkl}^{1/2} \tag{7.26}$$

The small-scale fading coefficients (h_{nmjkl}) are assumed to be zero-mean, unit-variance and piecewise constant over the frequency smoothness interval. The latter is defined as the reciprocal of the OFDM guard interval measured in subcarrier spacings. For channel estimation purposes, only one pilot per frequency smoothness interval is required. The large-scale fading coefficients (β_{jkl}) are the same for all subcarriers and for different antennas at the same BS since the geometric attenuation and the shadowing change slowly over space. The $M \times K$ channel matrix between all K users in the lth cell to the M antennas at the jth BS is expressed as

$$\mathbf{G}_{jl} = \begin{pmatrix} g_{n1j1l} & \cdots & g_{n1jKl} \\ \vdots & \ddots & \vdots \\ g_{nMj1l} & \cdots & g_{nMjKl} \end{pmatrix} = \mathbf{H}_{jl}\mathbf{D}_{jl}^{1/2} \tag{7.27}$$

where for notational simplicity we have dropped the subcarrier index, and $[\mathbf{H}_{jl}]_{mk} = h_{nmjkl}$, $\mathbf{D}_{jl} = diag(\beta_{j1l},\ldots,\beta_{jKl})$.

For a TDD-based massive MIMO system, pilot training sequences are transmitted by the users in the UL to assist channel estimation. Let $\boldsymbol{\psi}_{kl} = (\psi_{kl}^{[1]},\ldots,\psi_{kl}^{[\tau]})^T$ be the pilot sequence of kth user in lth cell, where τ denotes the length of the pilot sequence. The sequences employed by users within the same cell and in the neighboring cells should be orthogonal, that is for unit energy sequences (Lu et al. 2014),

$$\psi_{k_1 l_1}^{H} \psi_{k_2 l_2} = \delta[k_1 - k_2]\delta[l_1 - l_2] \qquad (7.28)$$

where $\delta[\cdot]$ is the Kronecker delta. However, the max number of orthogonal pilot sequences that can exist, is upper-bounded by the duration of the coherence interval divided by the channel delay spread. This limits the number of users that can be served, especially in a multicell system. In order to accommodate more users, one may allocate nonorthogonal sequences to neighboring cells, that is, $\psi_{k_1 l_1}^{H} \psi_{k_2 l_2} \neq 0$, or reuse the pilot sequences. A direct implication is that the channel vector to a user becomes correlated with the channel vectors of those users with nonorthogonal sequences. The effect of reusing pilots from one cell to another and the associated negative consequences of correlated channel vectors is termed *pilot contamination*. When the BS antenna array correlates its received pilot sequence with the sequence associated with a particular user, it derives a channel estimate that is contaminated by a linear combination of channels with other users that use the same sequence. During DL transmission, the beamforming signal that is based on the contaminated channel estimate results in interference directed at those users that share the same sequence. This directional interference will not disappear with the increase of M. A similar phenomenon occurs during the UL transmission.

Assume that the pilot sequences $\mathbf{\Phi} = (\psi_1, \psi_2, \ldots, \psi_K) \in \mathbb{C}^{\tau \times K}$ used in one cell are orthogonal, that is, $\mathbf{\Phi}^{H}\mathbf{\Phi} = \tau \mathbf{I}_K$, and the same pilot group is reused in other cells. Assume further that the pilot transmission from different cells is synchronized, which is actually the worst case. The conventional pilot assignment methods usually assign the sequence ψ_k to the kth user.

The received signal matrix at the jth BS, $\mathbf{Y}_j^P \in \mathbb{C}^{M \times \tau}$, is given by

$$\mathbf{Y}_j^P = \sqrt{\rho_p} \sum_{l=1}^{L} \mathbf{G}_{jl} \mathbf{\Phi}^T + \mathbf{N}_j^P = \sqrt{\rho_p} \sum_{l=1}^{L} \sum_{k=1}^{K} \mathbf{g}_{jkl} \psi_k^T + \mathbf{N}_j^P \qquad (7.29)$$

where ρ_p is the pilot transmit power and $\mathbf{N}_j^P \in \mathbb{C}^{M \times \tau}$ is the noise matrix at the jth BS during pilot transmission, whose components are i.i.d. circular complex Gaussian random variables with zero-mean and unit variance, and thus mutually uncorrelated, and uncorrelated with the channel matrices. It is reminded that the kth column of the matrix $\mathbf{G}_{jl}$ denoted by $\mathbf{g}_{jkl}$, is the channel vector between the kth user in the lth cell and the M antennas of the jth cell. The channel estimate of the kth user in the jth cell, is obtained at the jth cell by correlating the received pilot sequence with the corresponding sequence ψ_k

$$\hat{\mathbf{g}}_{jkj} = \frac{1}{\sqrt{\rho_p}\tau} \mathbf{Y}_j^P \psi_k^* = \sum_{l=1}^{L} \mathbf{g}_{jkl} + \frac{1}{\sqrt{\rho_p}\tau} \mathbf{N}_j^P \psi_k^* \qquad (7.30)$$

It is clear that the channel estimate is a linear combination of the channel vectors of the users with the same pilots (*k*th users) in all cells, a phenomenon referred to as *pilot contamination*. The estimate at the *j*th BS for all users is given by

$$\hat{\mathbf{G}}_{jj} = \frac{1}{\sqrt{\rho_p}\tau} \mathbf{Y}_j^p \mathbf{\Phi}^* = \mathbf{G}_{jj} + \sum_{l \neq j} \mathbf{G}_{jl} + \frac{1}{\sqrt{\rho_p}\tau} \mathbf{N}_j^p \mathbf{\Phi}^* \qquad (7.31)$$

There have been proposed many techniques to mitigate pilot contamination that fall in three main categories: optimization of pilot sequences allocation in multicell systems, where pilot reuse policies are followed to put mutually contaminated cells farther apart (Su and Yang 2015; Zhu et al. 2015), smart channel estimation techniques or even blind channel estimation (Ngo and Larsson 2012; Müller et al. 2014; Teeti et al. 2015), and pilot contamination precoding techniques that utilize cooperative transmission in multicell systems to nullify the directed interference that results from pilot contamination. Such precoding techniques are commented in the following paragraph. Recently, a novel pilot decontamination technique was proposed in Chen and Yang (2016), which combines a channel estimation and pilot assignment technique by exploiting the sparsity of correlated channels.

7.5.2 Orthogonality and Signal Processing Techniques

As already mentioned, the important result of increasing the number of antenna elements at the BS is the asymptotic orthogonality of channel vectors to different user terminals. These terminals are separated by many wavelengths and the random matrix theory implies that the effects of uncorrelated noise and small-scale fading are eliminated. Assuming single-antenna terminals, linear processing techniques are optimal and can be used, that is, maximum ratio combining on the UL (also called matched filtering—MF) and maximum ratio transmission on the DL. The signal model used to illustrate these results follows. It is based on a TDD with pilot reuse multicell system where the BSs are equipped with *M* antennas and there are *K* single-antenna user terminals in each cell.

7.5.2.1 Uplink Transmission

Assuming that channel estimation has been performed, the received signal at the *j*th BS is given by

$$\mathbf{y}_j^u = \sqrt{\rho_u} \sum_{l=1}^{L} \sum_{k=1}^{K} \mathbf{g}_{jkl} x_{kl}^u + \mathbf{n}_j^u = \sqrt{\rho_u} \sum_{l=1}^{L} \mathbf{G}_{jl} \mathbf{x}_l^u + \mathbf{n}_j^u \qquad (7.32)$$

where:

x_{kl}^u is the symbol from the kth user in the lth cell with $E\{|x_{kl}^u|^2\}=1$, $\mathbf{x}_l^u$ is the $K \times 1$ vector of symbols from all users in the lth cell

ρ_u is the uplink transmission power

$\mathbf{n}_j^u$ is the AWGN vector with $E\{\mathbf{n}_j^u(\mathbf{n}_j^u)^H\}=\sigma_n^2\mathbf{I}_M$

The elements of $\mathbf{n}_j^u$ are also uncorrelated with the channel matrices. When the matched filter detector is applied (maximum ratio combining), the BS processes the received signal by multiplying it by the conjugate transpose of the channel estimate. The detected symbol for the kth user in the jth cell is given by

$$\hat{x}_{kj}^u = \hat{\mathbf{g}}_{jkj}^H \mathbf{y}_j^u$$

$$= \left(\sum_{l_1=1}^{L}\mathbf{g}_{jkl_1}+\frac{1}{\sqrt{\rho_p}\tau}\mathbf{N}_j^p\boldsymbol{\psi}_k^*\right)^H\left(\sqrt{\rho_u}\sum_{l_2=1}^{L}\sum_{k'=1}^{K}\mathbf{g}_{jk'l_2}x_{k'l_2}^u+\mathbf{n}_j^u\right) \qquad (7.33)$$

$$= \sqrt{\rho_u}\left(\mathbf{g}_{jkj}^H\mathbf{g}_{jkj}x_{kj}^u+\sum_{l_2\neq j}\mathbf{g}_{jkl_2}^H\mathbf{g}_{jkl_2}x_{kl_2}^u\right)+v_{kj}^u$$

where v_{kj}^u denotes the intracell interference and uncorrelated noise, which can be significantly reduced by increasing the number of antennas at the BS. The second term in parenthesis constitutes a residual interference due to transmission from users in other cells that use the same kth pilot sequence. This expression contains inner products between M-component random vectors. As M grows without bound the L2-norms of the vectors grow proportional to M, while the inner products of uncorrelated vectors grow at a lesser rate (Marzetta 2010). For practical large values of M only the products of identical quantities remain significant.

The detected symbol vector for all users in the jth cell is given by the multiplication of the received signal vector with the conjugate transpose of the channel matrix estimate,

$$\hat{\mathbf{x}}_j^u = \hat{\mathbf{G}}_{jj}^H\mathbf{y}_j^u$$

$$= \left(\mathbf{G}_{jj}+\sum_{l_1\neq j}\mathbf{G}_{jl_1}+\frac{1}{\sqrt{\rho_p}\tau}\mathbf{N}_j^p\boldsymbol{\Phi}^*\right)^H\left(\sqrt{\rho_u}\sum_{l_2=1}^{L}\mathbf{G}_{jl_2}\mathbf{x}_{l_2}^u+\mathbf{n}_j^u\right) \qquad (7.34)$$

The same observations for the inner products apply to this equation. Using Equation 7.27, one may write

$$\frac{1}{M}\mathbf{G}_{jl_1}^H \mathbf{G}_{jl_2} = \mathbf{D}_{jl_1}^{1/2}\left(\frac{\mathbf{H}_{jl_1}^H \mathbf{H}_{jl_2}}{M}\right)\mathbf{D}_{jl_2}^{1/2} \tag{7.35}$$

As M grows without limit and assuming *favorable propagation conditions* we get the so called *asymptotic channel orthogonality*

$$\frac{1}{M}\mathbf{H}_{jl_1}^H \mathbf{H}_{jl_2} \to \mathbf{I}_K \delta[l_1 - l_2] \tag{7.36}$$

and the corresponding detected signal vector is written as

$$\frac{1}{M\sqrt{\rho_u}}\hat{\mathbf{x}}_j^u \to \mathbf{D}_{jj}\mathbf{x}_j^u + \sum_{l \neq j}\mathbf{D}_{jl}\mathbf{x}_l^u \tag{7.37}$$

and the kth component of the detected signal is written as

$$\frac{1}{M\sqrt{\rho_u}}\hat{x}_{kj}^u \to \beta_{jkj}x_{kj} + \sum_{l \neq j}\beta_{jkl}x_{kl} \tag{7.38}$$

Therefore, for a very large number of BS antennas the effects of uncorrelated noise and fast fading are eliminated completely, and the intracell interference is zeroed. However, inter-cell interference is present. The effective signal-to-interference ratio (SIR) is identical for all subcarriers but depends on the cell and the terminal, that is,

$$\text{SIR}_{jk}^u = \frac{\beta_{jkj}^2}{\sum_{l \neq j}\beta_{jkl}^2} \tag{7.39}$$

The important observation is that the SIR is constant along the frequency domain because the shadow-fading coefficients are independent of frequency. Moreover, the SIR is independent of the cell-size because the range dependence is the same in the numerator and the denominator.

7.5.2.2 Downlink Transmission

In the DL the BS uses the channel estimation obtained from UL and performs conjugate beamforming, that is, transmits a symbol vector through a precoding matrix which is proportional to the conjugate transpose of the channel estimate. The DL transmission from a BS to the kth user in the cell suffers from interference due to transmissions from BSs in neighboring cells to the respective kth users. The received symbol vector by the K users in the lth cell is given by

$$\mathbf{y}_i^d = \sqrt{\rho_d} \sum_{j=1}^{L} \mathbf{G}_{ji}^T \hat{\mathbf{G}}_{jj}^* \mathbf{x}_i^d + \mathbf{n}_i^d$$

$$= \sqrt{\rho_d} \sum_{j=1}^{L} \mathbf{G}_{ji}^T \left(\mathbf{G}_{jj} + \sum_{l \neq j} \mathbf{G}_{jl} + \frac{1}{\sqrt{\rho_p} \tau} \mathbf{N}_j^p \Phi^* \right)^* \mathbf{x}_i^d + \mathbf{n}_i^d \qquad (7.40)$$

where ρ_d is the DL transmission power. If the number of BS antennas increases without limit, then one may use Equations 7.35 and 7.36 to conclude for the detected signal vector that

$$\frac{1}{M\sqrt{\rho_d}} \hat{\mathbf{x}}_i^d \to \mathbf{D}_{ll} \mathbf{x}_i^d + \sum_{j \neq l} \mathbf{D}_{jl} \mathbf{x}_j^d \qquad (7.41)$$

and the kth user in the lth cell receives

$$\frac{1}{M\sqrt{\rho_d}} \hat{x}_{kl}^d \to \beta_{lkl} x_{kl} + \sum_{j \neq l} \beta_{jkl} x_{kj} \qquad (7.42)$$

The second term on the right hand of the equation implies that each BS transmits coherent interference to the users of other cells due to the pilot contamination effect. The effective SIR is given by

$$\mathrm{SIR}_{lk}^d = \frac{\beta_{lkl}^2}{\sum_{j \neq l} \beta_{jkl}^2} \qquad (7.43)$$

The aforementioned analysis has utilized the MF decoding in the UL and the conjugate beamforming in the DL. Other linear types of precoding and decoding like ZF or MMSE can be used with better results in high SNR regime, but with increased complexity (Jose et al. 2011). There are also several proposals in the literature for the case of multicell precoding where many BSs cooperate to jointly serve users in different cells. Three main techniques have been proposed depending on the overhead and the information exchange among BSs (Lu et al. 2014): single-cell processing, coordinated beamforming and network MIMO multicell processing. In single-cell processing a BS has channel knowledge only for the users in this cell and not for users in other cells. This technique reduces the overhead but cannot deal with the inter-cell interference (Hoydis et al. 2013). The network MIMO multicell processing provides the best results but with increased overhead (Huh et al. 2012). Coordinated beamforming achieves a trade-off between performance and required overhead (Zhang and Cui 2010).

7.5.3 Achievable Rates and Efficiency

The achievable rates for the uplink and the downlink in a single cell case are given by applying the asymptotic orthogonality condition to the corresponding MU-MIMO rates for the MAC and BC. The total throughput in uplink is written as

$$C^{UL} = \log_2 \det\left(\mathbf{I}_K + \rho_u \mathbf{G}^H \mathbf{G}\right)$$

$$\overset{M \gg K}{\approx} \log_2 \det\left(\mathbf{I}_K + M\rho_u \mathbf{D}\right) \qquad (7.44)$$

$$= \sum_{k=1}^{K} \log_2\left(1 + M\rho_u \beta_k\right)$$

In the downlink the channel is known at both ends of the link and it is possible for the BS to perform power allocation to maximize the sum rate. If $\mathbf{P}$ is a positive diagonal matrix with the power allocations $(p_1, \ldots, p_K)$ at its diagonal elements, and $\sum_{k=1}^{K} p_k = 1$, the sum rate is written as

$$C^{DL} = \max_{\mathbf{P}} \log_2 \det\left(\mathbf{I}_M + \rho_d \mathbf{G} \mathbf{P} \mathbf{G}^H\right)$$

$$\overset{M \gg K}{\approx} \max_{\mathbf{P}} \log_2 \det\left(\mathbf{I}_K + M\rho_d \mathbf{P} \mathbf{D}\right) \qquad (7.45)$$

The authors in Ngo et al. (2013) derived the power scaling law for massive-MIMO systems. The basic result is that one user can scale down the transmit power proportional to $1/M$ if perfect CSI is assumed and to $1/\sqrt{M}$ if imperfect CSI is assumed, to get the same performance as in the SISO case. This results applies even in multicell systems with pilot contamination. In the same publication the authors also studied the trade-off between energy and spectral efficiency for the uplink. They defined the *energy efficiency* as the ratio of the spectral efficiency and the transmit power. If perfect CSI is assumed, the energy efficiency decreases as the spectral efficiency increases. The same applies to the case of imperfect CSI and high transmit power region. On the contrary, in the low transmit power region and for imperfect CSI the energy efficiency increases with the spectral efficiency. The authors also studied different precoding techniques. At high spectral efficiency, ZF outperforms MRC. This is due to the fact that MRC is limited by the intracell interference, which is significant at high spectral efficiency. Thus, when transmitted power is increased, the spectral efficiency of MRC approaches a constant value, while the energy efficiency goes to zero. These observations are obtained from Figure 7.6 where all simulation parameters are the same with those given in Ngo et al. (2013). Indeed, the reference point for the relative efficiency is the case with $K = 1, M = 1$ and transmit power of 10 dB. For the multicell scenario seven (7) cells are assumed with an intercell interference factor equal to $\beta = 0.04$.

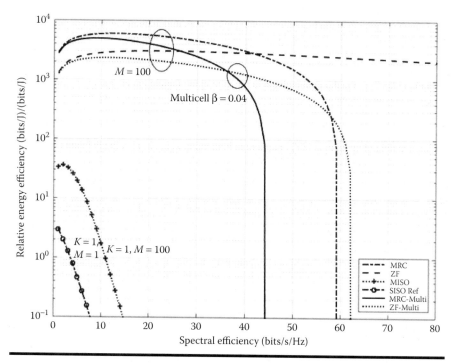

Figure 7.6 **Energy efficiency versus spectral efficiency for massive MIMO systems. ZF and MRC are compared for single-cell and multicell scenarios.**

7.5.4 Hardware Impairments

The deployment of the antenna array includes colocated and distributed scenarios. The antenna elements in colocated configurations may form 2D or 3D array structures in order to exploit three dimensional channel characteristics, that is, being able to resolve paths in azimuth and elevation. The compactness of these arrays implies the effect of mutual coupling, which calls for efficient matching networks. Another important issue related to the channel and the antenna arrays is the correlation of the signals received by the closely spaced elements. What really matters in massive MIMO systems is the effects of coupling and correlation on the asymptotic orthogonality of the channel vectors. It has been shown by simulation in Rusek et al. (2013) that there is a power loss and a capacity penalty due to these effects. In Björnson et al. (2015), the authors studied the effect of hardware imperfections at the BSs taking into account the multiplicative phase-drifts, additive distortion noise, noise amplifications, and intercarrier interference. It was concluded that only the phase drifts limit the achievable rates as the number of antennas grows to infinity. This implies that massive MIMO systems are quite robust to hardware imperfections due to the fact that distortions are uncorrelated with the useful signals. In the same publication the authors established scaling laws for the variance of the distortion noise and receiver noise, and showed that a circuit-aware design can make the total circuit power consumption of the M ADCs and LNAs increase as $\sqrt{M}$.

7.5.5 Testbeds and Measurement Results

The feasibility of massive MIMO systems has been questioned many times in the literature, especially with respect to its application to 5G cellular systems. There are few testbeds and measurement campaigns that attempt to answer the question on the feasibility and prove the advantages promised by massive MIMO concept (Hoydis et al. 2012; Martnez et al. 2014; Vieira et al. 2014; Gao et al. 2015). The first crucial assumption is the favorable propagation conditions and the corresponding asymptotic orthogonality of the channel vectors. This assumption has been validated in a great extent by outdoor measurements in Hoydis et al. (2012) where a scalable virtual antenna array of up to 112 elements was used. The orthogonality was examined using the correlation coefficient and the condition number as meaningful metrics. The same issue was addressed in Gao et al. (2015) where outdoor-to-outdoor measurements were performed at 2.6 GHz using two different arrays: a virtual uniform linear array (ULA) with a large aperture and a uniform cylindrical array (UCA) with 128 antenna ports. Three propagation environments were measured and the favorable propagation conditions were confirmed. Moreover, singular value spreads and sum-rate capacities are calculated for LOS and NLOS conditions as well as for closely located users. Again, the conclusions of this research are very promising for the application of massive MIMO systems. In Vieira et al. (2014) the authors presented the Lund University Massive MIMO testbed (LuMaMi) and performed indoor measurements with four single-antenna users. The testbed operated with 20 MHz bandwidth OFDM signals and with 100 antennas at the BS. The importance of this research activity lies on the implementation of massive MIMO system with off-the-shelf hardware components. In Martnez et al. (2014) the authors used three different arrays to investigate the impact of the array aperture on the performance of the massive MIMO systems. The arrays were equipped with 64 antennas and eight user terminals were used with two antennas each. The measurements were conducted in indoor environment under LOS and NLOS conditions, for various user terminals distribution and terminals proximity. The main result is that the performance is increased with the aperture, which directly affects the number of spatial degrees of freedom available in real scenarios.

7.6 Full-Dimension MIMO

FD-MIMO is an interim technology term used in 3GPP to denote the massive-MIMO capabilities of LTE-Advanced Pro standard (Release 13). Because the direct implementation of massive-MIMO systems is still not mature for the reasons mentioned in Section 7.5, 3GPP initiated a study (3GPP 2015) to propose a feasible way of integration of this technology to 4G systems. FD-MIMO was initially proposed in Nam et al. (2013) and utilizes a large number of antennas placed in a 2D rectangular antenna array at the BSs. The 2D grid allows for spatial separation not only in the azimuth domain but also in the elevation domain. A direct implication of the utilization of 2D arrays is that the radiation beam patterns produced by these arrays may be adapted in both elevation and azimuth planes providing 3D beamforming capabilities and consequently, offering many more degrees of freedom in the spatial separation of users in the cells (Razavizadeh et al. 2014). This is achieved by dynamic and adaptive precoding across all antenna elements (Kim et al. 2014). The 3D in beamforming, sometimes called *vertical beamforming*, allows for different power allocation to the beams that serve cell-edge and cell-centre regions, maintaining low intercell interference.

In the following the main issues related to the application of FD-MIMO to next-generation mobile communication systems are discussed.

7.6.1 CSI, Feedback, and Beamforming

One of the main challenges in implementing massive MIMO systems is the acquisition of instantaneous CSI at the BS, that is, on the UL. In TDD systems, one may exploit channel reciprocity to obtain the instantaneous CSI at BS via uplink training. If nonorthogonal pilots are employed in neighboring cells, channel estimation in a given cell will be impaired by the pilots in other cells, a phenomenon referred to as pilot contamination. Therefore, even in TDD mode, the measurement at the BS does not capture the DL interference from neighboring cells or coscheduled user terminals. In FDD systems, time variation and frequency response is measured via downlink pilot signals, called downlink reference signals (RS) in LTE, and the BS acquires the CSI through a feedback channel. If a large number of antennas is used at the BS, the feedback link is overwhelmed and it is quite demanding for the BS to get accurate CSI, especially for users with high mobility. An alternative approach is to exploit the second-order statistics of the channel and utilize slowly varying statistical CSI. This allows for a long-term feedback (Adhikary et al. 2013). Recently, the authors in Li et al. (2016) proposed a new downlink low complexity 3D beamforming space-division multiple access (SDMA) transmission algorithm for FDD FD-MIMO systems, where each user has access to its own perfect effective CSI, while the BS has only the statistical CSI of each user. They derived the optimal beamforming vector for each user, as well as the main guidelines for user scheduling, by maximizing a lower bound of the average signal-to-leakage-plus-noise ratio (SLNR). The term *leakage* in SLNR refers to the total power leaked from one user's beamforming direction to other user's channel direction. Numerical results presented in that paper showed that, even with statistical CSI at the BS, the achievable sum rate performance of the proposed algorithm is comparable to traditional ZF and MF algorithm with imperfect instantaneous CSI, while requires much less feedback overhead for FDD systems and is of low complexity. The algorithm is based on the observation that the BS should schedule the users that have orthogonal strongest horizontal or vertical statistical eigenmodes, and the optimal transmission strategy is to transmit signals along the strongest horizontal and vertical statistical eigenmode of each user, based on the maximization of a lower bound of average SLNR. Moreover, the scheduler should select users whose strongest horizontal and vertical statistical eigenmodes capture most of their channels' power.

7.6.1.1 Reference Signals in 3GPP

In LTE, Rel. 8 and 9 of 3GPP standards, channel training and data demodulation are based on the same RS called common RS (CRS). In LTE-Advanced, Rel. 10, two new RSs have been introduced: the CSI-RS to assist to CSI acquisition, and the demodulation RS (DM-RS) to perform demodulation of data channel (Lim et al. 2013). The first is common to all users in the cell and thus unprecoded, the latter is UE-specific and is precoded with the same vector as the data. In order to minimize overhead, CSI-RS is transmitted in a fraction of subframes, whereas the CRS, which is used for both demodulation and CSI measurements, needs to be transmitted every subframe. In FD-MIMO a new RS is proposed, the precoded or beamformed CSI-RS that provides lower UL feedback overhead, lower DL pilot overhead, and higher quality in RS (Ji et al. 2016). In order to support the beam-formed CSI-RS scheme, a new transmitter architecture called the transceiver unit (TXRU) architecture has been introduced, that is, a hardware connection between the baseband signal path and antenna array elements.

7.6.2 Antenna Issues and Architectures

One of the practical limiting factors in massive MIMO systems is the space needed in BSs for large antenna arrays at low operating frequencies. For a ULA with copolarized antenna elements, a horizontal space of about 4.7 m is required to fit 64 antenna elements with 0.5λ interelement distance for an LTE carrier frequency of 2.0 GHz. The requirement is relaxed if one uses cross-polarized elements. In the case of vertical sectorization with two or three sectors, the number of elements in the vertical plane may be small. This is not the case when dynamic adaptation of beams per UE is required. Therefore, in FD-MIMO 2D grids are used to place co- or cross-polar antenna elements leading to significantly reduced dimensions, for example, for an a 8×8 array with dual polarized elements an area of 1.0×0.5 m is needed. The model used for antenna radiation pattern is that introduced in (ITU-R 2009)

$$A(\theta,\phi) = G_{\max} - \min\left[A_H(\theta) + A_V(\phi), A_m \right] \qquad (7.46)$$

where:

G_{max} is the maximum antenna gain at the main beam direction (boresight)

θ and ϕ are the angles of the user direction from the boresight in the horizontal and the vertical plane, respectively

A_m is the side-lobe level attenuation of the antenna pattern, also called maximum attenuation

$A_H(\theta)$ is the relative antenna gain (dB) in the horizontal plane, $-180° \leq \theta \leq 180°$

$A_V(\phi)$ is the relative antenna gain (dB) in the vertical plain, $-90° \leq \phi \leq 90°$, given by

$$A_H(\theta) = \min\left[12\left(\frac{\theta}{\theta_{3dB}}\right)^2, A_m \right] \tag{7.47}$$

$$A_V(\phi) = \min\left[12\left(\frac{\phi - \phi_{tilt}}{\phi_{3dB}}\right)^2, A_m \right] \tag{7.48}$$

where:

ϕ_{tilt} represents the tilt angle of the main beam

θ_{3dB} and ϕ_{3dB} are the 3 dB beamwidth of the horizontal and vertical patterns, respectively

The angles are given in Figure 7.7a and b. Typical values for the parameters are $G_{max} = 17$ dBi, $A_m = 20 - 30$ dB, $\theta_{3dB} = 60° \sim 70°$, $\phi_{3dB} = 8° \sim 15°$.

We assume that the BS is deployed with an $M \times N$ uniform planar array (UPA) with N antennas in each row, each row forming a ULA, and a total number of M rows in the vertical dimension. Another useful parameter is the number of polarization dimensions P, which is equal to 1 for single-polarized antennas and equal to 2 for dual-polarized antennas. A typical configuration indicated in (3GPP 2015) is that of dual polarized antenna array ($P = 2$) with $M = 8$ with 0.8λ spacing in vertical dimension and $N = 4$ with 0.5λ spacing in horizontal dimension.

As already mentioned in previous paragraph, the TXRU architecture is responsible for the connection of baseband and antenna array. The active antenna elements incorporate phase shifters and variable gain amplifiers so that the BS can control the phase and the gain of all antenna elements. The TXRU architecture is the feeding network that brings the baseband signals from the TXRUs to the array elements. In (3GPP 2015) it is called *TXRU virtualization model*. A TXRU is only associated with antenna elements with the same polarization. According to Ji et al. (2016), depending on the CSI-RS transmission and feedback strategy, two representative options, *array partitioning* and *array connected* architecture, are suggested and shown in Figure 7.8. The former is for the conventional codebook scheme and the latter is for the beamforming scheme. In the array partitioning architecture, antenna elements are divided into multiple groups (subarrays) and each TXRU is connected to one of

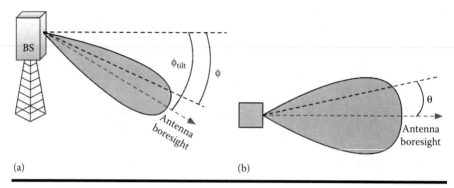

Figure 7.7   BS Antenna pattern planes: (a) vertical plane and (b) horizontal plane.

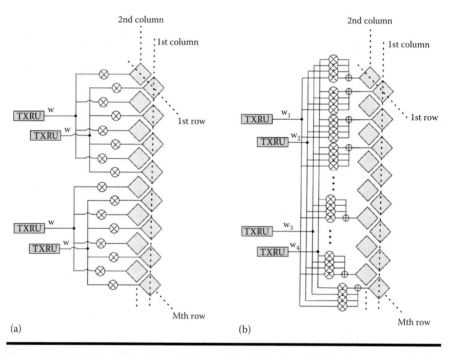

Figure 7.8   TXRU architectures: (a) array partitioning and (b) array connected.

them, whereas in the array connected structure, the RF signals of multiple TXRUs are delivered to a single antenna element. In the array partitioning an orthogonal CSI-RS is assigned to each group, that is, to each TXRU. In order to achieve the same coverage for all CSI-RSs an identical weight is applied to all groups (subarrays). In the array connected architecture, each antenna element is connected to more than one TXRUs and an orthogonal CSI-RS is assigned for each TXRU. A different 3D beamforming weight for each beam is applied to all antenna elements.

The CSI feedback mechanisms that proposed in (3GPP 2015) are the composite codebook and the beam index feedback. In the first scheme the overall codebook is divided into two codebooks one for vertical and one for horizontal plane, and these are fed back separately to the BS. This way the two codebooks may be different in size. The combination of the two codebooks is based on Kronecker product. In the beam index feedback the user terminal send back the index of the beam that maximizes the received power and the corresponding CQI. If dual-polarized antennas are used then extra information regarding the relative phase is required.

7.6.3 3D Channel Model

The performance evaluation of any novel technique that incorporates azimuth and elevation angles, requires a 3D channel model to characterize the propagation paths in both the horizontal and vertical directions along with time and frequency characteristics. The majority of channel models used in the literature of mobile communication systems are two dimensional models that neglect the elevation angle. Typical examples are the spatial channel model (SCM) proposed by 3GPP in (3GPP 2014), and the WINNER channel model proposed in (WINNER II WP1 2007). The latter has been recommended by ITU-R as a baseline for performance evaluation of various radio interface technologies in (ITU-R 2009). Nevertheless, especially in urban environments, a great percentage of energy is incident with elevation angles larger than 10° (Kuchar et al. 2000). Therefore, 3D channel models are required and in this direction WINNER+ as an extension to WINNER model was proposed in Hentila et al. (2010). Moreover, 3GPP initiated a study on 3D models in order to support the performance evaluation of FD-MIMO techniques (3GPP 2013). A critical issue is the implementation of these models with no limitations to antenna array structure and propagation environment. Some implementations of the 3GPP model are given in Kammoun et al. (2014) and Ademaj et al. (2015, 2016). This model is a 3D geometric stochastic model where the scatterers are represented by statistical parameters. Three environmental scenarios are defined, urban macrocell (UMa), urban micro cell (UMi), and UMa-high rise (UMa-H). In UMa and UMa-H scenarios a sector antenna height of 25 m is considered, that is, above the rooftop level, whereas in UMi scenario the height of the antenna is 10 m, that is, below the rooftop level. The propagation conditions specified are the LOS, the non-line-of-sight (NLOS), and the outdoor-to-indoor (O-to-I). For each of these conditions different parameters have been defined for the mean path loss, the large-scale, and the small-scale fading. The main features to be highlighted is that the LOS conditions, the path loss and the spread of departure angles are height and distance-dependent.

7.7 mm-Wave MIMO

The channel bandwidth availability and the significantly reduced size requirements to accommodate tens or even hundreds of antennas at the transceivers are the two main advantages offered by the use of mm-Wave frequencies. Although spatial multiplexing and beamforming techniques are efficiently exploited in many available wireless standards in sub-6 GHz bands, the design and implementation of these techniques in mm-Wave bands is not straightforward and is quite different from the lower frequencies. The main reasons for this deviation are the hardware constraints imposed by the high frequencies, the different channel propagation conditions and the large number of antennas at both ends of the links (Alkhateeb et al. 2014; Bai et al. 2014; Heath et al. 2016). The hardware constraints are related to the difficult practical implementation of mm-Wave arrays with one RF chain for each available antenna and to the high cost and power consumption of components like the power amplifier (PA) and the analog-to-digital converters (ADCs). Moreover, the increased bandwidth implies many gigasamples per second for each antenna-related digital conversion stage, which in turn implies increased power consumption for the signal processing stage. All these constraints make the conventional sub-6 GHz MIMO architecture with digital processing taking place entirely at baseband, infeasible. The channel in mm-Wave bands is quite different from that in lower frequencies (Rappaport et al. 2013a, 2013b). The first difference is related to LOS and NLOS conditions. The LOS signals propagate as in free space, whereas NLOS signals are much weaker. The mm-Wave frequencies are quite sensitive to blockage effects caused by walls, bodies, and foliage. The diffraction mechanism does not contribute much to the coverage, as in sub-6 GHz bands, and the scattering clusters are fewer. Furthermore, the angle spread of the multipaths in a cluster is small. These effects lead to sparse MIMO channel matrices affecting the rank of the channel and the corresponding capacity achieved. The large number of antennas will allow for large directivity gains and a more efficient interference mitigation. On the other hand, the increased number of antennas increases the complexity for channel estimation, precoding, combining and equalization. All these issues have led the research community to the proposal of different MIMO architectures. The available options to serve the beamforming and/or the spatial multiplexing operations are discussed in Sections 7.7.1 through 7.7.3.

7.7.1 Analog Beamforming

This architecture is based on a network of digitally controlled phase shifters connected to the antenna elements. The signals from the phase shifters are added and driven to a single RF chain. The weights of the phase shifters are adaptively adjusted and are designed to shape and steer the transmit and receive beams according to some criterion, for example, the SNR. The analog beamforming is proposed in IEEE 802.11ad and in IEEE 802.15.3c. The phase shifters may be active or passive (Heath et al. 2016) and the main issues are the insertion losses and the power consumption, which is related also to the resolution of the quantized phases. The drawbacks of analog beamforming are, (1) the support of single-user and single-stream transmission for a simple implementation with one RF chain behind the phase shifters network and (2) the beam steering is quite complicated requiring beam training and channel estimation.

7.7.2 Hybrid Analog/Digital Precoding/Combining

This technique is a compromise between purely analog or digital implementations. The initial idea of joint RF and baseband signal processing design may be found in Zhang et al. (2005), later on in Theofilakos and Kanatas (2007), where the concept of adaptive receive antenna subarray formation for MIMO systems was proposed, and in Venkateswaran and van der Veen (2010), where a multichannel beamformer was proposed in the RF domain, followed by a digital beamformer in baseband. The architecture proposed for mm-Wave frequencies (Heath et al. 2016) is shown in Figure 7.9. The DACs and ADCs are given in pairs since the sampling is performed in the in-phase and the quadrature components. The relationship among the number of streams, RF chains, and antenna elements is the following: $N_s < L_t$ (or L_r) $< N_t$ (or N_r). If we denote as $\mathbf{F}_{RF}$ and $\mathbf{F}_{BB}$ the analog and digital precoding matrices respectively, and as $\mathbf{W}_{RF}$ and $\mathbf{W}_{BB}$ the analog and digital combining matrices respectively, then the input-output relationship is given by

$$\mathbf{y} = \mathbf{W}_{BB}^H \mathbf{W}_{RF}^H \mathbf{H} \mathbf{F}_{RF} \mathbf{F}_{BB} \mathbf{x} + \mathbf{W}_{BB}^H \mathbf{W}_{RF}^H \mathbf{n} \tag{7.49}$$

There are different strategies for the design of the precoding and combining matrices which are briefly presented in Heath et al. (2016). Furthermore, the authors in Sohrabi and Yu (2016) show that hybrid beamforming can achieve the same performance of any fully digital beamforming scheme with much fewer RF chains. The required number of RF chains and the relationship to the number of data streams is also considered in that paper, as well as in Bogale et al. (2016), where an extensive study is presented for the number of RF chains and phase shifters for a downlink multiuser massive MIMO system in frequency selective channels.

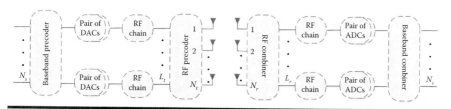

Figure 7.9 Hybrid analog/digital precoding/combining mm-Wave MIMO architecture.

7.7.3 Low Resolution ADCs

This alternative architecture is based on the replacement of the high-resolution ADCs, for example, 6-bit or higher resolution, with low resolution ADCs of 3, 2 or even 1-bit resolution. This replacement is proposed since the ADCs with high resolution and large sampling rates are considered to be costly and consume power of several Watts. Moreover, the interfacing cards connecting digital components to ADCs are also power-hungry and the consumption increases with the resolution. There are few drawbacks related to this new model with low-precision ADCs. The first is that the basic communication techniques with low-resolution ADCs are quite different. In Mezghani et al. (2012) the authors provide a modified minimum mean square error (MMSE) detector, extended to an iterative decision feedback equalizer, whereas in Choi et al. (2016), the authors present detectors and channel estimation methods for massive-MIMO systems with 1-bit ADCs. Nevertheless, the literature is not yet mature for multiuser applications. Next, the theoretical capacity achieved is limited even at high SNR regime (Mo and Heath 2015). There are expected many more contributions in the literature for 1-bit ADC architectures in the following years.

References

3GPP Technical Reports TR36.873. Study on 3D channel model for LTE. Technical Report.

3GPP. Spatial channel model for multiple input multiple output (MIMO) simulations v.12.0.0. TR 25.996, 3GPP, September 2014.

3GPP. Technical Report—Study on 3D channel model for LTE. TR 36.873 V.l.2.1, 3rd Generation Partnership Project (3GPP), September 2013.

3GPP. Technical Report—Study on elevation beamforming/full-dimension (FD) mimo for LTE. TR 36.897 V.1.0.1, 3rd Generation Partnership Project (3GPP), June 2015.

Ademaj, F., M. Taranetz, and M. Rupp. 3GPP 3D MIMO channel model: A holistic implementation guideline for open source simulation tools. *EURASIP Journal on Wireless Communications and Networking*, 2016(1):55, 2016.

Ademaj, F., M. Taranetz, and M. Rupp. Implementation, validation and application of the 3GPP 3D MIMO channel model in open source simulation tools. In *2015 International Symposium on Wireless Communication Systems (ISWCS)*, pp. 721–725, Belgium, Germany, August 2015.

Adhikary, A., J. Nam, J. Y. Ahn, and G. Caire. Joint spatial division and multiplexing—The large-scale array regime. *IEEE Transactions on Information Theory*, 59(10):6441–6463, 2013.

Alamouti, S. M. A simple transmit diversity technique for wireless communications. *IEEE Journal on Selected Areas in Communications*, 16(8):1451–1458, 1998.

Alexandropoulos, G. C., P. Ferrand, J. M. Gorce, and C. B. Papadias. Advanced coordinated beamforming for the downlink of future LTE cellular networks. *IEEE Communications Magazine*, 54(7):54–60, 2016.

Alkhateeb, A., J. Mo, N. Gonzalez-Prelcic, and R. W. Heath. MIMO precoding and combining solutions for millimeter-wave systems. *IEEE Communications Magazine*, 52(12):122–131, 2014.

Andrews, J. G., S. Buzzi, W. Choi, S. V. Hanly, A. Lozano, A. C. K. Soong, and J. C. Zhang. What will 5G be? *IEEE Journal on Selected Areas in Communications*, 32(6):1065–1082, 2014.

Bai, T., A. Alkhateeb, and R. W. Heath. Coverage and capacity of millimeter-wave cellular networks. *IEEE Communications Magazine*, 52(9):70–77, 2014.

Björnson, E., M. Matthaiou, and M. Debbah. Massive MIMO with non-ideal arbitrary arrays: Hardware scaling laws and circuit-aware design. *IEEE Transactions on Wireless Communications*, 14(8):4353–4368, 2015.

Bogale, T. E., L. B. Le, A. Haghighat, and L. Vandendorpe. On the number of RF chains and phase shifters, and scheduling design with hybrid analog-digital beamforming. *IEEE Transactions on Wireless Communications*, 15(5):3311–3326, 2016.

Caire, G. and S. Shamai. On the achievable throughput of a multiantenna Gaussian broadcast channel. *IEEE Transactions on Information Theory*, 49(7):1691–1706, 2003.

Chen, J. and V. K. N. Lau. Two-tier precoding for FDD multicell massive MIMO time-varying interference networks. *IEEE Journal on Selected Areas in Communications*, 32(6):1230–1238, 2014.

Chen, Z. and C. Yang. Pilot decontamination in wideband massive MIMO systems by exploiting channel sparsity. *IEEE Transactions on Wireless Communications*, 15(7):5087–5100, 2016.

Choi, J., D. J. Love, and P. Bidigare. Downlink training techniques for FDD massive MIMO systems: Open-loop and closed-loop training with memory. *IEEE Journal of Selected Topics in Signal Processing*, 8(5):802–814, 2014.

Choi, J., J. Mo, and R. W. Heath. Near maximum-likelihood detector and channel estimator for uplink multiuser massive MIMO systems with one-bit ADCS. *IEEE Transactions on Communications*, 64(5):2005–2018, 2016.

Cui, Q., H. Wang, P. Hu, X. Tao, P. Zhang, J. Hamalainen, and L. Xia. Evolution of limited-feedback comp systems from 4G to 5G: Comp features and limited-feedback approaches. *IEEE Vehicular Technology Magazine*, 9(3):94–103, 2014.

Foschini, G. J. Layered space-time architecture for wireless communication in a fading environment when using multi-element antennas. *Bell Labs Technical Journal*, 1(2):41–59, 1996.

Foschini, G. J., D. Chizhik, M. J. Gans, C. Papadias, and R. A. Valenzuela. Analysis and performance of some basic space-time architectures. *IEEE Journal on Selected Areas in Communications*, 21(3):303–320, 2003.

Foschini, G. J and M. J Gans. On limits of wireless communications in a fading environment when using multiple antennas. *Wireless Personal Communications*, 6(3):311–335, 1998.

Gao, X., O. Edfors, F. Rusek, and F. Tufvesson. Massive MIMO performance evaluation based on measured propagation data. *IEEE Transactions on Wireless Communications*, 14(7):3899–3911, 2015.

Gesbert, D., M. Kountouris, R. W. Heath Jr., C. B. Chae, and T. Salzer. Shifting the MIMO paradigm. *IEEE Signal Processing Magazine*, 24(5):36–46, 2007.

Goldsmith, A., S. A. Jafar, N. Jindal, and S. Vishwanath. Capacity limits of MIMO channels. *IEEE Journal on Selected Areas in Communications*, 21(5):684–702, 2003.

Gore, D., A. Paulraj, and R. Nabar. *Introduction to Space-Time Wireless Communications*. Cambridge University Press, Cambridge, UK, 2003.

Hassibi, B. and M. Sharif. Fundamental limits in MIMO broadcast channels. *IEEE Journal on Selected Areas in Communications*, 25(7):1333–1344, 2007.

Heath, R. W., N. González-Prelcic, S. Rangan, W. Roh, and A. M. Sayeed. An overview of signal processing techniques for millimeter wave MIMO systems. *IEEE Journal of Selected Topics in Signal Processing*, 10(3):436–453, 2016.

Hentila, L., T. Jamsa, E. Suikkanen, E. Kunnari, M. Narandzié, J. Meinila, P. Kyosti. D5.3: Winner+ final channel models. Technical Report, Wireless World Initiative New Radio—WINNER+, 2010.

Hochwald, B. M., C. B. Peel, and A. L. Swindlehurst. A vector-perturbation technique for near-capacity multiantenna multiuser communication—part II: Perturbation. *IEEE Transactions on Communications*, 53(3):537–544, 2005.

Hoydis, J., C. Hoek, T. Wild, and S. ten Brink. Channel measurements for large antenna arrays. In *2012 International Symposium on Wireless Communication Systems (ISWCS)*, pp. 811–815, Paris, France, August 2012.

Hoydis, J., S. ten Brink, and M. Debbah. Massive MIMO in the UL/DL of cellular networks: How many antennas do we need? *IEEE Journal on Selected Areas in Communications*, 31(2):160–171, 2013.

Huh, H., A. M. Tulino, and G. Caire. Network MIMO with linear zero-forcing beamforming: Large system analysis, impact of channel estimation, and reduced-complexity scheduling. *IEEE Transactions on Information Theory*, 58(5):2911–2934, 2012.

Hur, S., S. Baek, B. Kim, Y. Chang, A. F. Molisch, T. S. Rappaport, K. Haneda, and J. Park. Proposal on millimeter-wave channel modeling for 5G cellular system. *IEEE Journal of Selected Topics in Signal Processing*, 10(3):454–469, 2016.

ITU-R Report M.2135-1. Guidelines for evaluation of radio interface technologies for imt-advanced. Technical Report, ITU-R, 2009.

Ji, H., Y. Kim, J. Lee, E. Onggosanusi, Y. Nam, J. Zhang, B. Lee, and B. Shim. Overview of full-dimension MIMO in LTE-advanced pro. *IEEE Communications Magazine*, 49(2):102–111, 2016.

Jiang, X., F. Kaltenberger, and L. Deneire. How accurately should we calibrate a massive MIMO TDD system? In *2016 IEEE International Conference on Communications Workshops (ICC)*, pp. 706–711, Kuala Lumpur, Malaysia, May 2016.

Jose, J., A. Ashikhmin, T. L. Marzetta, and S. Vishwanath. Pilot contamination and precoding in multicell TDD systems. *IEEE Transactions on Wireless Communications*, 10(8):2640–2651, 2011.

Kammoun, A., H. Khanfir, Z. Altman, M. Debbah, and M. Kamoun. Preliminary results on 3D channel modeling: From theory to standardization. *IEEE Journal on Selected Areas in Communications*, 32(6):1219–1229, 2014.

Kim, Y., H. Ji, J. Lee, Y. H. Nam, B. L. Ng, I. Tzanidis, Y. Li, and J. Zhang. Full dimension mimo (FD-MIMO): The next evolution of MIMO in LTE systems. *IEEE Wireless Communications*, 21(3):92–100, 2014.

Kuchar, A., J. P. Rossi, and E. Bonek. Directional macro-cell channel characterization from urban measurements. *IEEE Transactions on Antennas and Propagation*, 48(2):137–146, 2000.

Larsson, E. G., O. Edfors, F. Tufvesson, and T. L. Marzetta. Massive MIMO for next generation wireless systems. *IEEE Communications Magazine*, 52(2):186–195, 2014.

Lee, J., Y. Kim, H. Lee, B. L. Ng, D. Mazzarese, J. Liu, W. Xiao, and Y. Zhou. Coordinated multipoint transmission and reception in LTE-advanced systems. *IEEE Communications Magazine*, 50(11):44–50, 2012.

Li, X., S. Jin, X. Gao, and R. W. Heath. 3D beamforming for large-scale FD-MIMO systems exploiting statistical channel state information. *IEEE Transactions on Vehicular Technology*, 65(11):8992–9005, 2016.

Lim, C., T. Yoo, B. Clerckx, B. Lee, and B. Shim. Recent trend of multiuser MIMO in LTE-advanced. *IEEE Communications Magazine*, 51(3):127–135, 2013.

Liu, Q., X. Su, J. Zeng, H. Gao, T. Lv, X. Xu, and C. Xiao. An improved relative channel reciprocity calibration method in TDD massive MIMO systems. In 24th *Wireless and Optical Communication Conference (WOCC) 2015*, pp. 98–102, Taipei, Taiwan, October 2015.

Lu, L., G. Y. Li, A. L. Swindlehurst, A. Ashikhmin, and R. Zhang. An overview of massive MIMO: Benefits and challenges. *IEEE Journal of Selected Topics in Signal Processing*, 8(5):742–758, 2014.

Luo, X. Multiuser massive MIMO performance with calibration errors. *IEEE Transactions on Wireless Communications*, 15(7):4521–4534, 2016.

Luo, X. and X. Wang. How to calibrate massive MIMO? In *2015 IEEE International Conference on Communication Workshop (ICCW)*, pp. 1119–1124, London, UK, June 2015.

Martnez, À O., E. De Carvalho, and J. Ø Nielsen. Towards very large aperture massive MIMO: A measurement based study. In *2014 IEEE Globecom Workshops (GC Wkshps)*, pp. 281–286, Austin, TX, December 2014.

Marzetta, T. L. Noncooperative cellular wireless with unlimited numbers of base station antennas. *IEEE Transactions on Wireless Communications*, 9(11):3590–3600, 2010.

Marzetta, T. L. Massive MIMO: An introduction. *Bell Labs Technical Journal*, 20:11–22, 2015.

Mezghani, A., M. Rouatbi, and J. A. Nossek. An iterative receiver for quantized MIMO systems. In *2012 16th IEEE Mediterranean Electrotechnical Conference*, pp. 1049–1052, Yasmine Hammamet, Tunisia, March 2012.

Mo, J. and R. W. Heath. Capacity analysis of one-bit quantized MIMO systems with transmitter channel state information. *IEEE Transactions on Signal Processing*, 63(20):5498–5512, 2015.

Müller, R. R., L. Cottatellucci, and M. Vehkaperä. Blind pilot decontamination. *IEEE Journal of Selected Topics in Signal Processing*, 8(5):773–786, 2014.

Nam, Y. H., B. L. Ng, K. Sayana, Y. Li, J. Zhang, Y. Kim, and J. Lee. Full-dimension MIMO (FD-MIMO) for next generation cellular technology. *IEEE Communications Magazine*, 51(6):172–179, 2013.

Ngo, H. Q. and E. G. Larsson. EVD-based channel estimation in multicell multiuser MIMO systems with very large antenna arrays. In *2012 IEEE International Conference on Acoustics, Speech and Signal Processing (ICASSP)*, pp. 3249–3252, Kyoto, Japan, March 2012.

Ngo, H. Q., E. G. Larsson, and T. L. Marzetta. Energy and spectral efficiency of very large multiuser MIMO systems. *IEEE Transactions on Communications*, 61(4):1436–1449, 2013.

Paulraj, A.J. and T. Kailath. Increasing capacity in wireless broadcast systems using distributed transmission/directional reception (DTDR), September 6 1994. US Patent 5,345,599.

Rappaport, T. S., F. Gutierrez, E. Ben-Dor, J. N. Murdock, Y. Qiao, and J. I. Tamir. Broadband millimeter-wave propagation measurements and models using adaptive-beam antennas for outdoor urban cellular communications. *IEEE Transactions on Antennas and Propagation*, 61(4):1850–1859, 2013a.

Rappaport, T. S., S. Sun, R. Mayzus, H. Zhao, Y. Azar, K. Wang, G. N. Wong, J. K. Schulz, M. Samimi, and F. Gutierrez. Millimeter wave mobile communications for 5G cellular: It will work! *IEEE Access*, 1:335–349, 2013b.

Razavizadeh, S. M., M. Ahn, and I. Lee. Three-dimensional beamforming: A new enabling technology for 5G wireless networks. *IEEE Signal Processing Magazine*, 31(6):94–101, 2014.

Roh, W., J. Y. Seol, J. Park, B. Lee, J. Lee, Y. Kim, J. Cho, K. Cheun, and F. Aryanfar. Millimeter-wave beamforming as an enabling technology for 5G cellular communications: Theoretical feasibility and prototype results. *IEEE Communications Magazine*, 52(2):106–113, 2014.

Rusek, F., D. Persson, B. K. Lau, E. G. Larsson, T. L. Marzetta, O. Edfors, and F. Tufvesson. Scaling up MIMO: Opportunities and challenges with very large arrays. *IEEE Signal Processing Magazine*, 30(1):40–60, 2013.

Sibille, A., C. Oestges, and A. Zanella, (Eds.). *MIMO: From Theory to Implementation*, 1st ed. Academic Press, Cambridge, MA, 2010.

Sohrabi, F. and W. Yu. Hybrid digital and analog beamforming design for large-scale antenna arrays. *IEEE Journal of Selected Topics in Signal Processing*, 10(3):501–513, 2016.

Su, L. and C. Yang. Fractional frequency reuse aided pilot decontamination for massive MIMO systems. In *2015 IEEE 81st Vehicular Technology Conference (VTC Spring)*, pp. 1–6, Glasgow, UK, May 2015.

Tarokh, V., N. Seshadri, and A. R. Calderbank. Space-time codes for high data rate wireless communication: Performance criterion and code construction. *IEEE Transactions on Information Theory*, 44(2):744–765, 1998.

Teeti, M., J. Sun, D. Gesbert, and Y. Liu. The impact of physical channel on performance of subspace-based channel estimation in massive MIMO systems. *IEEE Transactions on Wireless Communications*, 14(9):4743–4756, 2015.

Telatar, E. Capacity of multi-antenna Gaussian channels. *European Transactions on Telecommunications*, 10(6):585–595, 1999.

Theofilakos, P. and A. G. Kanatas. Capacity performance of adaptive receive antenna subarray formation for MIMO systems. *EURASIP Journal on Wireless Communications and Networking*, 2007(1):056471, 2007.

Venkatesan, S., H. Huang, and C. B. Papadias. *MIMO Communication for Cellular Networks*. Springer, New York, 2012.

Venkateswaran, V. and A. J. van der Veen. Analog beamforming in MIMO communications with phase shift networks and online channel estimation. *IEEE Transactions on Signal Processing*, 58(8):4131–4143, 2010.

Vieira, J., S. Malkowsky, K. Nieman et al. A flexible 100-antenna testbed for massive MIMO. In *2014 IEEE Globecom Workshops (GC Wkshps)*, pp. 287–293, Austin, TX, December 2014.

Vieira, J., F. Rusek, and F. Tufvesson. Reciprocity calibration methods for massive MIMO based on antenna coupling. In *2014 IEEE Global Communications Conference*, pp. 3708–3712, Austin, TX, December 2014.

Vishwanath, S., N. Jindal, and A. Goldsmith. Duality, achievable rates, and sum-rate capacity of Gaussian MIMO broadcast channels. *IEEE Transactions on Information Theory*, 49(10):2658–2668, 2003.

Viswanath, P. and D. N. C. Tse. Sum capacity of the vector Gaussian broadcast channel and uplink-downlink duality. *IEEE Transactions on Information Theory*, 49(8):1912–1921, 2003.

WINNER II WP1. Winner II channel models. Technical Report, IST-4-027756 WINNER II Deliverable D.1.1.2, 2007.

Wolniansky, P. W., G. J. Foschini, G. D. Golden, and R. A. Valenzuela. V-blast: An architecture for realizing very high data rates over the rich-scattering wireless channel. In *1998 URSI International Symposium on Signals, Systems, and Electronics. Conference Proceedings (Cat. No.98EX167)*, pp. 295–300, Pisa, Italy, September 1998.

Yu, W. and J. M. Cioffi. Trellis precoding for the broadcast channel. In *Global Telecommunications Conference, 2001. GLOBECOM'01. IEEE*, Vol. 2, pp. 1344–1348, San Antonio, TX, November 2001.

Yu, W., W. Rhee, S. Boyd, and J. M. Ciofli. Iterative water-filling for Gaussian vector multiple access channels. In *Information Theory, 2001. Proceedings. 2001 IEEE International Symposium On*, pp. 322, Washington, DC, 2001.

Zamir, R., S. Shamai, and U. Erez. Nested linear/lattice codes for structured multiterminal binning. *IEEE Transactions on Information Theory*, 48(6):1250–1276, 2002.

Zhang, R. and S. Cui. Cooperative interference management with MISO beamforming. *IEEE Transactions on Signal Processing*, 58(10):5450–5458, 2010.

Zhang, X., A. F. Molisch, and S.-Y. Kung. Variable-phase-shift-based RF-baseband codesign for MIMO antenna selection. *IEEE Transactions on Signal Processing*, 53(11):4091–4103, 2005.

Zhu, X., Z. Wang, L. Dai, and C. Qian. Smart pilot assignment for massive MIMO. *IEEE Communications Letters*, 19(9):1644–1647, 2015.

Chapter 8

Channel-Dependent Precoding for Multiuser Access with Load-Controlled Parasitic Antenna Arrays

Konstantinos Ntougias, Dimitrios Ntaikos, and Constantinos B. Papadias

Contents

In memory of our most humane professor, Dr. Philip Constantinou.

—Constantinos B. Papadias

Multiple-input multiple-output (MIMO) technology has been recently incorporated in cellular mobile radio communications systems as a means to increase the spectral efficiency (SE), in response to the exponential traffic growth noticed over the past few years [1] and the shortage in available spectrum [2]. In view of the enormous capacity requirements of future cellular networks [3], new variants of this communication paradigm that promise to further boost the achievable SE have been proposed [4,5].

The performance of the various MIMO transmission schemes studied in the literature depends on the degrees of freedom (DoF) provided by the antenna systems that are installed on the base stations (BSs) and the user terminals (UT) [6,7]. When conventional digital antenna arrays (DAA) are utilized, which make use of voltage-driven (i.e., active) antenna elements, this quantity equals the number of deployed antennas. However, cost, complexity, and energy consumption grow with the number of antennas, as each antenna element should be fed by a radio frequency (RF) chain [7]. Moreover, the interelement spacing should be above a predefined threshold, so that the electromagnetic coupling among the antennas is minimized [7]. These constraints place a limit on the number of antenna elements in such arrays and, therefore, on the performance gains of MIMO communication techniques [7]. Hence, there is a growing interest lately on antenna systems that address these issues.

Load-controlled parasitic antenna arrays (LC-PAA) constitute a representative example. These compact hybrid antenna systems employ a limited number of active antenna elements, surrounded by passive antenna elements that are terminated to tunable loads. Due to the strong mutual coupling, which is caused by the deliberately chosen small antenna spacing, the feeding voltages induce currents on the so-called parasitic antennas, thus enabling them to participate in the formation of the far-field radiation pattern. Hence, *by exploiting the mutual coupling, LC-PAAs manage to provide the same DoF as DAAs with more RF units or, alternatively, more*

DoF than DAAs with the same number of RF modules, thus leading to cost, complexity, and energy consumption savings or to performance improvement, respectively. Furthermore, by adjusting the impedance of the parasitic loads with the help of a low-cost digital control circuit, we can set the amplitude and phase of the currents that run on the passive antennas in a controllable manner, that is we can perform transmit beamforming (BF), even in the extreme case of single-fed LC-PAAs [8]. This is similar with the weighting of the currents that is performed in the baseband of DAAs in order to shape and steer beams.

A collection of recent studies demonstrates that the capability of single-RF LC-PAAs to reconfigure their radiation pattern allows them to perform also open-loop (i.e., channel-agnostic) MIMO [7], while [9] presents a method that enables the application of channel-aware precoding on such antenna systems—and, in extension, to multi-RF LC-PAAs as well. Nevertheless, the technique described in [9] cannot attain any given signal constellation or precoding scheme, since often the required load combinations result in system instability. In [10], an alternative is presented. However, this method is not robust. In addition, it presents high computational complexity.

The goal of this chapter is to introduce *a pragmatic approach that will facilitate the application of the LC-PAA technology in actual 5G networks.* More specifically, a low-complexity method for performing arbitrary channel-dependent precoding with such arrays in a variety of MIMO configurations is described. We have to note that the considered precoding methods include not only the conventional user-level precoding techniques but also symbol-level precoding schemes. The latter techniques improve the performance of multiuser MIMO systems that operate at low signal-to-noise ratio (SNR) levels [11]. Moreover, we should mention that this work focuses on both the cellular access and the centimeter-wave (cm-wave) portion of the radio spectrum (i.e., frequency bands in the 0.3–3 GHz and 3–30 GHz range, respectively), in accordance to the current trend of exploiting the available bandwidth (BW) in the spectral regions above 6 GHz in order to enhance the peak user rates and the system capacity [12]. The performance of the utilized transmission techniques is evaluated, in terms of the (average) per-user or system-wide downlink (DL) throughput, through analytic expressions, numerical simulations, and over-the-air (OTA) testbed-based experimentations. The design and implementation of the corresponding antenna arrays is presented as well.

8.1 Introduction

8.1.1 The 5G Era

Nowadays, we live in the dawn of the 5th-generation (5G) era. The 5G vision encompasses the provision of data-hungry services and the support of challenging use cases, such as the delivery of ultra-high-definition (UHD) video streams at locations with high user density (e.g., airports, train stations, shopping malls, stadiums). The characteristics of the envisioned applications and scenarios indicate that 5G cellular mobile radio communications systems should accommodate about 1000 times higher traffic volume per unit area than current 4G Long-Term Evolution (LTE)/LTE-Advanced (LTE-A) networks [3]. This statement refers to the cellular DL, where the BSs act as transmitters (TX), whereas the UTs play the role of the receivers (RX).

In principle, this objective could be met by allocating additional spectral resources to these bandwidth-demanding services. However, the sub-6 GHz frequency bands, which are typically utilized for wireless access purposes, are severely congested. This phenomenon is attributed to the increasing demand for wireless communications over the past decades, which led to the emergence of a plethora of services (e.g., TV and radio broadcasting), in conjunction with the adoption of the rather rigid and inefficient licensed-access spectrum management model, where license holders are granted the right to exploit exclusively specific parts of the spectrum at certain locations [2]. The scarcity and the subsequent high-acquisition cost of the radio spectrum prohibit the use of the aforementioned *brute-force* approach in practice. As a consequence, the industry and the academia agreed that the synergy between a number of alternatives is required, in order to address the spectrum crunch issue and achieve the capacity goal of 5G systems.

The area capacity (AC) of a cellular network (i.e., the sum-throughput per unit area) is given by

$$AC \text{ [bits/s/km}^2] = BW \text{ [Hz]} \times SE \text{ [bits/s/Hz]} \times CD \text{ [cells/km}^2] \qquad (8.1)$$

where:
BW is the system bandwidth
SE is the spectral efficiency (i.e., the aggregated data rate per unit of BW)
CD is the cell density (i.e., the number of cells per unit area)

Therefore, the typical strategies for meeting the 5G capacity target include [13]:

- The exploitation of additional spectrum in higher frequency bands, where there is an abundance of available BW, through the use of cm-wave and millimeter-wave (mm-wave) access technologies
- The utilization of techniques that increase the SE, for example, spectrum sharing methods, such as the Licensed Shared Access (LSA) paradigm
- The densification of the radio access network (RAN), that is, the deployment of a large number of small cells over the service area of interest, so that the available spectrum is reused more aggressively across different cells

8.1.2 *MIMO Wireless Communication Paradigms*

MIMO technology constitutes an integral component of the 5G ecosystem. MIMO communication techniques leverage the spatial dimension provided by the use of multiple antennas at the UTs or/and the BSs, in order to accomplish a substantial increase of the DL throughput at link or cell level [6]. This performance boosting is realized at no extra cost in terms of transmission power neither requires bandwidth expansion; it is instead an outcome of the increase in the dimensionality of the signal space. That is, it corresponds to the additional spatial DoF brought by the use of multiantenna nodes.

More specifically, MIMO transmission schemes involve the spatial multiplexing (SM) of a number of data signals (i.e., the concurrent transmission of multiple radio signals over the same frequency band) destined either to a single user or to a set of individual users [6]. The former paradigm is known as single-user MIMO (SU-MIMO) or point-to-point MIMO (PTP-MIMO), while the latter one is referred to as multiuser MIMO (MU-MIMO) or single-cell MIMO (SC-MIMO). Both technologies have been used extensively in LTE/LTE-A networks.

In SU-MIMO, the cell users are orthogonalized in the time domain through the utilization of the time-division multiple access (TDMA) scheme, that is, a single user is scheduled at each timeslot (TS). In MU-MIMO, on the other hand, which is sometimes called space-division multiple access (SDMA), a group of cell users is selected at each scheduling interval. These so-called active users share spatially the DL channel. Hence, in SU-MIMO both the serving BS and the UTs should be equipped with multiple antennas, for multistream communication to take place, whereas in MU-MIMO there is no such requirement for the user devices.

These multiantenna communication methods entail the employment of complex signal processing operations, in order to mitigate the resulting inter-stream or/and multiuser cochannel interference (CCI). The (pre)processing of the data signals that takes place at the BS in this regard prior to transmission is called precoding and is performed in the baseband (i.e., in the digital domain). Precoding exploits the availability of channel state information (CSI) at the TX (CSIT) (i.e., the knowledge of the composite DL channel at the BS), in order to spatially orthogonalize the transmitted cochannel signals. The DL channel is characterized by a matrix, whose entries are the gains of the scalar DL channels formed between each TX–RX antenna pair.

SU-MIMO exploits the multipath propagation, in order to enable the transmission of multiple data signals on a single time-frequency resource. To this end, linear precoding aligns these cochannel signals with spatially orthogonal scalar sub-channels created between the transmit and receive antennas, so that the inter-stream interference is eliminated [14]. In MU-MIMO, on the other hand, linear precoding constitutes a multistream variant of transmit BF which aims at steering the data signals toward the intended users, so that the received signal power is increased while the multiuser interference (MUI) caused by the energy leakage of each transmission at other users is limited [15]. Due to its inner-mechanics, MU-MIMO can be applied in both line-of-sight (LOS) and non-LOS communication scenarios, in contrast to SU-MIMO.

In view of the advent of 5G systems, the focus of the community has been shifted in recent years toward multicell MIMO (MC-MIMO). This technology is an extension of MU-MIMO that enables the adoption of universal frequency reuse, so that the system-wide capacity is further increased. This is accomplished by utilizing techniques that handle not only the intra-cell CCI but also the resulting inter-cell interference (ICI), which may degrade the performance of the cell-edge users if no countermeasure is taken.

Two MC-MIMO variants have been considered in the literature. Coordinated MIMO (cMIMO), which has been introduced in the LTE-A standard under the name of coordinated multipoint (CoMP), refers to a family of multiantenna transmission techniques that rely on the cooperation between neighboring nodes, so that their transmissions are coordinated and ICI is reduced or even eliminated [4]. Cooperation takes the form of the exchange of control information (e.g., user scheduling), CSI, user data, or combinations thereof between the corresponding BSs over the transport network. Massive MIMO (mMIMO), on the other hand, is based on the use of an excessive number of antennas at the BSs, in comparison with the number of active users. The excess of transmit antennas allows for serving a high number of users and controlling at the same time more effectively the resulting intra-cell and inter-cell CCI [5]. It is expected that both MC-MIMO technologies will be essential components of 5G systems. In this chapter, we focus on cMIMO.

8.1.3 Symbol-Level Precoding

The conventional user-level linear precoding methods that are utilized in MU-MIMO aim to reduce or even eliminate the CCI. In [11], though, it is demonstrated that at symbol level, CCI may be in some cases constructive instead of destructive, that is, it may increase the received SNR without the need to waste transmit power for this purpose. Moreover, in this work a symbol-level variant of the popular zero forcing (ZF) precoding scheme is studied. Numerical simulations show that this method outperforms its user-level counterpart.

8.1.4 Antenna Arrays

An antenna array is a device that consists of a group of collocated antenna elements that are connected to each other and arranged according to some topology in order to form a composite antenna. The cooperation of the antenna elements allows for advanced operations to take place, such as the shaping of complex radiation patterns and the concurrent transmission/reception of multiple data signals.

As we have already mentioned, the DoF provided by the antenna arrays installed on the BSs and the UTs determine the SM and interference management capabilities of MIMO communication techniques, that is, the number of cochannel signals that can be transmitted in parallel as well as, in the case of multiuser multiantenna communication setups, the achievable spatial directivity.

Conventional DAAs are comprised by active antenna elements that are placed sufficiently apart, so that the occurrence of mutual coupling among them is avoided [7]. Therefore, there is an one-to-one mapping between each driving voltage and the corresponding antenna current. This linear relationship facilitates the processing of the data signals in the digital domain and allows for full control of the antenna currents [7].

The array DoF of such antenna systems equal the number of antenna elements. This fact motivated the research on mMIMO. Similarly, the variant of cMIMO known as joint transmission (JT) or network MIMO takes advantage of the relationship between the number of antennas in each BS and the provided DoF to artificially increase the effective DoF. To this end, it enables the sharing of the user data among the cooperating BSs, so that the scheduled cell-edge users are served jointly by multiple nodes.

On the other hand, since in DAAs each antenna element is fed by a RF unit which consists of low-noise amplifiers (LNA), power amplifiers (PAs), filters, digital-to-analog/analog-to-digital converters (DAC/ADC), and so on, the cost, complexity, and energy consumption of the system grows proportionally to the number of antennas [7]. Moreover, the packing of a large number of antennas is prohibited by the requirement to eliminate the mutual coupling, which would degrade the radiation efficiency. These constraints limit the achievable SE of MIMO techniques.

LC-PAAs represent an alternative paradigm that solves these problems. These antenna systems exploit the mutual coupling among the antenna elements in order to increase the DoF that correspond to a given number of RF units or decrease the number of active antennas required to provide a target number of DoF. We distinguish between two types of LC-PAAs, namely, single-active multiple-passive (SAMP) arrays or single-fed arrays and multiple-active multiple-passive (MAMP) arrays. As we have described, these antenna systems calculate dynamically the values of the tunable parasitic loads, in order to generate the desired antenna currents for the application of interest (e.g., transmit BF, channel-dependent precoding). We should mention that single-RF LC-PAAs are of special interest, since they mimic the functionality of conventional antenna arrays while having a single voltage feed. Single-fed LC-PAAs with tunable loads are often referred to as single-RF electronically steerable passive array radiators (ESPAR).

8.2 Overview of MIMO Wireless Communication

8.2.1 Single-User MIMO

Consider a scenario where a BS communicates with a UT in the cellular DL over a PTP link established between them. We assume that CSI is available at both the TX and the RX. The communication is subject to a transmit power constraint P

and is contaminated by additive white Gaussian noise (AWGN) with variance σ_n^2. In addition, we consider a quasi-static frequency-flat fading channel. For convenience, we assume independent and identically distributed (i.i.d.) Rayleigh fading. We focus on a single realization of this fading channel, which corresponds to a time-invariant (TI) narrowband (NB) channel.

First, we assume that each node of this communication setup is equipped with a single antenna. The SE of this 5th single-input single-output (SISO) system (i.e., the maximum data rate per unit of BW for which reliable communication is possible over this scalar channel) for the given channel realization is [6]

$$C_{\text{SISO,CSIR}}^{\text{TI-NB}} = \log_2\left(1 + \frac{P|h|^2}{\sigma_n^2}\right) = \log_2\left(1 + \rho|h|^2\right)$$

$$= \log_2\left(1 + \text{SNR}_{\text{SISO}}\right) \quad [\text{bits/s/Hz}]$$

(8.2)

where:

h is the channel gain
$\text{SNR}_{\text{SISO}} = \left(P/\sigma_n^2\right)|h|^2 = \rho|h|^2$ is the receive SNR
$\rho = P/\sigma_n^2$ is the transmit SNR

We note from Equation 8.2 that the (BW-normalized) channel capacity C grows logarithmically with the transmit power P or, equivalently, with the receive SNR. This argument holds in the DoF-limited high-SNR regime as well, that is, every increase of 3 dB in transmit power leads to an additional bit/s/Hz of SE [6]. In the power-limited low-SNR regime, on the other hand, C increases linearly with P [6].

Next, we consider an equivalent $(1, N)$ single-input multiple-output (SIMO) system. The (BW-normalized) capacity of this system is expressed as [6]

$$C_{\text{SIMO,CSIR}}^{\text{TI-NB}} = \log_2\left(1 + \frac{P\|\mathbf{h}\|^2}{\sigma_n^2}\right) = \log_2\left(1 + \rho\sum_{i=1}^{N}|h_i|^2\right)$$

$$= \log_2\left(1 + \text{SNR}_{\text{SIMO}}\right) \quad [\text{bits/s/Hz}]$$

(8.3)

where the $(N \times 1)$ channel vector $\mathbf{h}$ holds the gains h_i of the scalar channels formed between the ith receive antenna and the transmit antenna and $\text{SNR}_{\text{SIMO}} = \left(P/\sigma_n^2\right)\|\mathbf{h}\|^2 = \rho\sum_{i=1}^{N}|h_i|^2$ is the receive SNR.

Similarly, the capacity of a $(M,1)$ multiple-input single-output (MISO) system is given by [6]

$$C_{\text{MISO,CSIT}}^{\text{TI-NB}} = \log_2\left(1 + \frac{P\|\mathbf{h}\|^2}{\sigma_n^2}\right) = \log_2\left(1 + \rho\sum_{j=1}^{M}|h_j|^2\right)$$

$$= \log_2\left(1 + \text{SNR}_{\text{MISO}}\right) \quad [\text{bits/s/Hz}]$$

(8.4)

where the $(M \times 1)$ channel vector $\mathbf{h}$ holds the gains h_j of the channels formed between the jth transmit antenna and the receive antenna and $\text{SNR}_{\text{MISO}} = \left(P/\sigma_n^2 \right) \|\mathbf{h}\|^2 = \rho \sum_{j=1}^{M} \left| h_j \right|^2$ is the receive SNR.

We notice that the use of multiple receive (transmit) antennas results in an increase of the receive SNR, that is, the quantity $\|\mathbf{h}\|^2$ represents a power gain. When the channel gain is normalized to unity, the power gain equals the number of receive (transmit) antennas. We also note that the performance benefit from the use of multiple receive (transmit) antennas is more apparent in the low-SNR regime, since then the power gain is translated into a linear capacity gain.

In the aforementioned setups, only single-stream transmission is possible. In a (M,N) multiple-input multiple-output (MIMO) system, on the other hand, multistream communication is also possible. The capacity of P2P-MIMO systems was studied extensively in the mid-1990s [16,17]. It has been shown that the MIMO channel, which is characterized by a $(N \times M)$ matrix $\mathbf{H}$ that holds the gains h_{ij} of the scalar channels formed between the ith receive antenna and the jth transmit antenna, can be decomposed into exactly $r = \text{rank}(\mathbf{H}) \le \min(M,N)$ parallel (i.e., non-interfering) SISO channels (or eigenmodes) with gain λ_i, where the quantities $\lambda_1, \ldots, \lambda_r$ represent the non-zero singular values of $\mathbf{H}$ arranged in descending order. The MIMO capacity is given by [6,18]

$$C_{\text{MIMO,CSIT}}^{\text{TI-NB}} = \max_{\substack{P_1, P_2, \ldots, P_r \\ \sum_i P_i = P}} \sum_{i=1}^{r} \log_2 \left(1 + \frac{P_i \lambda_i^2}{\sigma_n^2} \right) \tag{8.5}$$

The optimal power levels allocated to each eigenmode are computed according to the water-filling (WF) algorithm as

$$P_i^* = \left(\mu - \frac{\sigma_n^2}{\lambda_i^2} \right)^+, \quad i = 1, 2, \ldots, r \tag{8.6}$$

where $x^+ = \max(x, 0)$ and μ is chosen so that $\sum_{i=1}^{r} P_i^* = P$. In the medium-SNR regime, the WF algorithm allocates more power to the eigenmodes with larger gain. In the low-SNR regime, on the other hand, all power is allocated to the dominant eigenmode (i.e., the one with the higher gain)—that is, single-stream transmission takes place. Here, a power gain is noticed, as in the SIMO/MISO cases. Finally, at high SNR, the WF algorithm allocates approximately equal power to all eigenmodes. In this case, the channel capacity scales linearly with r, that is, an increase of 3 dB in the transmit power adds approximately r extra bits/s/Hz of SE. This is in contrast with the SISO, SIMO, and MISO cases, where we can only achieve a logarithmic increase of the capacity with the transmit power. The prelog factor r which determines the capacity increase of a MIMO system in comparison to an equivalent SISO system (i.e., the number of interference-free spatial sub-channels created between the transmit and transmit antennas or, equivalently, the number of data streams that can be spatially multiplexed) is referred to as the SM gain or the spatial DoF.

If the number of scatterers is limited or/and the inter-element distance at each side of the link is small, **H** may become ill-conditioned or/and rank-deficient (i.e., $r < \min[M, N]$), which results in a decrease of the channel capacity. On the other hand, under the assumption of i.i.d. Rayleigh fading, which implies a rich scattering environment as well as sufficient antenna separation at both the TX and the RX, **H** is full-rank and well-conditioned. Since in this case $r = \min(M, N)$, the capacity grows linearly with the minimum of the number of antennas installed at both sides of the link (or, in case of $M = N$, with the number of antennas at either side of the link). Hence, in order to double the capacity in a symmetric setup, we simply have to double the number of antennas at each node.

So far, we have focused on the capacity of a PTP system for a single channel realization. In order to fully characterize the performance of such a system, we typically use the average capacity, which is obtained by taking the expectation of the capacity per channel realization with respect to the distribution of the channel.

8.2.2 Multiuser MIMO

While SU-MIMO provides significant gains in terms of the per-user capacity, it presents also a number of drawbacks: (a) In SU-MIMO, a single user is scheduled at each TS, since there is no mechanism to handle the intra-cell interference. (b) The UTs should be equipped with multiple antennas. As we have already mentioned, this is seldom the case, due to size, cost, and energy consumption constraints. (c) SU-MIMO can be applied only in scattering environments. Even in virtual LOS environments, such as keyhole channels where the channel matrix has rank one, SM is not possible. And (d) fading correlation due to insufficient scattering or/and small antenna spacing results in throughput drop.

MU-MIMO addresses these issues. This paradigm enables a BS to serve concurrently multiple users over a single frequency band, so that the SE of the corresponding cell is increased. The resulting point-to-multipoint (PTMP) channel is referred to as the MIMO broadcast channel (MIMO-BC). In contrast to SU-MIMO, MU-MIMO does not require the use of multiantenna UTs or the existence of scattering. Moreover, it is more robust against propagation non-idealities than SU-MIMO. On the other hand, though, MU-MIMO requires the availability of CSIT, for the mitigation of the MUI that arises due to the parallel cochannel transmissions to be possible with the application of appropriate precoding techniques, while SU-MIMO systems may operate without CSIT as well.

Multiuser communication systems are described by their capacity region, which consists of all vectors of data rates $\mathbf{R} = (R_1, R_2, \ldots, R_K)$ that can be achieved simultaneously by the K active users, where R_k is the data rate of the kth user. A suitable performance measure is the sum-rate (SR) capacity, which is the maximum of the sum of the user rates over all possible data rate vectors [6]:

$$C_{\text{MIMO-BC}} = \max_{\mathbf{R} \in \mathcal{C}} \sum_{i=1}^{K} R_k \qquad (8.7)$$

The notation $\mathcal{C}$ refers to the capacity region of the MIMO-BC.

The average SR capacity of the MIMO-BC scales linearly with the minimum of the number of antennas installed at the BS and the total number of antennas at the UTs [6,7]. Hence, when there are at least as many users as transmit antennas, the average SR capacity grows with the number of service antennas, irrespective of the number of antennas installed at each UT [6,7]. These statements hold also in the high-SNR regime. In the low-SNR regime, on the other hand, the optimal strategy is to schedule a single user (the one with the best channel) and apply transmit BF [6].

Since both CSI and user data are available at the BS, the resulting MUI can be predicted in advance. The capacity-achieving transmission strategy is dirty paper coding (DPC), a multiuser encoding scheme that takes advantage of the noncausal knowledge of the MUI to subtract it prior to transmission [6]. However, the successive encodings and decodings involved in DPC turn this nonlinear preprocessing method practically infeasible, especially when the number of users is large [6].

Linear precoding constitutes a suboptimal alternative that offers a good trade-off between performance and complexity. This technique leverages the physical separation of the users to enable the spatial sharing of the channel through the application of multistream transmit BF. More specifically, each data stream is premultiplied by a different BF vector and then it is transmitted by all antennas. The BF weights determine the relative amplitude and phase of the signal at each antenna, so that the signal components are added constructively at the intended user and destructively at other users [15]. In other words, a beam is assigned to each user, so that the signal power is focused toward that user. The more the DoF provided by the antenna array installed on the BS are, the higher the spatial resolution and, therefore, the interference handling capability of the system will be.

The input-output relationship of a $(M,[K,1])$ MISO-BC channel formed between a BS with M transmit antennas and K single-antenna UTs, assuming that linear precoding is utilized, is given by

$$y_k = \mathbf{h}_k^\dagger \left(\sum_{m=1}^{K} \mathbf{w}_m \sqrt{p_m} s_m \right) + n_k, \quad k = 1, 2, \ldots, K \Leftrightarrow \mathbf{y} = \mathbf{HWP}^{1/2}\mathbf{s} + \mathbf{n} \qquad (8.8)$$

where:

y is a $(K \times 1)$ vector whose element y_k is the received signal at the kth user

H denotes the $(K \times M)$ channel matrix, whose rows $\mathbf{h}_k$ are $(1 \times M)$ vectors that hold the channels h_{km} between the kth user and each one of the M TX antennas

W represents the $(M \times K)$ precoding matrix, whose column $\mathbf{w}_k$ is the $(M \times 1)$ BF vector for the kth user

$\mathbf{P}$ is the $(K \times K)$ power allocation matrix, whose element p_k is the power allocated to the kth user

$\mathbf{s}$ refers to the $(K \times 1)$ symbol vector, with s_k being the data symbol intended for the kth user

$\mathbf{n}$ is the $(K \times 1)$ additive noise vector, whose elements n_k represent the noise at the kth RX

(Note that we have omitted the time index for convenience.) We should notice that the transmission is subject to a sum-power constraint P.

The signal-to-interference-plus-noise ratio (SINR) at the kth user is expressed as

$$\mathrm{SINR}_k = \frac{\left|\mathbf{h}_k^\dagger \mathbf{w}_k\right|^2 p_k}{\sum_{m \neq k} \left|\mathbf{h}_k^\dagger \mathbf{w}_m\right|^2 p_m + \sigma_n^2}, \quad k = 1, 2, \ldots, K \tag{8.9}$$

The data rate of the kth user is given by

$$R_k = \log_2\left(1 + \mathrm{SINR}_k\right) \tag{8.10}$$

and the SR throughput is

$$R = \sum_{i=1}^{K} R_k = \sum_{i=1}^{K} \log_2\left(1 + \mathrm{SINR}_k\right) \tag{8.11}$$

Several variants of linear precoding exist. Maximum ratio transmission (MRT) is a common example. This method utilizes BF vectors that match to the intended users' channel vectors, thus maximizing the signal power at these users [19]:

$$\mathbf{v}_k^{(\mathrm{MRT})} = \mathbf{h}_k, \quad k = 1, 2, \ldots, K \tag{8.12}$$

$$\mathbf{w}_k^{(\mathrm{MRT})} = \frac{\mathbf{v}_k^{(\mathrm{MRT})}}{\left\|\mathbf{v}_k^{(\mathrm{MRT})}\right\|}, \quad k = 1, 2, \ldots, K \tag{8.13}$$

MRT is optimal in the power-limited low-SNR regime, where typically a single user is selected at each TS, since in this case the power gain attributed to the application of transmit BF is translated into a linear capacity gain. However, its capacity floors in the DoF-limited high-SNR regime, where commonly SDMA is applied, since it cannot handle the MUI.

ZF precoding, on the other hand, makes use of BF vectors that are orthogonal to the subspace of other users' channel vectors in order to eliminate the MUI. That is, the inner product of a user's BF vector with other users' channel vectors is zero:

$$\left\|\mathbf{h}_k^\dagger \mathbf{w}_m^{(\mathrm{ZF})}\right\|^2 = 0, \quad k, m = 1, 2, \ldots, K, m \neq k \tag{8.14}$$

The ZF condition is translated into the use of the Moore-Penrose pseudo-inverse of the composite channel matrix as the precoding matrix [19]:

$$\mathbf{F}^{(ZF)} = \mathbf{H}^{+} = \mathbf{H}^{\dagger} \left(\mathbf{H} \mathbf{H}^{\dagger} \right)^{-1} \tag{8.15}$$

$$\mathbf{W}^{(ZF)} = \frac{\mathbf{F}^{(ZF)}(:,k)}{\| \mathbf{F}^{(ZF)}(:,k) \|}, \quad k = 1, 2, \ldots, K \tag{8.16}$$

This precoding scheme attains a significant portion of the DPC capacity in the high SNR regime, especially when single-antenna terminals are utilized [6]. Also, it approaches the capacity as the number of users grows toward infinity, since in this case user selection benefits from the abundance of spatial directions and the multiuser diversity effect (i.e., users that have both sufficient spatial separation and high-gain channels are scheduled). The main drawback of ZF precoding is that it is power-inefficient, since the BF vectors do not match to the users' channels as in MRT. Thus, ZF performs poorly at low SNR values.

Regularized ZF (RZF) precoding is an extension of ZF that introduces a controllable amount of MUI at the cell. The value of the coefficient that controls the level of the residual MUI is typically set such that the SINR at the users is maximized. More specifically, in RZF precoding we have [19]

$$\mathbf{v}_k^{(RZF)} = \mathbf{H}^{\dagger} \left(\frac{1}{p_k} \mathbf{I}_K + \mathbf{H} \mathbf{H}^{\dagger} \right)^{-1}, \quad k = 1, 2, \ldots, K \tag{8.17}$$

$$\mathbf{w}_k^{(RZF)} = \frac{\mathbf{v}_k^{(RZF)}}{\| \mathbf{v}_k^{(RZF)} \|}, \quad k = 1, 2, \ldots, K \tag{8.18}$$

RZF precoding is asymptotically optimal at both low and high SNR and performs reasonably well at intermediate SNR values [19]. Moreover, due to the regularization of the channel, it is more robust against pathological scenarios such as ill-conditioned channel matrices. However, the existence of residual MUI complicates the power allocation procedure.

We have to note that once a linear precoding scheme has been chosen, the optimization of the system's performance, in terms of the achieved average SR throughput, depends on the employed user selection and power allocation algorithms.

8.2.3 Coordinated MIMO

Contemporary cellular systems enable the reuse of frequencies at adjacent cells, so that the overall capacity is increased. Then, the transmissions are subject to ICI, which limits the system-wide SE. The effect of ICI is more prominent at the cell boundaries, where the data signals are attenuated whereas the interfering cochannel signals are strong, thus degrading the quality of service (QoS) of the cell-edge users.

MU-MIMO ignores the ICI (i.e., it treats it as noise). This fact motivated the study on cMIMO, which represents an extension of MU-MIMO whose goal is to mitigate the ICI. More specifically, cMIMO refers to a collection of MIMO transmission methods that are based on the cooperation between neighboring cells in order to control the ICI. BS cooperation may involve the exchange of control information, CSI, or user data over transport links. The various levels of cooperation differ on their transport capacity and latency requirements as well as on their performance gains over noncooperative MIMO.

In partial-cooperation variants, such as coordinated BF (CBF), the cooperating BSs exchange the channel gains associated with the direct and interfering links, possibly in conjunction with control information such as user scheduling or/and power allocation data, in order to design their precoders (or their beamformers, in the case of single-stream transmission) in such a way, so that the ICI is coordinated [20]. In full-cooperation schemes, on the other hand, these nodes share also the user data in order to further improve the performance of the system [20]. In the most promising flavor of the latter subset, which is known as JT or network MIMO, the cooperating BSs serve jointly the cell-edge users, thus forming essentially a composite antenna array. In this case, the ICI is actually exploited instead of mitigated, that is, the interfering signals are turned into data signals [20].

The precoding schemes employed in cMIMO are typically generalizations/extensions of the ones utilized in MU-MIMO.

8.3 Constructive-Interference Zero Forcing Precoding

Constructive-interference ZF (CI-ZF) precoding is a symbol-level variant of the conventional, user-level ZF precoding scheme. Similar to DPC, CIZF takes advantage of the availability of all data symbols at the BS prior to DL transmission to predict the interference and *zero-force* only the destructive interference, while leaving the CI unaffected [11,21].

Consider a K-user MU-MISO system. Let us define the $(K \times K)$ channel cross-correlation matrix $\mathbf{R}$ as [11]

$$\mathbf{R} = \mathbf{HH}^{\dagger} \tag{8.19}$$

The symbol-to-symbol CCI from s_k to s_m is expressed as

$$\mathrm{CCI}_{km} = s_k \rho_{km}, \quad k, m = 1, 2, \ldots, K, m \neq k \tag{8.20}$$

while the cumulative CCI on s_k from all symbols is given by

$$\mathrm{CCI}_k = \sum_{k=1}^{K} s_k \rho_{km}, \quad m = 1, 2, \ldots, K, m \neq k \tag{8.21}$$

where:

$$\rho_{km} = \frac{\mathbf{h}_k \mathbf{h}_m^\dagger}{\|\mathbf{h}_k\| \|\mathbf{h}_m\|} \tag{8.22}$$

is the (k,m)-th element of $\mathbf{R}$ that represents the cross-correlation factor between the kth user's channel and the mth transmitted data stream.

In CIZF, the precoding matrix has the following form [11]:

$$\mathbf{W}^{(\text{CIZF})} = \mathbf{W}^{(\text{ZF})}\mathbf{T} = \mathbf{H}^\dagger \mathbf{R}^{-1}\mathbf{T} \tag{8.23}$$

The received signal at the kth user is given by [21]

$$y_k = \tau_{kk}\sqrt{p_k}s_k + \sum_{m \neq k} CI_{km} + n_k, \quad k,m = 1,2,\ldots,K \tag{8.24}$$

where:

$CI_{km} = \tau_{km}\sqrt{p_m}s_m$ denotes the constructive CCI from the mth user to the kth user

τ_{km} is the (k,m) element of the $K \times K$ matrix $\mathbf{T}$

Then, the kth user's SINR is given by

$$\text{SINR}_k^{(\text{CIZF})} = \sum_{m=1}^{K} |\tau_{km}|^2 p_m, \quad k = 1,2,\ldots,K \tag{8.25}$$

$\mathbf{T}$ is calculated on a symbol-by-symbol basis as follows [11]: First, $\mathbf{R}$ is calculated according to Equation 8.19 and next, assuming for simplicity the use of binary phase-shift keying (BPSK) modulation (i.e., $s_k = \pm1$, $k = 1,2,\ldots,K$), the $(K \times K)$ matrix $\mathbf{G}$ is computed as

$$\mathbf{G} = \text{diag}(\mathbf{s})\text{Re}(\mathbf{R})\text{diag}(\mathbf{s}) \tag{8.26}$$

Then, $\tau_{kk} = \rho_{kk}$ and $\tau_{km} = 0$ if $g_{km} < 0$ or $\tau_{km} = \rho_{km}$ otherwise.

Since we do not have a Gaussian input but a finite-alphabet one, we do not calculate the SR capacity through the "$\log_2$" Shannon formula. Instead, we use the following relationship [21]:

$$R = (1 - \text{BLER})m \tag{8.27}$$

where $m = 1$ bit/symbol for BPSK and the block error rate (BLER) is given by $\text{BLER} = 1 - (1 - P_e)^{N_f}$, with P_e being the symbol/bit error rate (SER/BER) of BPSK and N_f being the frame size.

The generalization to higher-order modulation schemes is straightforward.

8.4 Robust Arbitrary Channel-Aware Precoding with Single-Fed LC-PAAs

Figure 8.1 shows the equivalent circuit diagram of a M-element single-fed LC-PAA whose $(M - 1)$ parasitic elements are connected to tunable loads with purely imaginary impedance (e.g., varactor diodes). The relation between the currents and voltages associated with the antenna elements is given by the generalized Ohm's law as [9]

$$\mathbf{i} = \left(\mathbf{Z} + \mathbf{Z}_L \right)^{-1} \mathbf{v} \tag{8.28}$$

where:

  $\mathbf{i}$ is the $(M \times 1)$ vector of the currents that run on the antenna elements

  $\mathbf{Z}$ is the $(M \times M)$ mutual coupling matrix whose diagonal entry Z_{mm} represents the self-impedance of the mth antenna element while the off-diagonal entry Z_{mk} denotes the mutual impedance between the mth and the kth antenna element

  $\mathbf{Z}_L$ is the $(M \times M)$ diagonal load matrix whose diagonal elements are the source resistance R_s and the impedances of the parasitic loads $jX_m (m = 2,3,\ldots,M)$, with $j = \sqrt{-1}$ denoting the imaginary unit

  $\mathbf{v}$ is the $(M \times 1)$ voltage vector that holds the sole feeding voltage v_s

Note that $\mathbf{Z}$ depends on the geometry of the array and it is typically measured or calculated with the help of appropriate computer software.

The system model of a (M, N) MIMO link established between a TX and a RX having M and N antennas, respectively, is given from an antenna perspective by [9]

$$\mathbf{y} = \mathbf{Hi} + \mathbf{n} \tag{8.29}$$

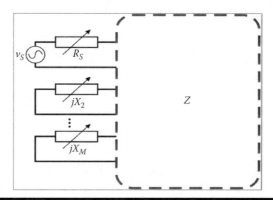

Figure 8.1 Equivalent circuit diagram of a single-fed load-controlled parasitic antenna array.

irrespective of the antenna array technology utilized at each end of the link. In Equation 8.29, **y** is the $(N \times 1)$ vector of open-circuit voltages at the receive antennas, **i** represents the $(M \times 1)$ vector of currents that run on the transmit antennas, **H** denotes the $(N \times M)$ channel matrix whose entry h_{nm} relates the mth input current with the nth open-circuit output voltage, and **n** constitutes a $(N \times 1)$ AWGN vector with covariance matrix $\mathbf{R}_n = \mathbb{E}(\mathbf{nn}^\dagger) = \sigma_n^2 \mathbf{I}_N$. Note that the same relation holds also for the case where the N receive antennas are shared by K users.

Assuming the application of channel-aware precoding, the input–output signal relationship in Equation 8.29 becomes

$$\mathbf{y} = \mathbf{HWs} + \mathbf{n} \tag{8.30}$$

where:
  W is the $(M \times M)$ precoding matrix
  s is the $(M \times 1)$ input signal vector

Hence, in order to apply channel-aware precoding to a single-fed LC-PAA, we have to map the precoded symbols to the antenna currents as follows [22]:

$$\mathbf{i} = \mathbf{Ws} \tag{8.31}$$

After calculating the required currents for the desired precoding scheme and the given input signal format according to Equation 8.31, we should compute the corresponding loading values according to Equation 8.28 under the constraint of a positive input resistance, that is, $\text{Re}\{Z_{in}\} > 0$, in order to ensure that the antenna system will not reflect power back [23]. The value of the input resistance depends on the impedances of the loads, which are determined by the antenna currents. The latter, in turn, depend on the precoded signals. Hence, it becomes apparent that this design condition cannot be met for any given input signal constellation or precoding scheme.

On the other hand, single-fed LC-PAAs can admit any input signal in transmit BF applications, since the array manifold required to shape the radiation pattern as desired does not depend on the format of the input signal. The role of the parasitic loads in this case is to generate currents with appropriate magnitude and phase shift, so that the desired beam is produced. This is similar with the functionality of BF weights in conventional antenna arrays. The only condition that has to be met is that the loadings should be tuned within a range of reasonable values.

Based on this remark, we suggest an alternative approach for performing *robust, low-complexity, arbitrary channel-aware precoding* with single-fed LC-PAAs:

1. First, we apply transmit BF using any valid method.
2. Then, we perform channel-aware precoding over the employed beam.

By taking advantage of the radiation pattern reconfiguration capabilities of single-fed LC-PAAs through the decoupling of the problem to a BF and a precoding part, we overcome the circuit stability and implementation complexity issues. This approach can be generalized to MAMP arrays as well.

8.5 Degrees of Freedom of LC-PAAs

The radiation pattern of any antenna array with M antenna elements is governed by the following equation [7,24]:

$$P(\phi,\theta) = \sum_{m=1}^{M} i_m a_m(\phi,\theta) = \mathbf{i}^T \mathbf{a}(\phi,\theta) \tag{8.32}$$

where:
  $i_m \in \mathbb{C}$ is the current that runs on the mth antenna element
  $a_m(\phi,\theta)$ is the response of that element at azimuthal angle ϕ and elevation angle θ
  $\mathbf{i}$ is the $(M \times 1)$ vector of the currents at the ports of the antenna elements
  $\mathbf{a}(\phi,\theta)$ is the $(M \times 1)$ steering vector of the array that holds the response of each antenna element

It has been shown that, assuming an arbitrary planar single-fed LC-PAA geometry, we can express the functions $a_m(\phi,\theta)$ $(m = 1,2,...,M)$ as a linear combination of a set of M orthonormal basis patterns $\Phi_n(\phi,\theta)$ $(n = 1,2,...,M)$ that span an M-dimensional space, so that $P(\phi,\theta)$ is represented at the beam-space domain as [7,24]:

$$P(\phi,\theta) = \sum_{m=1}^{M} i_m \underbrace{\sum_{n=1}^{M} q_{mn} \Phi_n(\phi,\theta)}_{a_m(\phi,\theta)}$$

$$= \sum_{n=1}^{M} \mathbf{i}^T \mathbf{q}_n \Phi_n(\phi,\theta) \tag{8.33}$$

$$= \sum_{n=1}^{M} w_n \Phi_n(\phi,\theta)$$

$$= \mathbf{\Phi}^T \mathbf{w}$$

where:
  $\mathbf{q}_n$ is the $(M \times 1)$ vector that holds the projections q_{mn} of all functions $a_m(\phi,\theta)$ onto $\Phi_n(\phi,\theta)$
  w_n is the weighting coefficient of $\Phi_n(\phi,\theta)$
  $\mathbf{\Phi}$ is the $(M \times 1)$ vector of the basis patterns
  $\mathbf{w}$ defines a coordinate vector at the beam-space domain which corresponds to a radiated pattern

By letting $s_n = w_n$, that is, by enabling the nth basis pattern $\Phi_n(\phi,\theta)$ to be modulated by the symbol $s_n = \mathbf{i}^T \mathbf{q}_n$, we can encode the $(M \times 1)$ vector of transmitted symbols $\mathbf{s}$ directly to a single radiation pattern $P(\phi,\theta)$. Hence, via the switching of different beam patterns in response to input data vector symbols, we are able to spatially multiplex M data streams using an M-element single-RF LC-PAA. Therefore, we can argue that *the aerial DoF (ADoF) of a single-RF LC-PAA equal the number of its elements* [7,24]. However, *this result represents only an upper bound for the effective DoFs (EDoFs) of the array*, since, depending on the geometry and the dimensions of the LC-PAA, the radiation efficiency of some basis patterns may be negligible [7,24].

In [25], the authors argue that the EDoF of an M-element single-RF LC-PAA equal M only if the impedances of the $(M-1)$ loads are complex (e.g., the parasitic loads are active circuits); otherwise (e.g., when varactor diodes are used), the ADoF equal $(M-1)/2+1$, since we can control $(M-1)$ real values (e.g., the reactance of the varactor diodes) and one complex value (i.e., the input voltage). This finding seems to comply with the examples in [7], where a 3-element and a 5-element single-RF LC-PAA using varactor diodes as parasitic loads provides 2 and 3 EDoF, respectively.

8.6 A Communication Protocol for Low-Mobility Scenarios

As the number of antennas in the BSs grows, the CSI feedback overhead grows as well and becomes a limiting factor in the system's design and performance. LC-PAAs solves this problem, since they provide the same DoF with equivalent DAAs while using a limited number of active antennas. However, the computational complexity related with the dynamic calculation of the parasitic loads' impedances discourages the application of the LC-PAA technology in practice. Moreover, regardless of the type of arrays used in MIMO systems, the beam tracking procedure is fairly complex, especially as we move to higher frequencies, where typically the beams are narrower.

In this Section, we describe a communication protocol for MU-MIMO/cMIMO LC-PAA-equipped systems, which can be utilized in low-mobility scenarios and addresses the aforementioned issues. More specifically, we assume that instead of tunable loads, the LC-PAAs employ a number of fixed loading sets. Each one of them corresponds to a predetermined radiation pattern (i.e., a beam). A simple RF switch allows us to connect the passive antenna elements to the desired loading set (Figure 8.2).

The system operation is divided in three phases [19]:

1. *Learning phase*: For each beam combination, the BS(s) sends a pilot signal. Then, the UTs measure their SINR or estimate the gain of the direct and cross channels and report back this channel quality metric.
2. *Beam-selection phase*: After switching through all possible beam combinations, the BS(s) selects the optimum one, in terms of the achieved SR throughput, based on the information reported by the UT.
3. *Transmission phase*: The BS(s) transmits over the selected beams.

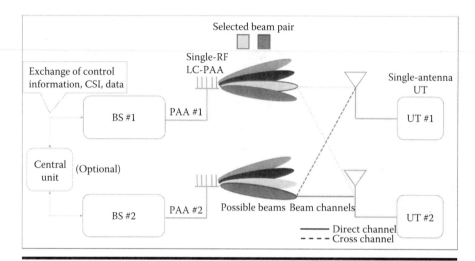

Figure 8.2   A cMIMO system comprised by 2 BSs with a single-fed LC-PAA and 2 single-antenna UTs. Each LC-PAA can generate at each time one out of four different beams. The best beam combination is selected jointly by the BSs, based on SINR or CSI feedback from the UTs. Then, transmission takes place over these beams.

The use of fixed beams reduces the complexity of load calculation and beam tracking. Moreover, the use of SINR feedback further reduces complexity, since it is commonly much easier to measure the SINR than to estimate the gain of a number of channels. After SINR-feedback-based beam selection, a CSI-feedback procedure for the selected composite beam-channel takes place, in order to enable the use of precoding.

8.7 Design of LC-PAAs

Designing LC-PAAs is an integral part of the wireless communication protocol outlined in the previous Section. In all the following cases, a commercially available electromagnetic analysis software was used. This software uses the finite element method (FEM) in order to simulate the desired antenna design and it provides a wide variety of results. Other similar software use the method of moments (MoM) or the finite-difference time-domain (FDTD) method, yielding very similar results. In our case, we are mainly interested in the scattering parameters (S-parameters) and the 3D far field radiation pattern of the antenna. The first is presented to verify that the antenna resonates at the desired frequency, while the latter is used to show where the electromagnetic radiation is directed and it is presented via the three main planes of the 3D far field radiation pattern, namely the X-Z, the X-Y and the Y-Z planes.

In our designs we used as dielectric board either the FR-4 or the Rogers RO4350 material. The FR-4 board has relative permittivity (ε_r) 4.4 and dielectric loss tangent (*tanδ*) 0.02 and it is used for frequencies up to 3 GHz. The Rogers RO4350 board which has relative permittivity (ε$_r$) 3.48 and dielectric loss tangent (*tanδ*) 0.004, is used for frequencies above the 10 GHz regime.

8.7.1 Cross-Like Parasitic Antenna Array at 2.5 GHz

The first design that was studied was a simple patch parasitic antenna array. It is known that by varying the length of the two sides of the patch, one can change its resonant frequency. Also, by varying the feeding point position of the patch, one can alter the input impedance of the patch in order to match the resistance of the feeding line (which in our case was a 50 Ohm coaxial cable with SMA connectors). In close vicinity to the active patch, about 2 mm far, four identical parasitic (nonfeeding) patches were placed. This design for the cross parasitic antenna array, is presented in Figure 8.3. In this cross-like configuration of the patch parasitic antenna array, the center element is the only one that is active, that is, connected to an RF chain, while the surrounding patch elements 1, 2, 3, and 4 are terminated to precalculated loads (capacitors in this case with impedance $[-jX]$). Thus, the array is considered an ESPAR antenna.

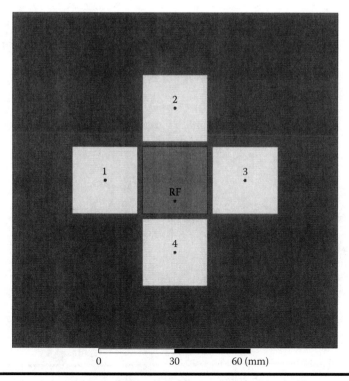

Figure 8.3   Initial prototype design of the cross parasitic antenna array.

The actual values of the loads that were used, along with their equivalent Ohmic resistance at 2.5 GHz are presented in Table 8.1.

In Figure 8.4 we present the scattering parameters versus frequency and the three main planes of the far field radiation pattern.

Table 8.1 Calculated Load Values for the Cross Parasitic Antenna Array

Element Number	Capacitor or Inductor Value	Impedance at 2.5 GHz
1	0.8pF	$X_C = 76.517$ Ohms
2	0.8pF	$X_C = 76.517$ Ohms
3	0.5pF	$X_C = 122.42$ Ohms
4	0.8pF	$X_C = 76.517$ Ohms

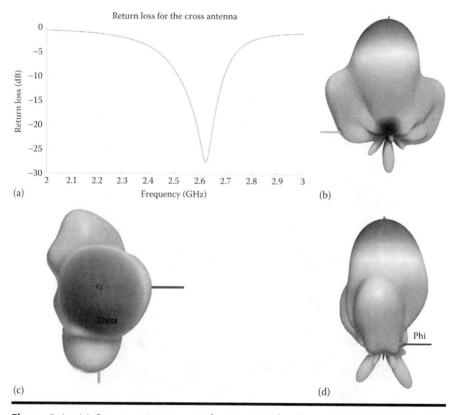

Figure 8.4 (a) S-parameters versus frequency. The three main plains of the far field radiation pattern. (b) x-z, (c) x-y, and (d) y-z.

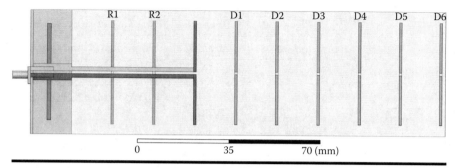

Figure 8.5 Initial prototype design of the parasitic Yagi-Uda planar antenna.

8.7.2 Yagi-Uda Parasitic Antenna Array at 2.6 GHz

The initial design of the prototype planar Yagi-Uda model is presented in the Figure 8.5. This parasitic antenna resonates at 2.6 GHz. The antenna consists of the active dipole (along with its microstrip feeding lines that stretch till the edge of the FR-4 dielectric board so that an SMA connector for feeding the antenna can be mounted), the six directors that are placed in front of the active dipole (on the same side of the FR-4 board as the active dipole, noted as D1–D6) and the two reflectors that are placed behind the active dipole (on the back side of the FR-4 board, noted as R1–R2). The overall FR4 board's dimensions are 16 cm by 10 cm.

The role of the directors is to concentrate the electromagnetic field and shape the radiation pattern in a desired way. Similarly, the role of the reflectors is to reflect the back-scattered radiation. Both the directors and the reflectors are implemented by small printed dipoles. Note that in the middle of each dipole there is a small gap, where the loads are soldered. The directors are loaded with capacitors, while the reflectors are loaded with inductors.

The actual values of the loads that were used, along with their equivalent Ohmic resistance at 2.6 GHz are presented in Table 8.2. In Figure 8.6 we present the scattering parameters versus frequency and the three main planes of the far field radiation pattern.

8.7.3 10.23 GHz Bowtie Parasitic Antenna Array

The first attempt to migrate to higher frequencies (i.e., paving the way to the 18 GHz and the 28 GHz regimes) is presented in this section. Initially, in Figure 8.7, we present a simple bow-tie patch antenna resonating at 10.23 GHz.

The actual values of the loads that were used, along with their equivalent Ohmic resistance at 10.23 GHz are presented in Table 8.3.

In Figure 8.8 we present the scattering parameters versus frequency and the three main planes of the far field radiation pattern. It should be noted that the main lobe has a half power beam width (HPBW) of around 56 and a gain of 7.4 dBi.

Table 8.2   Calculated Load Values for the Planar Yagi-Uda Parasitic Antenna

Element Number	Capacitor or Inductor Value	Impedance at 2.6 GHz
Reflector 1	1.6nH	$X_L = 23.122$ Ohms
Reflector 2	1.8nH	$X_L = 26.012$ Ohms
Director 1	1.3pF	$X_C = 53.229$ Ohms
Director 2	1.2pF	$X_C = 57.665$ Ohms
Director 3	1.1pF	$X_C = 62.907$ Ohms
Director 4	1.0pF	$X_C = 69.198$ Ohms
Director 5	0.9pF	$X_C = 76.886$ Ohms
Director 6	1.0pF	$X_C = 69.198$ Ohms

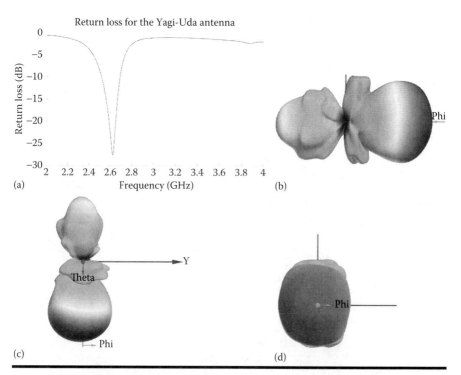

Figure 8.6   (a) S-parameters versus frequency. The three main plains of the far field radiation pattern. **(b)** x-z, **(c)** x-y, and **(d)** y-z.

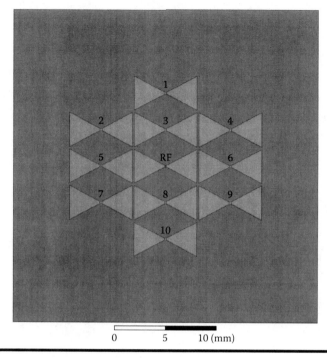

Figure 8.7   Initial prototype design of the bowtie planar antenna.

Table 8.3   Calculated Load Values for the 10.23 GHz Bowtie Parasitic Antenna

Element Number	Capacitor or Inductor Value	Impedance at 10.23 GHz
1	41.34pH	$X_L = 5.0$ Ohms
2	41.34pH	$X_L = 5.0$ Ohms
3	45.47pH	$X_L = 5.5$ Ohms
4	41.34pH	$X_L = 5.0$ Ohms
5	0.275pF	$X_C = 30$ Ohms
6	0.275pF	$X_C = 30$ Ohms
7	0.275pF	$X_C = 30$ Ohms
8	0.275pF	$X_C = 30$ Ohms
9	0.275pF	$X_C = 30$ Ohms
10	0.275pF	$X_C = 30$ Ohms

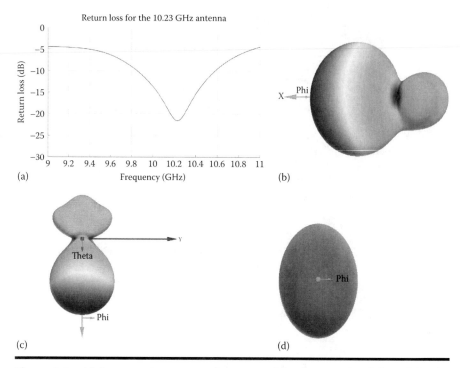

(a)

(b)

(c)

(d)

Figure 8.8 (a) S-parameters versus frequency. The three main plains of the far field radiation pattern. (b) x-z, (c) x-y, and (d) y-z.

Considering that the active element is similar to a dipole, thus having a donut-like far-field radiation pattern with a gain of 0 dBi, it is quite impressive that by carefully placing the loaded parasitic elements, we confined the far-field radiation pattern to less than 60, while at the same time boosting its gain to 7.4 dBi.

8.7.4 19.25 GHz Bowtie Parasitic Antenna

Migrating to even higher frequencies with the use of parasitic antennas is a fairly difficult task. Even a slight change in the dimensions has a large impact on the resonant frequency and/or the radiation pattern. This section presents our initial simulated results of a K-band (20 GHz) parasitic antenna. In Figure 8.9 we present the initial design of the bowtie parasitic antenna array, resonating at 19.25 GHz. It consists of one active and ten parasitic elements. The active element is located at the center of the dielectric board and it is surrounded by the parasitic elements. Note that in the middle of each parasitic element there is a small gap, where the loads (capacitors or inductors) are placed.

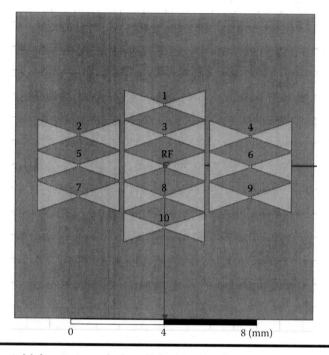

Figure 8.9   Initial prototype design of the bowtie planar antenna.

The actual values of the loads that were used, along with their equivalent Ohmic resistance at 19.25 GHz, are presented in Table 8.4.

In Figure 8.10 we present the scattering parameters versus frequency and the three main planes of the far field radiation pattern. It should be noted that the main lobe has a HPBW of around 38 and a gain of 9 dBi. Considering that the active element is similar to a dipole, thus having a donut-like far-field radiation pattern with a gain of 0 dBi, it is quite impressive that by carefully placing the loaded parasitic elements, we confined the far-field radiation pattern to less than 40, while at the same time boosting its gain to 9 dBi.

8.8 Numerical Simulations and Performance Evaluation

In this Section, we evaluate the performance of the proposed arbitrary precoding framework for various MIMO setups through a number of numerical simulations. We compare the case where the nodes are equipped with LC-PAAs against the scenario where equivalent DAAs are installed on them. The target SNR range is (0, 30 dB). The results refer to the average SR throughput obtained after 100 simulation runs. The simulation is based on a realistic scattering environment. The radiation patterns have been generated from appropriate antenna design software.

Table 8.4   Calculated Load Values for the 19.25 GHz Bowtie Parasitic Antenna

Element Number	Capacitor or Inductor Value	Impedance at 19.25 GHz
1	41.34pH	$X_L = 5.0$ Ohms
2	41.34pH	$X_L = 5.0$ Ohms
3	45.47pH	$X_L = 5.5$ Ohms
4	41.34pH	$X_L = 5.0$ Ohms
5	0.275pF	$X_C = 30$ Ohms
6	0.275pF	$X_C = 30$ Ohms
7	0.275pF	$X_C = 30$ Ohms
8	0.275pF	$X_C = 30$ Ohms
9	0.275pF	$X_C = 30$ Ohms
10	0.275pF	$X_C = 30$ Ohms

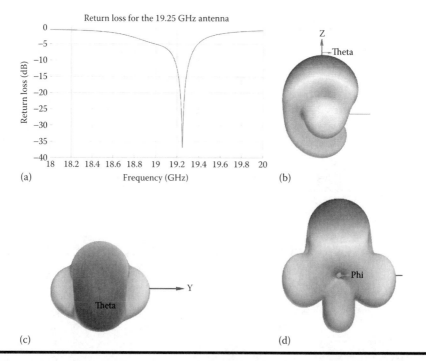

Figure 8.10   (a) S-parameters versus frequency. The three main plains of the far field radiation pattern. (b) x-z, (c) x-y, and (d) y-z.

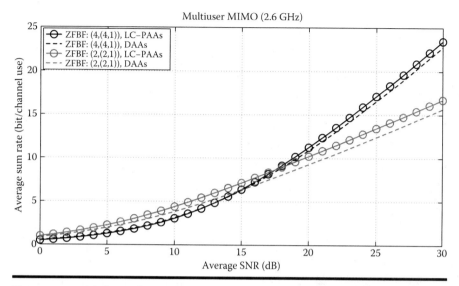

Figure 8.11 **SR throughput of (2,(2,1)) and (4,(4,1)) systems utilizing ZFBF. Each system is equipped with either LC-PAAs or DAAs and operates at the 2.6 GHz band. The ZFBF scheme is utilized.**

In Figure 8.11 the performance of (2,(2,1)) and (4,(4,1)) MU-MIMO systems equipped with LC-PAAs versus corresponding systems equipped with DAAs is illustrated. These systems operate at the 2.6 GHz band and utilize the ZFBF scheme. We note that the LC-PAA-equipped systems outperform their counterparts. The performance gain is more prominent in the (2,(2,1)) setup.

In Figure 8.12 the average SR throughput of two cMIMO systems operating at the 19 GHz band is shown. Each one of them is comprised by two BSs with two active antennas each, serving two users with a single-antenna UT each. One system is equipped with LC-PAAs, while the other makes use of DAAs. Again, the ZFBF scheme is utilized. The JT variant of cMIMO has been employed. Similar to the MU-MIMO case, the LC-PAAs setting is a few dB better than the DAAs one.

In Figure 8.13 the performance of the system setup considered in Figure 8.2 is depicted. The following scenarios are considered: (a) nonprecoding-based transmission, where the beam pair selection is based on SINR feedback; (b) ZF precoding, where the beam pair selection is based on SINR feedback; (c) MRT, ZF, and RZF precoding where the beam pair selection is based on CSI feedback; and (d) ZF precoding in an equivalent single-RF setup where each BS is equipped with a single omnidirectional antenna instead of a LC-PAA. The employed cMIMO variant is JT.

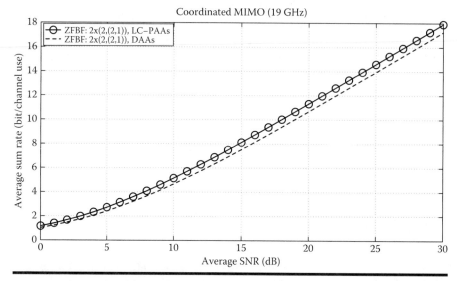

Figure 8.12 SR throughput of 2 cMIMO system. Each one is comprised by 2 BSs, having 2 active antennas each, and 2 single-antenna UTs served by each BS. One system is equipped with LC-PAAs, while the other makes use of DAAs. Both systems operate at the 19 GHz band. The ZFBF scheme is utilized. The JT variant of cMIMO has been employed.

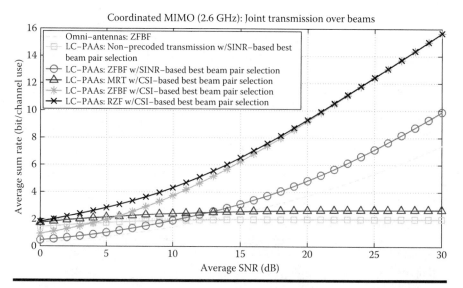

Figure 8.13 Performance of the communication protocol described in Section 8.6.

Some interesting remarks can be made based on Figure 8.13.

■ The average SR throughput of nonprecoding-based transmission and MRT floors at high SNR due to the residual ICI.

■ ZF precoding with LC-PAAs outperforms its omniantenna counterpart over the entire SNR range, even when the beam pair selection is based on SINR feedback, due to the power gain attributed to transmit BF.

■ Beam pair selection based on CSI feedback improves significantly the performance of ZF precoding, since in this case the beam pair selection and precoding tasks are performed jointly.

■ Nonprecoding-based transmission outperforms ZF precoding with omniantennas over the entire range of relevant SNRs (i.e., until the SR throughput flooring at the high SNR regime occurs)! In fact, nonprecoding-based transmission almost resembles MRT. Hence, we note that *we can significantly reduce complexity and still pay only a negligible penalty on performance.*

■ Finally, we note that MRT is optimal at low SNR, ZF with beam pair selection based on CSI feedback is optimal at high SNR, and RZF approaches these two extremes at the corresponding SNR regimes while it performs better at intermediate SNRs, as expected.

Finally, in Figure 8.14 the performance of CIZF against ZFBF for the setup of Figure 8.12 is presented, considering this time systems that operate at 2.6 GHz. The use of BPSK modulation is assumed. We note that CIZF performs much better than

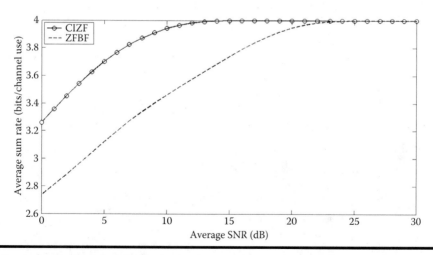

Figure 8.14 SR throughput of 2 cMIMO system. Each one is comprised by 2 BSs, having 2 active antennas each, and 2 single-antenna UTs served by each BS. One system is equipped with LC-PAAs, while the other makes use of DAAs. Both systems operate at the 2.6 GHz band. One system utilizes the CIZF scheme, whereas the other adopts the ZFBF scheme.

its user-level counterpart. The SR throughput of CIZF floors at the SNR level of 15 dB. At the SNR value of 24 dB, the performance of both precoding schemes converges.

8.9 Over-the-Air Demonstration

In this Section, we present the results of an OTA demonstration of the aforementioned robust channel-aware precoding method. For the purposes of the demo, three identical parasitic patch arrays as the ones illustrated in Figure 8.3 are placed on a single base board and in an angle of 45° with respect to each other, as seen in Figure 8.15. Each board is connected to a fixed set of loads. This gave us three narrow beams forming at these angles, which could be switched/selected simply by controlling an RF switch that has a low insertion low and fast switching time.

For setting up the testbed to demonstrate precoding using parasitic patch antennas, we used the WARP v.3 platform. WARP stands for Wireless Open-Access Research Platform and is a scalable and extensible programmable wireless platform, built from the ground up to prototype advance wireless networks. These modules integrate a high-performance field-programmable gate array (FPGA), two flexible RF interfaces and multiple peripherals to facilitate rapid prototyping of custom wireless designs. The central controller consists of a single host PC, which uses MATLAB to send data and control commands to the radio modules. WARP conveniently provides an open-source MATLAB-based framework called WARPLab, which allows users to control and configure the WARP boards and process transmit and receive data samples. This baseline framework is used for rapid physical layer prototyping that allows the coordination of arbitrary combinations of single and multiantenna transmit and receive nodes. The extensible framework gives users the flexibility to develop and deploy large arrays of nodes to meet the application or research need.

The testbed setup is configured for two BSs transmitting concurrently to a single user each, while having their antenna arrays placed within a distance of

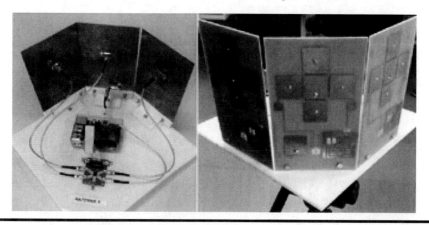

Figure 8.15 Antenna prototype and RF-switch control circuit for beam-switching.

few meters. For testing purposes, we have implemented an orthogonal frequency division multiplexing (OFDM) transceiver system that utilizes ZFBF, with special emphasis on the CSI feedback mechanism which is deemed crucial for a successful and adequate interference cancellation.

With the use of pilot signals at the transmitted signal, we can accurately track channel state information at the RX and provide feedback via the host PC to the TX node for precoding. We have assumed that the channel conditions in a short time period remain unchanged in respect to the surrounding environment, so we could tolerate a small delay in CSI feedback. To accomplish this, we used channel training at the preamble of the transmission. A short training sequence (STS) was sent in order to properly select gains, while a long training sequence (LTS) was sent and used by the RX to estimate and correct for carrier frequency offset (CFO) as well as timing alignment via a cross-correlator. Finally, the preamble was concluded with channel training symbols from each of the two transmit antennas. These symbols were used to generate the 2×2 channel matrix which was used to build the precoder matrix. The TX nodes select each time the best beam to transmit, based on the CSI feedback.

We connected two directional parasitic patch antenna arrays at each TX port and sent two independent spatial streams of data at the same time and in the same frequency band. We also used two omnidirectional monopole antennas at the RX ports to disentangle the received data symbols.

As is shown in Figure 8.16, the performance of the system approaches closely the theoretical performance. Also, as it was expected, ZFBF outperforms significantly raw transmission (i.e., nonprecoded communication).

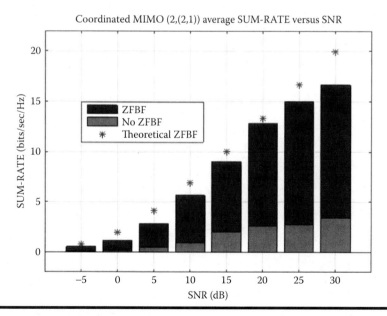

Figure 8.16 OTA demonstration results.

8.10 Summary and Conclusions

In this chapter, we presented a novel yet realistic technology framework that allows us to activate an increased number of spatial degrees-of-freedom in future wireless networks with manageable additional hardware complexity. The approach is based on the use of load-controlled parasitic antenna arrays, which we advocate as a technology contender for wireless nodes of either small (e.g., compact) or large (e.g., massive) size, operating in either single link or multiuser setups and over frequencies that may range from radio to cm- and even mm-wave.

After reviewing the key attributes of conventional multiantenna systems and load-controlled antenna arrays, we focused on the important case of multiuser downlink access. For this, we described a robust, low complexity, arbitrary channel-aware precoding framework for single-fed and multi-RF LC-PAAs.

Capitalizing on this framework, we outlined a communication protocol that further reduces the operation complexity of such systems, by avoiding the need for dynamic load computation, as well as by reducing drastically the number of required channel estimation coefficients. The approach is based on a combination of: (1) The use of fixed (precalculated) loads that shape certain possible radiation patterns (beams) destined for each user; (2) the choice of preferred combinations of the preshaped beams based on channel-dependent switching, thus reducing the effective channel dimensionality once the beam combination is chosen; and (3) the precoding of the user streams over the chosen beam combination. Due to the reduced dimensionality, this approach also reduces channel estimation complexity and CSI feedback overhead. The proposed protocol is not limited to LC-PAAs, being universally applicable to any kind of antenna arrays whose radiation patterns can be computed in an offline fashion and controlled/switched dynamically (e.g., it can be used with mechanically steered arrays, conventional DAAs, reflect arrays).

The performance of the proposed approach and protocol was evaluated under various MIMO setups (MU-MIMO, cMIMO) and frequency bands (2.6, 19 GHz), both via numerical simulations and OTA testbed-based experimentations. The results showcase the important performance gains of LC-PAAs over conventional antenna systems of similar complexity. Our studies included symbol-level ZF precoding, which has the potential to improve the performance of the system in scenarios where the SNR is low. Finally, we demonstrated various implementations of LC-PAAs operating either in cellular or in cm-wave frequencies.

Acknowledgments

The main part of this work has been supported by the EC FP7 project HARP (grant agreement no: 318489) and the EC H2020 Research Project SANSA (grant agreement no: 645047). The authors would like also to acknowledge Mr. Bobby Gizas,

Lab Engineer at Athens Information Technology (AIT), for his work on the implementation of the testbed and the conduction of the relevant over-the-air experiments.

References

1. Ericsson, Mobility report, *Technical Report*, June 2016.
2. M. Mueck et al., White paper: Novel spectrum usage paradigms for 5G, IEEE SIG CR in 5G, *Technical Report*, November 2014.
3. 5G-PPP, 5G vision, *Technical Report*, February 2015.
4. D. Lee et al., Coordinated multipoint transmission and reception in LTE-advanced: Deployment scenarios and operational challenges, *IEEE Communications Magazine*, 50(2), 148–155, 2012.
5. E. Larsonn et al., Massive MIMO for next generation wireless systems, *IEEE Communications Magazine*, 52(2), 186–195, 2014.
6. H. Huang et al., Eds., *MIMO Communication for Cellular Networks*. 1em plus 0.5em minus 0.4em. Springer-Verlag, New York, 2012.
7. A. Kalis et al., Eds., *Parasitic Antenna Arrays for Wireless MIMO Systems*. 1em plus 0.5em minus 0.4em. Springer-Verlag, New York, 2014.
8. T. Ohira and K. Gyoda, Electronically steerable passive array radiator antennas for low-cost analog adaptive beamforming, in *IEEE International Conference on Phased Array Systems and Technology*, 2000, pp. 101–104.
9. V. Barousis et al., A new signal model for MIMO communications with compact parasitic arrays, in *IEEE International Symposium on Communications, Control and Signal Processing*, Athens, Greece, May 21–23, 2014, pp. 109–113.
10. L. Zhou et al., Achieving arbitrary signals transmission using a single radio frequency chain, *IEEE Transactions on Communications*, 63(12), 4865–4878, 2015.
11. C. Masouros and E. Alsusa, Dynamic linear precoding for the exploitation of known interference in MIMO broadcast systems, *IEEE Transactions on Wireless Communications*, 8(3), 1396–1404, 2009.
12. White Paper: 5G spectrum recommendations, 4G Americas, *Technical Report*, August 2015.
13. White Paper: 5G radio access: Requirements, concept and technologies, NTT DOCOMO, *Technical Report*, July 2014.
14. A. Goldsmith, *Wireless Communications*. 1em plus 0.5em minus 0.4em. Cambridge University Press, Cambridge, UK, 2005.
15. E. Bjornson and E. Jorswieck, Optimal resource allocation in coordinated multi-cell systems, *Foundations and Trends in Communications and Information Theory*, 9(2–3), 113–381, 2013.
16. G. J. Foschini, Layered space-time architecture for wireless communication in fading environments when using multi-element antennas, Bell Labs, *Technical Report*, 1996.
17. E. Telatar, Capacity of multi-antenna Gaussian channels, *European Transactions on Telecommunications*, 10(6), 585–596, 1999.
18. A. Goldsmith et al., Capacity limits of MIMO channels, *IEEE Journal on Selected Areas in Communications*, 21(5), 684–702, 2003.
19. K. Ntougias et al., Coordinated MIMO with single-fed load-controlled parasitic antenna arrays, in *17th IEEE International Workshop on Signal Processing advances in Wireless Communications* (*SPAWC 2016*), Edinburgh, UK, July 3–6, 2016, pp. 1–5.

20. P. Marsch and G. Fettweis, Eds., *Coordinated Multi-Point in Mobile Communications: From Theory to Practice.* 1em plus 0.5em minus 0.4em. Cambridge University Press, Cambridge, UK, 2011.

21. K. Ntougias et al., Robust low-complexity arbitrary user- and symbol-level multi-cell precoding with single-fed load-controlled parasitic antenna arrays, in *23rd International Conference on Telecommunications (ICT 2016)*, Thessaloniki, Greece, May 16–18, 2016, pp. 1–5.

22. G. Alexandropoulos et al., Precoding for multiuser MIMO systems with single-fed parasitic antenna arrays, in *IEEE Global Communications Conference (GLOBECOM)*, Austin, TX, December 8–12, 2014, pp. 3897–3902.

23. V. Barousis and C. B. Papadias, Arbitrary precoding with single-fed parasitic arrays: Closed-form expressions and design guidelines, *IEEE Wireless Communications Letters*, 3(2), 229–232, 2014.

24. V. Barousis and A. G. Kanatas, Aerial degrees of freedom of parasitic arrays for single RF front-end MIMO transceivers, *Progress in Electromagnetics Research B*, 35, 287–306, 2011.

25. M. A. Sedaghat and R. Mueller, Large system analysis of low-cost MIMO transmitters, in *17th International ITG Workshop on Smart Antennas (WSA)*, Stuttgart, Germany, March 13–14, 2013, pp. 1–4.

Chapter 9

Spatial Modulation for 5G Systems

Konstantinos Peppas and Panagiotis Mathiopoulos

Contents

9.1 Introduction

Research and development for the 5th-generation (5G) wireless systems has been initiated several years ago [1–3]. Such systems, which are set for commercial use sometime around 2020, are expected to provide new types of enhanced user connectivity services, in terms of providing very high data rates, increased capacity, improved security, higher reliability, reduced latency, increased quality of service and availability, and energy efficiency (EE). According to the 5G standard such systems should provide higher data rates, for example, tens of Mb/s and accommodating tens of thousands of users providing data rates of 100 Mb/s for metropolitan areas. Furthermore, their spectral efficiency (SE) will increase significantly, as compared to the SE achieved by the 4th-generation (4G) wireless systems, their coverage will also improve and their latency will be reduced significantly as compared to Long-Term Evolution (LTE) [2].

Multiple-input multiple-output (MIMO) systems, incorporating multiple antennas at the transmitter and/or receiver, constitute a key technology for the design of current (4G) and future (5G) wireless communications systems. MIMO systems are capable of providing higher data rates and superior error performance as compared to their single antenna counterparts without increasing spectrum utilization and transmit power [4,5]. As such, MIMO has become an essential element of current wireless communication standards including the IEEE 802.11n (Wi-Fi), WiMAX (4G), and LTE (4G) [5].

The MIMO paradigm can be exploited in different ways to improve the overall performance by means of: (1) diversity, (2) antenna array (beamforming), and (3) spatial multiplexing. For the diversity case, system performance enhancements result from diversity gain, which reduces the bit error rate (BER) for the same SE, as the number of transmitting and/or receiving antennas increases. These systems can be designed to achieve full diversity gain with relatively low complexity receivers. Moreover, they can efficiently deal with channel imperfections that exist in real-time implementations of MIMO systems [6–9] (Table 9.1).

In the beamforming case, by exploiting knowledge of the channel at the transmitter and employing singular value decomposition, the channel matrix can be decomposed to a number of parallel single-input single-output (SISO) channels. The resulting unitary matrices can then be used as pre- and post-filters at the transmitter and receiver, respectively, to increase channel capacity by employing the so-called water-filling technique [10,11].

In the spatial multiplexing case, a layered space-time approach is employed to transmit multiple independent data streams over the antennas to increase capacity. A representative example of this approach is the so-called vertical Bell Labs layered space-time (VBLAST) architecture [4,12] in which a user data stream is demultiplexed into a number of substreams which are equal to the number of transmitting antennas. If the transmit-to-receive links are sufficiently independent, system throughput may be increased linearly with the number of antennas [4,12].

Table 9.1 Table of Acronyms

Acronym	Explanation
5G	Fifth generation
4G	Fourth generation
ABEP	Average bit error probability
ARQ	Automatic repeat request
ASM	Antenna subset modulation
AWGN	Additive white Gaussian noise
BER	Bit error rate
BPSK	Binary phase shift keying
BS	Base station
CFO	Carrier frequency offset
CPM	Continuous phase modulation
CSI	Channel state information
EE	Energy efficiency
GSM	Generalized spatial modulation
GSSK	Generalized space shift keying
HPA	High power amplifiers
IAS	Inter-antenna synchronization
ICI	Inter-channel interference
IFFT	Inverse fast Fourier transform
LTE	Long term evolution
MGF	Moment generating function
MIMO	Multiple-input multiple-output
ML	Maximum-likelihood
OFDM	Orthogonal frequency division multiplexing
OSTBC	Orthogonal space-time block coding

(*Continued*)

Table 9.1 (*Continued*)   Table of Acronyms

Acronym	Explanation
PA	Power amplifier
PAPR	Peak-to-average power ratio
PSK	Phase shift keying
QAM	Quadrature amplitude modulation
QPSK	Quadrature phase shift keying
RF	Radio-frequency
SC	Single carrier
SE	Spectral efficiency
SISO	Single-input single-output
SM	Spatial modulation
SSK	Space shift keying
VBLAST	Vertical bell labs layered space-time
WiMax	Worldwide interoperability for microwave access

Despite the aforementioned advantages of MIMO systems, their main drawback is the increased implementation complexity and cost. More specifically, as compared to their SISO counterparts, they are more costly to implement and require increased power consumption [13,14]. Their receiver implementation complexity is mainly because of their ability to mitigate *inter-channel interference* (ICI), which is introduced by coupling multiple symbols in both time and space, thus increasing the signal processing complexity. In addition, detection algorithms require that all symbols are transmitted at the same time and thus sophisticated *inter-antenna synchronization* (IAS) techniques should be employed to exploit the benefits of space time-coded and multiuser MIMO transmissions. Finally, *multiple RF chains* are required at the transmitter to simultaneously transmit several data streams. Unfortunately, current radio frequency (RF) electronics are expensive, difficult to implement, do not follow Moore's law and as such make the MIMO transmitter rather bulky [15].

As far as power consumption requirements are concerned, on the one hand, the electronic circuits utilized by MIMO systems, such as RF power amplifiers (PAs), mixers, synthesizers and filters, substantially increase the circuit power dissipation of the base stations. Recent studies, carried out at the European project EARTH [16], have suggested that EE decreases linearly with the number of active antennas and it depends mostly on the PAs. On the other hand, conventional MIMO communication systems aim at maximizing the capacity under peak or average power constraints. Such communication systems may transmit data with the maximum allowed power for long periods, and in that sense their design is not energy efficient. EE is commonly defined as the transmitted information bits per unit of transmit energy and has been studied from an information theoretical perspective [17,18]. It is well known that, for an additive white Gaussian noise (AWGN) channel, the channel capacity, C, is given by the well-known Shannon's formula, that is,

$$C = B\log_2\left(1 + \frac{P}{N_0 B}\right) = B\eta_{SE} \tag{9.1}$$

where:
$\eta_{SE} \triangleq \log_2\left(1 + P/N_0 B\right)$ is the SE in bits/s/Hz
P is the transmit power
B is the system bandwidth
N_0 is the noise power spectral density

According to [18], the EE is given by

$$\eta_{EE} = \frac{C}{P} = \frac{\eta_{SE}}{N_0(2^{\eta_{SE}} - 1)} \tag{9.2}$$

From Equation 9.2, it becomes evident that EE decreases monotonically with η_{SE}, that is, with SE.

Because of the abovementioned disadvantages of conventional MIMO systems, recent research efforts have focused on the design of less complex MIMO schemes with a limited number of active RF chains and relaxed IAS and ICI requirements. At the same time, these systems are still able to exploit all transmit-antenna elements for multiplexing and transmit-diversity gains. In this context, the so-called single-RF MIMO design has recently emerged as a promising alternative to conventional MIMO systems [15].

Figure 9.1 illustrates the block diagrams of a single-RF MIMO system and that of a conventional MIMO system. By comparing these two systems, it is noted that the single-RF MIMO can achieve the gains of MIMO communications by employing all available antenna elements. However only one antenna is typically activated (single-RF front-end) at the transmitter at any modulation instant [19]. The motivation behind single-RF paradigm originates from the fact that large numbers of transmitting antennas may be available at the base stations (BSs) [15,19], especially in the emerging millimeter-wave bands [20–23]. This happens since the complexity and power consumption/dissipation of MIMO communications are mainly determined by the number of the number of active RF chains [19].

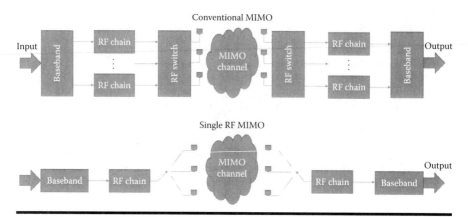

Figure 9.1   Block diagrams of conventional and single-RF MIMO systems.

Given the aforementioned considerations, an ideal transmission scheme for MIMO systems should have the following operational characteristics. First, it should employ one (or just a few) active RF chains but still being able to exploit all transmit-antenna elements for multiplexing and transmit-diversity gains. Second, it should offer maximum likelihood (ML) optimum decoding performance with single-stream decoding complexity. And third, it should work without the need of power inefficient linear modulation schemes, such as quadrature amplitude modulation (QAM), or it should allow the use of more power efficient modulation schemes, such as phase shift keying (PSK) or continuous phase modulations (CPM).

Recently, the so-called spatial modulation (SM) has been proposed as a promising transmission scheme, which reduces the complexity and cost of multiple-antenna schemes [14]. SM can achieve high data rates, improved system performance and EE, thus having the inherent potential to meet the aforementioned desired operational characteristics. It is also noted that SM-MIMO can outperform many state-of-the-art MIMO schemes, provided that a sufficiently large number of antenna elements is available at the transmitter [24]. These properties can be achieved by employing the following simple operational mechanism. At any signaling time instance only one transmit-antenna is activated for data transmission and therefore SM can avoid ICI. In this way, no IAS is required and only one RF chain is necessary for data transmission. Furthermore, SM employs a low-complexity single-stream receiver design for ML decoding. The spatial position of each transmit-antenna in the antenna-array is used as a source of information. This is realized by employing the so-called SM constellation diagram [25], where a one-to-one mapping between each antenna index and a block of information bits to be transmitted is performed. The resulting coding mechanism, also known as transmit-antenna index coded modulation, allows SM to achieve a spatial multiplexing gain with respect to SISO systems.

9.2 Working Principles

In this section, the basic working principles for SM MIMO will be presented. Let us consider a MIMO system having N_t and N_r transmitting and receiving antennas, respectively. The system employs either M-PSK or M-QAM where M is the cardinality of the signal constellation diagram. At the receiver, ML detection is employed. In Figure 9.2, the SM-MIMO concept is illustrated and compared with the orthogonal space-time block coding (OSTBC) scheme proposed by Alamouti [6] and the VBLAST scheme [4]. In the first case, two M-PSK/M-QAM symbols, S_1 and S_2 are encoded and simultaneously transmitted from a pair of transmitting antennas in two time slots. The spectral rate of the Alamouti scheme equals to $R_{\text{Alamouti}} = \log_2(M)$ bits per channel use (bpcu). In VBLAST, S_1 and S_2 are simultaneously transmitted from a pair of transmitting antennas in a single time slot and in this case the spectral rate of VBLAST equals to $R_{\text{VBLAST}} = N_t \log_2(M)$ bpcu.

The basic idea of SM is to map a block of information bits into two information carrying units, namely a symbol that is chosen from a complex signal-constellation diagram and a unique transmit-antenna index that is chosen from the set of transmit-antenna in the antenna array [14]. Essentially, SM is a hybrid modulation and MIMO technique in which the modulated signals belong to a 3D constellation diagram, which jointly combines signal and spatial information and generalizes the well-known 2D (complex) signal-constellation diagram of M-PSK/QAM schemes. It is noted that an additional 3rd dimension is provided by the antenna array, where

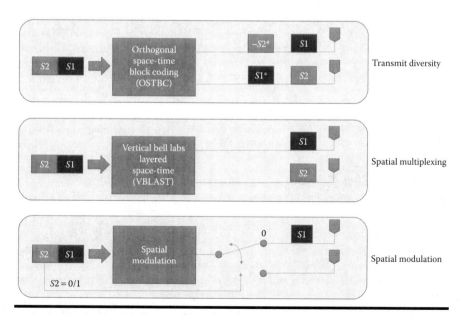

Figure 9.2   **The SM concept and its comparison to the OSTBC [6] and VBLAST [4] schemes.**

some of the bits are mapped to the transmitted antennas. The spectral rate of SM is given as $R_{SM} = \log_2(N_t) + \log_2(M)$ bpcu [25].

The signal model of a SM-MIMO system with N_t transmitting and N_r receiving antennas, assuming a frequency-flat fading channel model, can be mathematically expressed as [14]

$$\mathbf{y} = \mathbf{Hx} + \mathbf{n} \tag{9.3}$$

where:

$\mathbf{y} \in \mathbb{C}^{N_r \times 1}$ is the received complex signal vector
$\mathbf{H} \in \mathbb{C}^{N_r \times N_t}$ is the complex channel matrix
$\mathbf{n} \in \mathbb{C}^{N_r \times 1}$ is the complex additive white Gaussian noise at the receiver

Furthermore, $\mathbf{x} = \mathbf{e}s \in \mathbb{C}^{N_t \times 1}$ is the complex modulated vector with s being the complex PSK/QAM modulated symbol belonging to the signal-constellation diagram and $\mathbf{e} = [e_1, e_2, \ldots, e_{N_t}]^T$ is a $N_t \times 1$ vector belonging to the spatial-constellation diagram whose tth element, e_t, is given as

$$e_t = \begin{cases} 1 & \text{if the } t\text{th transmitting antenna is active} \\ 0 & \text{if the } t\text{th transmitting antenna is not active} \end{cases} \tag{9.4}$$

If $N_t = 1$, the SM-MIMO reduces to conventional SISO communications, where the information bits are encoded only onto the signal-constellation diagram. The achieved spectral rate in this case is $R_{SISO} = \log_2 M$ bpcu. On the other hand, for the special case of $M = 1$, the SM-MIMO reduces to the special case of the so-called space shift keying (SSK) modulation [26]. According to the operation of SSK, the incoming bit stream is encoded into the index of one unique antenna each time switched on for transmission while the others are switched off thus providing a spectral rate of $R_{SSK} = \log_2 N_t$ bcpu.

Figure 9.3 presents an example for the encoding mechanism of SM-MIMO for $N_t = 4$ and $M = 4$. Let us assume that the block of bits to be encoded is 1101. The first $\log_2 4 = 2$ bits, 11, determine the index of the active transmitting antenna, T×3, whereas the second $\log_2 4 = 2$ bits, namely 01, determine the transmitted point from a quadrature phase-shift keying (QPSK) constellation. Let us now consider a SM-MIMO system with $N_t = 4$ and $M = 2$, as shown in Figure 9.4. At the transmitter, the information source is divided into blocks containing $\log_2(N_t) + \log_2(M)$ bits each. Each block is then processed by a SM mapper, which splits each of them into two sub-blocks each having $\log_2(N_t)$ and $\log_2(M)$ bits. The index of the selected antenna that is switched on for data transmission is determined by the bits in the first sub-block. All other transmit antennas are kept silent during the current signaling time interval. The symbol in the signal constellation diagram is selected based on the bits in the second subblock. In the special case, where SSK modulation (instead of SM) is considered, each transmit-antenna, when switched on, will transmit exactly the same signal. Thus, the information is encoded only in the position within the antenna array.

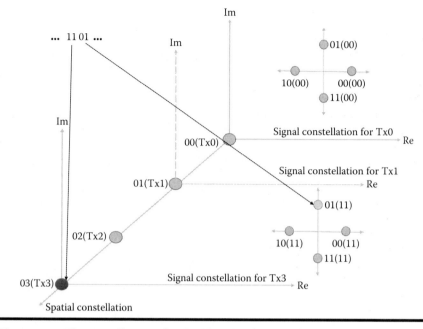

Figure 9.3 The encoding mechanism for SM-MIMO.

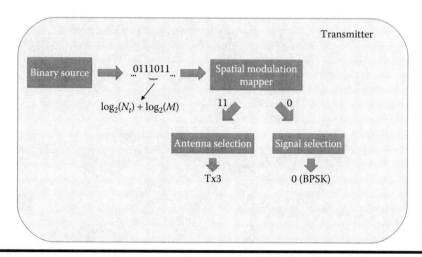

Figure 9.4 The encoding mechanism for SM-MIMO with $N_t = 4$ and M = 2.

The transmitted by the active antenna signal propagates through a generic wireless channel as shown in Figure 9.5. Because of the different spatial positions occupied by the transmit-antenna in the antenna array, the signal transmitted by each antenna will experience different propagation conditions due to different propagation paths. This represents the fundamental working principle of SM since the wireless channel plays

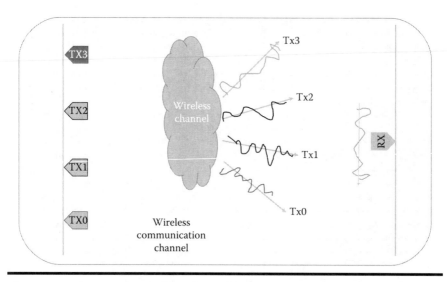

Figure 9.5   The generic wireless channel for SM-MIMO transmission.

the role of a *modulation unit*. It is implemented by introducing a distinct fingerprint that renders the signal transmitted by distinct transmit-antenna distinguishable at the receiver. Note that, if the transmit-to-receive wireless links are not sufficiently different, the signals emitted by the transmit-antenna will look approximately the same and hence the receiver will not be able to correctly detect the transmitted signals. The receiver exploits the randomness introduced by the wireless channel for signal detection. As illustrated in Figure 9.6, in order to detect the transmitted signal from the

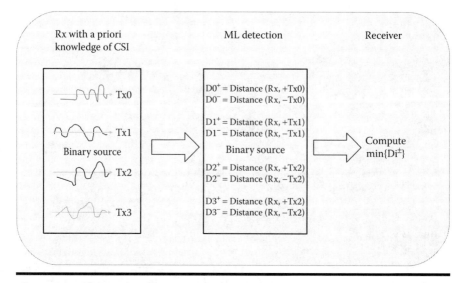

Figure 9.6   The receiver for SM-MIMO transmission.

noisy received signal, the receiver must have a priori knowledge of the channel state information (CSI) of all the transmit-to-receive wireless links. Such knowledge can be obtained by employing channel estimation techniques [27]. According to the ML principle, the receiver computes the Euclidean distance between the received signal and the set of all possible signals modulated by the wireless channel and chooses the smallest one. In this way, all the bits in the transmitted block can be decoded and the original bitstream is recovered [28].

9.3 Advantages and Disadvantages

In this section the advantages and disadvantages of SM-MIMO, as compared to state-of-the-art MIMO communications, will be discussed.

9.3.1 Advantages

The main operational advantages of the SM-MIMO, as compared to conventional MIMO schemes, such as VBLAST and the Alamouti space-time schemes, are the following:

1. SM entirely avoids ICI and IAS, and requires only a single RF chain at the transmitter. This is due to the simple working mechanism of SM where a single transmit-antenna is activated, at each time, for data transmission while all the other antennas are kept silent. In addition, as compared to the conventional SISO system, its 3D constellation diagram introduces a multiplexing gain in the spatial domain. This gain logarithmically increases with the number of transmit-antennas. When viewed as a coding technique, SM provides a high spectrally-efficient code with an equivalent code rate greater than one [25].

2. The receiver design is inherently simpler than the one employed by the V-BLAST scheme. This is due to the fact that complicated interference cancellation algorithms are not required since SM completely avoids ICI. Another advantage of SM is that, unlike conventional spatial-multiplexing methods for MIMO systems, SM can attain ML decoding via a simple single-stream receiver. Furthermore, SM can work effectively, even when the number of receiving antennas is less than the number of transmitting antennas, because the receive-antennas are used to achieve only a diversity gain rather than a multiplexing gain. In principle, since a single receive-antenna is needed to exploit the SM paradigm, it makes it suitable for downlink settings with low-complexity mobile units.

3. Because of its unique operation, SM has the ability to work well in various multiple-access scenarios. Due to the fact that different pairs of transmitters and receivers usually occupy different spatial positions, the channel impulse response of each pair of users is likely to be statistically different from other user pairs. If each intended receiver uses, for data detection,

the set of channel impulse responses of all the transmitters (i.e., multiuser detection), several users might share the same wireless resources for communication. In other words, with SM the wireless channel acts as a natural source of purely random signatures (i.e., in this case the channel impulse responses) for multiple-access.

4. SM provides larger capacities than conventional low-complexity coding methods for MIMO systems, such as OSTBCs because of the multiplexing gain achieved by exploiting the spatial domain to convey part of the information bits.

5. It is noted that SSK modulation can further reduce the receiver complexity, since only the spatial dimension of the 3D constellation diagram is utilized for encoding information bits to symbols. Thus, conventional modulation schemes such as M-PSK and/or M-QAM are not required for data transmission. The price to be paid for this complexity reduction is a loss in the achievable data rate [29].

9.3.2 Disadvantages

There are also some disadvantages associated with SM-MIMO, again with respect to conventional MIMO schemes, as follows.

1. The main disadvantage of SM-MIMO is that at least two transmit-antenna are required to exploit the SM concept. Moreover, if the transmit-to-receive wireless links are not sufficiently different, the SM paradigm might not be efficiently used and/or not yield in an adequate performance. This limitation is somehow similar to the conventional spatial multiplexing techniques, which require a rich-scattering environment to guarantee a significant boost in the achievable data rate [4].

2. Another disadvantage is that the receiver requires perfect CSI for data detection. This might pose complexity constraints on the channel estimation unit, as well as some overhead which are necessary for channel estimation.

3. When compared to more conventional MIMO systems, for example, V-BLAST, SM can offer only a logarithmic (instead of linear) increase of the data rate with the number of transmit-antenna. This might limit SM to achieve very high spectral rates for small numbers of antennas at the transmitter.

9.4 Transmission Techniques

In this section, various transmission techniques for SM-MIMO will be discussed. The simplest transmission technique for SM-MIMO is the SSK modulation. Let us assume that the source produces a random sequence of independent bits, $\mathbf{b} = [b_1, b_2, \ldots b_k]$. Groups of m bits are mapped to a constellation vector $\mathbf{x}_j = [0, \ldots, 0, 1, 0, \ldots, 0]$, whose jth element equals to 1. Table 9.2 depicts an example

Table 9.2   An Encoding Example for a SSK Scheme

$b = [b_1\ b_2]$	Symbol	Antenna Index j	$x = [x_1, x_2, x_3, x_4]^T$
[0 0]	0	1	$[1\ 0\ 0\ 0]^T$
[0 1]	1	2	$[0\ 1\ 0\ 0]^T$
[1 0]	2	3	$[0\ 0\ 1\ 0]^T$
[1 1]	3	4	$[0\ 0\ 0\ 1]^T$

for the encoding of SSK when $N_t = 4$ transmitting antennas are used, resulting in a spectral rate of 2 bpcu. As previously mentioned, SM and SSK are appealing because of their single RF configuration which greatly simplifies the transmitter structure as compared to the more complex MIMO schemes, such as the VBLAST. However, their corresponding spectral rates are $\log_2(N_t) + \log_2(M)$ bpcu and $\log_2(N_t)$ bpcu, respectively, as opposed to the VBLAST scheme which achieves a spectral rate of $N_t \log_2(M)$ bpcu. Nevertheless, rate and complexity can be traded off by allowing more than one active antennas in each time slot. The resulting schemes are known as generalized SSK (GSSK) [30] and generalized SM (GSM) [31].

GSSK uses n_t out of N_t available antenna indices to convey information. Therefore, there are $M' = \binom{N_t}{n_t}$ possible constellation points and the corresponding spectral rate is $R_{\text{GSSK}} = \lfloor \log_2 \binom{N_t}{n_t} \rfloor$. For example, with $n_t = 2$ and $N_t = 7$, the number of possible combinations is $M' = 21$. Since the constellation sizes are multiples of 2, only 16 out of the possible 21 combinations can be used. Although the set of antenna combinations, $\mathcal{X}$, may be chosen at random, there also exist optimal selection rules [30]. Table 9.3 presents an example of 8-ary GSSK with $N_t = 5$, $n_t = 2$, with randomly chosen $\mathcal{X}$.

Table 9.3   An Encoding Example for a GSSK Scheme with $N_t = 5$, $n_t = 2$

$b = [b_1\ b_2\ b_3]$	j	$x = [x_1, x_2, x_3, x_4]^T$
[0 0 0]	(1, 2)	$[1/\sqrt{2}\ \ 1/\sqrt{2}\ 0\ 0\ 0]^T$
[0 0 1]	(1, 3)	$[1/\sqrt{2}\ 0\ 1/\sqrt{2}\ 0\ 0]^T$
[0 1 0]	(1, 4)	$[1/\sqrt{2}\ 0\ 0\ 1/\sqrt{2}\ 0]^T$
[0 1 1]	(1, 5)	$[1/\sqrt{2}\ 0\ 0\ 0\ 1/\sqrt{2}]^T$
[0 0 0]	(1, 2)	$[0\ 1/\sqrt{2}\ \ 1/\sqrt{2}\ 0\ 0]^T$
[0 0 1]	(1, 3)	$[0\ 1/\sqrt{2}\ 0\ 1/\sqrt{2}\ 0\ 0]^T$
[0 1 0]	(1, 4)	$[0\ 1/\sqrt{2}\ 0\ 0\ 1/\sqrt{2}]^T$
[0 1 1]	(1, 5)	$[0\ 0\ 1/\sqrt{2}\ \ 1/\sqrt{2}\ 0]^T$

**Table 9.4 An Encoding Example for GSM with
$N_t = 5$, $N_u = 2$ and BPSK Modulation**

Grouped Bits	Antenna Combination ℓ	Symbol s
0000	(1, 2)	−1
0001	(1, 2)	+1
0010	(1, 3)	−1
0011	(1, 3)	+1
0100	(1, 4)	−1
0101	(1, 4)	+1
0110	(1, 5)	−1
0111	(1, 5)	+1
1000	(2, 3)	−1
1001	(2, 3)	+1
1010	(2, 4)	−1
1011	(2, 4)	+1
1100	(3, 5)	−1
1101	(3, 5)	+1
1110	(4, 5)	−1
1111	(4, 5)	+1

GSM uses more than one transmit antenna to send the same complex symbol in a similar fashion as GSSK. In this case, the number of possible antenna combinations is $M' = \binom{N_t}{N_u}$ where N_u is the number of active antennas at each instance. As with GSSK, the number of the allowed antenna combinations must be a power of two and, therefore, only 2^{m_l} combinations can be used, where $m_l = \lfloor \log_2 \binom{N_t}{N_u} \rfloor$. The resulting spectral rate is $R_{GSM} = \lfloor \log_2 \binom{N_t}{n_t} \rfloor + \log_2 M$. Table 9.4 depicts the encoding for a GSM scheme assuming $N_t = 5$, $N_u = 2$ and binary phase-shift keying (BPSK) modulation. In this case, $m_l = \lfloor \log_2 \binom{5}{2} \rfloor = 3$. Thus, the first three bits are mapped to the antenna combinations, and the remaining one bit is modulated using BPSK.

9.5 Performance Analysis of SM over Fading Channels

The performance of SM-MIMO has been extensively analyzed and studied during the last few years. Such representative past works can be found in [26, 27, 32–39], and references therein. On the one hand, these works have provided useful insights as to the factors affecting system performance, for example, the impact of wireless propagation on the end-to-end error probability and achievable rate, identifying propagation scenarios in which the adoption of SM-MIMO is useful and the achievable diversity order. On the other hand, such analytical results are useful to the system engineer for performance evaluation purposes as they can provide guidelines for system design and optimization. In this section, the performance analysis of SM-MIMO over fading channels will be discussed. It is noted that the analysis presented herein is valid for arbitrarily distributed fading envelopes and channel phases. In order to simplify the underlying mathematical analysis, a tight union bound for the evaluation of the average bit error probability (ABEP) will be presented.

9.5.1 Channel and System Model

An $N_t \times N_r$ MIMO system employing SM is considered, equipped with N_t transmit and N_r receive antennas, which can send digital information via M complex symbols, $\chi_\ell = |\chi_\ell| \exp(j\theta_\ell)$, with $\ell = 1,2,....,M$ and $|\cdot|$ denoting complex magnitude. A frequency-flat slowly-varying fading channel model is assumed where the channel impulse response of the wireless link from the n_tth transmit antenna to the n_rth receive antenna can be mathematically expressed as

$$h_{n_t,n_r}(\xi) = \alpha_{n_t,n_r}(\xi - \tau_{n_t,n_r}) \tag{9.5}$$

where:

$\alpha_{n_t,n_r} = |\alpha_{n_t,n_r}| \exp(j\phi_{n_t,n_r})$
τ_{n_t,n_r} is the propagation time delay

Throughout this analysis, it is assumed that the channel envelopes, $|\alpha_{n_t,n_r}|$, and the channel phases, ϕ_{n_t,n_r}, may follow any of the well-known distributions available in the open technical literature [40] and that the delays τ_{n_t,n_r} are known at the receiver. Furthermore, it is assumed that $\tau_{1,1} = \tau_{1,2} = ... = \tau_{N_t,N_r}$, which is a realistic assumption when the distance between transmitter and receiver is much larger than the spacing of the antenna elements [40].

9.5.2 Optimal Detector

Let $\mu(\tilde{n}_t, \chi_{\tilde{\ell}})$ be the transmitted message. The signal received by the n_rth receive antenna, if $\mu(\tilde{n}_t, \chi_{\tilde{\ell}})$ is transmitted, can be expressed as [37, Equation 9.1]

$$r(t) = s_{ch,n_r}[t \,|\, \mu(\tilde{n}_t, \chi_{\tilde{\ell}})] + \eta_{n_r}(t) \tag{9.6}$$

where:

$s_{ch,n_r}[t \,|\, \mu(\tilde{n}_t, \chi_{\tilde{\ell}})] = \sqrt{E_m}\, a_{\tilde{n}_t,n_r} \chi_{\tilde{\ell}} w(t)$ for $\tilde{n}_t = 1, 2, ..., N_t$, $n_r = 1, 2, ..., N_r$,
$\tilde{\ell} = 1, 2, ..., M$
E_m is the average energy per transmitted symbol

Moreover, the noise $\eta_{n_r}(t)$ at the input of the n_rth receive antenna is a complex AWGN stochastic process with single-sided power spectral density N_0.

Equation 9.6 is a general $N_t \times M$ hypothesis detection problem in AWGN when conditioning upon fading channel statistics. Thus, the ML optimum detector with full CSI and perfect time synchronization at the receiver is given by [34,37]

$$(\hat{n}_t, \chi_{\hat{\ell}}) = \underset{\text{for } n=1,2,\ldots,N_t \text{ and } \ell=1,2,\ldots,M}{\arg\max} \left\{ D(n_t, \chi_\ell) \right\} \tag{9.7}$$

where:

$$D(n_t, \chi_\ell) = \sum_{n_r=1}^{N_r} \left[\int_{T_m} z_{n_r}(t) s^*_{\text{ch},n_r}(t \mid \mu(\tilde{n}_t, \chi_{\tilde{\ell}})) dt - \frac{1}{2} \int_{T_m} \mid s^*_{\text{ch},n_r}(t \mid \mu(\tilde{n}_t, \chi_{\tilde{\ell}})) \mid^2 dt \right] \tag{9.8}$$

T_m is the transmission time slot of each message

$(\cdot)^*$ denotes complex conjugate

The estimated message $\mu(\hat{n}_t, \chi_{\hat{\ell}})$ is obtained by solving Equation 9.7. Thus, the whole block of bits is successfully decoded by the receiver, if and only if, $\mu(\hat{n}_t, \chi_{\hat{\ell}}) = \mu(\tilde{n}_t, \chi_{\tilde{\ell}})$, that is, $\hat{n}_t = \tilde{n}_t$ and $\chi_{\hat{\ell}} = \chi_{\tilde{\ell}}$.

9.5.3 A Union Bound for the Average Error Probability

The exact ABEP of the detector in Equation 9.7 can be computed in closed-form for arbitrary fading channels and modulation schemes as [37, Equation 9.3]

$$\text{ABEP} = \mathbb{E}_\alpha \left\{ \frac{1}{N_t M} \frac{1}{\log_2(N_t M)} \right.$$

$$\left. \sum_{n_t=1}^{N_t} \sum_{\ell=1}^{M} \sum_{\tilde{n}_t=1}^{N_t} \sum_{\tilde{\ell}=1}^{M} [N_H((\tilde{n}_t, \chi_{\tilde{\ell}})) \rightarrow (n_t, \chi_\ell)] \Pr\{(\hat{n}_t, \chi_{\hat{\ell}}) = (n_t, \chi_\ell) \mid (\tilde{n}_t, \chi_{\tilde{\ell}})\} \right\} \tag{9.9}$$

where:

α denotes the set of $N_t \times N_r$ complex channel gains, that is, α_{n_t, n_r} for $n_t = 1, 2, \ldots, N_t$ and $n_r = 1, 2, \ldots, N_r$

$N_H((\tilde{n}_t, \chi_{\tilde{\ell}}) \rightarrow (n_t, \chi_\ell))$ is the Hamming distance of the messages $\mu(\tilde{n}_t, \chi_{\tilde{\ell}})$ and $\mu(n_t, \chi_\ell)$

$\mathbb{E}_\alpha$ is the expectation computed over all the fading channels

However, the estimation of $\Pr\{(\hat{n}_t, \chi_{\hat{\ell}}) = (n_t, \chi_\ell)\}$ is very complicated, especially for large values of N_t and M, as it requires the computation of multidimensional integrals. Therefore, it is common practice to exploit union-bound methods

[41] to obtain such ABEP performance. Such a tight union bound is the following [37, Proposition 1]

$$\text{ABEP} \leq \text{ABEP}_{\text{signal}} + \text{ABEP}_{\text{spatial}} + \text{ABEP}_{\text{joint}} \tag{9.10}$$

where $\text{ABEP}_{\text{signal}}$, $\text{ABEP}_{\text{spatial}}$, and $\text{ABEP}_{\text{joint}}$ show how the error performance of SM is affected by the signal constellation, the spatial constellation and the interaction of both signal and space constellation, respectively.

The term $\text{ABEP}_{\text{signal}}$ is given by

$$\text{ABEP}_{\text{signal}} = \frac{1}{N_t} \frac{\log_2 M}{\log_2 MN_t} \sum_{n_t=1}^{N_t} \text{ABEP}_{\text{MOD}}(n_t) \tag{9.11}$$

where:

$$\text{ABEP}_{\text{MOD}(n_t)} = \frac{1}{M} \frac{1}{\log_2 M} \sum_{\ell=1}^{M} \sum_{\tilde{\ell}=1}^{M} \left[N_H(\chi_{\tilde{\ell}} \to \chi_\ell) \mathbb{E}_{\alpha(n_t)} \left\{ \Pr \left\{ \chi_{\tilde{\ell}} = \chi_\ell \mid \chi_{\tilde{\ell}} \right\} \right\} \right] \tag{9.12}$$

$N_H(\chi_{\tilde{\ell}} \to \chi_\ell)$ is the Hamming distance of the bits transmitted via the signal constellation.

$\mathbb{E}_{\alpha(n_t)}\{\cdot\}$ is the expectation with respect to the fading channel from the n_tth transmit antenna to the N_r receive antenna.

As it can be observed, every term in the sum is the ABEP of a conventional modulation scheme having constellation points that belong to the signal constellation diagram of the employed SM and are transmitted only through the n_tth transmit antenna. Thus, the corresponding error probabilities, $\text{ABEP}_{\text{MOD}}(n_t)$, depend only on the Euclidean distance of the points in the signal constellation diagram.

The term $\text{ABEP}_{\text{spatial}}$ is given by

$$\text{ABEP}_{\text{spatial}} = \frac{1}{M} \frac{\log_2 N_t}{\log_2 MN_t} \sum_{\ell=1}^{M} \text{ABEP}_{\text{SSK}}(\ell) \tag{9.13}$$

where:

$$\text{ABEP}_{\text{SSK}(\ell)} = \frac{1}{N_t} \frac{1}{\log_2 N_t} \sum_{n_t=1}^{N_t} \sum_{\tilde{n}_t=1}^{N_t} \left[N_H(\tilde{n}_t \to n_t) \Psi_\ell(n_t, \tilde{n}_t) \right] \tag{9.14}$$

$N_H(\tilde{n}_t \to n_t)$ is the Hamming distance of the bits transmitted via the spatial constellation, $\Psi_\ell(n_t, \tilde{n}_t) = (1/\pi) \int_0^{\pi/2} \mathcal{M}_{\gamma_{n_t, \tilde{n}_t}}(\overline{\gamma}\kappa_{\tilde{\ell}}^2/2\sin^2\theta)\,d\theta$, with $\overline{\gamma} = E_m/4N_0$, $\gamma_{n_t,\tilde{n}_t} = \sum_{n_r=1}^{N_r} |\alpha(n_t, n_r) - \alpha(\tilde{n}_t, n_r)|^2$ and $\mathcal{M}_X(\cdot)$ is the moment generating function

(MGF) of the random variable X. As it can be observed, $ABEP_{spatial}$ is the summation of M terms $ABEP_{SSK}(\ell)$. These terms correspond to the ABEP of an equivalent SSK MIMO scheme where the average SNR, $\bar{\gamma}$, is replaced by $\kappa^2\bar{\gamma}$. Consequently, these probabilities again depend only on the Euclidean distance of the points in the spatial constellation diagram.

The term $ABEP_{joint}$ is given by

$$
ABEP_{joint} = \left\{ \frac{1}{N_t M} \frac{1}{\log_2 N_t M} \right.
$$

$$
\left. \times \sum_{n_t=1}^{N_t} \sum_{\ell=1}^{M} \sum_{\substack{\tilde{n}_t \neq n_t=1}}^{N_t} \sum_{\ell \leq \ell=1}^{M} \left[N_H(\tilde{n}_t \to n_t) + N_H(\chi_{\tilde{\ell}} \to \chi_\ell) \right] \mathcal{Y}(n_t, \chi_\ell, \tilde{n}_t, \chi_{\tilde{\ell}}) \right\} \tag{9.15}
$$

where $\mathcal{Y}(n_t, \chi_\ell, \tilde{n}_t, \chi_{\tilde{\ell}}) = (1/\pi) \int_0^{\pi/2} \mathcal{M}_{\gamma_{n_t,\chi_\ell,\tilde{n}_t,\chi_{\tilde{\ell}}}} (\bar{\gamma}/2\sin^2\theta) d\theta$ and $\gamma_{n_t,\chi_\ell,\tilde{n}_t,\chi_{\tilde{\ell}}} = \sum_{n_r=1}^{N_r} |\alpha(n_t, n_r)\chi_\ell - \alpha(\tilde{n}_t, n_r)\chi_{\tilde{\ell}}|^2$. Clearly, the structure of this term is more complicated than the previous two and depends on the Euclidean distance of points belonging to signal and spatial constellation diagrams.

Based on Equation 9.10, the ABEP of SM can be evaluated numerically for arbitrary fading channels and modulation schemes provided that the MGFs of the signal-to-noise ratios (SNRs) $\gamma_{n_t} = \sum_{n_r=1}^{N_r} |h_{n_t,n_r}|^2$, $\gamma_{n_t,\tilde{n}_t}$, and $\gamma_{n_t,\chi_\ell,\tilde{n}_t,\chi_{\tilde{\ell}}}$ are available in closed-form. For arbitrary correlated fading channels, analytical expressions for the MGFs of γ_{n_t} are available in [40]. For correlated Rice and Nakagami-m fading channels, analytical expressions for the MGFs of $\gamma_{n_t,\tilde{n}_t}$ and $\gamma_{n_t,\chi_\ell,\tilde{n}_t,\chi_{\tilde{\ell}}}$ have been derived in [33,37]. Finally, a general framework for the evaluation of these MGFs for independent channels and assuming uniform phases of $\alpha(n_t, n_r)$ has been proposed in [35].

9.6 SM-MIMO for 5G Wireless Communications

In this section, we focus on the application of SM-MIMO for future 5G wireless communications. Firstly, the relationship between SM-MIMO and massive MIMO is discussed [42]. Then, GSM is introduced as an efficient way to utilize large-scale transmit antennas available in massive MIMO systems. The combination of both orthogonal frequency division multiplexing (OFDM) and single carrier (SC) with SM-MIMO is also considered as a means of improving the performance of 5G systems in frequency selective fading channels. We close this discussion by mentioning some MIMO transmission schemes closely related to the SM-MIMO paradigm, such as single RF MIMO schemes based on compact parasitic architectures, the incremental MIMO and the antenna subset modulation (ASM) schemes.

9.6.1 SM and Massive MIMO

The motivation behind the use SM-MIMO for the design of spectral- and energy-efficient future generation wireless communication systems is two-fold, as follows:

- Firstly, to minimize, under performance constraints, the number of active antenna elements in order to increase the EE by reducing the circuit power consumption.
- Secondly, to maximize, under implementation and size constraints, the number of passive antenna elements in order to increase both SE and EE by reducing the transmit power consumption [14].

The design of massive MIMO schemes that retain the benefits of multiple-antenna transmission while having a single active RF element is another recently introduced but important trend for current and future MIMO R&D activities. Massive MIMO are systems that use antenna arrays with an order of magnitude more elements than in currently available MIMO systems. They are designed to employ a large number of antennas, typically not less than 100, at the base station. Clearly such systems entail an unprecedented number of antennas simultaneously serving a much smaller number of terminals, with each antenna unit using extremely low power (typically of the order of mW). One of their advantages is that several expensive and bulky items, such as long coaxial cables, can be eliminated altogether. Massive MIMO designs can be made extremely robust, since when one or even a few of the antenna units fail, their performance will not be significantly altered. The main effect of increasing the number of antennas is that uncorrelated thermal noise and fast fading can be averaged out and vanish so that the system performance is predominantly limited by interference from other transmitters. As the number of antennas tends to infinity, uncorrelated interference, noise, and CSI errors vanish [42–44]. In addition to the earlier, the achieved spectral efficiency is independent of bandwidth [42,43].

For all the previous reasons, SM seems to be an appealing transmission technology for high-rate and low-complexity MIMO implementations that exploit the massive MIMO paradigm. In this respect, SM can be regarded as a low-complexity modulation scheme exploiting the massive MIMO idea with the use of a single active RF chain.

9.6.2 GSM for Massive MIMO

Recent research efforts have focused on increasing the efficiency of conventional SM for future wireless communication systems by combing spatial modulation and spatial multiplexing. The previously discussed GSM is quite attractive because of its ability to work well with smaller numbers of transmit RF chains as compared to other more traditional spatial multiplexing, for example the VBLAST system [31].

As previously mentioned, in GSM the transmitted information is conveyed in the activated combination of transmit antennas whereas the transmitted symbol is chosen from a signal constellation. Transmitting the same data symbol from more than one antenna at a time, retains the key advantage of SM, which is the complete avoidance of ICI at the receiver. Moreover, GSM offers spatial diversity gains and increases the reliability of the wireless channel, by providing replicas of the transmitted signal to the receiver. Because of these reasons, GSM is regarded as a promising technique for massive MIMO in 5G wireless systems [45].

A main drawback of GSM is that the increase of possible antenna combinations results in an increase of the detection complexity at the receiver. Thus research efforts have focused on the development of low-complexity near-ML detectors which can be used to incorporate GSM-MIMO into more practical system implementations [46].

9.6.3 Spatial Modulation-Orthogonal Frequency Division Multiplexing

Although most of current research efforts assume flat fading for spatial modulation-orthogonal frequency division multiplexing (SM-MIMO), the influence of frequency selective fading cannot be neglected in 5G broadband wireless channels. OFDM is an efficient way to combat frequency selective fading and has been adopted by most current wireless communication standards. Note that the combination of SM-MIMO with OFDM is regarded as an attractive technique for future wireless systems [47–49].

As compared to current MIMO-OFDM techniques, some of the advantages of the SM-OFDM are the following.

- *Lower peak-to-average power ratio (PAPR)*: High PAPR of the transmitted signals is one of the main limitations of OFDM systems. Due to the special structure of SM, SM-OFDM transmit signals are sparse in the frequency domain, thus facilitating the construction of lower PAPR transmit signals to improve the efficiency of high power amplifiers (HPA) [50].
- *Low-complexity parameter estimation*: In SM-OFDM systems, the estimation of timing offset, carrier frequency offset (CFO) and channel coefficient are very important for signal detection. The structure of SM-MIMO reduces the interference among antennas and thus such parameter estimations can be significantly simplified estimation [45,47].

Note also that SM-OFDM systems have also limitations which should be fully considered in future system design. For example, after the inverse fast Fourier transform (IFFT) transform process takes place, all transmitting antennas of SM-OFDM will be occupied by time-domain signals. In this way, the single-antenna activation feature is lost in SM-OFDM and thus its complexity is greater than that of the conventional SM.

9.6.4 Spatial Modulation—Single Carrier

SM-SC is a technique for dealing with frequency selective fading in a similar fashion as OFDM and as such is an attractive transmission scheme for 5G communications [51–54]. As compared to SM-OFDM, SM-SC has several advantages with the most important one being that it can maintain the single-antenna-activation feature as the conventional SM while combating frequency selective fading. Therefore, it can be used to efficiently implement large-scale SM-MIMO systems by reducing the number of RFs chains. Furthermore, it can achieve lower PAPR compared to conventional SM-OFDM [51,54].

Despite these advantages, there exist several challenges in SM-SC system design, for example, reducing the high complexity of their detector. On the one hand, ML detection is essential for improving the performance, while on the other hand, hardware complexity inherent to the design of a ML detector for such systems is prohibitively high. Therefore, several practical problems related to complexity, performance, and practical implementation, should be addressed before SM-SC can be utilized in 5G communications.

Research efforts in this field has focused on several directions, including the efficient design and implementation of the following:

- *Frequency-domain signal detection techniques* [52]. Research efforts have focused on the design of near-ML detectors for performing a near ML detection for SM-SC signals [52]. However, the computational complexity still remains large in practical implementations.
- *Time-domain signal detection techniques* [54]. Such techniques seem to have a better trade-off between complexity and performance.
- *Turbo detection schemes*, for which combination of soft-in soft-out (SISO) equalizer and decoder can efficiently improve the system performance of SM-SC with relatively low computational complexity [52,54].

Moreover, comparison studies between SM-ODFM and SM-SC should be carried out as was done for SC-FDMA and OFDMA satellite up-links [55].

9.6.5 Miscellaneous Transmission Schemes Related to SM-MIMO

9.6.5.1 MIMO Designs Based on Compact Parasitic Architectures

Multiple antenna designs based on compact parasitic architectures that enable multiplexing gains with a single active RF element and many passive antenna elements

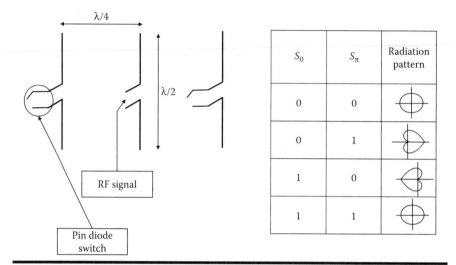

Figure 9.7   Parasitic single-RF MIMO.

have been proposed in [19]. As shown in Figure 9.7, the key idea is to change the radiation pattern of the antenna array at each symbol time instance, and to encode independent information streams onto angular variations of the far-field in the wave-vector domain.

In Figure 9.7 a switched parasitic array is depicted, consisting of one central active element driven by the PA of the device, and two colinear parasitic elements that may be either open or short circuited using high speed p-i-n diodes and digital control signals. When both peripheral elements are open-circuited, the radiation pattern is omnidirectional in the azimuth plane. When one peripheral element is open circuited and the other short circuited, then a cardioid pattern is produced with a maximum toward the direction of the short-circuited element. SM-MIMO employs a single-RF chain in a similar fashion as the parasitic MIMO scheme but the information is encoded onto the transmit antenna switching mechanism, rather than onto the radiation pattern of the antenna array.

9.6.5.2 Incremental MIMO

This is another MIMO related scheme which jointly combines multiple-antenna transmission and automatic repeat request (ARQ) feedback to avoid keeping all available antennas active, thus enabling MIMO gains with a single RF chain and a single PA [56]. As shown in Figure 9.8, the main idea of this so-called incremental MIMO scheme, is to reduce complexity and to improve EE by activating one antenna at a given time slot. It also exploits ARQ feedback to randomly cycle through the available antennas at the transmitter in case of incorrect data reception. SM-MIMO has the characteristic of being an open-loop scheme, while the incremental single-RF MIMO is a closed-loop scheme using ARQ feedback.

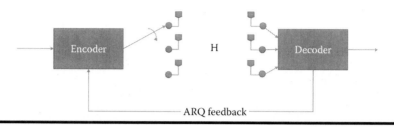

Figure 9.8   Incremental MIMO.

9.6.5.3 Antenna Subset Modulation

New directional modulation schemes for mm-Wave frequencies have been also proposed to enable secure and low-complexity wireless communications. Such a system employs the so-called ASM [57]. As shown in Figure 9.9, the main idea in ASM is to modulate the radiation pattern at the symbol rate by driving only a subset of antennas in the array. While the random switching antenna subsets do not affect the symbol modulation for a desired receiver along the main direction, it effectively randomizes the amplitude and phase of the received symbol for an eavesdropper along a side lobe.

This feature gives to the ASM scheme an advantage in supporting secure communications while the SM-MIMO provides higher bit rates. Moreover, ASM uses directional beamforming to mitigate the impact of path loss, atmospheric absorption, and high noise levels present at millimeter-wave frequencies, whereas SM-MIMO schemes are not designed to provide beamforming gains.

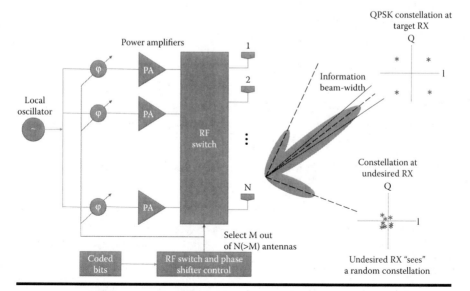

Figure 9.9   Antenna subset modulation.

References

1. IMT vision—Framework and overall objectives of the future development of IMT for 2020 and beyond, Rec. ITU-RP M.2083-0. International Telecommunication Union (ITU-R), Radio-communication Sector of ITU, (09/2015), Electronic Publication, Geneva, 2015.

2. A. Osseiran et al., Scenarios for 5G mobile and wireless communications: The vision of the METIS project, *IEEE Commun. Mag.*, 52(5), 26–35, 2014.

3. J. G. Andrews et al., What 5G will be, *IEEE J. Sel. Areas Commun.*, 32(6), 1065–1082, 2014.

4. P. W. Wolniansky, G. J. Foschini, G. D. Golden, and R. A. Valenzuela, V-BLAST: An architecture for realizing very high data rates over the rich-scattering wireless channel, in *Proceedings of the ISSSE-98*, Pisa, Italy, September 1998, pp. 295–300.

5. D. Gesbert, M. Shafi, D. Shiu, P. Smith, and A. Naguib, From theory to practice: An overview of MIMO space-time coded wireless systems, *IEEE J. Sel. Areas Commun.*, 21(3), 281–302, 2003.

6. S. Alamouti, A simple transmit diversity technique for wireless communications, *IEEE J. Sel. Areas Commun.*, 16(8), 1451–1458, 1998.

7. V. Tarokh, N. Seshadri, and A. Calderbank, Space-time codes for high data rate wireless communication: Performance criterion and code construction, *IEEE Trans. Inf. Theory*, 44(2), 744–765, 1998.

8. H. E. Gamal, On the robustness of space-time coding, *IEEE Trans. Signal Process.*, 50(10), 2417–2428, 2002.

9. M. Godavarti, T. Marzetta, and S. Shamai, Capacity of a mobile multiple antenna wireless link with isotropically random Rician fading, *IEEE Trans. Inf. Theory*, 49(12), 3330–3334, 2003.

10. P. Viswanath, D. Tse, and V. Anantharam, Asymptotically optimal water filling in vector multiple-access channels, *IEEE Trans. Inf. Theory*, 47(1), 241–247, 2001.

11. G. Raleigh and J. Cioffi, Spatio-temporal coding for wireless communication, *IEEE Trans. Commun.*, 46(3), 357–366, 1998.

12. G. J. Foschini, Layered space-time architecture for wireless communication in a fading environment when using multi-element antennas, *Bell Labs Tech. J.*, 1(2), 41–59, 1996.

13. M. D. Renzo, H. Haas, and P. Grant, Spatial modulation for multiple antenna wireless systems: A survey, *IEEE Commun. Mag.*, 49(12), 182–191, 2011.

14. M. D. Renzo, H. Haas, A. Ghrayeb, S. Sugiura, and L. Hanzo, Spatial modulation for generalized MIMO: Challenges, opportunities, and implementation, *Proc. IEEE*, 102(1), 56–103, 2014.

15. A. Mohammadi and F. M. Ghannouchi, Single RF front-end MIMO transceivers, *IEEE Commun. Mag.*, 49(12), 104–109, 2011.

16. G. Auer et al., D2.3: Energy efficiency analysis of the reference systems, areas of improvements and target breakdown. EARTH: Energy aware radio netw. technol. January 2012, available: https://bscw.ict-earth.eu/pub/bscw.cgi/d71252/EARTH_WP2_D2.3_v2.pdf.

17. F. Heliot, M. A. Imran, and R. Tafazolli, On the energy efficiency-spectral efficiency trade-off over the MIMO Rayleigh fading channel, *IEEE Trans. Commun.*, 60(5), 1345–1356, 2012.

18. Y. Chen, S. Zhang, S. Xu, and G. Y. Li, Fundamental tradeoffs on green wireless networks, *IEEE Commun. Mag.*, 49(6), 30–37, 2011.

19. A. Kalis, A. G. Kanatas, and C. B. Papadias, A novel approach to MIMO transmission using a single RF front end, *IEEE J. Sel. Areas Commun.*, 26(6), 972–980, 2011.
20. F. Khan and J. Pi, Millimeter-wave mobile broadband: Unleashing 3300 GHz spectrum, in *Proceedings of the IEEE Wireless Communications and Network Conference (WCNC)*, March 2011, pp. 1–6.
21. T. S. Rappaport, J. Murdock, and F. G. Jr., State of the art in 60-GHz integrated circuits and systems for wireless communications, *Proc. IEEE*, 99(8), 1390–1496, 2011.
22. S. Rajagopal, S. Abu-Surra, Z. Pi, and F. Khan, Antenna array design for multi-Gbps mmWave mobile broadband communication, in *Proceedings of the IEEE Global Communication Confererence (Globecom)*, December 2011, pp. 1–6.
23. T. S. Rappaport et al., Millimeter wave mobile communications for 5G cellular: It will work!, *IEEE Access*, 1, 335–349, 2013.
24. M. D. Renzo and H. Haas, On transmit-diversity for spatial modulation MIMO: Impact of spatial-constellation diagram and shaping filters at the transmitter, *IEEE Trans. Veh. Technol.*, 62(6), 2507–2531, 2013.
25. M. D. Renzo, H. Haas, and P. Grant, Spatial modulation for multiple-antenna wireless systems: A survey, *IEEE Commun. Mag*, 49(12), 182–191, 2011.
26. J. Jeganathan, A. Ghrayeb, L. Szczecinski, and A. Ceron, Space shift keying modulation for MIMO channels, *IEEE Trans. Wireless Commun.*, 8(7), 3692–3703, 2009.
27. M. D. Renzo and H. Haas, Space shift keying (SSK) modulation with partial channel state information: Optimal detector and performance analysis over fading channels, *IEEE Trans. Commun.*, 58, 3196–3210, 2010.
28. J. Jeganathan, A. Ghrayeb, and L. Szczecinski, Spatial modulation: Optimal detection and performance analysis, *IEEE Commun. Lett.*, 12(8), 545–547, 2008.
29. S. Song et al., A channel hopping technique I: Theoretical studies on band efficiency and capacity, in *IEEE International Conference on Communications, Circuits and Systems*, June 2004, pp. 229–233.
30. J. Jeganathan, A. Ghrayeb, and L. Szczecinski, Generalized space shift keying modulation for MIMO channels, in *Proceedings of the IEEE Symposium Personal, Indoor Mobile Radio Communications (PIMRC)*, September 2008, pp. 1–5.
31. A. Younis, N. Serafimovski, R. Mesleh, and H. Haas, Generalized spatial modulation, in *Proceedings of Asilomar Conference on Signals, Systems, and Computers*, November 2010, pp. 1498–1502.
32. R. Y. Mesleh, H. Haas, S. Sinanovic, C. W. Ahn, and S. Yun, Spatial modulation, *IEEE Trans. Veh. Technol.*, 57(4), 2228–2241, 2008.
33. M. D. Renzo and H. Haas, A general framework for performance analysis of space shift keying (SSK) modulation for MISO correlated Nakagami-*m* fading channels, *IEEE Trans. Commun.*, 59(9), 2590–2603, 2010.
34. M. D. Renzo and H. Haas, Space shift keying (SSK) MIMO over correlated Rician fading channels: Performance analysis and a new method for transmit-diversity, *IEEE Trans. Commun.*, 59(1), 116–129, 2011.
35. K. Peppas, M. Zamkotsian, F. Lazarakis, and P. Cottis, Unified error performance analysis of space shift keying modulation for MISO and MIMO systems under generalized fading, *IEEE Wireless Commun. Lett.*, 2(6), 663–666, 2013.
36. K. P. Peppas, M. Zamkotsian, F. Lazarakis, and P. G. Cottis, Asymptotic error performance analysis of spatial modulation under generalized fading, *IEEE Wireless. Comm. Lett.*, 3(4), 421–424, 2014.

37. M. D. Renzo, H. Haas, and P. Grant, Bit error probability of SM-MIMO over generalized fading channels, *IEEE Trans. Veh. Technol.*, 61(3), 1124–1144, 2012.

38. K. P. Peppas and P. T. Mathiopoulos, Free space optical communication with spatial modulation and coherent detection over H-K atmospheric turbulent channels, *IEEE J. Lightw. Technol.*, 33(20), 4221–4232, 2015.

39. K. P. Peppas, P. Bithas, G. Efthymoglou, and A. Kanatas, Space shift keying transmission for intervehicular communications, *IEEE Trans. Intell. Transp. Syst.*, 17(12), 3635–3640, 2016.

40. M. K. Simon and M.-S. Alouini, *Digital Communication over Fading Channels*, 2nd ed. New York: Wiley, 2005.

41. J. G. Proakis, *Digital Communications*, 4th ed. New York: McGraw-Hill, 1995.

42. T. L. Marzetta, Noncooperative cellular wireless with unlimited numbers of base station antennas, *IEEE Trans. Wireless Commun.*, 11(9), 3590–3600, 2010.

43. E. G. Larsson, F. Tufvesson, O. Edfors, and T. L. Marzetta, Massive MIMO for next generation wireless systems, *IEEE Commun. Mag.*, 52(2), 186–195, 2014.

44. G. Wright, Greentouch initiative: Large scale antenna systems demonstration, in *Proceedings of the Spring Meeting*, Seoul, Korea, 2011.

45. Y. Xiao, L. Xiao, L. Dan, and X. Lei, Spatial modulation for 5G MIMO communications, in *Proceedings of the 19th International Conference on Digital Signal Processing*, Hong Kong, August 2014, pp. 847–851.

46. Y. Xiao, Z. Yang, L. Dan, P. Yang, L. Yin, and W. Xiang, Low complexity signal detection for generalized spatial modulation, *IEEE Commun. Lett.*, 18(3), 403–406, 2014.

47. R. Mesleh, H. Haas, C. Ahn, and S. Yun, Spatial modulation OFDM, in *Proceedings of the 11th International OFDM Workshop*, Hamburg, Germany, 2006, pp. 288–292.

48. F. Yu, X. Lei, L. Peng, Y. Xiao, P. Wei, and X. Wen, Performance analysis of spatial modulation OFDM system with n-continuous precoder, in *IEEE 83rd Vehicular Technology Conference (VTC Spring)*, 2016, pp. 1–5.

49. B. T. Vo, H. H. Nguyen, and N. Quoc-Tuan, Spatial modulation for OFDM with linear constellation precoding, in *International Conference on Advanced Technologies for Communications (ATC)*, 2015, pp. 226–230.

50. V. Dalakas, A. A. Rontogiannis, and P. T. Mathiopoulos, A time domain constellation technique for PAPR reduction, *IET Commun.*, 3(7), 1144–1152, 2009.

51. P. Yang et al., Single-carrier SM-MIMO: A promising design for broadband large-scale antenna systems, *IEEE Commun. Surv. Tut.*, 18(3), 1687–1716, 2016.

52. B. Zhou, Y. Xiao, P. Yang, J. Wang, and S. Li, Spatial modulation for single carrier wireless transmission systems, in *Proceedings of the ICST Conference on Harbin*, August 2011, pp. 11–15.

53. P. Som and A. Chockalingam, Spatial modulation and space shift keying in single carrier communication, in *Proceedings of the IEEE Symposium Personal, Indoor Mobile Radio Communications (PIMRC)*, September 2012, pp. 1962–1967.

54. R. Rajashekar, K. Hari, and L. Hanzo, Spatial modulation aided zeropadded single carrier transmission for dispersive channels, *IEEE Trans. Commun.*, 61(6), 2318–2329, 2013.

55. V. Dalakas, P. T. Mathiopoulos, F. D. Cecca, and G. Gallinaro, A comparative study between SC-FDMA and OFDMA schemes for satellite uplinks, *IEEE Trans. Broadcast*, 58, 370–378, 2012.

56. P. Hesami and J. N. Laneman, Incremental use of multiple transmitters for low-complexity diversity transmission in wireless systems, *IEEE Trans. Commun.*, 60(9), 2522–2533, 2013.

57. N. Valliappan, A. Lozano, and R. W. Heath, Antenna subset modulation for secure millimeter-wave wireless communication, *IEEE Trans. Commun.*, 61(8), 3231–3245, 2013.

Chapter 10

Device-to-Device Communication Aspects for 5G Cellular Networks

---

Petros S. Bithas and George P. Efthymoglou

Contents

10.1 Introduction

Device-to-device (D2D) communication is recognized as one of the technology components of the evolving 5G architecture by the European Union project METIS [1]. METIS stands for mobile and wireless communications enablers for the twenty-twenty information society. The METIS project is currently evaluating the role that D2D technology can play in various scenarios such as vehicle-to-vehicle (V2V) communication, national security, and public safety, cellular network offloading, or service advertisement. In D2D communication, device user equipment (DUE) transmit data signals to each other over a direct link using the cellular resources instead of through the evolved node B (eNB), which differs from femtocell where users communicate with the help of small low-power cellular base stations (BSs).

A survey of the many proposals for D2D communication can be found in [2,3]. A main distinction among existing proposals is between underlay and overlay systems [4]. Most attention draws D2D communications as an underlay to the cellular. In this scenario, DUE communicates directly using the same resources as the cellular system. Therefore, the potential of improving spectral utilization has promoted much work in recent years, which shows that D2D can improve system performance by reusing cellular resources [5]. The reason to reuse the licensed instead of the unlicensed spectrum is that the former can operate in a controlled (by the BS) environment whereas the latter may not provide quality of service (QoS) guarantees because of uncontrollable interference in the unlicensed spectrum.

In D2D underlay systems, the level of control exercised by the eNB may vary depending on the QoS constraints imposed to the cellular and the D2D users. Most works consider that the eNB controls the resources used for cellular communications and by the D2D link. The eNB can set constraints on the transmit power of D2D transmitters (Txs) to limit the interference experienced by the cellular receivers. Therefore, interference coordination mechanisms enable underlay D2D communications [6].

In D2D overlay systems, a part of the cellular spectrum is assigned to D2D transmissions. Therefore, no interference occurs between cellular users equipment (CUE) and DUE. However, reusing the same subchannels (SCs) by different D2D pairs will result in cochannel interference for the D2D links. It must be noted that most of the results on D2D link performance are based on simulation [7], in order to capture (1) the random locations of devices and cellular users, (2) the available

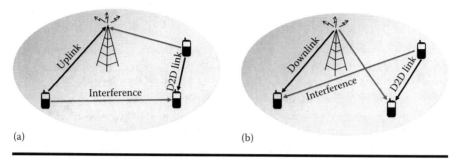

(a) (b)

Figure 10.1 Interference for scenarios where DUEs use the same frequency resources, (a) for the uplink and (b) for the downlink.

resources of the eNB based on traffic load, and (3) fading and path loss models for the cellular users and D2D links. Nevertheless, analytical results can also be obtained on D2D performance metrics to capture the impact of various system parameters [8,9].

One fundamental design consideration for D2D neighbor discovery and communications is the choice of resources to use for transmission. In principle, it should be possible to use either the uplink (UL) or downlink (DL) resources for D2D transmissions. In the case where UL resources are used for D2D transmissions in underlay systems, as shown in Figure 10.1a, the D2D transmissions cause intercell interference to neighbor BSs; pretty much in the same way as in current systems. The D2D receivers (Rxs) are on the other hand interfered by inter-cell nearby cellular devices transmitting to the eNB. On the other hand, in the case where DL resources are used for D2D transmission, as shown in Figure 10.1b, the D2D transmitting devices interfere to other CUEs receiving normal communications on the DL resources. Conversely, the eNBs in that case interfere with the D2D transmissions. The impact of interference from CUEs and DUEs using the same subchannels on the establishment of a new D2D link is a major issue. Moreover, the system designer needs to determine the main design parameters, taking also into consideration spectrum sharing criteria and D2D network connectivity. Finally, the impact of D2D links to the cellular performance needs to be investigated.

DUE may operate in both cellular and D2D modes. Mode selection is usually based on measurements to check (1) if the D2D devices are in communication range and (2) if D2D communication actually offers higher throughput than cellular communication. However, in order to limit the interference of D2D connections to the cellular network, the eNB should be able to control the maximum transmit power of D2D transmitters [5]. Interference coordination resembles the one found in underlay cognitive radio systems, where the transmit power of a secondary user in constrained by the interference imposed to the primary receiver.

10.2 Spectrum Sharing and Interference

In conventional cellular networks, interference emanates from fixed BS locations in the DL and worst case interference scenarios with the CUE located close to the cell edge is usually considered. In heterogeneous multitier wireless networks, stochastic geometry is employed to model the randomly located interference sources within a macrocell. Using homogeneous Poisson point process (HPPP) to model the random locations of interfering sources, the aggregate interference at a receiver located at the origin of a two-dimensional (2D) circular plane is known to follow the alpha-stable distribution. Using the HPPP model, various performance results for heterogeneous K-tier wireless networks can be obtained, such as signal-to-interference-plus-noise ratio (SINR) statistics, association probability to a k-tier, $k = 1,..., K$ BS, coverage probability, bit error rate performance and throughput, as a function of the node density for the k-tier network. It follows that stochastic geometry modeling is also applicable to model the aggregate interference generated by CUEs and D2D links.

In underlay-inband systems, the D2D links can employ either UL or DL frequency resources of a frequency division duplexing (FDD) system. The introduction of D2D links inside a macro-cell creates additional interference to the CUEs when using DL resources and to the macro BS when using UL resources. It has recently been decided in 3GPP to use UL resources for the study item work [10]. The use of UL frequencies for D2D links will impact the SINR at the macro BS, which will be better when D2D pairs are farther away from the BS. On the other hand, using UL resource sharing, the strength of interference experienced by D2D links depends not only on the position of D2D users, but also on the position of the CUE.

In overlay-inband systems, there is dedicated bandwidth for D2D transmissions which can be further partitioned into SCs with each D2D transmitting in only one of them. However, the possibility of one SC used by many D2D links can be allowed. For this scenario, an uncoordinated (probabilistic) scheduling scheme is the simplest option where each D2D Tx randomly and independently selects one of the available SCs for transmission. However, coordinated network-assisted scheduling can also be employed in order to reduce the interference induced to D2D links. In [8], a random time-frequency hopping channel access scheme for D2D transmissions is proposed and its performance is analyzed. Moreover, power control schemes can limit the transmit power of D2D Txs which will lead to D2D SINR degradation but will decrease the interference induced to CUEs or the BS.

For both underlay and overlay systems, the appearance of uncoordinated D2D transmissions inside a macro-cell with potential D2D Txs located randomly inside the cell has forced researchers to use stochastic geometry in order to accurately model the induced cochannel interference. A commonly adopted interference model in many ad-hoc and cognitive radio networks is based on the assumption that the interferers are scattered according to a spatial HPPP and operate asynchronously in a wireless environment that is subject to path-loss, fading, and shadowing. It can be shown that the aggregate interference in these networks follows an α-stable distribution with well-specified characteristic function [11–13].

10.2.1 Distribution of the Aggregate Interference

Let us consider a wireless system where the desired signal of a given user is corrupted by both a number of interfering signals and additive white Gaussian noise (AWGN) with single-sided power spectral density N_0. The number of interfering signals is a Poisson distributed discrete random variable (RV) and belongs to a HPPP $\mathcal{K}$. This implies that if the average number of transmitting terminals per unit area around a given node is λ, then the number of transmitting nodes in an area A is a Poisson RV with parameter $A\lambda$.

The instantaneous output SINR can be expressed as

$$\gamma = \frac{P_s a_s^2 d_s^{-v}}{N_0 + P_I \sum_{i \in \mathcal{K}} \xi_{si}^2 r_i^{-v}} \tag{10.1}$$

where:

P_s is the transmitting power of the desired signal
d_s is the distance between the desired transmitter and the receiver
a_s is the Rayleigh-faded amplitude of the desired signal
ξ_i is the fading amplitude of the ith interfering signal

It is assumed that all interfering signals are transmitted with the same power P_I [14], but experience mutually independent path loss and Rayleigh fading. Moreover, it is assumed that the interferers are independent and identically distributed (i.i.d.). Also, r_i is the distance from the interfering terminal i to the receiver, while v ($v > 2$) is the path loss exponent in the environment surrounding the receiver.

The SINR in Equation 10.1 may be expressed as $\gamma = \gamma_s/(1 + \mathcal{Z})$ where $\gamma_s = d_s^{-v}(P_s/N_0)a_s^2$ and $\mathcal{Z} = \sum_{i \in \mathcal{K}} \gamma_i r_i^{-v}$ with $\gamma_i = (P_I/N_0)\xi_i^2$. Assuming that a_s and ξ_i are Rayleigh distributed with scale parameters equal to one, γ_s and γ_i are exponentially distributed RVs with inverse scale parameters $\overline{\gamma}_s = d_s^{-v}(P_s/N_0)$ and $\overline{\gamma}_I = P_I/N_0$, respectively.

When the interfering signals sum incoherently, the aggregate interference can be modeled as an α-stable RV. Thus, with an appropriate choice of the system parameters and assuming that r_i is uniformly distributed in an unbounded region around the receiver, it can be shown that the moments generating function of the aggregate interference in a wireless network can be expressed in terms of the system parameters as [14,15]

$$\mathcal{M}_{\mathcal{Z}}(s) = \exp(-Bs^\alpha) \tag{10.2}$$

where $B \triangleq \pi\lambda\Gamma(1 - \alpha)\Gamma(1 + \alpha)(\mathbb{E}\langle\gamma_i\rangle)^\alpha$, $\alpha = 2/v$, $\mathbb{E}\langle\cdot\rangle$ denotes expectation and $\Gamma(\cdot)$ denotes the gamma function [16, Eq. (8.310/1)]. Since γ_i are exponentially distributed RVs, $\mathbb{E}\langle\gamma_i\rangle = \overline{\gamma}_I$.

10.2.2 Performance Analysis

The cumulative distribution function (CDF) of the instantaneous SINR, γ, can be obtained as

$$F_\gamma(\gamma) = \Pr\left(\frac{\gamma_s}{1+\mathcal{Z}} \leq \gamma\right) = \int_0^\infty \Pr\left(\gamma_s \leq \gamma(1+\mathcal{Z}) \mid \mathcal{Z} = z\right) f_\mathcal{Z}(z)\,dz$$

$$= 1 - e^{-\gamma/\bar{\gamma}_s} \int_0^\infty e^{-\gamma z/\bar{\gamma}_s} f_\mathcal{Z}(z)\,dz = 1 - e^{-\gamma/\bar{\gamma}_s} \mathcal{M}_\mathcal{Z}\left(\frac{\gamma}{\gamma_s}\right) \tag{10.3}$$

$$= 1 - e^{-\gamma/\bar{\gamma}_s} e^{-B(\gamma/\bar{\gamma}_s)^\alpha}$$

Using the CDF-based approach, the average bit error probability (ABEP) of binary modulation schemes can be expressed as

$$\bar{P}_E = \frac{a^b}{2\Gamma(b)} \int_0^\infty \gamma^{b-1} \exp(-a\gamma) F_\gamma(\gamma)\,d\gamma \tag{10.4}$$

where:

$F_\gamma(\gamma)$ is the CDF of γ

a and b are the parameters that account for different modulation/demodulation schemes

Specifically, $a = 1$, $b = 1$ for binary differential phase shift keying (BDPSK), $a = 1$, $b = 0.5$ for coherent binary phase shift keying (BPSK), and $a = 0.5$, $b = 0.5$ for coherent binary frequency shift keying (BFSK).

Substituting Equation 10.3 to Equation 10.4, expressing the exponential in terms of the Fox's H-function and evaluating the integral with the help of [17, Eq. (2.25.1.1)] gives the ABEP [18]

$$\bar{P}_E = \frac{1}{2} - \frac{1}{2\Gamma(b)} \left(\frac{a\bar{\gamma}_s}{a\bar{\gamma}_s + 1}\right)^b H_{1,1}^{1,1}\left[\frac{B}{(a\bar{\gamma}_s + 1)^\alpha} \middle| \begin{array}{c} (1-b,\alpha) \\ (0,1) \end{array}\right] \tag{10.5}$$

For $v = n$, with n being integer, Equation 10.5 can be expressed in terms of the Meijer's G-function as [18]

$$\bar{P}_E = \frac{1}{2} - \frac{\sqrt{n}\,2^{-\frac{n+3}{2}+b}}{\pi^{\frac{n}{2}}\Gamma(b)} \left(\frac{a\bar{\gamma}_s}{a\bar{\gamma}_s + 1}\right)^b G_{2,1}^{n,2}\left[\frac{4B^n}{n^n(a\bar{\gamma}_s + 1)^2} \middle| \begin{array}{c} \frac{2-b}{2}, \frac{1-b}{2} \\ \Xi_n \end{array}\right] \tag{10.6}$$

where $\Xi_n = \left\{0, \dfrac{1}{n}, \ldots, \dfrac{n-1}{n}\right\}$.

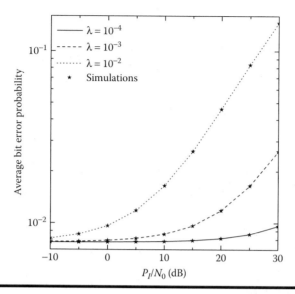

Figure 10.2 **ABEP of BPSK as a function of P_I/N_0 for various values of λ with $d_s = 1$, $P_s/N_0 = 15$ dB and $v = 3$.**

10.2.3 Numerical and Computer Simulation Results

In this section, we illustrate the derived expressions and study the impact of the spatial density of interferers on the ABEP of BPSK. Figure 10.2 depicts the ABEP as a function of P_I/N_0 for various values of λ, assuming $d_s = 1$, $v = 3$, and $P_s/N_0 = 15$ dB. As it is expected, the ABEP increases as P_I/N_0 increases. Moreover, the impact of P_I/N_0 on the ABEP depends also on λ and as it can be observed, it is more severe for high values of the interferer's density. Finally, the numerically evaluated results match perfectly with equivalent ones obtained using Monte Carlo simulations.

10.3 D2D Neighbor Discovery

The work in [4] investigates a distance-based D2D mode selection: cellular mode is used if the distance to a D2D Rx is greater than a mode selection threshold; otherwise, D2D mode is selected. If we assume that the received signal power (averaged over fast fading) is only a function of distance and pathloss exponent, distance-based D2D mode selection is equivalent to the average received-signal-power or average SNR based mode selection. Once users realize high D2D speeds are possible, more local sharing is likely to occur. So far, no commonly agreed upon D2D distance distribution has appeared in the literature. In the absence of such an accepted model, similar to [4], we assume that each potential D2D Rx is randomly and independently placed around its associated potential D2D Tx according to a

HPPP. In view of this scenario, we consider a large circular area of radius R centered around the D2D Tx and assume that DUEs are uniformly distributed over the two-dimensional area $A = \pi R^2$ according to a HPPP with density λ being the average number of DUEs per unit area.

For a DUE located at distance d from the D2D Tx of interest, the conditional probability density function (PDF) of the received SNR, is given by

$$f_X(x|d) = \left(\frac{m_s}{\Omega_s}\right)^{m_s} \frac{d^{m_s-1}}{\Gamma(m_s)} \exp\left(-\frac{m_s}{\Omega_s}x\right) \tag{10.7}$$

where m_s and Ω_s are the distribution's shaping and scaling parameters. When the path-loss exponent follows the decaying power law, the average SNR Ω_s at distance d is given by

$$\Omega_s = \frac{P_t \cdot K \cdot d^{-v}}{N} = \tilde{P}_t \cdot d^{-v} \tag{10.8}$$

where:
P_t is the transmit power
K is a constant that depends on the antenna characteristics and free-space path-loss up to distance $d_0 = 1m$
d is a RV
v is the path loss exponent with values in the range [2,6]
N is the receiver noise power
$\tilde{P}_t$ is the transmit SNR

Assuming a DUE located randomly over a circular coverage area of radius R centered around the D2D Tx, the distance d is a RV $d \in [0, R]$ with marginal PDF.

$$f_d(r) = \frac{2r}{R^2} \tag{10.9}$$

The unconditional distribution of the received SNR can be obtained by averaging Equation 10.7 over the PDF of the distance given in Equation 10.9. Making the change of variables $u = (d/R)^v$, the PDF of the received SNR X is given in [19, Eq. (10.7)] using $i = 1$, $\beta_1 = 1$, and $B_1 = 2$, as

$$f_X(x) = \frac{1}{v\Gamma(m_s)}\left(\frac{m_s}{\Omega_0}\right)^{-\left(\frac{2}{v}\right)} x^{-\left(\frac{2}{v}+1\right)} \gamma\left(m_s + \frac{2}{v}, \frac{m_s x}{\Omega_0}\right) \tag{10.10}$$

where:
$\Omega_0 = \tilde{P}_t R^{-v}$ is the received SNR at distance R
$\gamma(v,z) = \int_0^z x^{v-1} e^{-x} dx$ is the lower incomplete gamma function of the first kind [16, Eq. (8.350.1)]

The CDF of the received SNR X is then given by [19]

$$F_X(x) = \frac{1}{\Gamma(m_s)}\left[\gamma\left(m_s, \frac{m_s x}{\Omega_0}\right) - \left(\frac{m_s x}{\Omega_0}\right)^{-\frac{2}{v}} \gamma\left(m_s + \frac{2}{v}, \frac{m_s x}{\Omega_0}\right)\right] \tag{10.11}$$

It follows that for a D2D Tx located at the origin the CDF of the received SNR at a random location inside the cell is given by

$$F_{\text{SNR}}(\beta) \underline{\triangleq} \Pr\{X \le \beta\} = F_X(\beta) \tag{10.12}$$

10.3.1 Impact of Interference on System Performance

In this section, we consider the impact of cochannel interference generated by a CUE on the association performance of a D2D pair. We assume the presence of a Nakagami-m faded interfering signal at the D2D receiver. The PDF of the instantaneous interference is given by

$$f_I(x) = \left(\frac{m_I}{\Omega_I}\right)^{m_I} \frac{x^{m_I-1}}{\Gamma(m_I)} \exp\left(-\frac{m_I}{\Omega_I}x\right) \tag{10.13}$$

where:

m_I is the Nakagami fading parameter for the interference channel
Ω_I is the average received interference-to-noise ratio (INR) of the interfering signal

We will investigate the effect of INR on the association probability.

10.3.1.1 Impact of Interference Only

When the interference is dominant, the effect of noise may be neglected in the outage analysis. The CDF of the received signal-to-interference ratio (SIR) is given in [19], as

$$\begin{aligned}
F_{\text{SIR}}(\beta) = {} & \frac{(m_I)_{m_s}}{\Gamma(m_s)}\left(\frac{m_s}{m_s + \frac{\Lambda m_I}{\beta}}\right)^{m_s}\left(\frac{m_I}{m_I + \frac{m_s\beta}{\Lambda}}\right)^{m_I} \\
& \times \left[\frac{1}{m_s}\,{}_2F_1\left(1, m_s + m_I; m_s + 1; \frac{m_s}{m_s + \frac{\Lambda m_I}{\beta}}\right)\right. \\
& \left. - \left(\frac{1}{m_s + \frac{2}{v}}\right){}_2F_1\left(1, m_s + m_I, m_s + \frac{2}{v} + 1; \frac{m_s}{m_s + \frac{\Lambda m_I}{\beta}}\right)\right]
\end{aligned} \tag{10.14}$$

where:

$_2F_1(\cdot,\cdot;\cdot;\cdot)$ is the Gauss hypergeometric function defined in [16, Eq. (9.100)]

$\Lambda = \Omega_0/\Omega_I$ is the average SIR per interferer

10.3.1.2 Impact of Interference and Noise

In the presence of interference and AWGN, the CDF of the received SINR is given in [19], as

$$
F_{\mathrm{SINR}}(\beta) = \frac{1}{\Gamma(m_s)}\left(\frac{m_s\beta}{\Omega_0}\right)^{m_s}\sum_{j=0}^{m_s}\binom{m_s}{j}\frac{(m_I)_j}{\left(\dfrac{m_I}{\Omega_I}\right)^j}
$$

$$
\times\left[\frac{1}{m_s}\Phi_1\left(m_s, m_I + j, m_s + 1; -\frac{m_s\beta}{m_I\Lambda}; -\frac{m_s\beta}{\Omega_0}\right)\right.
$$
(10.15)

$$
\left.-\left(\frac{1}{m_s+\dfrac{2}{v}}\right)\Phi_1\left(m_s+\frac{2}{v}, m_I + j, m_s+\frac{2}{v}+1; -\frac{m_s\beta}{m_I\Lambda}; -\frac{m_s\beta}{\Omega_0}\right)\right]
$$

where $(z)_q = \Gamma(z+q)/\Gamma(z)$ is the Pochhammer symbol and the confluent bivariate hypergeometric function $\Phi_1(\cdots, \cdots, \cdots; \cdots; \cdots)$ is defined in integral form as [16, Eq. (3.385)]

$$
\Phi_1(a, b; c; x; y) = \frac{\Gamma(c)}{\Gamma(a)\Gamma(c-a)}
$$

$$
\times\int_0^1 t^{a-1}(1-t)^{c-a-1}(1-xt)^{-b}e^{yt}\,dt,
$$
(10.16)

$$
|x| < 1, \mathrm{Re}(a) > 0, \mathrm{Re}(c-a) > 0
$$

10.3.2 Probability of D2D Association

We consider a circle of radius R centered around a potential D2D Tx and assume that other DUEs are uniformly distributed over a two-dimensional area $A = \pi R^2$ according to a homogeneous PPP with density λ. The probability that $Q = q$ DUEs exist within area A is a discrete RV with PDF

$$
P_Q(q) = \frac{\left(\lambda\pi R^2\right)^q}{q!}e^{-\lambda\pi R^2}, \quad q = 0, 1, \ldots
$$
(10.17)

whereas the average number of DUEs within the area A is given by

$$\bar{Q}(R,\lambda) = \lambda \pi R^2 \tag{10.18}$$

The probability that at least N DUEs can associate to the D2D Tx of interest can be obtained in terms of the incomplete beta function $I_p(r, n-r+1) = \sum_{i=r}^{n} \binom{n}{i} p^i (1-p)^{n-i}$ [16, Eq. (8.392)] as follows

$$
\begin{aligned}
P_{A,N}^{\text{SNR}} &= \sum_{q=N}^{\infty} \Pr[\text{at least N SNRs exceed } \beta] \Pr[Q=q] \\
&= \sum_{q=N}^{\infty} \sum_{k=N}^{q} \binom{q}{k} \left[1 - F_{\text{SNR}}(\beta)\right]^k \left[F_{\text{SNR}}(\beta)\right]^{q-k} \Pr[Q=q] \\
&= \sum_{q=N}^{\infty} I_{1-F_{\text{SNR}}(\beta)}(N, q-N+1) \frac{\bar{Q}^q(R,\lambda)}{q!} e^{-\bar{Q}(R,\lambda)}
\end{aligned}
\tag{10.19}
$$

Using the integral definition of the incomplete beta function $I_x(a,b) = \Gamma(a+b)/\Gamma(a)\Gamma(b) \int_0^x t^{a-1}(1-t)^{b-1} dt$ and substituting $k = q - N$, the association probability of at least N DUEs is obtained in closed form as

$$
\begin{aligned}
P_{A,N}^{\text{SNR}} &= \frac{e^{-\bar{Q}(R,\lambda)}}{\Gamma(N)} \int_0^{1-F_{\text{SNR}}(\beta)} t^{N-1}(1-t)^{-N} \sum_{k=0}^{\infty} \frac{\left[(1-t)\bar{Q}(R,\lambda)\right]^{N+k}}{k!} dt \\
&= \frac{\left[\bar{Q}(R,\lambda)\right]^N}{\Gamma(N)} \int_0^{1-F_{\text{SNR}}(\beta)} t^{N-1} e^{-\bar{Q}(R,\lambda)t} dt \\
&= \frac{1}{\Gamma(N)} \gamma\left(N, [1-F_{\text{SNR}}(\beta)]\bar{Q}(R,\lambda)\right), \quad N=1,2,\dots
\end{aligned}
\tag{10.20}
$$

Moreover, for the special case of $N=1$ we obtain the closed form expression

$$
\begin{aligned}
P_{A,N=1}^{\text{SNR}} &= \frac{1}{\Gamma(N)} \gamma\left(N, [1-F_{\text{SNR}}(\beta)]\bar{Q}(R,\lambda)\right)\Big|_{N=1} \\
&= 1 - \exp\left([F_{\text{SNR}}(\beta)-1]\bar{Q}(R,\lambda)\right)
\end{aligned}
\tag{10.21}
$$

The previous expressions show that the association probability depends on the outage probability of each DUE receiver and the average number of potential DUE receivers within the communication area. The probability that zero D2D receivers can connect to the D2D Tx of interest can be obtained as $P_{out}^{SNR} \triangleq 1 - P_{N=1}^{SNR}$. Note that this result can also be obtained as follows

$$
\begin{aligned}
P_{out}^{SNR} &= \sum_{q=0}^{\infty} \Pr\left[\max_{q}(SNR_q) \leq \beta\right] \Pr[Q = q] \\
&= \sum_{q=0}^{\infty} F_{SNR}^q(\beta) \frac{\bar{Q}^q(R,\lambda)}{q!} e^{-\bar{Q}(R,\lambda)} \qquad (10.22) \\
&= \exp\left[\left(F_{SNR}(\beta) - 1\right)\bar{Q}(R,\lambda)\right]
\end{aligned}
$$

It follows that by replacing the SNR association metric with the SIR or SINR association metrics, the previous results in Equations 10.20 through 10.22 hold by just replacing $F_{SNR}(\beta)$ with $F_{SIR}(\beta)$ and $F_{SINR}(\beta)$, respectively.

10.3.3 Numerical and Computer Simulation Results

Some representative examples illustrate the formulations derived here. A large circular area with radius $R = 600$ m is used to approximate the infinite area of the PPP in the simulation model. All plots consider Nakagami-m fading with $m_s = 2$ for the desired signal and Rayleigh fading (i.e., $m_I = 1$) with $\Omega_I = 10$ dB for the interference signal at the device receiver. Figure 10.3 portrays the probability of finding at least one DUE Rx to connect to as a function of P_t under SINR/SNR/SIR association metrics, assuming $R = 600$ m, $\nu = 3.5$, $\lambda = 10^{-4}$ (corresponds to average number of devices equal to 113 for the area considered), and $\beta = 5$ dB. The value of $\tilde{P}_t$ was obtained using the relationship $P_t = \tilde{P}_t + K - N$, assuming $N = -105$ dBm and $K = -35$ dB. The figure shows the impact of the association metric employed on the probability of finding at least one neighbor device receiver to connect to. We observe that analytical results match our computer simulation ones.

Figure 10.4 depicts the association probability to at least one DUE Rx as a function of DUE density λ, assuming $P_t = 10$ dBm, $\nu = 4$, and $\beta = 5$ dB. As it is expected, association probability increases as the number of DUE Rxs randomly located inside the circular area increases. Finally, comparisons of numerically evaluated results with equivalent ones obtained using Monte Carlo simulations, clearly show that there is practically no distinction between the simulated and theoretical curves.

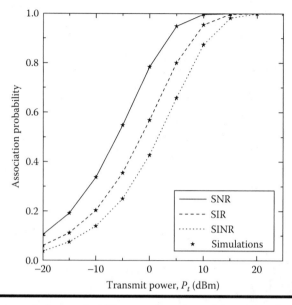

Figure 10.3   Association probability as a function of the transmit power.

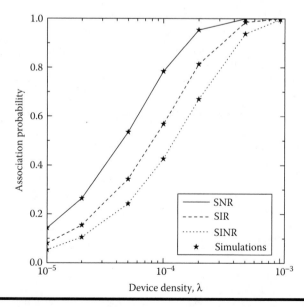

Figure 10.4   Association probability as a function of the device density λ.

10.4 D2D Link Performance with Power Control

In this section we consider the scenario where D2D users communicate underlay to CUEs using uplink resources with the goal of protecting the BS. One way of such protection is to limit the transmission power of D2D users. Therefore, all D2D transmissions are allowed as long as the generated interference on the BS remains below an interference threshold I [20]. Thus, the transmit power of a D2D Tx should be constrained as follows [21]

$$P_s = \min\left\{\frac{I}{\left(|h_{DB}|^2\right)}, P_{max}\right\}$$ (10.23)

where:

h_{DB} denotes the channel gain between the DUE and the BS

P_{max} denotes the maximum allowable transmit power

I denotes the maximum tolerable interference level at which BS can still maintain reliable communication

In addition, due to the simultaneous transmission from a CUE, the received signal at the D2D receiver is also subject to interfering effects. The corresponding received instantaneous SINR is given by

$$\gamma_{D2D} = \frac{P_s |h_{DD}|^2}{|h_{CD}|^2 + N_0}$$ (10.24)

where:

h_{DD}, h_{CD} denote the channel gains between DUEs and BS and DUE, respectively

N_0 denotes the noise power

It is assumed here that $|h_{DB}|^2, |h_{DD}|^2, |h_{CD}|^2$ follow the gamma distribution with PDF given by

$$f_{|h_X|^2}(y) = \frac{m_X^{m_X} y^{m_X-1}}{\bar{\gamma}_X^{m_X} \Gamma(m_X)} \exp\left(-\frac{m_X}{\bar{\gamma}_X} y\right)$$ (10.25)

where:

$m_X, \bar{\gamma}_X$ denote the shaping parameter and the mean value of the distribution, with $X \in \{DB, DD, CD\}$

Assuming integer values for m_X, the corresponding CDF can be expressed as

$$F_{|h_X|^2}(y) = 1 - \exp\left(-\frac{m_X}{\bar{\gamma}_X} y\right) \sum_{k=0}^{m_X-1} \left(\frac{m_X}{\bar{\gamma}_X}\right)^k \frac{y^k}{k!}$$ (10.26)

Next, an analytical expression for the CDF of γ_{D2D} will be provided. Based on it, the OP, which is defined as the probability that the SINR falls below a predefined threshold γ_{th}, can be directly evaluated as $P_{out} = F_{\gamma_{D2D}}(\gamma_{th})$.

10.4.1 Outage Probability Analysis

In this section, we investigate the CDF of γ_{D2D}. We start by defining the following new RV

$$U = \min\left\{ \frac{I}{\max\left(|h_{DB}|^2\right)}, P_{max} \right\} |h_{DD}|^2 \tag{10.27}$$

The CDF of U can be expressed as

$$F_U(x) = \underbrace{\Pr\left(\frac{I}{|h_{DB}|^2}|h_{DD}|^2 \leq x, |h_{DB}|^2 > \frac{I}{P_{max}} \right)}_{\mathcal{F}_1(x)} \tag{10.28}$$
$$+ \underbrace{\Pr\left(P_{max}|h_{DD}|^2 \leq x, |h_{DB}|^2 \leq \frac{I}{P_{max}} \right)}_{\mathcal{F}_2(x)}$$

In the following analysis, we have assumed i.i.d. fading conditions and integer values of m_{DB}. Thus, in Equation 10.28, $\mathcal{F}_1(x)$ can be expressed as

$$\mathcal{F}_1(x) = \int_{I/P_{max}}^{\infty} f_{|h_{DB}|^2}(y) F_{|h_{DD}|^2}\left(\frac{xy}{I} \right) dy \tag{10.29}$$

Assuming also integer values of m_{DD}, substituting Equations 10.26 and 10.25 in 10.29 and employing the definition of the upper incomplete gamma function [16, Eq. (3.350/2)], yields the following closed form expression

$$\mathcal{F}_1(x) = \frac{\Gamma\left(m_{DB}, \dfrac{Im_{DB}}{P_{max}\overline{\gamma}_{DB}} \right)}{\Gamma(m_{DB})} - \sum_{k=0}^{m_{DD}-1} \frac{\left(\dfrac{m_{DD}}{\overline{\gamma}_{DD}} \right)^k}{k!} \frac{m_{DB}^{m_{DB}}\left(\dfrac{x}{I} \right)^k}{\overline{\gamma}_{DB}^{-m_{DB}}\Gamma(m_{DB})} \tag{10.30}$$
$$\times \left(\frac{m_{DB}}{\overline{\gamma}_{DB}} + \frac{m_{DD}x}{I\overline{\gamma}_{DD}} \right)^{-k-m_{DB}} \Gamma\left(m_{DB}+k, \frac{Im_{DB}\overline{\gamma}_{DD}+m_{DD}\overline{\gamma}_{DB}x}{P_{max}\overline{\gamma}_{DB}\overline{\gamma}_{DD}} \right)$$

In addition $\mathcal{F}_2(x)$ can be easily evaluated as

$$\mathcal{F}_2(x) = F_{|h_{DD}|^2}\left(\frac{x}{P_{max}} \right) F_{|h_{DB}|^2}\left(\frac{I}{P_{max}} \right) \tag{10.31}$$

The CDF expression for the output SINR, γ_{D2D} is then given by

$$F_{\gamma_{D2D}}(\gamma) = \int_0^\infty F_U\left[(N_0 + x)\gamma\right] f_{|h_{CD}|^2}(x)dx \tag{10.32}$$

Substituting Equations 10.25, 10.30, and 10.31 in Equation 10.32, using [16, Eq. (8.310/1)], [16, Eq. (8.352/2)], and [22, Eq. (2.3.6/9)], and after some mathematical manipulations, yields the following closed-form expression for γ_{D2D}

$$
F_{\gamma_{2D}}(\gamma) = \frac{\Gamma\left(m_{DB}, \dfrac{Im_{DB}}{P_{max}\bar{\gamma}_{DB}}\right)}{\Gamma(m_{DB})} - \sum_{k=0}^{m_{DD}-1}\sum_{m=0}^{m_{DB}-1}\sum_{p=0}^{k} \frac{(k+m_{DB}-1)!}{m!k!} \frac{\Gamma(m_{CD}+p)N_0^{k-p}}{\Gamma(m_{CD})\Gamma(m_{DB})\gamma^A}\binom{k}{p}
$$

$$
\times \exp\left[-\left(\frac{Im_{DB}}{P_{max}\bar{\gamma}_{DB}}+\mathcal{D}N_0\gamma\right)\right]\frac{I^{m_{DB}}}{P_{max}^m}\frac{m_{DB}^{m_{DB}}m_{CD}^{m_{CD}}m_{DD}^{-A}}{\bar{\gamma}_{DB}^{-m_{DB}}\bar{\gamma}_{CD}^{-m_{CD}}\bar{\gamma}_{DD}^{-m-m_{DB}}}\left(\frac{m_{DB}I\bar{\gamma}_{DD}}{\bar{\gamma}_{DB}}+m_{DD}N_0\gamma\right)^{A-m_{DB}+m}
$$

$$
\times U\left(m_{CD}+p, A+1-m_{DB}+m, \left(\mathcal{D}\gamma+\frac{m_{CD}}{\bar{\gamma}_{CD}}\right)\left(\frac{m_{DB}I\bar{\gamma}_{DD}}{\bar{\gamma}_{DB}m_{DD}\gamma}+N_0\right)\right)
$$

$$
\times F_{|h_{DB}|^2}\left(\frac{I}{P_{max}}\right)\left[1-\sum_{k=0}^{m_{DD}-1}\sum_{t=0}^{k}\frac{N_0^{k-t}}{k!}\frac{m_{CD}^{m_{CD}}\Gamma(m_{CD}+t)}{\bar{\gamma}_{CD}^{-m_{CD}}\Gamma(m_{CD})}\binom{k}{t}\frac{(\mathcal{D}\gamma)^k\exp(-\mathcal{D}N_0\gamma)}{\left(\dfrac{m_{CD}}{\bar{\gamma}_{CD}}+\mathcal{D}\right)^{m_{CD}+t}}\right]
$$

$$\tag{10.33}$$

where:

$\mathcal{A} = m_{CD} + p - k, \mathcal{D} = m_{DD}/(\bar{\gamma}_{DD}P_{max})$

$U(\cdot,\cdot,\cdot)$ denotes the confluent hypergeometric function [16, Eq. (9.210/2)]

10.4.2 Numerical Results

In this section, using the previously derived expression for the CDF of the instantaneous output SINR, the OP of the D2D underlay scheme will be studied. For obtaining Figure 10.5, we have assumed, an outage threshold $\gamma_{th} = -5$ dB, $N_0 = 0$ dB, $m_{DD} = m_{DB} = m_{CD} = 2$, $\bar{\gamma}_{DD} = 5$ dB, $\bar{\gamma}_{DB} = 0$ dB, and $\bar{\gamma}_{CD} = 1$ dB. Under these assumptions and based on Equation 10.33, the OP is plotted as a function of the maximum allowed transmission power, P_{max}, and for different values of the maximum tolerable interference level I. It is shown that the performance improves with the increase of P_{max}, reaching in all cases a floor for higher values of P_{max}. The performance also improves for higher values of I.

For obtaining Figure 10.6, we have assumed the same values for the shaping and scaling parameters, $M = N = 3$, $I = 5$ dB. Under these assumptions, the OP is plotted as a function of γ_{th} for different values of the maximum allowable transmission

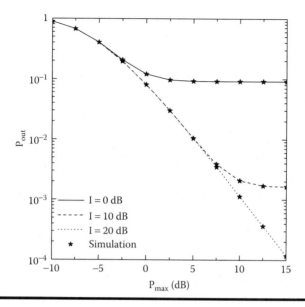

Figure 10.5 Outage probability as a function of the maximum transmission power.

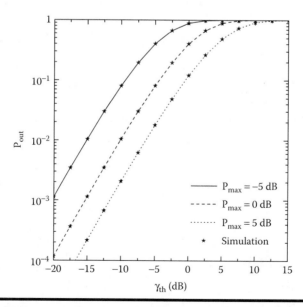

Figure 10.6 Outage probability as a function of the normalized outage threshold.

power P_{max}. It is depicted that the performance improves with the increase of P_{max} and/or the decrease of γ_{th}. For comparison purposes, computer simulation performance results are also included in all figures, verifying the validity of the proposed theoretical approach.

10.5 Effect of Mobility

In static wireless networks, a uniform distribution is usually adopted to model the node distribution in order to study network capacity and connectivity properties [23,24]. However, the presence of mobility which is an inherent feature of many wireless networks, requires a more realistic and nonuniform model for the spatial distribution of the mobile nodes. With mobile receiving nodes in which the transmit power remains constant but the propagation distance is a RV because of node mobility, the conditional PDF of the received SNR, given the transmitter-receiver distance r, is given by Equation 10.7. The unconditional distribution of the received SNR can be obtained by averaging Equation 10.7 over the PDF of distance of a particular node from the access point over the unit space [0,1], that is,

$$f_X(x) = \int_0^1 f_X(x|r) f_r(r) dr \tag{10.34}$$

Several mobility models are available to characterize the distribution of the transmitter-receiver distances in a mobile environment. In the random waypoint (RWP) mobility model, it is usually assumed that the receiving nodes are located at randomly selected coordinate points in the service area, which depends on the network topology. For a one-dimensional (1D) topology, we consider a line with the transmitter or access point being located at the origin. The two-dimensional (2D) topology is assumed to be a circle of unit radius, while a three-dimensional (3D) topology is a spherical network with unit radius. In both the 2D and 3D network topologies, it is assumed that the transmitter is located at the origin. The receiving node is located at a randomly selected location within the communication range of the transmitter. The steady state spatial node distributions for the RWP mobility models are polynomials in the transmitter-receiver distance r. Therefore, for each of these models, the PDF of the distance for a realistic mobility scenario with $0 \le r \le D$, is given by [25]

$$f_r(r) = \sum_{i=1}^{n} B_i \frac{r^{\beta_i}}{D^{\beta_i+1}}, \quad 0 \le r \le D \tag{10.35}$$

where the parameters n, B_i, and β_i depend on the number of dimensions considered in the topology and are summarized in [25, Table 1]. Substituting Equations 10.7 and 10.35 in Equation 10.34, and making the change of variables $u = (r/D)^v$, the unconditional PDF of the received SNR is given by [25]

$$f_X(x) = \left(\frac{m_s}{\Omega_0}\right)^{m_s} \frac{x^{m_s-1}}{v\Gamma(m_s)} \sum_{i=1}^{n} B_i \int_0^1 u^{\left(m_s + \frac{\beta_i+1}{v}\right)-1} \exp\left(-\frac{m_s x}{\Omega_0} u\right) du$$

$$= \frac{1}{v\Gamma(m_s)} \sum_{i=1}^{n} B_i \left(\frac{m_s}{\Omega_0}\right)^{-\left(\frac{\beta_i+1}{v}\right)} x^{-\left(\frac{\beta_i+1}{v}+1\right)} \gamma\left(m_s + \frac{\beta_i+1}{v}, \frac{m_s x}{\Omega_0}\right)$$

(10.36)

where:

$\Omega_0 = \tilde{P}_t D^{-v}$ is the received SNR at distance D

The PDF in Equation 10.36 may be alternatively expressed in terms of the confluent hypergeometric function [16, Eq. (8.351.2)] as

$$f_X(x) = \left(\frac{m_s}{\Omega_0}\right)^{m_s} \frac{x^{m_s-1}}{\Gamma(m_s)} \sum_{i=1}^{n} \frac{B_i}{(m_s v + \beta_i + 1)}$$

$$\times {}_1F_1\left(m_s + \frac{\beta_i+1}{v}; m_s + \frac{\beta_i+1}{v} + 1; -\frac{m_s x}{\Omega_0}\right)$$

(10.37)

10.5.1 Outage Probability

The CDF of the received SNR X is given by [19]

$$F_X(x) = \int_0^x f_X(u) du$$

$$= \left(\frac{m_s}{\Omega_0}\right)^{m_s} \frac{1}{\alpha\Gamma(m_s)} \sum_{i=1}^{n} B_i \int_0^1 u^{\left(m_s + \frac{\beta_i+1}{\alpha}\right)-1} \int_0^x x^{m_s-1} \exp\left(-\frac{m_s}{\Omega_0} ux\right) dx du$$

(10.38)

$$= \frac{1}{\alpha\Gamma(m_s)} \sum_{i=1}^{n} B_i \int_0^1 u^{\left(\frac{\beta_i+1}{\alpha}\right)-1} \gamma\left(m_s, \frac{m_s xu}{\Omega_0}\right) du$$

We can show, via integration by parts, that

$$\int_0^1 x^{v-1} \gamma(m, ax) dx = v^{-1}\left[\gamma(m, a) - a^{-v}\gamma(v + m, a)\right], v > 0$$

(10.39)

Using Equation 10.39, the integral in Equation 10.38 may be evaluated to give

$$F_X(x) = \frac{1}{\Gamma(m_s)} \sum_{i=1}^{n} \left(\frac{B_i}{\beta_i+1}\right)\left[\gamma\left(m_s, \frac{m_s x}{\Omega_0}\right) - \left(\frac{m_s x}{\Omega_0}\right)^{-\frac{\beta_i+1}{v}} \gamma\left(m_s + \frac{\beta_i+1}{v}, \frac{m_s x}{\Omega_0}\right)\right]$$

(10.40)

Commonly used as a performance metric in many wireless communications systems, the outage probability is the probability that the received signal falls below a preset threshold θ. It can then be obtained as $P_{\text{out}} = F_X(\theta)$.

10.5.2 Average Bit Error Probability

The exact ABEP for binary modulations is given by

$$\bar{P}_b = \int_0^\infty \frac{\Gamma(b, ax)}{2\Gamma(b)} f_X(x)dx = \frac{a^b}{2\Gamma(b)} \int_0^\infty x^{b-1} e^{-ax} F_X(x)dx \qquad (10.41)$$

where $\Gamma(\cdot,\cdot)$ is the upper incomplete gamma function [16, Eq. (8.350.2)] and parameters $a, b \in (1/2, 1)$ depend on the type of binary modulation/demodulation employed [27]. Moreover, Equation 10.41 can closely approximate the average BER of M-ary modulations with conditional BER given by a linear combination of terms $P_b(\gamma) = Q\left(\sqrt{2g_M\gamma}\right)$, where $Q(x) = \left(1/\sqrt{2\pi}\right)\int_x^\infty \exp\left(-t^2/2\right)dt$ is the Gaussian Q-function, by substituting $a = g_M$ and $b = 1/2$ in Equation 10.41. Substituting Equation 10.40 in Equation 10.41, the ABEP is given by [19]

$$\bar{P}_b = \frac{\Gamma(m_s + b)}{2\Gamma(b)\Gamma(m_s)} \left(\frac{a\Omega_0}{m_s + a\Omega_0}\right)^b \left(\frac{m_s}{m_s + a\Omega_0}\right)^{m_s}$$

$$\sum_{i=1}^n \left(\frac{B_i}{\beta_i + 1}\right) \left[\frac{1}{m_s} \, {}_2F_1\left(1, m_s + b; m_s + 1; \frac{m_s}{m_s + a\Omega_0}\right)\right. \qquad (10.42)$$

$$\left. -\left(\frac{\nu}{m_s\nu + \beta_i + 1}\right) {}_2F_1\left(1, m_s + b; m_s + \frac{\beta_i + 1}{\nu} + 1; \frac{m_s}{m_s + a\Omega_0}\right)\right]$$

10.5.3 Impact of Interference and Noise

In the presence of interference and AWGN, the outage probability is given by

$$P_{\text{out}}^{\text{SINR}} = \Pr\left\{\frac{X}{I_a + 1} \leq \theta\right\} \qquad (10.43)$$

where I_a is the aggregate INR at the receiver emanating from static sources in the network. The outage probability is then given by

$$P_{\text{out}}^{\text{SINR}} = \mathbb{E}_{I_a}\left\langle \Pr\left(X \leq \theta[I_a + 1]\right)\right\rangle = \mathbb{E}_{I_a}\left\langle F_X\left(\theta[I_a + 1]\right)\right\rangle = \frac{1}{\Gamma(m_s)} \sum_{i=1}^n \left(\frac{B_i}{\beta_i + 1}\right)$$

$$\times \mathbb{E}_{I_a}\left\langle \gamma\left(m_s, \frac{m_s\theta(I_a + 1)}{\Omega_0}\right) - \left(\frac{m_s\theta(I_a + 1)}{\Omega_0}\right)^{-\frac{\beta_i + 1}{\nu}} \gamma\left(m_s + \frac{\beta_i + 1}{\nu}, \frac{m_s\theta(I_a + 1)}{\Omega_0}\right)\right\rangle \qquad (10.44)$$

The final expression for the outage probability in the presence of interference and noise is given in [19], as

$$
P_{\text{out}}^{\text{SINR}} = \frac{1}{\Gamma(m_s)} \left(\frac{m_s \theta}{\Omega_0} \right)^{m_s} \sum_{i=1}^{n} \left(\frac{B_i}{\beta_i + 1} \right) \sum_{j=0}^{m_s} \binom{m_s}{j}
$$

$$
\times \frac{(m_I L)_j}{\left(\frac{m_I}{\Omega_I} \right)^j} \left[\frac{1}{m_s} \Phi_1 \left(m_s, m_I L + j, m_s + 1; -\frac{m_s \theta}{m_I \Lambda}; -\frac{m_s \theta}{\Omega_0} \right) \right.
$$

(10.45)

$$
\left. - \left(\frac{\alpha}{m_s \nu + \beta_i + 1} \right) \Phi_1 \left(m_s + \frac{\beta_i + 1}{\nu}, m_I L + j, m_s + \frac{\beta_i + 1}{\nu} + 1; -\frac{m_s \theta}{m_I \Lambda}; -\frac{m_s \theta}{\Omega_0} \right) \right]
$$

10.5.4 Numerical and Computer Simulation Results

In this section, we assume a 2D topology with a D2D Rx moving around the D2D Tx according to the RWP mobility model over a maximum distance of $D = 50$ m. In Figure 10.7, we examine the effect of the path-loss exponent for the desired signal and cochannel interference with average INR $\Omega_I = 10$ dB on the D2D link performance of a mobile DUE Rx. We plot the outage probability versus transmit power P_t of the DUE Tx for path loss exponents $\nu = \{3,4\}$ for the desired signal assuming Nakagami-m fading channels with $m_s = 2$ and $m_I = 1$, and threshold $\theta = 5$ dB.

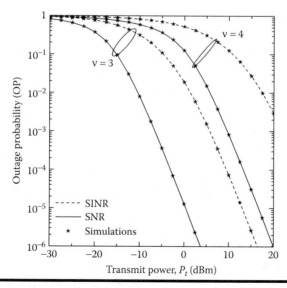

Figure 10.7 **Outage Probability vs transmit power for $\nu = \{3,4\}$ assuming noise-only and noise-plus-interference cases.**

For the noise-only case considered, the outage probability is found to be considerably lower for smaller values of ν, due to higher values of average received SNR. However, in the presence of interference, we observe much higher outage probabilities than for the noise-only case for both values of ν, with the performance gap to increase considerably as ν decreases.

10.6 User Selection in Mobile Device-to-Device Communications

The vehicle-to-everything (V2X) communications are considered to be an integral part of the 5G networks. Based on these systems, important improvements are expected in road safety, traffic efficiency, and comfort to both the drivers and the passengers. These systems' performance depends, mainly, on the channel model, which is sufficiently different from the classical cellular one. Reasons for this include, the equal heights of the transmitter and the receiver, their movement, the surrounding scatterers, the highly dynamic propagation conditions and so on. Well established distributions that have been widely employed to model V2V channel conditions are the ones that are based on the multiple scattering radio propagation channels [28]. Using this generic model, and under the assumption that only double scattering effects are present, new families of distributions are proposed, namely double-Rayleigh, double-Nakagami, and double-Rice [29–31]. A common assumption in all these models is that both the transmitter and the receiver are in motion.

Toward enabling V2X communication in 5G systems, several requirements should be satisfied, including (1) low latency, (2) high reliability, and (3) high throughput. In this context, cooperative relaying is expected to play an important role for improving spatial diversity in V2X communication networks, by extending wireless network coverage with low energy budget. Moreover, by applying cooperative relaying to multiuser communication scenarios, a new network architecture has been introduced known as multiuser relay network (MRN) architecture. Investigating new distributed and centralized user selection techniques is therefore an important topic for research in V2X communications [32,33].

10.6.1 System Model

We consider a two-phase multiuser cooperative V2V communication network, where M mobile users-sources U_j communicate with the destination (D) with the help of a relay node (R) as it is shown in Figure 10.8. The relay and destination nodes are equipped with single antennas, while no direct link between the mobile users and the destination exists, due to severe shadowing. The relay node is assumed

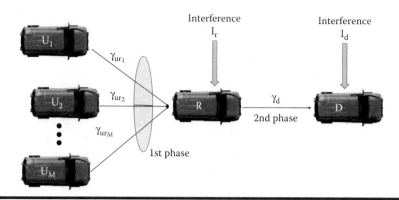

Figure 10.8   Mobile device-to-device system model.

to operate in an amplify and forward (AF) half-duplex mode, based on the variable gain approach. Moreover, the relay node and the destination are subject to AWGN and interfering effects. Finally, all links, including the interfering ones, are subject to double-Nakagami (DN) fading.

In the first phase, a user j transmits a signal to the relay. In the second phase, the relay transmits a modified version of the received signal to the destination. In the AF scheme, the relay nodes cannot differentiate between source and interference signals. The instantaneous end-to-end SIR at the destination is given by [34]

$$\gamma_{\text{urd}_j} = \frac{\gamma_{\text{ur}_j}^{\text{ef}} \gamma_{\text{rd}}^{\text{ef}}}{\gamma_{\text{ur}_j}^{\text{ef}} + \gamma_{\text{rd}}^{\text{ef}} + 1} \tag{10.46}$$

where $\gamma_{\text{ur}_j}^{\text{ef}}, \gamma_{\text{rd}}^{\text{ef}}$ are the effective SIRs at the relay and destination, respectively, defined as

$$\gamma_{\text{ur}_j}^{\text{ef}} = \frac{\gamma_{\text{ur}_{\text{sel}}}}{\gamma_{I_r}} \quad \text{and} \quad \gamma_{\text{rd}}^{\text{ef}} = \frac{\gamma_d}{\gamma_{I_d}} \tag{10.47}$$

where:

$\gamma_{\text{ru}_{\text{sel}}}, \gamma_d$ are the desired instantaneous received SNRs at the R and D, respectively

γ_{I_r} and γ_{I_d} are the instantaneous INR, received at R and D, respectively

As far as the user selection scheme is concerned, here the best user selection approach is adopted, which results to the following instantaneous received SNR at the relay

$$\gamma_{\text{ur}_{\text{sel}}} = \max\left\{\gamma_{\text{ur}_1}, \gamma_{\text{ur}_2}, \ldots, \gamma_{\text{ur}_M}\right\} \tag{10.48}$$

Based on this approach the CDF of $\gamma_{u_{\text{rsel}}}$ can be expressed as

$$F_{\gamma_{u_{\text{rsel}}}}(\gamma) = \prod_{j=1}^{M} F_{\gamma_{u_{\text{r}_j}}}(\gamma) \tag{10.49}$$

It should be noted that since DN fading is considered, the PDF of $X \in \{\gamma_{u_{\text{r}_j}}, \gamma_{I_r}, \gamma_d, \gamma_{I_d}\}$ is given by

$$f_X(y) = \frac{1}{\Gamma(m_{X,1})\Gamma(m_{X,2})y} G_{0,2}^{2,0}\left(\frac{m_{X,1}m_{X,2}y}{\overline{\gamma}_{X,1}\overline{\gamma}_{X,2}}\middle|\begin{array}{c} - \\ m_{X,1}, m_{X,2} \end{array}\right) \tag{10.50}$$

while the corresponding CDF expression is given by

$$F_{\gamma_X}(x) = \frac{1}{\Gamma(m_{X,1})\Gamma(m_{X,2})} G_{1,3}^{2,1}\left(\frac{m_{X,1}m_{X,2}x}{\overline{\gamma}_{X,1}\overline{\gamma}_{X,2}}\middle|\begin{array}{c} 1 \\ m_{X,1}, m_{X,2}, 0 \end{array}\right) \tag{10.51}$$

where:

$m_{X,i}, \overline{\gamma}_{X,i}$ are the distribution's shaping and scaling parameters
$G_{p,q}^{m,n}[\cdot|\cdot]$ denotes the Meijer's G-function [16, Eq. (9.301)].

10.6.2 CSI Model

In a V2V communication scenario, the fading behavior changes rapidly. Hence, the CSI of the $U_j - R$ links is assumed to be outdated, due to the delay between the user selection and data transmission phases as well as the fast time varying nature of the wireless medium. For this reason we consider the model for the outdated CSI that is used by many authors in the past, for example, [35]. More specifically, the imperfection between the actual SNR of the *j*th user at the data reception instance, $\gamma_{u_{\text{r}_j}}$, and $\tilde{\gamma}_{u_{\text{r}_j}}$, which is available during the selection phase, can be measured based on a correlation coefficient. In that case, the PDF of the actual received SNR (based on the *selected* user) at the data reception instance can be expressed as [35]

$$f_{\gamma_{u_{\text{rsel}}}}(y) = \int_{0}^{\infty} f_{\gamma_{u_{\text{r}_j}}, \tilde{\gamma}_{u_{\text{r}_j}}}(y,x) \frac{f_{\tilde{\gamma}_{u_{\text{rsel}}}}(x)}{f_{\gamma_{u_{\text{r}_j}}}(x)} dx \tag{10.52}$$

For evaluating Equation 10.52, $f_{\gamma_{\mathrm{ur}_j}, \tilde{\gamma}_{\mathrm{ur}_j}}(y, x)$ is required. Based on the results presented in [31], the following expression for the joint PDF between γ_{ur_j} and $\tilde{\gamma}_{\mathrm{ur}_j}$ can be obtained

$$
f_{\gamma_{\mathrm{ur}_j}, \tilde{\gamma}_{\mathrm{ur}_j}}(x, y) = \sum_{h,q=0}^{\infty} \frac{\dfrac{4\rho_1^h \rho_2^q}{\left[\Gamma(m_{Y,1})\Gamma(m_{Y,2})\right]} \left(\displaystyle\prod_{i=1}^4 \bar{\gamma}_{Y,i}\right)^{-\frac{m_{Y,1}+m_{Y,2}+q+h}{2}}}{\Gamma(m_{Y,1}+h)\Gamma(m_{Y,2}+q)h!q!} \frac{(xy)^{\frac{m_{Y,1}+m_{Y,2}+h+q}{2}-1}}{(1-\rho_1)^{m_{Y,2}+h+q}(1-\rho_2)^{m_{Y,1}+h+q}}
$$

$$
\times K_{m_{Y,1}-m_{Y,2}+h-q}\left[\frac{2x^{1/2}}{\sqrt{\bar{\gamma}_{Y,1}\bar{\gamma}_{Y,3}\hat{\rho}}}\right] K_{m_{Y,1}-m_{Y,2}+h-q}\left[\frac{2y^{1/2}}{\sqrt{\bar{\gamma}_{Y,2}\bar{\gamma}_{Y,4}\hat{\rho}}}\right]
\tag{10.53}
$$

where $\hat{\rho} = (1-\rho_1)(1-\rho_2)$, $p_1 = h+q+m_1+m_2$, $p_2 = h-q+m_1-m_2$, $Y \equiv \mathrm{ur}_j$, and $K_v(\cdot)$ is the modified Bessel function of the second kind and vth order [16, Eq. (8.432/1)]. Moreover, in Equation 10.53, $0 \leq \rho_1, \rho_2 < 1$ are the power correlation coefficients of the underlying fading processes of the first and second bounces, respectively. Based on the previous expression, in the next section the SIR statistics for the scheme under consideration will be studied.

10.6.3 SIR Statistics

As far as the first link is concerned, based on the approach presented in [36], the following exact expression for the CDF of $\gamma_{\mathrm{ur}_j}^{\mathrm{ef}}$ has been derived

$$
F_{\gamma_{\mathrm{ur}_j}^{\mathrm{ef}}}(\gamma) = \sum_{n=1}^{M}\left[\prod_{j=1 \, j\neq n}^{M} \frac{\sqrt{\pi}\,\Gamma(2m_{Y,1})2^{1-2m_{Y,1}}}{\Gamma(m_{Y,1})\Gamma(m_{Y,2})}\right]\sum_{h,q=0}^{\infty}\sum_{g_1,g_2=0}^{\left|h-\frac{1}{2}-q\right|-\frac{1}{2}}\pi^{\frac{3}{2}}\mathcal{A}_1
$$

$$
\times \frac{\rho_1^h \rho_2^q (1-\rho_2)^{\frac{1}{2}}\Gamma(p_{4,2}+2m_{Z,1})}{h!q!\Gamma(m_{Z,1}+h)\Gamma(m_{Z,2}+q)p_{4,2}}\frac{(\bar{\gamma}_{I_d,1}\bar{\gamma}_{I_d,2})^{p_{4,2}}}{2^{p_{4,2}+2}}\frac{(\mathcal{D}_1+\mathcal{B}_1\mathcal{D}_2)}{\Gamma(m_{I_d,1})\Gamma(m_{I_d,2})}
\tag{10.54}
$$

$$
\times \gamma^{p_{4,2}/2}\,{}_2F_1\left(p_{4,2}, p_{4,2}+2m_{Z,1}; p_{4,2}+1; -\left(\frac{\gamma\bar{\gamma}_{I_d,1}\bar{\gamma}_{I_d,2}}{\bar{\gamma}_{u_d,2}\bar{\gamma}_{Z,4}\hat{\rho}}\right)^{1/2}\right)
$$

where:

$$\mathcal{A}_1 = \prod_{j=1}^{2} \frac{\left(g_j + \left|h - q - \dfrac{1}{2}\right| - \dfrac{1}{2}\right)!}{g_j! \left(-g_j + \left|h - q - \dfrac{1}{2}\right| - \dfrac{1}{2}\right)!} \frac{(1-\rho_j)^{\frac{h+q+m_{Z,1}-(g_1+g_2)}{2}}}{2^{2g_j} \left(\overline{\gamma}_{Z,j}\overline{\gamma}_{Z,j+2}\right)^{\frac{p_{4,j}}{2}} \Gamma(m_{Z,j})}$$

$$\mathcal{B}_1 = \sum_{\substack{k=1}}^{M-1}(-1)^k \sum_{\substack{\ell_1=1 \\ \ell_1 \neq i}}^{M-k+1} \sum_{\substack{\ell_2=\ell_1+1 \\ \ell_2 \neq \ell_1}}^{M-k+2} \cdots \sum_{\substack{\ell_k=\ell_{k-1}+1 \\ \ell_k \neq \ell_{k-1}}}^{M} \sum_{d_{k,1}=0}^{2m_{\ell_1,1}-1} \sum_{d_{k,2}=0}^{2m_{\ell_2,1}-1} \cdots \sum_{d_{k,k}=0}^{2m_{\ell_k,1}-1} \prod_{t=1}^{k} \frac{(2\sqrt{p_{5,t}})^{d_{k,t}}}{d_{k,t}}$$

$$\mathcal{D}_1 = f\left(\frac{p_{4,1}}{2} - 1, \frac{1}{\sqrt{\overline{\gamma}_{Z,1}\overline{\gamma}_{Z,3}\hat{\rho}}}\right)$$

$$\mathcal{D}_2 = f\left(\sum_{t=1}^{k} \frac{p_{5,t} + p_{4,1}}{2} - 1, \sum_{t=1}^{k} p_{5,t}^{\frac{1}{2}} + \frac{1}{\overline{\gamma}_{Z,1}\overline{\gamma}_{Z,2}\hat{\rho}}\right)$$

with $f(x,y) = \Gamma(2+2x)/(2^{2x+1}y^{2(x+1)})$, $p_{4,j} = 2m_{r_n,1} + h + q - g_j$, $p_{5,i} = 1/(\overline{\gamma}_{\ell_i,1}\overline{\gamma}_{\ell_i,2})$, and $Z \equiv \mathrm{ur}_n$.

As far as the second link is concerned, substituting the CDF of γ_d and the PDF γ_{I_d}, given by Equations 10.51 and 10.50, respectively, in $F_{\gamma_{rd}^{ef}}(\gamma) = \int_0^{\infty} F_{\gamma_d}(x\gamma)f_{\gamma_{I_d}}(x)dx$, using [37, Equation 10.14], and after some mathematical manipulations, the CDF of γ_{rd}^{ef} is given by

$$F_{\gamma_{rd}^{ef}}(\gamma) = \frac{1}{\Gamma(m_{d,1})\Gamma(m_{d,2})} \frac{1}{\Gamma(m_{I_d,1})\Gamma(m_{I_d,2})}$$

$$G_{3,3}^{3,2}\left(\frac{m_{I_d,1}m_{I_d,2}\overline{\gamma}_{d,1}\overline{\gamma}_{d,2}x}{m_{d,1}m_{d,2}\overline{\gamma}_{I_d,1}\overline{\gamma}_{I_d,2}}\middle|\begin{matrix}1-m_{d,1},1-m_{d,2},1\\ m_{I_d,1},m_{I_d,2},0\end{matrix}\right) \tag{10.55}$$

Employing the exact expression for the end-to-end SINR $\gamma_{\mathrm{urd}j}$, provided in Equation 10.46, will result to a cumbersome mathematical analysis. A mathematically more convenient approach is to use, instead, a tight upper bound of $\gamma_{\mathrm{urd}j}$, as in [34], that is,

$$\gamma_{\mathrm{urd}j} \leq \gamma_{\mathrm{up}} = \min\{\gamma_{\mathrm{ur}j}^{ef}, \gamma_{rd}^{ef}\} \tag{10.56}$$

The CDF of γ_{up} is then given by

$$F_{\gamma_{\mathrm{up}}}(\gamma) = \Pr\{\gamma_{\mathrm{up}} \leq \gamma\} = F_{\gamma_{\mathrm{ur}j}^{ef}}(\gamma) + F_{\gamma_{rd}^{ef}}(\gamma) - F_{\gamma_{\mathrm{ur}j}^{ef}}(\gamma)F_{\gamma_{rd}^{ef}}(\gamma) \tag{10.57}$$

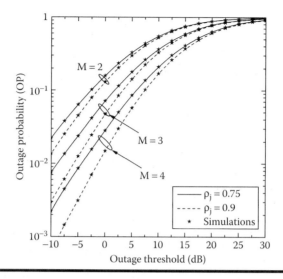

Figure 10.9 **Outage probability as a function of the outage threshold.**

Therefore, using Equations 10.54 and 10.55 in Equation 10.57, an tight upper bound for the CDF of the output SIR of the considered scheme is finally obtained.

10.6.4 Performance Analysis and Numerical Results

The performance of the system will be studied using the criterion of outage probability (OP), which using Equation 10.57 can be evaluated as $P_{out} = F_{\gamma_{up}}(\gamma_T)$, where γ_T denotes the predetermined threshold. In Figure 10.9, the OP is plotted as a function of the outage threshold γ_T. For obtaining this figure, the following assumptions have been made: $m_{Y,1} = 1$, $m_{Y,2} = 1.5$, $\bar{\gamma}_{Y_1,p_1} = 5/m_{Y,1}$ dB (with $p_1 \in \{1,2\}$), $\bar{\gamma}_{Y_1,p_2} = 5/m_{Y,2}$ dB (with $p_2 \in \{3,4\}$), $\bar{\gamma}_{Y_2,p_1} = 7/m_{Y,1}$ dB, $\bar{\gamma}_{Y_2,p_2} = 7/m_{Y,2}$ dB, $\bar{\gamma}_{Y_3,p_1} = 9/m_{Y,1}$ dB, $\bar{\gamma}_{Y_3,p_2} = 9/m_{Y,2}$ dB, $\bar{\gamma}_{Y_4,p_1} = 11/m_{Y,1}$ dB, $\bar{\gamma}_{Y_4,p_2} = 11/m_{Y,2}$ dB, (with $Y_i \equiv ur_i$) $m_{I_r,1} = 1$, $m_{I_r,2} = 1.5$, $\bar{\gamma}_{I_r,1} = \bar{\gamma}_{I_r,2} = 3$ dB, $m_{d,1} = 1.1$, $m_{d,2} = 1.5$, $\bar{\gamma}_{d,1} = 10/m_{d,1}$ dB, $\bar{\gamma}_{d,1} = 10/m_{d,2}$ dB, $m_{I_d,1} = 1.5$, $m_{I_d,2} = 2$, $\bar{\gamma}_{I_d,1} = \bar{\gamma}_{I_d,2} = 5$ dB. In Figure 10.9, it is shown that the performance improves as γ_T decreases and/or M increases. It is also shown that the performance improves as the correlation coefficients ρ_1, ρ_2 increase, that is, the SNR at the selection instance approaches the one at the reception instance.

References

1. H. Tullberg, P. Popovski, Z. Li, M. A. Uusitalo, A. Hoglund, O. Bulakci, M. Fallgren, and J. F. Monserrat, The METIS 5G system concept: Meeting the 5G requirements, *IEEE Commun. Mag.*, 132–139, 2016.
2. A. Asadi, Q. Wang, and V. Mancuso, A survey on device-to-device communication in cellular networks, *IEEE Commun. Surv. Tut.*, 16(4), 1801–1819, Fourthquarter 2014.

3. J. Liu, N. Kato, J. Ma, and N. Kadowaki, Device-to-device communication in LTE-advanced networks: A survey, *IEEE Commun. Surv. Tut.*, 17(4), 1923–1940, Fourthquarter 2015.

4. X. Lin, J. G. Andrews, and A. Ghosh, Spectrum sharing for device-to-device communication in cellular networks, *IEEE Trans. Wirel. Commun.*, 13(12), 6727–6740, December 2014.

5. P. Janis, Device-to-device communication underlaying cellular communications systems, *Int'l. J. Commun., Network and Sys. Sci.*, 2(3), 169–178, 2009.

6. Y. Xu, R. Yin, T. Han, and G. Yu, *Interference-Aware Channel Allocation for Device-to-Device Communication Underlaying Cellular Networks*, Beijing, China, August 2012, pp. 422–427.

7. R. K. Mungara, X. Zhangy, A. Lozano, and R. W. Heath, *Performance Evaluation of ITLinQ and FlashLinQ for Overlaid Device-to-Device Communication*, London, UK: IEEE, June 2014, pp. 5245–5250.

8. Q. Ye, M. Al-Shalash, C. Caramanis, and J. G. Andrews, Resource optimization in device-to-device cellular systems using time-frequency hopping, online: http://arxiv.org/abs/1309.4062.v3, pp. 1–14, March 2014.

9. H. ElSawy, E. Hossain, and M. S. Alouini, Analytical modeling of mode selection and power control for underlay D2D communication in cellular networks, *IEEE Trans. Commun.*, 62(11), 4147–4161, 2014.

10. Y. Zhao, B. Pelletier, P. Marinier, and D. Pani, *D2D Neighbor Discovery Interference Management for LTE Systems*, Atlanta, GA: IEEE, December 2013, pp. 556–560.

11. K. Gulati, B. L. Evans, J. G. Andrews, and K. R. Tinsley, Statistics of co-channel interference in a field of Poisson and Poisson-Poisson clustered interferers, *IEEE Trans. Signal Process.*, 58(12), 6207–6222, 2010.

12. M. D. Renzo, C. Merola, A. Guidotti, F. Santucci, and G. E. Corazza, Error performance of multi-antenna receivers in a Poisson field of interferers: A stochastic geometry approach, *IEEE Trans. Commun.*, 61(5), 2025–2047, 2013.

13. X. Yang and A. P. Petropulu, Co-channel interference modeling and analysis in a Poisson field of interferers in wireless communications, *IEEE Trans. Signal Process.*, 51(1), 64–76, 2003.

14. M. Z. Win, P. C. Pinto, and L. A. Shepp, A mathematical theory of network interference and its applications, *Proc. IEEE*, 97(2), 205–230, 2009.

15. P. Cardieri, Modeling interference in wireless ad hoc networks, *IEEE Commun. Surveys Tuts.*, 12(4), 551–572, 2010.

16. I. S. Gradshteyn and I. M. Ryzhik, *Table of Integrals, Series, and Products*. New York: Academic Press, 1980.

17. A. P. Prudnikov, Y. A. Brychkov, and O. I. Marichev, *Integrals and Series vol. 3: More Special Functions*, 1st ed. New York: Gordon and Breach Science Publishers, 1986.

18. V. Aalo, K. Peppas, G. Efthymoglou, M. Alwakeel, and S. Alwakeel, Evaluation of average bit error rate for wireless networks with alpha-stable interference, *IET Electr. Lett.*, 50(1), 47–49, 2014.

19. V. Aalo, C. Mukasa, and G. Efthymoglou, Effect of mobility on the outage and BER performances of digital transmissions over Nakagami-m fading channels, *IEEE Trans. Veh. Technol.*, 65(4), 2715–2721, 2016.

20. J. Lee, H. Wang, J. G. Andrews, and D. Hong, Outage probability of cognitive relay networks with interference constraints, *IEEE Trans. Commun.*, 10(2), 390–395, 2011.

21. A. Ghasemi and E. Sousa, Fundamental limits of spectrum-sharing in fading environments, *IEEE Trans. Wireless Commun.*, 6(2), 649–658, 2007.
22. A. P. Prudnikov, Y. A. Brychkov, and O. I. Marichev, *Integrals and Series vol. 1: Elementary Functions*, 1st ed. New York: Gordon and Breach Science Publishers, 1986.
23. M. Grossglauser and D. N. C. Tse, Mobility increases the capacity of ad-hoc networks, *IEEE/ACM Trans. Networking*, 10(4), 477–486, 2002.
24. D. Miorandi, E. Altman, and G. Alfano, The impact of channel randomness on the coverage and connectivity of ad hoc and sensor networks, *IEEE Trans. Wirel. Commun.*, 7(3), 1062–1072, 2008.
25. K. Govindan, K. Zeng, and P. Mohapatra, Probability density of the received power in mobile networks, *IEEE Trans. Wireless Commun.*, 10(11), 3613–3619, 2011.
26. M. Abramowitz and I. Stegun, *Handbook of Mathematical Functions*. New York: Dover Publications, 1970.
27. M. K. Simon and M.-S. Alouini, *Digital Communication over Fading Channels*, 2nd ed. New York: Wiley, 2005.
28. J. Salo, H. M. El-Sallabi, and P. Vainikainen, Statistical analysis of the multiple scattering radio channel, *IEEE Trans. Antennas Propag.*, 54(11), 3114–3124, 2006.
29. R. Shakeri, H. Khakzad, A. Taherpour, and S. Gazor, Performance of two-way multi-relay inter-vehicular cooperative networks, in *IEEE Wireless Communications and Networking Conference* (WCNC), April 2014, pp. 520–525.
30. H. Ilhan, Performance analysis of cooperative vehicular systems with co-channel interference over cascaded Nakagami-m fading channels, *Wireless Pers. Commun.*, 83(1), 203–214, 2015.
31. P. Bithas, G. Efthymoglou, and A. Kanatas, A cooperative relay selection scheme in V2V communications under interference and outdated CSI, in *IEEE International Symposium on Personal, Indoor and Mobile Radio Communications*, September 2016.
32. M. Boban, K. Manolakis, M. Ibrahim, S. Bazzi, and W. Xu, Design aspects for 5G V2X physical layer, in *2016 IEEE Conference on Standards for Communications and Networking* (CSCN), October 2016, pp. 1–7.
33. B. Aygun, C. W. Lin, S. Shiraishi, and A. M. Wyglinski, Selective message relaying for multi-hopping vehicular networks, in *2016 IEEE Vehicular Networking Conference* (VNC), December 2016, pp. 1–8.
34. S. Ikki and S. Aissa, Performance analysis of dual-hop relaying systems in the presence of co-channel interference, in *IEEE Global Telecommunications Conference*, December 2010, pp. 1–5.
35. Y. Gu, S. Ikki, and S. Aissa, Opportunistic cooperative communication in the presence of co-channel interferences and outdated channel information, *IEEE Commun. Lett.*, 17(10), 1948–1951, 2013.
36. P. S. Bithas, G. P. Efthymoglou and A. G. Kanatas, A cooperative relay selection scheme in V2V communications under interference and outdated CSI, in *2016 IEEE 27th Annual International Symposium on Personal, Indoor, and Mobile Radio Communications* (PIMRC), Valencia, Spain, 2016.
37. V. S. Adamchik and O. I. Marichev, The algorithm for calculating integrals of hypergeometric type functions and its realization in REDUCE system, in *International Conference on Symbolic and Algebraic Computation*, Tokyo, Japan, 1990, pp. 212–224.

Chapter 11

Management of Resources in Virtual Radio Networks

Luis M. Correia, Luisa Caeiro, and Filipe Cardoso

Contents

11.1 Introduction

11.1.1 Initial Considerations

Network virtualization is an abstraction process aiming at separating the logical network functionalities from the underlying physical network resources. It enables the aggregation and provision of the network by combining different physical networks into a single virtual one, or splitting a physical network into multiple virtual ones, which are isolated from each other. Network virtualization has been introduced as a tool for large scale experimental networks, for example, PlanetLab [1] or GENI [2], but it is also proposed as an approach for the future Internet architecture and for the 5th-generation (5G) of mobile communications, [3–5]. By enabling a plurality of diverse network architectures to coexist on a shared physical substrate, virtualization mitigates the ossifying forces in the current architectures and allows the continuous development of innovative network technologies [6].

Network virtualization covers aspects like resource virtualization and slicing. The virtualization of the physical resources consists of implementing multiple instances of a required logical resource on a single machine/node within the same or different set of physical resources allocated to the Virtual Network (VNet), the slice. When compared to wired ones, wireless resources introduce some new challenges to virtualization, due to the specific characteristics of the wireless environment. On one hand, the isolation of traffic cannot be guaranteed due to the scarcity of the radio spectrum, which cannot be overprovisioned, while on the other hand, the radio signal propagation is a very node-specific property, being difficult to control, which has a significant impact on most VNets. Slicing consists of allocating a coherent subset of physical resources to a specific VNet. The slicing process in wireless networks has also some specific issues derived from the characteristics of the medium; the provisioning of slices to multiple VNets with different radio links requires the capability to share radio resources, while at the same time avoiding interference among the different VNets [7].

Sharing radio resources on multiple access schemes has been intensively investigated for wireless systems concerning the separation of the radio links for different end-users of the same system. Still, with the introduction of Mobile Virtual Network Operators (MVNOs), radio resource sharing investigation has been confined to the same system. In the context of network virtualization, the target is to manage radio resources sharing for the VNet's aggregated link, abstracting the involved wireless systems. Therefore, this approach must be extended to the separation of different VNets.

To overcome the scarcity of the radio spectrum, several cognitive radio techniques have also been proposed recently in literature, for example, the smart cognitive radio in [8], which can be explored as additional measures to be integrated in wireless access virtualization context.

The presented approach for wireless access virtualization considers a broader perspective of virtual resources as an aggregated connectivity resource abstracted from a group of radio resources of different technologies, overcoming the limited bandwidth availability of wireless technologies by managing the radio resources across the diverse technologies. Instead of looking at the wireless virtualization from the perspective of the instantiation of virtual machines in the wireless nodes, our view is the virtualization of the wireless access to provide a required capacity to the virtual resource in order to serve its end-users. This approach is then agnostic to the point where the virtual node instantiation takes place, being possible to have the virtual nodes in each physical wireless node, or somewhere in the cloud, requesting virtual access over a given geographic area covered by a set of wireless nodes. It is worthwhile noting that this capacity can be modified on demand without manually changing the configuration of the network. The main target is the management of the radio resources sharing to provide the contracted amount of capacity to a VNet of a given type. Aspects like quality of service (QoS) and fairness are taken into account only at the VNet level, being delegated to physical (PHY)/medium access control (MAC) schedulers who are related to end-users.

The generalization of the problem as a cooperative radio resources management (CoRRM) problem with an additional level of abstraction, the virtual RRM level, allows following an approach of integration of the several levels of RRM, which needs to be adapted, but that actively participates in the process to achieve the main target of provision of the contracted level of service for all MVNOs operating over the common infrastructure. Naturally, the added virtual RRM level needs to assume the coordination role of all the underlying RRM levels, as it is aware of VNets requirements and has the responsibility to satisfy them. Still, the specific algorithms to implement the needed functionality at underlying RRM levels can evolve without overthrowing the outlined approach.

The next sections are organized as follows. In the Sections 11.1.2 and 11.1.3, a brief overview of the relevant aspects of RRM and CoRRM is provided, and radio resource sharing and wireless-virtualization-related work is presented and discussed. In Section 11.2, the network architecture, the model description, the strategies and algorithms for VNet radio resource allocation are presented, and the metrics for evaluation are defined. The analysis of the most relevant results is provided in Section 11.3; finally, conclusions are drawn in Section 11.4.

11.1.2 RRM and Cooperative RRM

One of the objectives of a network operator is to deploy a network that is able to support its customers with the required QoS. Focusing on the radio component, the output of radio network planning should be the provision of radio resource units (RUs) along the service area, by means of a certain radio network topology and a given configuration of the cell sites [9]. However, the amount of RUs to be provisioned varies with service penetration and usage profile, which change in time and space. The most basic way to overcome these issues is by means of network over dimensioning, and RUs overprovision, in order to guarantee QoS to end-users, but, radio resources are limited, and this is not a cost-efficient alternative. The challenge is to be able to provide the desired QoS level with minimum resources, therefore, minimizing operator's investment while meeting network design requirements.

Wireless communications are dynamic in nature due to several varying conditions, including propagation, traffic generation, and interference, among others. Hence, the management of the provisioned RUs should be also dynamic, in order to maintain end-users' QoS. RRM allocates and manages the RUs provided by the radio network.

RRM functions are responsible for taking decisions regarding the setting of different parameters influencing air-interface behavior. The overall behavior of the air interface at any given time results from the decisions taken by different RRM functions. However, consistency needs to be ensured among the different actions that will be undertaken by the different functions and mechanisms to solve conflicts deriving from contradictory actions/reactions. The correct design of RRM functions considers that some functionalities rely on actions/reactions of other functionalities to achieve a global performance.

RRM functions gather information and measurements related to the general radio environment and QoS. This can include signal-to-noise ratio (SNR), throughput, delay of radio bearers, handover, and admission statistics, and technology-dependent values, such as, channel allocation, orthogonal coding, and intra- and inter-cell interference values. Different RRM functions target different radio interface elements and effects, hence, they can be classified according to the time scales they use to be activated and executed. A set of RRM functions with the corresponding typical time scales between consecutive activations of the different algorithms are:

- Inner-loop power control, for example, 1 slot (less than 1 ms) in CDMA
- Packet scheduling and MAC algorithms, in around 1 frame
- Admission control, handover, congestion control, outer-loop power control in CDMA transmission, and from tens to thousands of frames

The most relevant RRM functions in the context of wireless access virtualization are handover, admission control, and scheduling algorithms.

Nowadays, mobile communications networks are composed of various types of radio access technologies (RATs) that constitute a global heterogeneous wireless network. New RATs may appear in future generation mobile communications, namely 5G, enforcing the need for cooperation among them, in order to provide users the best connectivity anytime and anywhere. The heterogeneous wireless networks concept is intended to propose a flexible and open architecture for a large variety of different wireless access technologies, for applications and services with different QoS demands, and different protocols. The main goal is to make the heterogeneous network transparent to users, a secondary one being to design an architecture that is independent of the wireless access technology.

In order to accomplish these objectives, and to optimize the global radio resources utilization, cooperation among the specific RRMs of each air-interface technology is needed. The complementary characteristics of the different RATs allow achieving a more efficient use of the overall resources with CoRRM, rather than with the usage of the various RRMs independently, the so-called trunking gain. CoRRM must take into consideration the overall resources in all available RATs, and dynamically select the best RAT, to guarantee, at each moment, the most efficient use of the available radio resources. A vertical handover procedure must be considered to enable a number of necessary features: avoiding disconnections due to lack of coverage in the current RAT, avoiding blocking due to overload in the current RAT, improvement of QoS by changing RAT, and supporting user's and operator's preferences in terms of usage or load balance among RATs. Inter-RRM signaling among RATs should also be required, in order to transfer information among RRM entities upon which resource allocation and admission control decisions can be taken.

A number of architectures and algorithms to implement CoRRM have been studied and proposed in the past few years: A Common RRM (CRRM) approach was proposed by 3GPP to enable the cooperation in between UMTS and GSM [9]; joint RRM (JRRM) was introduced by the European IST-SCOUT project for inter-working between high performance local area network (HIPERLAN/2) and UMTS [10]; a multilayered RRM scheme was introduced by the European IST-MIND project [11] for the cooperation among various RATs [12]; the European IST ambient networks (AN) project [13] defined the multiradio access (MRA) architecture; although not being a strict CoRRM architecture, the 3GPP system architecture evolution (SAE) integrates different access technologies into a common packet core network, allowing for inter-system handover [14]; more recently, the European IST mobile cloud networking (MCN) project exploited cloud computing as infrastructure for future mobile network deployment and operation. The proposed architecture allows the integrated operation of several radio access technologies, mobile core networks, as well as data centers [15].

11.1.3 Radio Resource Sharing and Wireless Virtualization

In the current mobile communications marketplace, functionalities that enable various forms of network sharing are becoming more and more important. Wireless network sharing is a way for operators to share the heavy deployment costs for mobile networks, not only during the launching phase, but also during the operation phase, through the optimization of resources utilization.

Sharing is an important topic for Long-Term Evolution (LTE), standards on network sharing being already in place [16]. Two architectures have been identified to be supported: gateway core network and multioperator core network. In the former, besides sharing the radio access network nodes, core network operators also share the core network nodes, while in the latter, multiple core network nodes operated by different operators are connected to the same radio access network nodes. In both architectures, the radio access network is shared; still the standard does not specify how capacity is shared among the several core network operators competing for radio access. Besides national roaming, in which a standard roaming agreement is established among operators, or the passive sharing, where only sites are shared, an active sharing is needed to support those network sharing architectures. In active sharing, the radio access network is common to several MVNOs, and these networks are shared among them. One proposal for radio access network sharing is made in [17], taking spectrum usage, QoS, and capacity sharing into account, among other aspects. Although it is a wide-range and very interesting approach for radio access network sharing, the main drawbacks are the direct mapping of the amount of radio resources reservation for capacity provision and the static way of configuration, at the eNodeB level, of the several strategies proposed for capacity sharing.

In the previous presented cases, operators are forced to use similar network functions, as defined by 3G specifications, hence, the possibility of having different multiple VNets with their own functions and communication protocols, isolated from each other, cannot be achieved. Still, without having an integrated perspective relative to multiple RATs, the abstraction of the wireless access is only partially made, avoiding one to take advantage of all available wireless infrastructures. Furthermore, the several models proposed for radio resources sharing are not based on capacity demand, the allocation of radio resources being more or less fixed and not dynamically adapted to the network state, in order to satisfy the requested capacity. This may lead to situations in which the VNets (or rather the logical networks, as the notion of VNet is not present) are running out of contract, denying service to their end-users even when some radio resources are available.

Wireless virtualization can introduce some additional promising aspects relative to wireless network sharing; in fact, besides the sharing of physical resources, network virtualization main targets are the possibility of running simultaneously different network protocols over multiple VNets, which are isolated from each other and with independent management functions. In this way, by setting the VNet type of service with adequate requirements, for example, minimum data rate and/or maximum delay, the VNets may satisfy their end-users' QoS. It should be noted that the amount of radio resources is fixed and capacity is not directly increased through network virtualization. However, by sharing the radio resources network, providers can optimize their utilization, and by managing their allocation to different virtual, operators can provide the requested capacity and QoS at each moment.

Wireless virtualization for specific wireless and mobile networks has been more recently addressed in literature, being an important topic nowadays for 5G as it is demonstrated by the presented virtualization strategies of radio access network [18], in addition to advanced computational platforms, such as cloud computing.

A framework for the efficient radio resources sharing without interference among different virtual radio networks is presented in [7]. In this work, a multiple access scheme to allocate radio resources to each VNet is proposed, establishing a virtual radio resource for each particular VNet. Within each VNet, a further multiple access scheme is then applied to distribute the resources of the VNet to the different end-users of that VNet.

In [19], the authors chose LTE as a case study to extend network virtualization into the wireless component. Their proposal is to add a hypervisor to the LTE eNodeB, in order to map the scheduling of physical resources onto virtual ones. Each operator expresses its contract in terms of the number of radio resources, in this case the physical resource blocks (PRBs), and is responsible for the estimation of the amount of these blocks needed for a time interval.

WiMAX [20] addresses the challenges of resources virtualization by proposing an architecture that enables shared use by multiple independent slice users, for example, MVNOs, each with possibly distinct flow types and network layer protocols. The design and implementation of a network virtualization substrate (NVS) for effective virtualization of wireless resources in cellular networks are presented in [21]. NVS introduces a slice scheduler that allows the existence of slices with bandwidth-based and resource-based reservations, simultaneously; it also includes a generic framework for efficiently enabling customized flow scheduling within the base station (BS) on a per-slice basis.

Concerning wireless local area networks (WLANs), the SplitAP architecture is proposed in [22]. The problem of sharing uplink airtime across groups of users is addressed by extending the idea of network virtualization, allowing the deployment of different algorithms for enforcing uplink airtime fairness across different client groups. A different approach to support full WLAN functionalities inside virtual machines is described in [23], named virtual WiFi. Each virtual machine establishes its own connection with self-supplied credentials, and multiple separate WLAN connections are supported through one physical WLAN network interface.

The majority of these approaches mainly address wireless resources virtualization, which is not the focus here, and only some of them tackle the management of radio resources to be shared among the several virtual resources. However, in these approaches, the assignment of radio resources to VNet end-users is handled within one physical resource in which the virtual resources are instantiated. Still, besides NVS, they do not address the allocation of radio resources based on the capacity required to the virtual resources, but based on a required amount of radio resources, which may perform differently according to the wireless medium conditions, possibly not providing the requested capacity.

Rather, the approach presented in what follows is based on an algorithm that manages the allocation of capacity to the virtual resources, adapting the allocation of radio resources to the wireless medium conditions, that is, the amount of radio resources allocated to the virtual resource is not an issue, as the amount of contracted capacity is provided to the virtual resource. Furthermore, instead of considering the sharing of radio resources within one physical resource, a set of physical resources from several RATs should be considered, independently abstracted by their RUs, for example, resource blocks in LTE, codes in UMTS, or time-slots in GSM, to the point where the virtual resources are instantiated [24]. In fact, as mentioned earlier, besides resource virtualization, a main issue in wireless virtualization is slicing, that is, how to share the available wireless resources to extend VNets to the wireless access, maintaining isolation and allowing the use of different communication protocols per VNet. This leads to the question: Where should the boundary for the instantiation of virtual nodes in the wireless access be? One can think about several alternatives for the instantiation of a virtual node: in each wireless node, in one node within a set of wireless nodes serving the same area, or even in a core node the wireless component being totally abstracted. However, independently of the place where the virtual node is instantiated, the key problem is how to manage the allocation of radio resources to the virtual resources in order to satisfy the amount of contracted capacity.

11.2 Models and Algorithms for Wireless Access Virtualization

11.2.1 Network Architecture

In a traditional business model, the end-user has a business relationship with a network operator and can only connect to the network via its infrastructure, if roaming is not considered. The network operator can be the owner of the infrastructure or also the physical network operator, providing added-value services to end-users.

MVNOs have appeared in the mobile telecommunications market many years ago, using a given infrastructure to provide specific services to their clients. In this case, end-users have a contract with the MVNO, which in turn has a contractual relationship with a network operator. Hence, MVNO clients can connect only via the infrastructure of the network operator with whom they have a contract.

With the introduction of network virtualization, new business roles are foreseen. In particular, the roles of the traditional network operator have been split into the virtual network operator, the MVNO within this context, which will most probably use several infrastructures, and the infrastructure providers (InP), which in turn will be shared by various MVNOs. In this case, MVNO clients may connect via the infrastructure of any InP providing network resources for the VNet in the area where they are located.

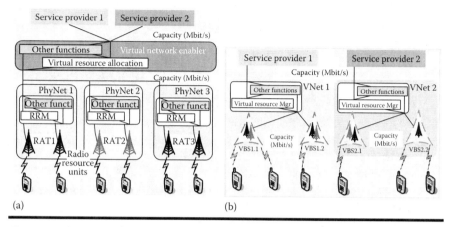

Figure 11.1 Network architecture: (a) physical view and (b) logical view.

The network architecture, Figure 11.1, refers to the virtualization of the wireless access as an integral part of VNets, being based on a generic network virtualization environment, for example, [3]. Hence, the considered network environment envisages the existence of multiple VNets created by a VNet enabler. Service providers, which can also accumulate the MVNO role, will use these VNets, formed on demand to satisfy their service requirements, in order to deliver services to their customers. In this way, the physical infrastructure owned by InPs is shared among several VNets, providing services with different requirements, and to multiple service providers.

The physical network architecture is depicted in Figure 11.1a. The following elements are represented:

- *Service provider*: Entity that provides/delivers services/applications over the (virtual) network for a set of clients.
- *VNet enabler*: Entity enabling network virtualization. Several functions are under the responsibility of this entity, for example, receiving and processing the requirements for virtualization, negotiation with InPs to use their physical resources for the provision of capacity, and VNets' creation. Although the operation and maintenance of VNets can be performed by an external entity, it is considered within this block for simplicity. Virtual resource allocation, besides other type of resource allocation (e.g., computational resources), includes the algorithm that manages the allocation of radio resources from the physical networks to the virtual resources created by the virtualization process. As one is considering different wireless access technologies, with different physical meanings for the RUs, a level of abstraction is added, enabling a common approach to manage all radio resources. It is considered that each wireless link is generically composed of RUs, which varies in number and capacity according to the wireless technology involved.

■ *Physical network* (*PhyNet*): Set of physical resources from each RAT, for example, the BSs or other nodes of the wireless system architecture, owned by InPs. The physical resources should allow instantiation of virtual resources, that is, should be capable of sharing their physical components. RRM is the set of specific mechanisms of each physical network performing the well-known radio-management functionalities, for example, admission control, scheduling, radio resource allocation, and handover, among others.

For simplicity, service providers' requests are illustrated as being only a capacity demand, although their requirements are not limited to it, for example, it can be also delay or reliability. Based on the request for capacity and infrastructure availability, the VNet enabler defines a VNet adequate for that service delivery, performing the virtual resource allocation. Virtual resources composing the VNet are then created on top of the network infrastructure, by sharing the available physical network capacity.

Within this scope, only the virtual resources deployed over wireless infrastructures are considered, from now on designated as virtual base stations (VBSs). It is worthwhile to note that though the instantiation of a VBS involves the virtualization of processing and memory resources, in order to run the inherent functions of a physical BS, within the scope of this work one considers only the radio part of the VBS, that is, the set of radio resources allocated to the VBSs. In this sense, VBSs are assumed to be implemented on top of a group of BSs from heterogeneous wireless networks, serving a given geographic area over which the capacity demand is requested. This group of BSs serving a delimited geographical area is designated as a cluster. The requested capacity may be split over one or several VBSs by the virtual resource allocation function. In case several VBSs coexist, a partial capacity requirement is established for each one.

VBSs' capacity is provided by the allocation of RUs over the several BSs deployed in the cluster, being coordinated across the cluster to satisfy the VBSs' contracted capacity. Although the number of RUs per BS can be different for BSs of the same RAT, in the scenario under evaluation it was assumed that all the BSs of the same RAT have the same number of RUs, for the sake of simplicity. The RU is the minimum radio resource unit that can be assigned to an end-user in a physical BS, hence, depending on the RAT, for example, a time-slot in TDMA (time division multiple access), a code in CDMA (code division multiple access), or a resource block (group of sub-carriers) in OFDMA (orthogonal frequency division multiple access). The RU is then abstracted by its capabilities, namely, the achievable data rates according to the modulation and coding schemes allowed in each RAT.

Figure 11.1b depicts the VNet operators' view of the network, the logical view. The following elements are considered:

■ *Virtual base station* (*VBS*): Virtual resources created to provide the capacity required by a service provider over a given geographical area. VBSs capacity is collected from the available radio resources of all BSs in that area. The VBS can implement the functionality of a physical BS being part of one virtual network.

- *Virtual resource management*: Process that manages the utilization of VBS's capacity, enabling to perform resource management functions, for example, to adapt the capacity required to the VBSs utilization or to an increase of clients in the area. Scheduling end-users within the VBS is also one of its responsibilities.
- *Virtual networks* (*VNets*): Entity characterized by the type of contract, the amount of required capacity and other kind of requirements, like location and topology, among others. They are composed of several virtual resources, which are the VBSs in the scope of this work, that is, virtual resources sharing wireless physical infrastructures.

MVNOs are the players that manage and operate VNets, including their VBSs, to satisfy service providers' requests. They know only the virtual resources that are part of the VNet with their associated capacity, the set of physical resources being hidden from them.

In order to make use of a service, the end-user connects physically to the BSs, Figure 11.1a, but the connection to the VNet providing the service is logically made via a VBS, through a virtual link, as illustrated in Figure 11.1b. The physical link is the group of RUs allocated to the end-user, whereas the virtual link is the capacity, in bit/s, allocated by the VBS. The mapping between the physical and virtual links is essential to compute the VBS-aggregated capacity, allowing monitoring of the contract satisfaction, and, consequently, the trustiness between the MVNO and InPs.

Several types of VNets may be considered to represent different service level agreements for the VNets. However, for the sake of proof of concept only two types of VNets are considered: GRT (guaranteed) and BE (best effort) VNets. The former ensures that the requested constraints, capacity in this case, will not be violated at any time. The latter provides a best-effort service, that is, no guarantees at all are given when data will be delivered.

11.2.2 Model Description

An analytical model is presented next, to obtain VNet's capacity for the network architecture. From the physical viewpoint, a cluster with a set of BSs from various RATs is considered as the small management unit in terms of virtual radio resource allocation (VRRA):

$$\mathrm{BS}_{\mathrm{RAT}_i}^{\mathrm{Cl}} = \left\{ \mathrm{BS}_1^{\mathrm{RAT}_i}, ..., \mathrm{BS}_{N_{\mathrm{BS}}^{\mathrm{RAT}_i}}^{\mathrm{RAT}_i} \right\} \tag{11.1}$$

where:

$N_{\mathrm{BS}}^{\mathrm{RAT}_i}$ is the total number of BSs of RATi in the cluster

N_{RAT} is the total number of RATs in the cluster, defined as

$$\mathrm{RAT}^{\mathrm{Cl}} = \left\{ \mathrm{RAT}_1, ..., \mathrm{RAT}_{N_{\mathrm{RAT}}} \right\} \tag{11.2}$$

The BS characterization is made from the viewpoint of the RAT it belongs to and the relation to the end-users connected through it. Concerning the RAT, besides the number of RUs specific of that RAT, the BS is characterized by its maximum capacity or *Maximum BS Data Rate*, that is, the total capacity (bit/s) provided by the RUs of any given BS, from now on designated as data rate, when the most favorable modulation and coding scheme is applied. Hence, the *Maximum BS Data Rate* is given by

$$R_{\max}^{\mathrm{BS}j}\left[\mathrm{bit/s}\right] = N_{\mathrm{RU}}^{\mathrm{RAT}_i} \cdot R_{\mathrm{RU}_{\max}}^{\mathrm{RAT}_i}\left[\mathrm{bit/s}\right] \tag{11.3}$$

where:

$N_{\mathrm{RU}}^{\mathrm{RAT}_i}$ is the total number of RUs per BS of RATi

$R_{\mathrm{RU}_{\max}}^{\mathrm{RAT}_i}$ is the data rate of the RU of RATi can provide, if the most favorable modulation and coding scheme is applied

Regarding the relationship between BS and end-users, the BS is characterized by the *BS Serving Data Rate* computed from:

$$R_{\mathrm{serv}}^{\mathrm{BS}j}\left[\mathrm{bit/s}\right] = \sum_{n=1}^{N_{\mathrm{EU}}^{\mathrm{BS}j}} R_{\mathrm{serv}\,n}^{\mathrm{EU}}\left[\mathrm{bit/s}\right] \tag{11.4}$$

where:

$N_{\mathrm{EU}}^{\mathrm{BS}j}$ is the number of end-users connected to BS$_j$

$R_{\mathrm{serv}_n}^{\mathrm{EU}}$ is the *End-user Served Data Rate*, that is, the data rate with which end-user n is being served; it depends on the number of RUs assigned to the end-user and the data rate the RUs are achieving, being obtained by

$$R_{\mathrm{serv}_n}^{\mathrm{EU}}\left[\mathrm{bit/s}\right] = N_{\mathrm{RU}}^{\mathrm{EU}_n} \cdot R_{\mathrm{MCSm}}\left[\mathrm{bit/s}\right] \tag{11.5}$$

where:

$N_{\mathrm{RU}}^{\mathrm{EU}_n}$ is the number of RUs assigned to end-user n

R_{MCSm} is the data rate achieved by each RU assigned to the end-user, according to the applied modulation and coding scheme m (for the sake of simplicity, it is considered that all assigned RUs achieve the same data rate)

It is assumed that the distribution of end-users among BSs is uniform. The cluster, being a group of BSs, can inherit the BS characterization, that is, it can be described by its maximum capacity and serving data rate. Hence, two other parameters have been defined: *Maximum Cluster Data Rate* and *Cluster Serving Data Rate*. The *Maximum Cluster Data Rate* is the maximum capacity of the cluster, that is, the sum of the *Maximum BS Data Rate* of all the BSs of that cluster:

$$R_{max}^{Cl}\left[bit/s\right] = \sum_{n=1}^{N_{BS}^{Cl}} R_{max}^{BS_n}\left[bit/s\right] \qquad (11.6)$$

where:

$R_{max}^{BS_n}$ is the maximum data rate for BS_n

N_{BS}^{Cl} is the total number of BSs within the cluster

The *Cluster Serving Data Rate* is the sum of the serving data rates of all BSs composing the cluster, being computed from:

$$R_{serv}^{Cl}\left[bit/s\right] = \sum_{n=1}^{N_{EU}^{Cl}} R_{serv\,n}^{EU}\left[bit/s\right] \qquad (11.7)$$

where N_{EU}^{Cl} is the total number of end-users in the cluster.

Concerning the virtual network, several VBSs from various VNets may exist in the cluster, being identified by

$$VBS^{Cl} = \left\{VBS_1,..., VBS_{N_{VBS}^{Cl}}\right\} \qquad (11.8)$$

where N_{VBS}^{Cl} is the number of VBSs in the cluster.

The VBS can be defined according to the contracted capacity and the capacity used by end-users. Three VBS data-rate-related parameters are identified:

- *Minimum contracted data rate*: $R_{min}^{VBS_j}\left[bit/s\right]$ is the data rate contracted by the MVNO as the minimum value InPs should provide when requested.
- *Reference contracted data rate*: $R_{ref}^{VBS_j}\left[bit/s\right]$ is the data rate contracted by the MVNO as a reference value to be provided by InPs to the MVNO.
- *VBS serving data rate*: $R_{serv}^{VBS_j}\left[bit/s\right]$ is the data rate provided to all end-users connected to the VBS:

$$R_{serv}^{VBS_j}\left[bit/s\right] = \sum_{n=1}^{N_{EU}^{VBS_j}} R_{serv\,n}^{EU}\left[bit/s\right] \qquad (11.9)$$

where $N_{EU}^{VBS_j}$ is the total number of end-users in VBS_j.

To depict the relation between MVNOs and InPs, which allows evaluating the established serving level agreement (SLA), two parameters have been defined:

- *Penalty*: p is the amount the InP should pay to the MVNO when the VBS is operating out of contract, that is, when SLAs are not satisfied
- *Time frame*: Δt_{TF} is the interval of time of the same order of magnitude of the time scale defined for cooperative radio resource management algorithms

Concerning the description of the VBSs according to the two types considered in this work, GRT and BE, the GRT VBS, VBS_{GRT}, is characterized by a *Minimum Contracted Data Rate*, $R_{\text{min}}^{\text{VBS}_i}$ [bit/s], which should be guaranteed for all time frames, and a *Penalty* computed as the total number of time frames the VBS is out of contract.

$$p = \sum_{i=1}^{N_{\text{TF}}} p_i \tag{11.10}$$

where:

N_{TF} is the total number of time frames in the observation interval

p_i is the penalty in time frame i, according to:

$$\begin{cases} p_i = 0, & R_{\text{serv}}^{\text{VBS}_j} \geq R_{\text{min}}^{\text{VBS}_j} \text{ in } \Delta t_{\text{TF}_i} \\ p_i = 1, & R_{\text{serv}}^{\text{VBS}_j} < R_{\text{min}}^{\text{VBS}_j} \text{ in } \Delta t_{\text{TF}_i} \end{cases} \tag{11.11}$$

The BE VBS, VBS_{BE}, is defined by a *Reference Contracted Data Rate*, $R_{\text{ref}}^{\text{VBS}_i}$ [bit/s], which is indicative and should be followed in a percentage $P_{R_{\text{ref}}}$ of the total number of time frames, that is, the minimum fraction of time frames InPs should make available the reference contracted data rate to the MVNO in order to avoid penalties. An associated *Penalty* accounts for the number of time frames the VBS is out of contract above $P_{R_{\text{ref}}}$ percentage of the total.

$$p = \sum_{i=1}^{N_{\text{TF}}} p_i - P_{R_{\text{ref}}} \cdot N_{\text{TF}} \tag{11.12}$$

subject to

$$\frac{\sum_{i=1}^{N_{\text{TF}}} p_i}{N_{\text{TF}}} \geq P_{R_{\text{ref}}} \tag{11.13}$$

where p_i is the penalty in time frame i, according to:

$$\begin{cases} p_i = 0, & R_{\text{serv}}^{\text{VBS}_j} \geq R_{\text{ref}}^{\text{VBS}_j} \text{ in } \Delta t_{\text{TF}_i} \\ p_i = 1, & R_{\text{serv}}^{\text{VBS}_j} < R_{\text{ref}}^{\text{VBS}_j} \text{ in } \Delta t_{\text{TF}_i} \end{cases} \tag{11.14}$$

In order to account for the global profit, the targets of maximizing the serving data rate of the cluster, Equation 11.7,

$$\max\left(R_{\text{serv}}^{\text{Cl}} \text{ [bit/s]}\right) = \max\left(\sum_{n=1}^{N_{\text{EU}}^{\text{Cl}}} R_{\text{serv}_n}^{\text{EU}} \text{ [bit/s]}\right) \tag{11.15}$$

and minimizing the penalties, Equations 11.10 and 11.12,

$$\min p = \min\left(\sum_{n=1}^{N_{\text{VBS}}^{\text{Cl}}} p_n\right) \tag{11.16}$$

are considered through an adequate allocation of RUs to the VBSs. The former considers that MVNOs pay the service based on used capacity. The latter assumes that an amount of money must be paid back to the MVNO if the contract is not fulfilled.

11.2.3 Strategies and Algorithms

The main target of VRRA is to provide the required capacity to VBSs, optimizing radio resources utilization. The OnDemandVRRA algorithm presented here is a heuristic one, which manages the allocation of RUs among VBSs only when they are requested by VNet end-users. The management of radio resources allocation from VBSs is coordinated, to provide different levels of service to the various MVNOs or service providers. This is achieved by taking the variability of the wireless medium and the diversity of the existing RATs into account.

It is important to ensure consistency among the decisions taken at the different levels of RRM, namely, intra-RAT (RRM), inter-RAT (CoRRM), and among VNets, to achieve an overall coherent behavior. Given that at the VNet level one should have the perspective of the several RATs, the OnDemandVRRA algorithm takes decisions at a time scale that is defined for CoRRM. This time scale is taken as the major common denominator of all RATs, for the sake of simplicity.

The OnDemandVRRA is responsible for dynamically (re)allocating RUs, satisfying the *Minimum Contracted Data Rate* for GRT VNets Equation 11.17, and aiming at the *Reference Contracted Data Rate* for BE VNets Equation 11.18:

$$R_{\text{serv}}^{\text{VBS}_i}\left[\text{bit}/\text{s}\right] \geq R_{\text{min}}^{\text{VBS}_i}\left[\text{bit}/\text{s}\right], \ \forall \ \text{VBS}_i \equiv \text{VBS}_{\text{GRT}} \tag{11.17}$$

$$\min\left(R_{\text{ref}}^{\text{VBS}_j} - R_{\text{serv}}^{\text{VBS}_j}\right)\left[\text{bit}/\text{s}\right], \ R_{\text{ref}}^{\text{VBS}_j}\left[\text{bit}/\text{s}\right] > R_{\text{serv}}^{\text{VBS}_j}\left[\text{bit}/\text{s}\right], \ \forall \ \text{VBS}_j \in \text{VBS}_{\text{BE}} \tag{11.18}$$

subject to

$$R_{\text{req}}^{\text{VBS}_i}\left[\text{bit}/\text{s}\right] \geq R_{\text{min}}^{\text{VBS}_i}\left[\text{bit}/\text{s}\right], \ \forall \ \text{VBS}_i \equiv \text{VBS}_{\text{GRT}} \tag{11.19}$$

$$\sum_{n=1}^{N_{\text{VBS}}} R_{\text{serv}}^{\text{VBS}_n}\left[\text{bit}/\text{s}\right] < \sum_{n=1}^{N_{\text{RAT}}} N_{\text{RU}}^{\text{RAT}_n} \cdot R_{\text{RU}_{\max}}^{\text{RAT}_n}\left[\text{bit}/\text{s}\right] \tag{11.20}$$

where $R_{\text{req}}^{\text{VBS}_i}\,[\text{bit/s}]$ is the *VBS Requested Data Rate*, that is, the total data rate requested by end-users in VBSi given by

$$R_{\text{req}}^{\text{VBS}_i}\,[\text{bit/s}] = \sum_{n=1}^{N_{\text{EU}}^{\text{VBS}_i}} R_{\text{req}\,n}^{\text{EU}}\,[\text{bit/s}] \tag{11.21}$$

where $R_{\text{req}\,n}^{\text{EU}}$ is the data rate requested by end-user n.

One should note that if a GRT VBS is not using all the contracted capacity, its end-users must be served with the capacity they are requesting, that is, if a given GRT VBS serving data rate is below the contracted capacity, the RUs allocated to its end-users must correspond to the data rate requested by them.

The optimization of radio resources utilization is indirectly achieved by allowing the allocation of RUs to any VBS, after all other VBSs in the cluster have their contracted capacity satisfied. This means that all available RUs in the cluster are allocated to any VBS as long as they have been requested, avoiding the waste of radio resources, for example, due to a previous allocation to VBSs that did not use them. In fact, one is not directly dealing with the scheduling of the radio resources to the end-users, but rather indirectly, by enforcing the decisions taken from the cluster viewpoint to be considered by RRM and CoRRM algorithms.

To cope with these objectives, the OnDemandVRRA algorithm is supported by a VNet priority scheme and a data-rate reduction strategy, besides the access selection mechanism.

Concerning access selection, end-users are connected to the different VBSs according to the requested service and their contract with the MVNO(s). The physical connection is established through one of the BSs of existing RATs in the coverage area, based on a list of preferences related to the requested service, the available capacity, and the strategy defined for resource evaluation. This strategy, for example, minimum load, minimum cost, and/or minimum energy state, is based on the BS cost, where several key performance indicators are weighted.

The VNet priority scheme running at cluster level assumes a coordination role and enables to set differentiated end-users according to the type of VBS and the *VBS Serving Data Rate*. VBSs are initialized to be handled with priority, all BSs in the cluster being informed of this, to activate the data-rate reduction process. When the *Minimum Contracted Data Rate* is reached, that is, the *VBS Serving Data Rate*≥*VBS Contracted Data Rate*, the priority to be given to end-users who wish to connect to this VBS is deactivated. This priority scheme based on the *VBS Serving Data Rate*, allows one to implement a data-rate reduction strategy whenever the GRT VBSs have priority, preventing starvation on BE VBSs when the contracted data rate in GRT VBSs is reached.

The data-rate reduction strategy is essential to compensate possible end-user data-rate decrease due to degradation of wireless medium conditions. It is applied to services with a minimum required data rate, for example, video streaming, while the VBS operates within the contracted capacity, that is, when the VBS priority is activated. Although another data-rate reduction strategy could be applied, the adopted data-rate reduction strategy is as follows. Whenever the VBS priority is activated for a GRT VBS, and the end-user tries to connect to a BS in which there are not enough RUs for his/her service, BE end-users connected to the BS are reduced according to:

- The out-of-contract rate of the VBS they are connected to, in order to mini-mize penalties, introducing some degree of fairness among BE VBSs
- The QoS priority class of the performed service [25]; end-users performing services with lower priority being the first to be reduced
- Their SINR, end-users with lower SINR being reduced first to allow optimiz-ing radio resource utilization

Still, if there are not enough RUs to reach the requested data rate, the evaluation of colocated BSs is performed, in order to select the one with enough RUs available and with the minimum cost to handover end-users.

A brief description of the end-users handling process is presented next. After the access is selected, the assignment of RUs to end-users is constrained by the level of contract fulfillment of all the VBSs in the cluster, which determine the RUs avail-ability. The state of operation of the VBS in relation to the contracted, that is, the priority of the VBS, is then checked to allow serving end-users of VBSs operating within contract. If it is the case, that is, the VBS has priority, the data-rate reduc-tion process is evoked in order to find enough RUs to assign to the end-user without violating the established contracts with other VBSs.

It is worthwhile to note the difference between OnDemandVRRA and the radio resource allocation and adaptation mechanisms at the PHY/MAC level, which deal with end-user performance instead of the VBS one. OnDemandVRRA acts mainly as a coordinator that enforces its VRRA decisions onto RRM func-tions, namely, scheduling and admission control, for the wireless systems within the cluster. However, these RRM functions should be adapted to receive these set-tings and extended to deal with it, for example, through the set of functions pro-posed for OnDemand end-users' handling.

11.2.4 Metrics for Evaluation

The metrics used for assessment are defined as follows.

The *Average Serving Data Rate* allows evaluating the algorithm ability to allocate the adequate quantity of RUs to the VBS, in order to satisfy the VBS-contracted data rate. It is the average of *VBS Serving Data Rate* over the total number of time frames in the observation time interval:

$$\overline{R_{\text{serv}}^{\text{VBS}}}\left[\text{bit/s}\right] = \frac{\displaystyle\sum_{n=1}^{N_{\text{TF}}} R_{\text{serv}_n}^{\text{VBS}}\left[\text{bit/s}\right]}{N_{\text{TF}}} \tag{11.22}$$

where $R_{\text{serv}_n}^{\text{VBS}}$ is the VBS serving data rate in time frame n.

The *Average Cluster Data Rate* is a metric to evaluate the performance of the overall cluster, allowing one to observe the impact of using VRRA algorithms for different use cases. It is defined as the average of the *Cluster Serving Data Rate* over the total number of time frames in the observation time interval:

$$\overline{R_{\text{serv}}^{\text{CI}}}\left[\text{bit/s}\right] = \frac{\displaystyle\sum_{n=1}^{N_{\text{TF}}} R_{\text{serv}_n}^{\text{CI}}\left[\text{bit/s}\right]}{N_{\text{TF}}} \tag{11.23}$$

where $R_{\text{serv}_n}^{\text{CI}}$ is the cluster serving data rate for time frame n.

The *Average Cluster Utilization* is a measure of the RUs utilization within the cluster. This metric should be analyzed together with the *Average Cluster Data Rate*, as the efficiency of the use of the RUs is as important as maximizing their use. It is defined as the average of the *Cluster Utilization* over the total number of time frames in the observation time interval:

$$\overline{\eta_{\text{CI}}} = \frac{\displaystyle\sum_{n=1}^{N_{\text{TF}}} \eta_{\text{CI}_n}}{N_{\text{TF}}} \tag{11.24}$$

where η_{CI_n} is the ratio between the maximum data rate corresponding to the RUs occupied by end-users and the maximum data rate the cluster can provide in time frame n, given by

$$\eta_{\text{CI}_n} = \frac{\displaystyle\sum_{i=1}^{N_{\text{RAT}}}\left(N_{\text{RU}_{\text{occ}}\,n}^{\text{RAT}} \cdot R_{\text{RU}_{\text{max}}}^{\text{RAT}_i}\left[\text{bit/s}\right]\right)}{\displaystyle\sum_{i=1}^{N_{\text{RAT}}}\left(N_{\text{RU}}^{\text{RAT}} \cdot R_{\text{RU}_{\text{max}}}^{\text{RAT}_i}\left[\text{bit/s}\right]\right)} \tag{11.25}$$

where $N_{\text{RU}_{\text{occ}}\,n}^{\text{RAT}}$ is the number of RUs occupied by end-users in each RAT in time frame n, being subjected to:

$$N_{\text{RU}_{\text{occ}}}^{\text{RAT}_i} \leq N_{\text{RU}_{\text{max}}}^{\text{RAT}_i}, \quad \forall\,\text{RAT}_i \in \text{RAT}^{\text{CI}} \tag{11.26}$$

It should be highlighted that the metrics defined for evaluation are not directly related to fairness, as it is considered that this is under the responsibility of the MVNOs, while the network operators of each VNet provide fairness among end-users within the same VNet. On the other hand, giving the same priority to VNets of the same type, and operating in a similar state relative to the contracted capacity, implies MAC schedulers to evenly handle end-users of different VNets, given that they are in the same priority level, allowing to provide fairness among VNets.

11.3 Analysis of Results

In order to analyze the proposed VRRA algorithm from different perspectives, the assessment was done by starting from a reference scenario over which several changes were applied, by varying a set of relevant parameters. The reference physical cluster is composed of 2 TDMA, 1 CDMA, 4 OFDMA, and 8 OFDM BSs, Figure 11.2. Two MVNOs are considered, each one providing a different set of services, one is GRT with a minimum guaranteed data rate and another BE with a reference data rate. The total data rate contracted for both VNets is on the average cluster capacity. It is considered that 8,000 end-users are uniformly distributed within the cluster, allowing to depict a situation in which the total data rate requested by end-users is above the maximum cluster data rate. Concerning the service profile, it is assumed that 4% of end-users use VoIP, 35% video, 3% file sharing, and 58% Web derived from the service penetration of mobile data traffic forecast for 2016 made by Cisco [26]. VoIP corresponds to traffic from retail VoIP services and PC-based VoIP, while for video streaming is considered, Web depicts Web traffic (excluding file sharing), and file sharing corresponds to peer-to-peer traffic.

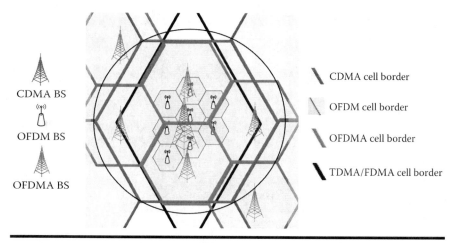

Figure 11.2   Reference physical cluster.

The signal quality of the end-users is changing over time; the modulation and coding scheme (MCS) is being adapted to this variation.

The strategy used for resource evaluation to select the best BS to connect the end-user is the maximum absolute capacity available. Although the number of RUs per BS can be different for BSs of the same RAT, being an input of our model, in the scenario under evaluation, it was assumed that all the BSs of the same RAT have the same amount of RUs for the sake of simplicity. The time frame common to all RATs is taken as 1 s.

In the following, the most relevant results from a set of use cases defined to analyze different perspectives are presented and discussed. Although the results for each use case are not exhaustively presented, further results and analysis can be found in [27].

11.3.1 Non-Virtual versus Virtual Network Operators Model

Two scenarios have been defined to study the benefits that can be achieved with the introduction of virtualization in wireless heterogeneous networks: *Standard* (without virtualization) and *Virtual* (with virtualization). These two scenarios allow comparing current operators versus virtual network operators model.

In the *Standard* scenario, two network providers are considered, one having a contract with 80% of the end-users and the other with 20%. Concerning the *Virtual* scenario, two different use cases were defined, *VNet-Limited* and *VNet-UnLimited*, both having one BE VBS and one GRT VBS. For the *VNet-UnLimited* use case, the BE VBS has a low-reference data rate, meaning that it does not impose any additional constraint to the GRT VBS, because the reference data rate is always served. For the *VNet-Limited* one, the reference data rate is defined according to the expected BE VBS request data rate, being a GRT VBS restriction, as most of the time it cannot be reached due to the maximum data rate of the cluster. For the rest of the parameters, the related reference scenario is used.

Observing the results for the *Average Cluster Data Rate*, $\overline{R_{\text{serv}}^{\text{Cl}}}$, obtained from Equation 11.23, and depicted in Figure 11.3a, one can see that for the *VNet-Limited* use case the cluster achieves the highest value, $\overline{R_{\text{serv}}^{\text{Cl}}} \approx 2.6\,\text{Gbit/s}$. On the other hand, the highest *Average Cluster Utilization*, η_{Cl}, computed from Equation 11.24, is achieved for *VNet-UnLimited*, which denotes a less efficient use of the overall resources than in *VNet-Limited*, as the $\overline{R_{\text{serv}}^{\text{Cl}}}$ for the former is less than the one for the latter, Figure 11.3b. In fact, there are more end-users from GRT VBS entering the network and the RUs may be assigned to end-users in poor wireless performance conditions, because for GRT services, a minimum service data rate should be satisfied. For the *Standard* scenario, the lowest value of η_{Cl} is observed, as there is an additional limitation arising from the NetProv's physical infrastructure partition, avoiding to share the overall RUs within the cluster.

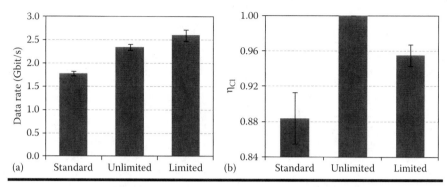

Figure 11.3 (a) Average cluster data rate and (b) average cluster utilization for operators' use cases.

By comparing the wireless access virtualization supported by the proposed OnDemandVRRA with the standard approach, in which there are multiple network operators (each owning part of the physical infrastructure), it can be said that the former allows achieving a better performance of a cluster of BSs from different RATs, enabling the provision of contracted capacity for GRT VNets. It is demonstrated by simulation that, in *Virtual* use cases, the *Average Cluster Serving Data Rate* may increase by approximately 46% and the utilization by 13%. On the other hand, for *Virtual* use cases, the *Average VBS Serving Data Rate* of GRT VBSs is always greater than the contracted minimum, being constrained by the BE VBSs reference data rate, which tends to be followed. For the *Standard* scenario, the values achieved for *Average VBS Serving Data Rate* are the lowest for BE services, but can be greater than in *Virtual* use cases for GRT services. On one hand, this denotes the limitation due to the split of the total cluster capacity by the two operators, and on the other hand, it denotes the uncoordinated allocation of radio resources as end-users are independently handled.

11.3.2 Physical versus Virtual Capacities

Situations in which the total amount of capacity contracted by VNOs, under- and overbooked, are discussed herein. It is considered that an underbooking situation, *Under* use case, occurs when the amount of contracted data rate by all the VBSs instantiated in the cluster is less than the average cluster capacity, that is, the data rate the cluster can provide when the modulation and coding schemes applied to all the RUs within the cluster is between the second and third higher data rates. Two overbooking situations were considered, *GRTOver* and *BEOver* use cases, in which the total contracted data rate is greater than the average cluster capacity. Finally, an *Average* use case is considered to depict the situation when the contracted capacity is near the average cluster capacity.

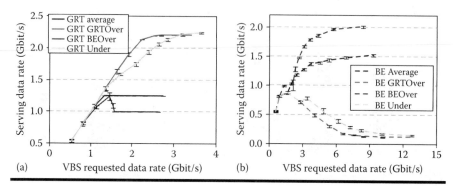

Figure 11.4 **Average VBS serving data rate for average, over and under use cases: (a) GRT VBS and (b) BE VBS.**

For all use cases, the reference scenario is considered, the quantity of end-users increasing from 1,000 to 15,000 to allow simulating situations in which the total data rate requested is below and above the maximum cluster data rate.

When the total capacity is on the average capacity of the cluster, the *Average* use case, both VBSs can achieve the contracted data rate since the GRT VBS is limited, Figure 11.4. However, when the GRT VBS requested data rate is near the contracted one, the data rate of BE end-users is reduced to allow the GRT VBS to reach the contracted data rate and the BE VBS becomes out of contract. It is worth to note that the average data rate per GRT end-user is almost constant, independently of number of end-users in the cluster, because the number of end-users admitted to the GRT VBS is limited by the contracted data rate of both VBSs.

For overbooking situations, *GRTOver* and *BEOver*, the results obtained for the average capacity of the cluster are extremely different depending on the type of VBS that is contracting more capacity.

For the GRT VBS, *GRTOver* use case, the behavior is to some extent similar to *Under*, as the number of GRT end-users is also high, though for diverse reasons. In this case, the priority given to GRT end-users, while the *VBS Serving Data Rate* is lower than the contracted capacity, is high, forcing the data rate reduction of BE end-users, to satisfy the contracted data rate. Concerning the *Average Cluster Data Rate*, and *Average Cluster Utilization*, it can be observed that, like in *Under*, there is an inefficiency in the use of RUs in this case, related to the value for contracted capacity of the GRT VBS, which is about 85% of the average cluster capacity, leading most of the connected end-users to be in GRT VBS. Also because of that, the contracted data rate of a BE VBS has only influence for situations in which both VBSs are heavy loaded and the GRT VBS is already operating on extra capacity requested, removing the priority of GRT end-users. In the analyzed case, due to overbooking, the cluster can never provide the BE VBS contracted capacity and the VBS operates out of contract as soon as the requested data rate is greater than the contracted.

While the BE VBS, *BEOver*, requested data rate is less than the contracted data rate, the GRT VBS achieves *Serving VBS Data Rate* above the minimum contracted capacity, which is forced to be the contracted data rate as soon as the BE VBS requested data rate becomes above its contracted data rate. The BE VBS though does not reach the contracted data rate due to cluster capacity limit is increasingly tending to it. The best RU efficiency for high values of cluster requested data rate is achieved in this use case, that is, the ratio of *Average Cluster Data Rate*, and *Average Cluster Utilization*, is greater than for the other use cases. Although the average data rate per GRT end-user is almost constant, a decrease of about 17% is observed when the *GRT VBS Serving Data Rate* is above the VBS contracted data rate and the BE VBS requested data rate is less than the contracted VBS, because the data rate reduction process is deactivated.

11.3.3 Diverse Combinations of VBSs from Different Types

The main target here is to analyze several combinations of different VNet types created in the physical cluster. The total number of VBSs in the cluster was set to four, the three use cases considered being characterized by the number of VNets of each type: 1GRT_3BE, 2GRT_2BE, and 3GRT_1BE.

The total data rate contracted by each type of VBSs is maintained for all use cases, that is, the total data rate contracted for GRT VBSs has always the same percentage relative to the total data rate contracted within the physical cluster. The *Average* use case was selected, representing one of the most favorable use cases for the total data rate contracted within the average cluster serving one. All the VBSs of the GRT type provide both VoIP and video services, FS, and Web services being provided by BE VBSs.

For all use cases, the reference scenario is considered. This allows depicting the situation in which the total data rate requested, approximately 5.5 Gbit/s on average, is above the average cluster data rate.

From the results obtained by changing the VBSs' type mixing, one can conclude the following. The total *Average VBS Serving Data Rate* of GRT VBSs is basically the same for all use cases and the individual VBS achieved value is always above the *Minimum Contracted Data Rate*, independent of the use case, Figure 11.5. It is verified that the total *Average VBS Serving Data Rate* of GRT VBSs has a maximum deviation among use cases around 0.2%, increasing slightly with the number of GRT VBSs. To achieve this, the OnDemandVRRA algorithm manages the allocation of RUs, causing the reduction of BE end-users and preventing the BE VBSs to get their *Reference Contracted Data Rate*. The total *Average VBS Serving Data Rate* of BE VBSs is under the *Reference Contracted Data Rate* for all use cases, decreasing approximately 1.8% from 1GRT-3BE to 3GRT-1BE. Each GRT *Average VBS Serving Data Rate* is always above the *Minimum Contracted Data Rate* independent of the use case.

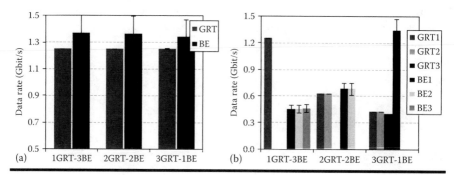

Figure 11.5 *Average VBS serving data rate* for VNet type mixing: (a) aggregated by VBS type and (b) per VBS.

Regarding the *Average Cluster Serving Data Rate*, it is observed that it decreases slightly with the number of GRT VBSs, reflecting the analysis done so far for the *Average VBS Serving Data Rate*. In fact, for 1GRT-3BE use case the Average Cluster Serving Data Rate achieve the maximum value, $R_{serv}^{CI} \approx 0.7\%$ above to the 3GRT-1BE. In accordance with this behavior, the *Average Cluster Utilization* decrease with the number of GRT VBSs is not significant.

11.3.4 Variation of the Quantity of VBSs Deployed on the Physical Cluster

The main target of the following analysis is to test the algorithm when the number of VBS in the cluster increases. The reference scenario is used for all use cases. The VBS-contracted data rate is typified according to the typical data rate of the service and the selected service profile: $R_{min}^{VBS} = 10$ Mbit/s for VoIP, $R_{min}^{VBS} = 500$ Mbit/s for video, $R_{ref}^{VBS} = 500$ Mbit/S for Web, and $R_{ref}^{VBS} = 100$ Mbit/s for file sharing. Three situations were explored: (1) the number of VBSs increases for all services in the same way, one VBS per service; (2) the number of GRT VBSs increases by one for each GRT service, while the number of BE VBSs maintains with two VBSs for each service; (3) the number of GRT VBSs maintains with two VBSs for each service, and the number of BE VBSs increases by one for each BE service. Concerning the relation between contracted and cluster capacity, use cases have been classified as in Table 11.1, allowing the assessment of the conclusions made earlier.

The results for *Average Cluster Data Rate* and *Average Cluster Utilization* are presented in Figure 11.6, as an example. Observing both metrics simultaneously, the best RU efficiency is achieved for *10-BE* and *12-BE*, which is when the strategy for the overall capacity provision is to limit the capacity contracted by GRT VNets, overbooking the capacity contracted by BE VNets. The worst RUs efficiency is observed for *12-GRT*, in which $\eta_{CI} = 1$, meaning that all RUs are assigned, and $R_{serv}^{CI} \approx 2.2 \, \text{Gbit/s}$.

Table 11.1   Strategy for Data Rate Contracted for Use Case and VNet Quantity

Use Case	Qty of VBSs	Total Contracted Data Rate [Mbit/s]	Strategy for Contracted Data Rate	Comment
4-Harmo	4	GRT VBSs: 510 BE VBSs: 600	Under	*Harmonized* equal number of VBSs providing each service
8-Harmo	8	GRT VBSs: 1020 BE VBSs: 1200	Under/Average	
10-GRT	10	GRT VBSs: 1530 BE VBSs: 1200	Average	*GRT Based* more VBSs providing GRT services
12-GRT	12	GRT VBSs: 2040 BE VBSs: 1200	GRT Over	
10-BE	10	GRT VBSs: 1020 BE VBSs: 1800	Average/BE Over	*BE Based* more VBSs providing BE services
12-BE	12	GRT VBSs: 1020 BE VBSs: 2400	BE Over	

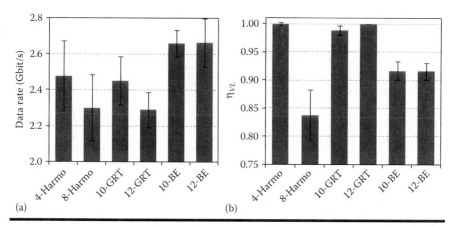

Figure 11.6   *Average cluster serving data rate* and *Average cluster utilization* for VNet quantity use cases: (a) aggregated by VBS type and (b) per VBS.

As a summary, one can say that changing the quantity of created VBSs as well as the contracted data rate in the cluster, GRT VBSs continue to achieve their minimum contracted data rate, though the *Average Cluster Serving Data Rate* can decrease if the number of GRT VBSs is higher than the number of BE VBSs. It is observed that the best RU efficiency is achieved when the strategy for the overall capacity provision is to limit the capacity contracted by GRT VNets, overbooking the capacity contracted by BE VNets. The worst RUs efficiency is observed when the number of GRT VBSs is the highest, the GRT overbooking situation being considered as the worst case or the limit situation for virtual wireless access implementation. Furthermore, contracting more data rate for BE VBSs over the same physical capacity does not cause the increase of their *Average VBS Serving Data Rate*, as the main target is to guarantee the minimum contracted data rate of GRT VBSs, BE VBSs only using the remaining capacity of the cluster.

11.4 Conclusions

The concept of network virtualization has been considered as the basis to address the problem of sharing the wireless infrastructure for provision of capacity to VNets. Following this approach, it is intended that the basic principles of network virtualization, such as the isolation between virtual resources and the possibility to deploy different protocols to take the diverse service requirements into account, can be applied.

The allocation of transmission resources is a challenging problem in virtualized environments, where they are shared among the different virtual resources, and there is the need to fulfill contracted capacity requirements. In wireless networks, the problem is even more challenging due to the inherently limited resources. In fact, the available radio resources are scarce with variable performance, and there is a lack of spare spectrum.

A reference network architecture for the virtualization of the wireless access based on the generic network virtualization environment is proposed. Both physical and virtual perspectives are considered, and the main stakeholders are considered. In terms of physical infrastructure, one considers a set of different RATs, which are abstracted by the specific RUs of each one. This allows the management of radio resources by the coordination of a pool of RUs, each having particular capabilities. Concerning the virtual resources, they can be differently defined by setting their type, GRT or BE, allowing to differentiate end-users handling, according to the VNet they belong to.

A new tier of RRM is proposed for inter–VNet RRM, designated by cooperative VNet RRM, managing how radio resources are allocated to the several VNets in order to satisfy the contracted VNets' capacity. This new level of management is proposed to interact with CoRRM and RRM, which is considered to be an intra–VNet RRM, thus, under the responsibility of VNOs. Moreover, the generalization of the inter–VNet RRM as a CoRRM problem with an additional level of abstraction, the virtual RRM level, allows following an approach of integration of the several levels of RRM, which needs to be adapted, but that actively participates in the process to achieve the main target of provision of the contracted level of service for all the VNOs operating on the common infrastructure. Naturally, the added virtual RRM level assumes the coordination role of all the underlying ones, as it is aware of VNets requirements and has the responsibility to satisfy them. Still the specific algorithms to implement the needed functionality at underlying RRM levels can evolve without overthrowing the outlined approach. The functionalities proposed for the initial VNet selection and VNet handover support are essential to provide CVRRM with the set of functionalities assigned to CoRRM, thus, allowing it to be considered as a transposition of CoRRM to the virtualization environment, though they are not further implemented. The VRRA function, being considered indispensable for the virtualization of the wireless access, is described in more detail, two different algorithms being proposed.

The two novel proposed VRRA algorithms, according to the type of guarantees of VBSs and the amount of contracted capacity and VBSs' utilization, take the variability of the wireless medium into account. RRM mechanisms, namely admission control and MAC scheduling, are continuously updated, to be aware of the VBSs' state relative to their service level agreement. Instead of looking at the wireless virtualization from the perspective of the instantiation of virtual machines into wireless nodes, our view is the virtualization of the wireless access to provide a contracted capacity to the VNet, in order to serve its end-users. Our approach is then agnostic to the point where the virtual node instantiation takes place, being possible to have the virtual nodes in each physical wireless node, or somewhere in the cloud requesting virtual access over a given geographic area covered by a set of wireless nodes. It is worthwhile noting that this capacity can be modified on demand, without manually changing the configuration of the network.

Comparing the wireless access virtualization, supported by the proposed OnDemandVRRA, with the standard approach, in which there are more than one network operator, each owning part of the physical infrastructure, it can be concluded that the former allows achieving a better performance from a cluster of BSs of several radio access technologies, enabling the provision of contracted capacity for GRT VNets. It is demonstrated by simulation that in virtual scenarios the *Cluster Serving Data Rate* may increase by approximately 46% and the utilization by 13%. On the other hand, for *Virtual* scenarios, the serving data rate of GRT VBSs is always greater than the minimum contracted being constrained by the defined BE VBSs reference data rate, which tends to be followed. The values achieved for the serving data rate are the lowest for BE services for the standard approach, but they can be larger for GRT services in the virtual approach, denoting, on one hand, the limitation arising from the split of the total cluster capacity by two operators, and on the other hand, the consequences of an uncoordinated allocation of radio resources when end-users are independently handled.

The proposed algorithm has been analyzed for different strategies for capacity provision, several usage profiles, diverse combinations of VBSs from different types, and several quantities of VBSs deployed on the physical cluster.

It is concluded that a limit for the percentage of GRT VBSs' contracted data rate should be defined as a function of the average capacity of the cluster and of the BE VBSs' contracted capacity. From the comparison between the virtual access approach and the standard one, it can be said that the setting of minimum and reference values for VNet's contracted data rate allows end-users to have a better network experience, considering the data rate as the main parameter to evaluate it.

Concerning the different strategies for capacity provision, that is, when the amount of contracted capacity by VBSs is over, on average, or under the physical capacity, it is concluded that a limit for the contracted data rate by GRT VBSs should be established in order to allow an efficient use of RUs among all the VBSs deployed within the cluster. It is verified that the *Cluster Serving Data Rate* may increase by approximately 20% if the amount of contracted capacity by BE VBS is 85% of the average capacity of the cluster, compared to the use case where the contracted capacity by the GRT VBS is the one with 85% of the cluster average data rate.

When the service profile is changing, OnDemandVRRA allows achieving isolation among the virtual resources, since the requested data rate of a VBS does not prevent the other to achieve the contracted data rate, if they are GRT. It is verified that even when 80% of end-users are requesting service in the BE VBS, the GRT VBS reaches the contracted data rate as soon as the requested data rate is greater or equal to it.

When changing the percentage of VBSs of each type instantiated in a cluster of BSs, only minor differences are perceived in the values obtained for the defined metrics, namely, VBS and *Cluster Average Serving Data Rate*, when the total number of VBSs is fixed and the relation between GRT and BE data rate contracted is maintained. The total *Average VBS Serving Data Rate* has a maximum deviation among the defined use cases around 0.2% for GRT VBS and approximately 1.8% for BE VBSs. According to this, also small variations are verified for the *Average Cluster Serving Data Rate*, increasing around 0.7% when the number of GRT VBSs increases.

By varying the quantity of created VBSs, as well as the contracted data rate in the cluster (although GRT VBSs are maintained within contract), the average cluster data rate can decrease if the number of GRT VBSs (hence, the contracted capacity) is higher than the number of BE VBSs. It is observed that the best RU efficiency is achieved when the strategy for the overall capacity provision is to limit the capacity contracted by GRT VNets, overbooking the capacity contracted by BE VNets. The worst RUs efficiency is observed when the number of GRT VBSs is the highest, which depicts the GRT overbooking situation as the worst case or the bound for virtual wireless access implementation. Furthermore, contracting more data rate for BE VBSs over the same physical capacity does not increase BE VBSs' serving data rate, as the implemented strategy is to guarantee the minimum data rate of GRT VBSs, BE VBSs only using the remaining capacity of the cluster.

References

1. Bavier, A., Bowman, M., Chun, B., Culler, D., Karlin, S., Muir, S., Peterson, L., Roscoe, T., Spalink, T., and Wawrzoniak, M., Operating system support for planetary-scale network services, in *Proceedings of the USENIX NSDI'04—1st Symposium on Networked Systems Design and Implementation*, San Francisco, CA, March 2004.
2. Sanjoy, P. and Srini, S., *GENI: Global Environment for Network Innovations, Technical Document on Wireless Virtualisation*, GENI project, Wireless Working Group, Document GDD-, Cambridge, MA, September 6–17, 2006 (http://www.geni.net).
3. Schaffrath, G., Werle, C., Papadimitriou, P., Feldmann, A., Bless, R., Greenhalgh, A., Wundsam, A., Kind, M., Maennel, O., and Mathy, L., Network virtualisation architecture: Proposal and initial prototype, in *Proceedings of VISA'09—1st ACM SIGCOMM Workshop on Virtualized Infrastructure Systems and Architectures*, Barcelona, Spain, August 2009.
4. Zhu, Y., Zhang-Shen, R., Rangarajan, S., and Rexford, J., Cabernet: Connectivity architecture for better network services, in *Proceedings of ReArch'08—ACM Workshop on Re-Architecting the Internet*, Madrid, Spain, December 2008.
5. Chih-Lin, I., Han, S., Xu, Z., Sun, Q., and Pan, Z., 5G: Rethink mobile communications for 2020+, *Philosophical Transactions of the Royal Society of London A—Mathematical, Physical and Engineering Sciences*, 374(2062), 20140432, 2016.
6. Anderson, T., Peterson, L., Shenker, S., and Turner, J., Overcoming the internet impasse through virtualisation, *IEEE Computer Magazine*, 38(4), 34–41, 2005.
7. Sachs, J. and Baucke, S., Virtual radio: A framework for configurable radio networks, in *Proceedings of WICON'08—4th Annual International Conference on Wireless Internet*, Maui, HI, November 2008.
8. Hu, S., Yao, Y., and Yang, Z., MAC protocol identification approach for implement smart cognitive radio, in *Proceedings of the ICC 2012—IEEE International Conference on Communications*, Ottawa, Canada, June 2012.
9. Perez-Romero, J., Sallent, O., Agusti, R., and Diaz-Guerra, M., *Radio Resource Management Strategies in UMTS*, John Wiley & Sons, Chichester, UK, 2005.
10. Hooli, K., Lara, J., Pfletschinger, S., Sternad, M., Thilakawardana, S., and Yutao, Z., *WINNER Spectrum Aspects: Assessment Report*, Deliverable D6.3, EC IST-WINNER Project, December 2005 (www.ist-winner.org).
11. Mobile IP based Network Developments (MIND), EC IST Project 28584, November 2002 (www.ist-world.org).
12. Sumaryo, S., Hepworth, E., Higgins, D., and Siebert, M., A Radio resource management architecture for a beyond-3G network, in *Proceedings of the International Workshop IST-MIND*, Budapest, Hungary, November 2002.
13. Ambient Networks (AN), EC FP6-IST Project 027662, January 2007 (www.ist-world.org).
14. 3GPP, *3GPP System Architecture Evolution (SAE): Report on Technical Options and Conclusions*, Technical Report TR 23.882 v8.0.0, September 2008 (http://www.3gpp.org).
15. Mobile Cloud Networking (MCN), EC FP7-ICT Project 318109, 2012–2016, January 2017 (http://www.mobile-cloud-networking.eu).
16. 3GPP, *3GPP Technical Specification Universal Mobile Telecommunications System (UMTS); LTE; Network Sharing; Architecture and Functional Description (Release 11)*, Technical Specification TS 23.251 V11.4.0, January 2013 (http://www.3gpp.org).

17. Alcatel-Lucent, *Network Sharing in LTE: Opportunities & Solutions*, Technology White Paper, July 2012 (http://alcatellucentmediaroom.files.wordpress.com/2010/07/lte_network_sharing_en_techwhitepaper1.pdf).

18. Huawei Technology, *5G: New Air Interface and Radio Access Virtualisation*, Technology White Paper, April 2015 (http://www.huawei.com/minisite/has2015/img/5g_radio_whitepaper.pdf).

19. Zaki, Y., Liang, Z., Goerg, C., and Timm-Giel, A., LTE wireless virtualisation and spectrum management, in *Proceedings of the WMNC'10—Wireless and Mobile Networking Conference*, Budapest, Hungary, October 2010.

20. Bhanage, G., Seskar, I., Mahindra, R., and Raychaudhuri, D., Virtual basestation: Architecture for an open shared WiMAX framework, in *Proceedings of the VISA'10—Virtualized infrastructure systems and architectures*, New Delhi, India, September 2010.

21. Kokku, R., Mahindra, R., Zhang, H., and Rangarajan, S., Cellular wireless resource slicing for active RAN sharing, in *Proceedings of the COMSNETS 2013—5th International Conference on Communication Systems and Networks*, Bangalore, India, January 2013.

22. Bhanage, G., Vete, D., Seskar, I., and Raychaudhuri, D., SplitAP: Leveraging wireless network virtualisation for flexible sharing of WLANs, in *Proceedings of the GLOBECOM 2010—Global Communication Conference*, Miami, FL, December 2010.

23. Xia, L., Kumar, S., Yang, X., Gopalakrishnan, P., Liu, Y., Schoenberg, S., and Guo, X., Virtual WiFi: Bring virtualisation from wired to wireless, in *Proceedings of the VEE'11—International Conference on Virtual Execution Environments*, Newport Beach, CA, March 2011.

24. Caeiro, L., Cardoso, F.D., and Correia, L.M., OnDemand virtual radio resource allocation for wireless access, *Wireless Personal Communications*, 82(4), 2431–2456, 2015.

25. IEEE, *IEEE Standard for Local and Metropolitan Area Networks: Virtual Bridged Local Area Networks*, 802.1 WG, IEEE Std. 802.1Q-2005, December 2005 (http://standards.ieee.org/getieee802/download/802.1Q-2005.pdf).

26. Cisco Systems, *Cisco Visual Networking Index: Global Mobile Data Traffic Forecast Update, 2011–2016*, Cisco® Visual Networking Index (VNI), February 2012 (http://www.cisco.com).

27. Caeiro, L., *Common Radio Resource Management in Virtual Heterogeneous Networks*, PhD Thesis, IST—University of Lisbon, Lisbon, Portugal, 2014.

Chapter 12

Relays and Cooperative Techniques for 5G Systems

Nikolaos Nomikos and Demosthenes Vouyioukas

Contents

12.1 Introduction

The 5th-generation (5G) of wireless networks aims, among others, at achieving the goals that have been set by regulatory authorities regarding the seamless connectivity of all types of devices. In various studies it is stated that next-generation networks should allow smart objects ranging from sensors and actuators to user equipment (UE) and vehicles, to connect and communicate in the Internet of things (IoT) [1]. Through the IoT, novel applications, such as smart cities, smart energy-grids, vehicle-to-vehicle communications, and health-oriented body area networks, are expected to become a reality [2,3].

In addition, the attention that cooperative relaying (CR) received in the past decade resulted in numerous contributions from both academia and industry, which outlined the gains that can be harvested. Through CR, network performance can be improved due to path-loss reduction, multipath diversity, and shadowing mitigation. At the same time, wireless networks have evolved as well, migrating toward denser heterogeneous deployments demanding:

1. Seamless integration of novel networking technologies, such as CR in wireless topologies.
2. Improved spectral efficiency in order to satisfy the needs of billions of users and devices that are connected in the IoT.
3. Consideration of the increased level of interference arising from the huge demand for wireless data and development of intelligent interference mitigation techniques.
4. Exploitation of the potential provided by heterogeneous networking in order to achieve the targets of the 5G wireless networks.
5. Support of the Green Communications paradigm aiming to enhance the sustainability of future wireless networks through improved power efficiency.

In the currently deployed 4th-generation (4G) networks, new techniques were introduced to wireless networking, such as coordinated multipoint transmission/reception, hierarchical and heterogeneous networks, and cooperative relaying. In the 802.16j and LTE-Advanced protocols, relay standards were detailed, including in-band and out-band relaying, as well as transparent and nontransparent relay connectivity with the users [4,5]. However, it became obvious that the role of relays in these networks was limited due to a number of reasons. First, hardware constraints held back the deployment of full-duplex (FD) relaying as the loop interference (LI) among the relay antennas cannot be easily mitigated. Also, mobile relays that provide the potential of ubiquitous connectivity were not adequately studied. In addition, cognitive relaying and device-to-device (D2D) relaying were partly investigated but their implementation was left for the next generation of wireless networks.

Moving forward, novel relay concepts must be developed and employed. As various types of UE and smart objects are to be connected to the IoT, it is of great importance that the network provides improved access to support the increased traffic demand. As node density increases, various devices can act as relays to forward traffic from the end-nodes to the core network and vice versa. These include, on one hand, small cells installed through uncontrolled network deployment by users to provide improved indoor coverage. These small cells are connected to the network backbone through wired technologies and serve either a closed group of users or all the users that are located in their coverage area as long as wireless channels are available.

On the other hand, by adopting the D2D paradigm, using UE as relays for all kinds of traffic enhances traditional cellular deployments with ad-hoc elements. In D2D topologies, users can communicate with each other either through the assistance of the wireless infrastructure or autonomously with direct communication of signalling and data [6,7]. In a similar way, vehicles, either private or public, may act as moving relays in order to avoid coverage holes in urban settings. Last but not the least, fixed relays with increased processing capabilities are destined to provide improved connectivity to groups of nodes that cannot directly communicate with the base station by employing advanced interference mitigation techniques [8–10].

At the heart of this chapter, there are unique challenges that have to be investigated in order to provide solutions applicable to various network topologies, while being easily adapted for specific wireless protocols. More specifically, CR is a technology that can be considered in an abstract way to provide additional coverage or improve the end-to-end quality of service (QoS) in a multitude of networks ranging from wireless personal area networks (WPANs) and wireless sensor networks (WSNs) to cellular topologies. Conducting research and providing algorithms with increased diversity allows CR to maximize its potential and provide robustness to communications facing challenges in environments with increased propagation losses and excessive fading.

This chapter serves as a useful manual by providing important definitions for relay networks and addressing practical challenges that must be tackled.

Various relaying strategies are studied, such as opportunistic relay selection (ORS), where after a selection phase one relay or relay-pair performs end-to-end communication. Also, successive relaying (SuR) is presented where a source and one relay concurrently transmit to mimic FD operation. In SuR, inter-relay interference (IRI) arises between the receiving and the transmitting relaying and mitigation techniques, such as interference cancellation (IC) and interference avoidance (IA), can be employed. Another interesting relay strategy is FD relaying in network where multiantenna relays are available. However, loop-interference has to be reduced as the output antenna of the relay interferes with the reception of the input antenna. To facilitate interference mitigation and improve the overall network performance, buffer-aided (BA) relays are used in the scenarios due to the increased freedom that they provide to transmission scheduling. Furthermore, a discussion about possible relay types for 5G networks is included and the ability of each type to serve applications with different requirements is studied.

It is considered that 5G relaying must exploit and support a multitude of devices and applications, respectively. So, it is important to adjust relay functionality on an application basis and toward this end, relay selection policies are presented aiming at high performance and low coordination overhead. More specifically, this chapter includes three different relay selection policies, which are tailored to different types of applications:

1. A reduced channel estimation overhead policy for high mobility scenarios.
2. A delay minimization policy for delay critical applications.
3. A low-power policy for power-efficient communications.

In addition, since multirelay networks must be efficiently coordinated, both centralized and distributed network coordination approaches are analyzed. For these schemes sketch diagrams are presented for message exchange and derive the implementation complexity order for relay selection.

The structure of this chapter is as follows. In Section 12.2, an in-depth discussion on CR is given with useful definitions of various relay strategies. Then, in Section 12.3, different interference cases are described, as well as respective mitigation techniques. In Section 12.4, a detailed description of possible relay classes is provided for 5G wireless networks and the capability of each class to support a specific application. Next, in Section 12.5, the system model is presented and the preliminaries necessary for understanding the contributions of this chapter. Subsequently, the three relay selection policies are presented in Section 12.6, based on the application's targets and centralized and distributed network coordination is analyzed in Section 12.7. In Section 12.8, performance evaluation and comparisons between relay classes are performed for each selection policy. Finally, Section 12.9 outlines some important future research directions and Section 12.10 provides the conclusions drawn from this chapter.

12.2 Cooperative Relaying

The notion of cooperation has a significant and positive impact on the physical world. Two or more entities tend to cooperate and overcome difficulties and burdens in order to achieve goals that otherwise would require much more effort or would even be impossible. It is natural that human-like behavior, such as cooperation, are inherited by wireless communication networks aiming to improve their performance. A technique that has received numerous contributions during the past years is CR where sources of information communicate with their corresponding destinations through relays.

Relays play the role of intermediate nodes offering three distinct gains to the network. First, they improve the characteristics of a wireless link through increased diversity [11] as additional and independent paths are available for the signal to propagate, thus reducing errors at the receiver. Second, as the transmission is broken into hops, the transmitter is closer to the receiver, resulting in reduced pathloss. Third, in cluttered environments, intelligent relay positioning and mobile relays provide resiliency against shadowing. In this relaying paradigm, two or more hops are required for the signal to reach the destination, thus resulting in a trade-off between harvesting network gains and increasing resources expenditure due to the use of more channels or more time-slots.

Various types of relays are used in wireless networks and the two main classes are amplify-and-forward (AF) and decode-and-forward (DF) based on the processing of relays on the signal. In AF relaying, the relay receives the source's signal and performs amplification in the analog domain, as well as phase steering before forwarding the signal to the destination. In DF relaying, the relay receives the source's signal, decodes it, and reencodes it prior to transmission toward the destination. In general, AF relays are more simple to implement as they are repeaters that do not perform advanced processing to the signal, as is the case in DF relaying where error correction and different modulation schemes can be used [11]. Moreover, AF relays not only amplify the desired signal but also the noise, in contrast to DF relays that decode the signal and try to remove the noise before reencoding takes place. So, due to the more complex signal processing performed by DF relays, increased delay compared to AF relays is introduced.

In a different spirit, depending on the ability of a relay to perform or not concurrent reception and transmission, the relays are characterized as FD or half-duplex (HD). When a relay operates in FD mode, only one channel is used for the end-to-end transmission, as the relay is forwarding the signal that is currently receiving from the source. However, this results in increased hardware complexity and LI from the relay's output antenna to its input antenna. If an HD relay is employed, orthogonal frequency channels have to be used, thus leading in reduced spectral efficiency; a loss imposed by the half-duplex constraint (HDC) of HD relays. Interesting relaying strategies, such as SuR [12], involve two or more HD relays and select one relay to receive the source's signal while another relay is forwarding a previously received signal to the destination. SuR mimics FD relaying using HD relays and offers increased spectral efficiency, as long as the IRI introduced by the simultaneous transmission is taken into consideration.

Depending on the physical characteristics of the relay node, there are many examples involving different relay types. The classic relay is a low-cost, power-supplied node that is set by the mobile operator in order to provide relay gains without requiring additional infrastructure, such as wired backhaul, which limits the practicality of small base, that is, micro and pico-cells [13]. However, predictions for the future wireless networks [6] bring in that the density of the users is rapidly increasing, thus leading to the increase of the number of the relays as well. In scenarios where numerous relays are available, battery-dependent relays either fixed or mobile, for example, on top of public vehicles [14], offer additional degrees of freedom compared to power supplied relays. Furthermore, similarly to the cases of WSNs and peer-to-peer (P2P) communications, users will play the role of relays for other users, thus giving rise to the D2D paradigm. Consequently, the extensive use of battery-dependent relays require the use of power-efficient techniques in order to avoid power outages, which threaten the diversity gains of CR.

Focusing on the number of relays that are employed to forward the signal to the destination in each hop, the primary categorization is between multiple relay transmission (MRT) and opportunistic relay transmission (ORT). For simplicity, let us consider a two-hop network where the signal reaches the destination after two phases and there is no direct link between the source and the destination. In MRT, the source broadcasts the signal to the relays that try to either amplify the signal in the analog domain or decode and reencode the signal prior to forwarding. As more than one relays participate in the second hop, more resources are needed either in time or in frequency in order to avoid interference when the destination is receiving each relay's signal. On the contrary, in ORS or best relay selection (BRS), the source either broadcasts or activates one relay to receive the signal through a selection process that involves channel state information (CSI) [15] and the characteristics of the relay, such as available energy [16,17] or buffer status [18,19]. Then, only one relay transmits to the destination using one channel, thus saving precious spectral and temporal resources. The most notable characteristic of ORT is that it exhibits the same diversity performance as MRT as proven in [15].

12.3 Interference Mitigation

Wireless networking is the major component of the ubiquitous access that is enjoyed by users and an enabling technology for the IoT that connects all kinds of smart objects into a network that is envisioned to offer new and exciting applications ranging from Smart Cities to Health Monitoring. As fast deployment and increased coverage is offered through wireless technologies, wireless networking is preferred for the access part of the network over wired networks. However, due to the broadcast nature of wireless propagation, when two or more signals use the same frequency channel, interference occurs leading in performance degradation as the receiver cannot distinguish the desired signal from the interference signal.

As denser deployments of wireless networks are performed by the mobile operators and users, interference mitigation techniques or exploitation have to be considered, otherwise network performance will be compromised. Moreover, as spectral-efficient relaying techniques operate in FD or mimic FD using HD relays, LI and IRI arise, respectively, at the relay's reception. Unless intelligent algorithms are employed, interference becomes a bottleneck that reduces the end-to-end performance of CR.

In FD relay networks, LI is the main limitation in achieving the optimal spectral efficiency as the relay's transmission causes interference while it receives the source's signal. Various techniques were presented including antenna isolation, power adaptation, IA, IC, and precoding. More specifically, IA is based on the multiple relay nodes available and the selection of one or more nodes, which provide links with the best end-to-end signal-to-interference-plus-noise ratio (SINR). In addition, IC is a strategy that aims at decoding the interference signal at the receiver prior to the decoding of the desired signal. Thus, interference-free reception can be achieved if the interference-to-signal-plus-noise ratio (ISNR) is above a decoding threshold, which is linked to the transmission rate employed by the source.

Antenna isolation was studied in [8] where natural isolation of the transmit and receive antennas provides an efficient way of reducing the LI level. The use of objects, such as building and shielding plates, is mentioned and isolation is further elaborated to include techniques, such as directional antennas and orthogonal polarization. However, in most cases LI residuals remain and additional techniques are employed to reduce LI to noise levels. A useful technique is a power adaptation that exploits CSI knowledge and calculates the required power level for successful transmission by matching it to the required signal-to-noise ratio (SNR) to support a target rate. In [20,21] the LI is significantly reduced as the power level in the relay-destination (RD) link is adapted to the fading conditions, thus improving the SINR in the first hop of two-hop relay networks. In [9,21], the availability of multiple FD relays is exploited by using ORS to choose the FD relay with the least amount of LI in each time slot. An interesting characteristic of FD relaying is that the relays know a priori the LI signal and interference cancellation can be applied with increased chances of success. In [8], time-domain cancellation (TDC) is proposed offering advantages, such as resiliency of the desired signal and keeping the same number of input and output antennas on the relay. On the downside, TDC is prone to channel estimation errors while it does not exploit the spatial domain and may offer worse isolation. The authors in [22], examine hardware constraints of FD relay receivers and the effect of imperfect CSI, which result in an amount of residual LI that cannot be canceled. Precoding techniques develop filters aiming to reduce LI level making use of CSI of the LI channel. In [8] various filters are designed for the following cases: (1) antenna selection at the receiver, (2) beam selection,

(3) null-space projection filters that enable the relay to receive and transmit in different subspaces, and (4) minimum mean square error (MMSE), which minimizes the distortion of the desired signal caused by the other three filter designs.

Contrary to FD relaying, in SuR with HD relays, it is possible that relays do not know a priori the interference signal and for this reason interference cancellation is more difficult to perform. Trying to mitigate the IRI interference, various techniques were proposed similar in spirit to those for LI mitigation but taking into consideration the different topology of SuR. In [23], a two-hop relay network is considered where IRI is decoded and subtracted when an interference cancellation threshold is satisfied. Using repetition coding, the source signal is transmitted after being decoded either IRI-free if IC was performed or if its SINR was above the decoding threshold. Extending this work to the case of multiple available relays [24,25] proposed the combination of ORS and SuR by selecting a relay-pair that achieves the maximum end-to-end SNR either by applying IC or by choosing the pair with the minimum IRI. Other works [26], propose the use of the IRI signal through superposition coding in order to achieve increased diversity gain when the receiver decodes the superposition of the IRI and the desired signal, while already having knowledge of the IRI signal from a previous reception.

A further extension for SuR was given in [10,18,27] where BA HD relays were employed in order to provide increased freedom when selecting the relay-pair, as relays with packets residing in their buffers were considered in the selection process. In this way, IRI avoidance is further improved and the end-to-end QoS target can be achieved. Many studies present power adaptation algorithms, which exploit CSI knowledge optimally setting the power level at the source and the transmitting relay to minimize the IRI and maintain the QoS of the source-relay-destination (SRD) communication. In [10], the optimal relay-pair is selected by first calculating the power level of relays that can transmit to the destination, that is, relays having non-empty buffers and then examining all possible pairs by evaluating the fading conditions in the inter-relay (IR) and source-relay (SR) channels.

12.4 Relay Classes and Applications

This section discusses in detail different relay classes and the potential of each class to support various applications.

12.4.1 Relay Classification

The classification of 5G relays depends foremost on the type of devices, which are to be deployed. Table 12.1 includes the capabilities of each relay class, ranging from low to high for various characteristics.

Table 12.1   Relay Classes and Their Distinct Capabilities

Relay Class/ Capability	Processing	MIMO	Storage	Channel Estimation	Density
User equipment	Low/ Medium	Low/ Medium	Low/ Medium	Low	High
Battery-dependent mobile relay	Medium/ High	Medium	Medium	Low/ Medium	Low/ Medium
Battery-dependent fixed relay	Medium/ High	High	High	Medium/ High	Low/ Medium
Power-supplied fixed relay	High	High	High	High	Low

First, processing capabilities vary significantly, as relay classes range from UE, such as smartphones and tablets, to battery-dependent mobile and fixed relays, as well as power-supplied fixed relays. Although UE's processing power has dramatically increased in recent years, their battery-dependence and small size puts them at a disadvantage against the other relay classes.

In addition, multiple-input multiple-output (MIMO) capability is expected to be a basic element in 5G setups. Again, as relay size increases, it becomes easier to fit multiple antennas on each relay. A subject related to MIMO is the potential for FD relaying, where the same channel is used to relay the signal in both hops at the expense of LI from the relay's output to its input. On one hand, the small size of UE prohibits them from employing isolated antennas, while directional antennas cannot be used due to the need for isotropic radiation. In the same spirit, the antenna installation on mobile relays may face difficulties compared to fixed relays that are easier to configure, as fading conditions change faster and this reflects on the performance of LI mitigation as well [8].

Lately, the storage capability and buffer management has been discussed in various works [10,18,21,27,28], as buffer-aided relays are able to increase the diversity of the network, as long as a specific amount of delay can be tolerated. It is to be expected that mobile relays, should employ medium-sized buffers, as packet expiration increases due to frequent handovers and disconnections. On the other hand, fixed relays can employ larger queues as their relative positions to UE and other devices remain constant for extended periods.

Another critical characteristic for 5G networks is the accurate and fast channel estimation, which provides the means for efficient resource utilization as the relay's behavior is adapted to its environment. As spectrum and channel condition sensing depend on the quality of the antenna sensors and hardware, small-sized UE provide decreased performance while more advanced mobile and fixed devices exhibit improved behavior.

The last characteristic that motivated the use of all kinds of devices as potential relays is node density. Here, UE have the advantage compared to operator-deployed relays and their efficient usage is the main target in D2D communications. Regarding the other classes, battery-dependent mobile and fixed relays can be installed easier compared to their power supplied counterparts, and their density is expected to be larger.

12.4.2 Application Performance Classification

Table 12.2 contains applications and the level of performance that each relay class can achieve. It is obvious that each application can aim at multiple performance goals that require separate investigation.

Taking into consideration that ubiquitous communication is to be provided to mobile UE and objects, high mobility should be efficiently supported in 5G networks. As the relative position among the relay and the end-nodes changes fast, channel estimation is performed harder and relays with accurate and fast sensing offer superior performance. Moreover, the increased coverage offered by power-supplied fixed relays results in less handovers and possible outages. However, the increased density of UEs and mobile relays results in outage avoidance if efficient relay selection schemes are employed. On the downside, the small transmit power and coverage offered by UE, lower their potential in high mobility applications.

In many cases, delay is the most critical concern, for example, emergency applications, high-precision industrial applications and various types of

Table 12.2   Relay Classes and Their Performance Capability on an Application Basis

Relay Class/Application	High Mobility	Delay Critical	Green Communications
User equipment	Low	Low/Medium	Medium
Battery-dependent mobile relay	Medium	Low/Medium	Medium
Battery-dependent fixed relay	Medium	Medium	Medium/High
Power-supplied fixed relay	Medium/High	Medium/High	High

smart-grid applications. The delay is denoted as the time required for a packet to reach the destination after it is transmitted by the source. For a fixed packet size, the average delay scales with the achieved throughput [29]. To this end, the relay device must be capable of handling delays by increasing the throughput. So, relays that offer low-latency processing, full-duplexity, and efficient buffer management, have an advantage in these types of applications. On the other hand, as stated in various works [10,18,27,28], when the number of relays increases, so does the average delay. As a result, the increased density of UE might lead in the distribution of various packets in many nodes, thus leading to packet expiration. Similarly, mobile relays that offer frequent handovers with the served devices, coupled with their reduced FD potential, might not offer the same performance as fixed infrastructure relays.

Regarding economic and environmental sustainability, Green Communications is a major research topic. In this field, relay selection aiming at efficient power usage, leads to two distinct gains. On one hand, carbon footprint reduction of telecommunication networks leads to environmentally friendly communications and decreased power expenses by mobile operators and users. On the other hand, reduced public exposure to electromagnetic fields (EMF) is achieved. Low EMF exposure has been a top priority for researchers, regulatory authorities, and operators in forthcoming international research initiatives (e.g., Horizon 2020 [30]). As a result, techniques that rely on smart-power adaptation based on channel conditions are of utmost importance. In this area, exploiting the increased density of UE might leverage their lower processing potential and offer extra channels, which can route the signals with more efficient power consumption as the offered diversity is greater. Still, more advanced nodes with increased MIMO and estimation capabilities might support this type of application, at the expense of reduced diversity and costlier deployments.

12.5 System Model and Preliminaries

12.5.1 System Model

Consider a two-hop system where a source S communicates with a destination D via a cluster of relays C. As various classes of relays with diverse characteristics are available, their specifications for processing, MIMO (or FD capability), channel estimation, and transmit power differ to a significant degree.

The UE relay class consists of single-antenna terminals and no FD capability. Moreover, they use the lowest transmit power equal to $0.2P$, where P is the maximum transmit power that can be used by a network node. Then, the battery-dependent mobile relays are considered to be MIMO equipped and FD capable. However, due to the varying fading environment and difficulties in antenna isolation compared to fixed relays, increased amount of

LI affects the FD operation, while their power output is equal to $0.8P$ due to battery limitations. For the class of battery-dependent fixed relays, MIMO and FD operation is assumed, as well as LI is reduced due to optimized antenna isolation. Their transmit power is equal to that of mobile battery-dependent relays. Finally, the power-supplied relay class supports multiple antennas and has FD characteristics, instantaneous channel estimation, and the highest level of transmit power P. For the transmit power ratios, the output power values of LTE-Advanced [4] are used while lowering the power of battery-dependent nodes in order to justify power-supply limitations.

Multiple antenna relays are equipped with $N = 2$ antennas, one for reception and one for transmission. Each relay R_k can operate in HD and in FD mode except for the UE, which operate only in HD. In FD mode, information can be transmitted to and from a point in two directions simultaneously on the same physical channel [8]. In this topology, ideal FD operation is not considered, but instead, LI at the relay is present. As an alternative to FD operation, successive relaying is adopted, where in each time slot two relays are used simultaneously. In this strategy, one relay receives the source's signal while another forward a previously received signal to the destination. In this way, FD operation can be mimicked but at the cost of IRI.

Each relay R_k holds a buffer (data queue) Q_k of length L_k (number of data elements) where it can store source data that has been decoded at the relay and can be forwarded to the destination. The parameter $l_k \in \mathbb{Z}_+$, $l_k \in [0, L_k]$ denotes the number of data elements that are stored in buffer Q_k; at the beginning, each relay buffer is empty (i.e., $l_k = 0$ for all k). In line with Table 12.1, it is assumed that for UE $L_k = 0.25L$, for battery-dependent mobile relays $L_k = 0.5L$ and for the two types of fixed relays $L_k = L$.

Figure 12.1 depicts the network topology where a pool of relays support the communication between a source and a destination each time. Also, the different types of interference are denoted, IRI for SuR transmission and LI for FD transmission. These relaying strategies are employed to satisfy the requirements of three specific applications targeting different performance metrics. More specifically, the delay critical application targets delay minimization and the transmitters use fixed power levels, while the number of transmitted codewords is matched to the capacity that is achieved in each time slot. The high mobility application adopts fixed power levels, due to CSI acquisition difficulties where CSI at the receivers is assumed. Also, a SNR threshold Z is set in order to avoid constant relay selection when SNR higher than the outage threshold γ_0 is desired. Finally, the green communication application uses power adaptation targeting a specific SNR threshold γ_0 in order to achieve power reduction. In this power minimization scenario, the threshold Z can be expressed by a target power value that must not be surpassed. As a result, when the power value Z is satisfied, there is no need to perform relay or relay-pair selection and the previously selected relay(s) continue to serve the end-to-end communication.

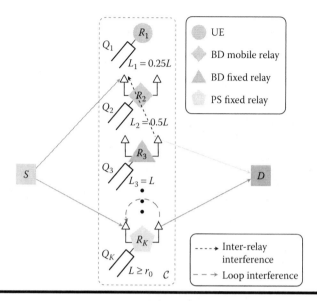

Figure 12.1   A network where three different applications take place and end-to-end communication is established via relay selection.

12.5.2 SNR Expressions

In our model, a reception is successful when the SNR at the receiver is above a SNR threshold γ_0. Hence, the instantaneous SNR from S to R_i and the instantaneous SINR from R_j to D when relay R_i is selected for reception and relay R_j is selected for transmission (in full-duplex $i = j$), are expressed as

$$\gamma_{SR_i} = \frac{g_{SR_i} P_{S_i}}{g_{R_j R_i} P_{R_j} + n} \geq \gamma_0 \tag{12.1}$$

and

$$\gamma_{R_j D} = \frac{g_{R_j D} P_{R_j}}{n} \geq \gamma_0 \tag{12.2}$$

respectively. Note that Equation 12.2 does not consider interference from the source, as it is assumed that there does not exist a direct SD link. When HD or out-of-band relaying takes place, in the SR_i link, the SINR expression in Equation 12.1 becomes

$$\gamma_{SR_i} = \frac{g_{SR_i} P_{S_i}}{n} \geq \gamma_0 \tag{12.3}$$

It is noted that when fixed power is used at the source and the transmitting relay, their corresponding power levels P_{S_i} and P_{R_j} are used.

Regarding the possibility of interference cancellation for the cases of IRI and LI, the factor $\mathbb{I}(R_t, R_r)$ is used indicating whether interference cancellation is possible and it is described by

$$\mathbb{I}(R_j, R_i) = \begin{cases} 0, & \text{if } \dfrac{g_{R_j R_i} P_{R_j}}{g_{SR_i} P_S + n_{R_i}} \geq \gamma_0 \\ 1, & \text{otherwise} \end{cases} \tag{12.4}$$

As a result, Equation 12.1 will vary according to Equation 12.4:

1. For $\mathbb{I}(R_j, R_i) = 0$: $\gamma_{SR_i} = \dfrac{g_{SR_i} P_{S_i}}{g_{R_j R_i} P_{R_j} + n} \geq \gamma_0$

2. For $\mathbb{I}(R_j, R_i) = 1$: $\dfrac{g_{R_j R_i} P_{R_j}}{n_{R_i}} \geq \gamma_0$

12.5.3 Reference Relay Selection Policies

12.5.3.1 Max–Min Relay Selection in Successive Relaying

The max–min relay selection policy implemented in a successive relaying network takes a different form if the IRI can be canceled at the relays. In [24,25,31] a reactive relay selection policy is proposed. More specifically, instead of considering only $\{S \rightarrow R\}$ and $\{R \rightarrow D\}$ channel gains, the feasibility of IC is also examined. In this way, with a very simple IC condition the following two cases for relay selection are given.

1. If candidate relay R_k can perform IC, then it may be selected to receive from the source based on the following value:

$$R^* = \arg\max_{R_k \in C} \min\{g_{SR_k}, g_{R_k D}\} \tag{12.5}$$

2. On the other hand, if R_k cannot perform IC, then it can be selected after competing with the rest of the relays as shown:

$$R^* = \arg\max_{R_k \in C} \min\left\{\dfrac{g_{SR_k}}{g_{R_t R_k}}, g_{R_k D}\right\} \tag{12.6}$$

It is obvious that having two simultaneous transmissions by the source and the transmitting relay reduces the diversity of the network, as R^* cannot participate in the selection process due to the half-duplex constraint. The lack of buffering at the relays results in the coupling of relay selection with the previous transmission phase where the currently transmitting relay was chosen among $K - 1$ candidates as one relay was forwarding to the destination and did not receive the source's frame. Additional diversity loss is introduced when IRI is not effectively mitigated but when IC cancellation or avoidance is feasible, a full-duplex behavior can be achieved and the half-duplex loss is leveraged.

12.5.3.2 Max–Max Relay Selection

The max–max relay selection (MMRS) [19] is the first relay selection policy that studied the diversity behavior of buffer-aided relay nodes. Given that the relay nodes are equipped with buffers and thus can store the data received from the source, the max–max policy splits the relay selection decision in two parts and selects the relay with the best source-relay link for reception and the relay with the best relay-destination link for transmission. The max–max selection policy follows the conventional two-slot cooperative transmission where the first slot is dedicated for the source transmission and the second slot for the relaying transmission, but the relay node may not be the same for both phases of the protocol. The max–max relay selection policy can be written as

$$R_r^* = \arg\max_{R_k \in \mathcal{C}}\{g_{SR_k}\} \tag{12.7}$$

$$R_t^* = \arg\max_{R_k \in \mathcal{C}}\{g_{R_kD}\} \tag{12.8}$$

where R_r^* and R_t^* denote the relay selected for the first phase and the second phase of the cooperative protocol, respectively. It has been proven that the max–max relay selection policy also ensures full diversity equal to the number of the relays and provides a significant coding gain in comparison to the conventional max–min selection scheme. However, it is worth noting that the previous selection strategy assumes that no relay's buffer can be empty or full at any time and thus all relays have always the option of receiving or transmitting [19, Sec. III. C].

12.5.3.3 Max-Link Relay Selection

The previous relay selection schemes are associated with a two-slot cooperative protocol where the schedule for the source and relay transmission is fixed a priori. In [28], this limitation is relaxed and each slot is allowed to be dynamically allocated to the source or a relay transmission, according to the instantaneous quality of the links and the status of the relays' buffers. More specifically, the proposed max-link relay selection scheme fully exploits the flexibility offered by the buffers at the relay nodes and at each time selects the strongest link for transmission (source or relay transmission) among the available links. A source-relay link is considered to be available when the corresponding relay node is not full and therefore can receive data from the source, while a relay-destination link is considered to be available when the relay node is not empty and thus can transmit the source's data toward the destination. The proposed scheme compares the quality of the available links and adjusts the relay selection decision and the time-slot allocation to the strongest link. If a source-relay link is the strongest link, the source transmits and the corresponding relay is selected for reception; on the other hand, if a relay-destination link is the strongest link, the corresponding relay is selected for transmission. The max-link relay selection policy can be analytically expressed as follows:

$$R^* = \arg\max_{R_k \in C} \left\{ \bigcup_{R_k \in C: \Psi(Q_k) \neq L} \{g_{SR_k}\}, \bigcup_{R_k \in C: \Psi(Q_k) \neq 0} \{g_{R_k D}\} \right\} \tag{12.9}$$

where R^* denotes the selected relay (either for transmission or reception) and the function $0 \leq \Psi(Q_k) \leq L$ gives the number of data elements that are stored in buffer Q_k.

12.5.3.4 Space Full-Duplex MMRS

The Space Full-Duplex MMRS (SFD-MMRS) [27] is based on the idea of choosing different relays for reception and transmission, according to the quality of the channels, so that the relay selected for reception and the relay selected for transmission can receive and transmit at the same time (i.e., in one time slot). The case where successive transmissions are performed is illustrated in Figure 12.1. In contrast to [19] in which reception and transmission have to be performed in successive time slots (as the same relay may be selected for reception and transmission), here two relays are simultaneously activated for reception and transmission. The relay pair (R^*, T^*) selected is given by

$$(R_r^*, R_t^*) = \begin{cases} (R_{r_1}, R_{t_1}) & \text{if } r_1 \neq t_1 \\ (R_{r_2}, R_{t_1}) & \text{if } r_1 = t_1 \text{ and } \min(g_{SR_{r_2}}, g_{R_{t_1}D}) \\ 0 & > \min(g_{SR_{r_1}}, g_{R_{t_2}D}) \\ (R_{r_1}, R_{t_2}) & \text{otherwise} \end{cases}$$

where:

$$R_{r_1} = \arg\max_{R_k \in C} g_{SR_k}$$
$$R_{t_1} = \arg\max_{R_k \in C} g_{R_k D}$$
$$R_{r_2} = \arg\max_{R_k \in C, R_k \neq R_{r_1}} g_{SR_k}$$
$$R_{t_1} = \arg\max_{R_k \in C, R_k \neq R_{t_1}} g_{R_k D}$$

SFD-MMRS provides the best performance in terms of both capacity and outage probability, compared with the existing schemes in the literature. However, as in [19], the strategy assumes that no relay's buffer can be empty or full at any time and in addition the IRI is assumed to be negligible.

12.5.3.5 Min-Power Relay Selection

The selection scheme of [10] called min-power merges the relay-pair selection algorithm of 18 with max-link selection of [28]. In this selection scheme, the target is to minimize the overall power expenditure; thus power adaptation is adopted. The min-power scheme is associated with a one-slot cooperative protocol (similar to the min-power and SFD-MMRS in [27] where, however, IRI and power minimization are not considered), contrary to protocols where the selection algorithm's operation spans two consecutive time slots (as in [19]). At each time slot, the source S attempts to transmit data to a selected relay with a nonfull buffer (i.e., $R_r \in \mathcal{A}$), and at the same time another relay with a nonempty buffer (i.e., $R_t \in \mathcal{T}$, $R_t \neq R_r$) attempts to transmit data to the destination D.

Denoting by $\mathcal{P}$ the set of all possible relay-pairs in the relay network, and by $|\mathcal{P}|$ its cardinality the best relay-pair (R_r, R_t) is selected as

$$b^{(SuR)} = \arg \min_{r,t \in \mathcal{P}} (P_{S_r}^* + P_{R_t}^*) \tag{12.10}$$

where $P_{S_r}^*$ and $P_{R_t}^*$ are the minimum power levels for the source and the transmitting relay when successive transmissions are performed and IRI is present and maybe cancelled according to Equation 12.4.

If $|\mathcal{P}| = 0$ (i.e., none of the relay-pairs can provide power levels below the power limits), then min-power switches to the max-link relay selection policy. Denoted by $\mathcal{M}$, the set of *links* that can be employed by the max-link relay selection policy; this set consists of the following two subsets: subset $\mathcal{T} \subseteq \mathcal{M}$, which includes all the relays for which their buffer is not empty and hence able to transmit to the destination, and subset $\mathcal{A} \subseteq \mathcal{M}$, which includes all the relays for which their buffer is not full and they are available to receive a packet from the source. Note that $\mathcal{A} \cup \mathcal{T} = \mathcal{M}$.

$$b^{(ML)} = \arg \min_{k \in \mathcal{M}} \min \left(P_{S_k}^{\dagger}, P_{R_k}^{\dagger} \right) \tag{12.11}$$

where $P_{S_k}^{\dagger}$ and $P_{R_k}^{\dagger}$ are the minimum power levels for the source and the transmitting relay when transmissions are based on adaptive link selection where IRI is absent.

12.5.3.6 Suboptimal Joint Relay-Pair Selection

The work in [32] presented a suboptimal scheme adopting DSSC [33,34] to select the best relay-pair using only partial CSI. To this end, a rate threshold Z is set at the start of the network's operation and is made globally known to its nodes. The relay-pair $b_{R_t R_r}^{(v)}$ in time slot v is selected as

$$
b_{R_t R_r}^{(v)} = \begin{cases} b_{R_t R_r}^{(v-1)}, & \text{if } r_{b_{R_t R_r}}^{(v)} \geq Z \\ b_{opt}^{(v)}, & \text{otherwise} \end{cases}
\tag{12.12}
$$

where $r_{b_{R_t R_r}}^{(v)}$ is the achievable rate in time slot v of the optimal pair of time slot $v-1$. Thus, in each time slot the previously selected relay-pair is examined first to check whether or not it can support a transmission rate above or equal to Z. If this is the case, CSI acquisition and exchange are avoided and switching between relay branches is reduced. On the contrary, when $b_{R_t R_r}^{(v-1)}$ cannot achieve Z, a new round of relay-pair selection is triggered.

12.6 5G Relay Selection

Here, the three relay selection policies are given which are tailored to a specific performance metric, set by the application taking place. The proposed policies combine the following three strategies:

- Max-link selection [28], where only one link is selected among the available SR and RD links, resulting in improved diversity. In [10], max-link with power adaptation was proposed to lower the power consumption.
- Successive opportunistic relaying (SOR), where two selected relays mimic FD operation, as one receives the source's signal while the other forward a previously received signal to the destination, providing spectral efficiency. Fixed-power SOR has been studied in [18], for throughput maximization. Moreover, SOR with power adaptation was the subject of [10], for power minimization.
- FD opportunistic relaying, where one selected relay transmits and receives on the same channel using its two antennas, offering throughput enhancement. FD with power adaptation was studied in [21], to reduce the power expenditure of the network.

As FD and SOR lead to FD operation, these schemes are denoted by FD, while max-link by *ML*. In the following, relay selection policies are introduced that can be employed in (1) high mobility, (2) delay critical, and (3) green communications scenarios.

12.6.1 High Mobility

In high mobility scenarios, constant CSI acquisition and processing is difficult, and fast relay selection must be performed in order to avoid disconnections and outages. For this reason, CSI is assumed to be available at the receiver-side only and thus, the transmitters use fixed power levels. SOR is omitted due to the incurred complexity of maintaining connection with two relays. As a result, only FD and max-link schemes are adopted, which offer increased throughput and robustness, respectively. Here, one best relay, denoted by R_b is selected and in subsequent transmissions it is examined first on whether or not it satisfies a SNR threshold denoted by Z, for FD transmission. So, in the uth time slot

$$b_{FD}^v = b^{v-1}, \text{ if } \min\{\gamma_{SR_b^v}, \gamma_{R_b^v D}\} \geq Z \qquad (12.13)$$

Otherwise, max-link selection is performed, where depending on the relay's buffer status, one or both its SR and RD links are investigated as follows:

$$b_{ML}^v = b^{v-1}, \text{ if } \max\{\gamma_{SR_b^v}, \gamma_{R_b^v D}\} \geq Z \qquad (12.14)$$

If Z is not satisfied, then all the available relays send feedback for their SR and RD links and the search for a FD relay transmission is initiated

$$b_{FD}^v = \arg \max_{i \in C} \min\{\gamma_{SR_i^v}, \gamma_{R_i^v D}\} \geq \gamma_0 \qquad (12.15)$$

If no FD relay can satisfy the SNR outage threshold $\gamma_0 \leq Z$, max-link is employed

$$b_{ML}^v = \arg \max_{i \in C} \max\{\gamma_{SR_i^v}, \gamma_{R_i^v D}\} \geq \gamma_0 \qquad (12.16)$$

12.6.2 Delay Critical

In many applications, low delay is the main target and relay selection policy must be adjusted toward delay minimization. As the average delay scales with the achieved throughput [29] for constant packet size, relay selection searches for a FD relay or a relay-pair, which operates in successive mode, in order to maximize the end-to-end throughput. This selection policy in a time slot v is formulated as

$$b_{FD}^v = \arg \max_{i,j \in C} \min\{\log_2(1 + \gamma_{SR_i^v}), \log_2(1 + \gamma_{R_j^v D})\} \qquad (12.17)$$

where $i = j$ if one FD relay is selected and $i \neq j$ when a relay-pair is selected.

When these modes fail, max-link is used as it improves the robustness of communication through increased diversity

$$b_{ML}^v = \arg \max_{i \in \mathcal{C}} \max\{\log_2(1+\gamma_{SR_i^v}), \log_2(1+\gamma_{R_i^v D})\} \quad (12.18)$$

12.6.3 Green Communications

As wireless networks migrate toward power-efficient deployments, power minimization is targeted. In these cases, power adaptation leads to the adjustment of the power level at the transmitters' side and defines relay selection. As proposed in [10], relay selection should take into account the minimum sum of powers required in a time slot v if full-duplex or successive relaying takes place. The power levels are calculated in such a way so as to fulfill a SNR threshold γ_0 set by the application. In this scenario, the source and the transmitting relay are able to perform power adaptation and the relay which transmits to the destination defines its power level as

$$P_{R_j} = \frac{\gamma_0 n}{g_{R_j D}} \quad (12.19)$$

Then, in the case of interference at the relay, the source power level is found for a specific SNR threshold γ_0 as in Equation 12.1. Thus, solving for P_{S_i}

$$P_{S_i} = \frac{\gamma_0(g_{R_j R_i} P_{R_j} + n)}{g_{SR_i}} \quad (12.20)$$

For the case of interference-free reception at the relay, the source power level is found by solving Equation 12.21 where the transmitting relay's power level does not affect the source's output power

$$P_{SR_i} = \frac{\gamma_0 n}{g_{SR_i}} \quad (12.21)$$

It is noted that Equation 12.21 holds when max-link is adopted.

So, for FD or SOR transmissions best relay(s) selection is defined by

$$b_{FD}^v = \arg \min_{i,j \in \mathcal{C}}(P_{S_i^v}^* + P_{R_j^v}^*) \quad (12.22)$$

When FD operation is infeasible, max-link selection exhibits power adaptation

$$b_{ML}^v = \arg \min_{i \in \mathcal{C}} \min\{P_{S_i^v}^*, P_{R_i^v}^*\} \quad (12.23)$$

It must be remarked here that power adaptation improves the performance of FD and SOR compared to fixed-power schemes as LI and IRI levels decrease due to the optimal power levels in the RD links.

12.7 Selection Coordination

It is worth mentioning that the majority of relevant works in the area of relay selection do not provide detailed descriptions of their implementations. There are several works that are related to the classic relay selection whose distributed implementation is presented in [15], and selection with outdated CSI, whose implementation is presented in [35]. In order to address practicality concerns on BA relay selection, in this part, both centralized and distributed implementations are discussed. For both approaches, it is assumed that the duration of pilot symbols is very small compared to the channel's coherence time of one time slot (see also discussion in [15]).

12.7.1 Centralized Coordination

First, in the centralized approach, the source is employed to select the relay-pair or FD relay in each time slot, while keeping track of the buffer sizes for all relays. At the start of each time slot, in phase 1, the destination transmits a pilot sequence to the K relays and the $\{D \to R_i\}$ CSI is estimated by each relay R_i. By assuming that the reciprocity property of antennas holds, relays can estimate the $\{R_i \to D\}$ CSI; similar arguments hold for the rest of the estimates, whenever necessary. Next, in the second phase, the relays take turns according to their indices, and broadcast pilots sequences to the source and the other $K-1$ relays. Thus, the source acquires the $\{S \to R_i\}$ CSI for all $R_i \in C$ and each relay R_j acquires the $\{R_i \to R_j\}$ CSI for all $R_i \in C$; also, in this step, the LI channels are estimated, that is, each relay R_j acquires the $\{R_j \to R_j\}$ CSI. The third phase starts by employing each relay R_j to sequentially inform the source on its buffer state and the CSIs of their respective $\{R_i \to R_j\}$ for all $R_i \in C$, as well as the $\{R_j \to D\}$ CSI. When all the relays have finished transmitting, the required information for centralized relay-pair selection has been gathered at the source.

For schemes that are based on power adaptation, selecting the relay-pair or relay that requires the least power expenditure involves the use of all the CSI that was acquired in the three phases. Compared to the cases where relays are assumed isolated and IRI is negligible as in [27], an additional overhead comes from the acquisition of the $\{R \to R\}$ and LI CSI, which are required in order to perform power adaptation and minimize the IRI and LI. The three phases of the centralized selection are shown in Figure 12.2.

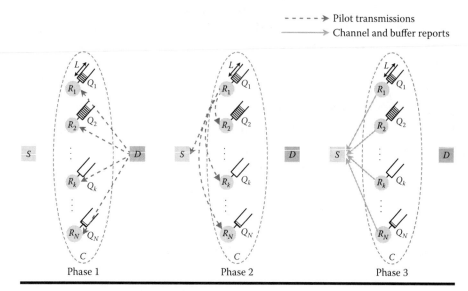

Figure 12.2   The three phases of centralized CSI and buffer status exchange.

12.7.2 *Distributed Coordination*

The distributed approach for the relay-pair selection process is based on the use of synchronized timers, as proposed in [15] for single relay selection. This approach was adopted in the distributed joint relay-pair selection of [32], where the idea of DSSC [33,34] was extended for the BA SOR scenario. In the first phase, the source broadcasts a pilot sequence and the K relays estimate the $\{S \to R\}$ CSI. Then, in the second phase, the destination sends pilot signals to the relays, which extract the $\{R \to D\}$ CSI. During the third phase, each relay takes turns in transmitting pilots to the other $K - 1$ relays that calculate the $\{R \to R\}$ CSI, while simultaneously, the LI CSI is estimated. The final, fourth phase requires that each relay notifies the rest $K - 1$ of its buffers status. Then, each relay sets its timer to be inversely proportional to the level of power minimization that can be achieved according to the total transmission power that is required. The four phases of the distributed selection are depicted in Figure 12.3.

Both centralized and distributed selection can be extended with DSSC in order to avoid constant rounds of CSI acquisition. In this case, the relay or relay-pair that was selected to perform an FD transmission, is examined first on whether or not it can satisfy the predetermined SNR threshold Z corresponding to the desired transmission rate. For the scenario of green communications, the threshold Z corresponds to a power value that must not be surpassed while achieving the desired transmission rate. If Z is not surpassed then the estimation of the CSI of the other

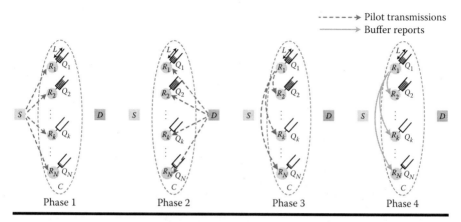

Figure 12.3   The four phases of distributed CSI and buffer status exchange.

relays is avoided. Also, one must note that the complexity of both implementations is $\mathcal{O}(K^2)$ as there will be $|\mathcal{P}| = K \times (K - 1)$ possible combinations.

As a final remark, when comparing the two approaches, although at first, the extra phase of distributed selection might be considered as an additional overhead, it is the third phase of the centralized approach that notifies the source of the various CSIs and buffer status that will introduce the largest amount of performance degradation, as pilot symbols are only a tiny fraction of a time slot's coherence time.

12.8 Numerical Results

In this part, the performance evaluation of the proposed relay selection policies is presented. More specifically, results are given for the average throughput for the high mobility application, the average delay for the delay critical application and the power reduction for the green communications application.

To examine the impact of the heterogeneous relay classes, three different pools of relays are compared. The first consists of eight UE, the second is a mixture of three UE, two battery-dependent mobile relays, two battery-dependent fixed relays, and one power-supplied fixed relay, denoted as MIX in the figures, while the last pool provides four fixed power-supplied (FPS) relays. The buffer size for a UE relay is set at $L = 4$, for a battery-dependent mobile relay $L = 8$ and for the fixed relays $L = 16$.

12.8.1 Average Throughput

In the first set of comparisons, the average throughput performance of the high mobility relay selection policy is examined. A fixed-rate transmission value equal

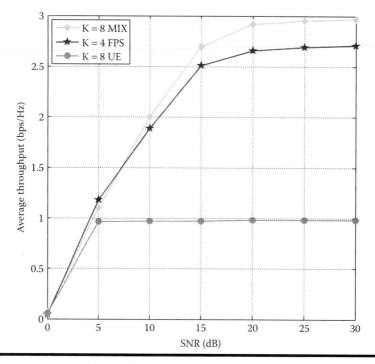

Figure 12.4   Average throughput for the high mobility scenario for increasing transmit SNR values.

to $r_0 = 2$ bits per channel use (bpcu) is imposed that changes to $r_0 = 3$ bpcu when the equivalent SNR threshold Z is satisfied. The results are depicted in Figure 12.4.

It is clear that the worst behavior is exhibited by the case when $K = 8$ UE are available as relays and only D2D communications are performed. This relay class is able to adopt only the max-link strategy, as FD capability falls victim to the devices' small size antenna isolation and low processing functionality, while, SOR is omitted in this scenario. The mixed pool of relays sports the best performance, mainly due to the battery-dependent relays, which transmit with lower power compared to the power-supplied fixed relay. As fixed power levels are considered in this scenario, LI is lower compared to the case of $K = 4$ FPS relays.

12.8.2 Average Delay

Next, Figure 12.5 shows results regarding the relay selection policy, which targets delay minimization. One may observe that the worst performance is achieved by the UE only the pool of relays. Although at the lowest transmit SNR value it has slightly better performance than the case of mixed relays, this can be attributed to the smaller buffer size of the UE. When transmit SNR increases,

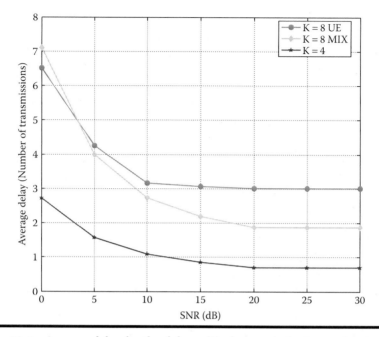

Figure 12.5 **Average delay for the delay critical scenario for increasing transmit SNR values.**

the set of mixed relay achieves better performance as it has an extra transmission mode, that is, FD relaying to compliment max-link and SOR, thus allowing for more packets to be transmitted from the relays' buffers. The best performance is observed by the four fixed relays. Although all the relays have larger buffers than UE and mobile relays, their number is smaller and they use all three different relaying modes.

12.8.3 Power Reduction

In the last comparisons, the power reduction performance of the Green Communications' relay selection policy is depicted in Figure 12.6. It must be outlined that when FD or SOR is employed, the overall power reduction achieved by the source and the transmitting relay is calculated. On the contrary, when max-link is used, only the power reduction of the source or the transmitting relay is calculated. Again, the UE pool of relays performs worse than the other two cases, as max-link is used in more instances. The other two cases perform similarly, with the case of $K = 4$ FPS relays offering slightly improved results. In this case, a trade-off can be seen between employing less relays offering better performance compared to more relays in the other two cases, at the cost of increased relay complexity and capital expenditure.

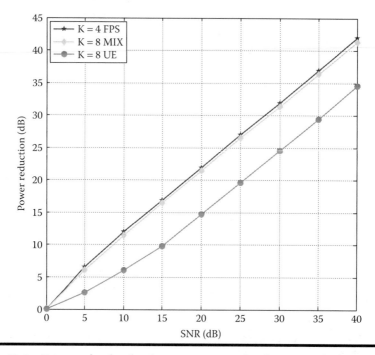

Figure 12.6 **Power reduction for the green communications scenario for increasing transmit SNR values.**

12.9 Open Challenges

The proposed algorithms address several challenges of wireless relay networks under specific assumptions regarding the symmetry of the network and the CSI. However, the area of CR is quite demanding and the continuous transformation and introduction of novel networking paradigms results in a dynamic and fertile research field. Thus, there are several future directions toward spectral-efficient heterogeneous networks.

Starting from implementation complexity, novel relaying algorithms providing near optimal performance with low CSI overhead are important. Such schemes should be developed under CSI at the receiver-only availability to compensate for fast changes in network topologies and relays with low processing and channel-estimation capabilities. Toward this end, more accurate CSI estimation and prediction techniques could decrease the cost of constant CSI acquisition and processing. Also, distributed synchronization should be examined as relay networks consist of multiple nodes which coordinate to achieve specific performance goals.

It is quite common in the literature that i.i.d. channels are considered aiming to simplify the performance analysis of the relay network. However, in reality i.n.i.d.

topologies are dominant and relay selection should be adjusted to this fact. So, networks with asymmetric channels should be further examined, especially for BA relaying where efficient scheduling algorithms should prioritize relays whose buffers tend to become empty or full in order to avoid diversity losses.

Next, interference mitigation is a field where contributions are expected to enable FD operation either using a single relay or relay-pairs as in the SuR paradigm. Network-coded interference cancellation can significantly reduce the impact of interference by exploiting a priori known interference signals at the receivers. Also, novel LI cancellation algorithms are needed to alleviate the HD constraint even for relays with lower processing capabilities, such as UE. Also, the feasibility of fast power adaptation should be examined as it is unrealistic at this point, to assume power adaptive transmissions per time slot especially for small channel coherence time.

Since UE are expected to be a major cooperative relaying type in future networks, D2D communications are a very important research field. Various scenarios are envisioned ranging from simple relaying of data to/from the core network, to direct D2D communications without any signalling from the overlay cellular network. Mobile providers are increasingly interested in D2D due to improved frequency reuse and better off-loading. However, to enable D2D, reliable CSI estimation from the UE is required in order to avoid interference toward other UE that communicate with nearby BSs. Furthermore, the incurred power expenditure from CSI acquisition and distributed coordination from the UE's side should be addressed by developing low-complexity, low-power techniques.

Another interesting technique relates to energy harvesting of renewable energy resources and radio-frequency signals. Depending on the type and capabilities of the relays, various levels of energy harvesting efficiency can be achieved. Toward this end, relaying algorithms should consider the energy harvesting profile of each relay, thus exploiting the potential of each relay type. From the operator's side BSs and infrastructure-based relay with energy harvesting capabilities could reduce the OPEX of the network, while lifetime maximization could be achieved for the UE, which adopt energy harvesting.

Finally, spectrum bands that were not exploited in previous generation of wireless networks are expected to play a significant role in overcoming the spectrum crunch of the currently used bands. Recently mm-Wave, free-space optical, and visible light bands were considered as possible solutions to increase the capacity of the fronthaul and backhaul of wireless networks. However, each band has unique characteristics that should be taken into account. Mm-Wave is prone to rain, while free-space optical exhibits reduced performance in heavy fog conditions. Thus, hybrid backhauling-fronthauling solutions should be developed to overcome such difficulties. Last, visible light communications can compliment the other spectrum bands and network engineers should harvest its potential by examining scenarios where such bands could be employed, for example, in indoor office environments where lighting and data transmission are merged.

12.10 Conclusions

In this chapter, the potential of Cooperative Relaying in 5G networks where different types of devices may act as relays was investigated. In such networks, the degrading role of interference was discussed and possible countermeasures, such as Interference Cancellation and Interference Avoidance, to improve network performance were given. Moreover, different relaying strategies were presented, such as opportunistic relay selection, successive relaying, full-duplex relaying, and buffer-aided relaying and their combination, led to the formulation of useful relay selection policies. Furthermore, a classification of relays based on their distinct characteristics was provided. In a similar manner, the level that each relay device can support different kinds of applications, which are of interest in 5G networks, that is, high mobility, delay critical, and Green Communications were discussed. Toward this end, three relay selection policies were presented targeting to improve the performance of each application. To alleviate possible practicality concerns, detailed descriptions for centralized and distributed network coordination were provided. To show the efficiency of the proposed algorithms, numerical results on the throughput, delay and power reduction performance comparing different relay types were included. Finally, interesting future directions were given as it is believed that the area of CR is open for contributions.

References

1. Demestichas, P., Georgakopoulos, A., Karvounas, D. et al., 5G on the horizon: Key challenges for the radio-access network, *IEEE Veh. Tech. Mag.* 8(3), (2013): 47–53.
2. Niyato, D. and Wang, P., Cooperative transmission for meter data collection in smart grid, *IEEE Commun. Mag.* 50(4), (2012): 90–97.
3. Zheng, K., Hu, F., Wang, W. et al., Radio resource allocation in LTE-advanced cellular networks with M2M communications, *IEEE Comm. Mag.* 50(7), (2012): 184–192.
4. 3GPP TR 36.814 v.9.0.0 Technical specification group radio access network; evolved universal terrestrial radio access (E-UTRA); Further advancements for E-UTRA physical layer aspects, (2010).
5. IEEE 802.16j IEEE standard for local and metropolitan area networks Part 16: Air interface for broadband wireless access systems amendment 1: Multihop relay specification, (2009).
6. Heath, R. W., Heath, R., Lozano, A. et al., Five disruptive technology directions for 5G, *IEEE Commun. Mag.* 52(2), (2014): 74–80.
7. Pahlevani, P., Hundeboll, M., Pedersen, M. V., Lucani, D., Charaf, H., Fitzek, F. H. P., Bagheri, H. and Katz, M., Novel concepts for device-to-device communication using network coding, *IEEE Commun. Mag.* 52(4), (2014): 32–39.
8. Riihonen, T., Werner, S. and Wichman, R., Mitigation of loopback self-interference in full-duplex MIMO relays, *IEEE Trans. Signal Process.* 59(12), (2011): 5983–5993.
9. Krikidis, I., Suraweera, H. A., Smith, P. J. et al., Full-duplex relay selection for amplify-and-forward cooperative networks, *IEEE Trans. Wireless Commun.* 11(12), (2012): 4381–4393.

10. Nomikos, N., Charalambous, T., Krikidis, I. et al., A buffer-aided successive opportunistic relaying selection scheme with power adaptation and inter-relay interference cancellation for cooperative diversity systems, *IEEE Trans. on Commun.* 63(5), (2015): 1623–1634.

11. Laneman, J. N., Tse, D. N. C. and Wornell, G. W., Cooperative diversity in wireless networks: Efficient protocols and outage behavior, *IEEE Trans. Inform. Theory* 50, (2004): 3062–3080.

12. Rankov, B. and Wittneben, A., Spectral efficient protocols for half-duplex fading relay channels, *IEEE J. Select. Areas Commun.* 8(2), (2007): 379–389.

13. Hoymann, C., Chen, W., Montojo, J. et al., Relaying operation in 3GPP LTE: Challenges and solutions, *IEEE Commun. Mag.* 50(2), (2012): 156–162.

14. Wang, C.-X., Haider, F., Gao, X. et al., Cellular architecture and key technologies for 5G wireless communication networks, *IEEE Commun. Mag.* 52(2), (2014): 122–130.

15. Bletsas, A., Khisti, A., Reed, D. et al., A simple cooperative diversity method based on network path selection, *IEEE J. Select. Areas Commun.* 24, (2006): 659–672.

16. Huang, W. J., Peter Hong Y. W. and Jay Kuo, C. C., Lifetime maximization for amplify-and-forward cooperative networks, *IEEE Trans. Wireless Commun.* 7, (2008): 1800–1805.

17. Ke, F., Feng, S. and Zhuang, H., Relay selection and power allocation for cooperative networkbased on energy pricing, *IEEE Commun. Lett.* 14, (2010): 396–398.

18. Nomikos, N., Vouyioukas, D., Charalambous, T. et al., Joint relay-pair selection for buffer-aided successive opportunistic relaying, *Wiley-Blackwell Trans. Emerg. Telecom. Techn.*, 25, (2014): 823–834.

19. Ikhlef, A., Michalopoulos, D. S. and Schober, R., Max-max relay selection for relays with buffers, *IEEE Trans. Wireless Commun.* 11, (2012): 1124–1135.

20. Riihonen, T., Werner, S. and Wichman, R., Hybrid full-duplex/half-duplex relaying with transmit power adaptation, *IEEE Trans. Wireless Commun.* 10(9), (2011): 3074–3085.

21. Nomikos, N., Charalambous, T., Krikidis, I. et al., Hybrid cooperation through full-duplex opportunistic relaying and max-link relay selection with transmit power adaptation, *Proceedings of the IEEE International Conference on Communications (ICC)*, (2014).

22. Day, B. P., Margetts, A. R., Bliss, D. W. et al., Full-duplex MIMO relaying: Achievable rates under limited dynamic range, *IEEE J. Select. Areas Commun.* 30(8), (2012): 1541–1553.

23. Fan, Y., Wang, C., Thompson, J. S. et al., Recovering multiplexing loss through successive relaying using repetition coding, *IEEE Trans. Wireless Commun.* 6, (2007): 4484–4493.

24. Tannious, R. and Nosratinia, A., Spectrally-efficient relay selection with limited feedback, *IEEE J. Select. Areas Commun.* 26, (2008): 1419–1428.

25. Nomikos, N. and Vouyioukas, D., A successive opportunistic relaying protocol with inter-relay interference mitigation, *Proceedings of the IEEE Wireless Communications and Mobile Computing Conference (IWCMC)*, (2012), pp. 228–333.

26. Wang, C., Fan, Y., Krikidis, I. et al., Superposition-coded concurrent decode-and-forward relaying, *Proceedings of the IEEE International Symposium Information Theory (ISIT)*, (2008), pp. 2390–2394.

27. Ikhlef, A., Junsu, K. and Schober, R., Mimicking full-duplex relaying using half-duplex relays with buffers, *IEEE Trans. Vehicular Tech.* 61, (2012): 3025–3037.

28. Krikidis, I., Charalambous, T. and Thompson, J. S., Buffer-aided relay selection for cooperative diversity systems without delay constraints, *IEEE Trans. Wireless Commun.* 11, (2012): 1957–1967.

29. El Gamal, A., Mammen, J., Prabhakar, B. et al., Optimal throughput-delay scaling in wireless networks—Part II: Constant-size packets, *IEEE Trans. on Inf. Theory*, 52(11), (2006), 5111–5116.

30. Horizon, Work programme 2014–2015, (2020), http://ec.europa.eu/research/participants/ portal/doc/call/h2020/common/1587758-05i._ict_wp_2014-2015_en.pdf (accessed April 20, 2016).

31. Bletsas, A., Dimitriou, A. G. and Sahalos, J. N., Interference-limited opportunistic relaying with reactive sensing, *IEEE Trans. Wireless Commun.* 9(1), (2010): 14–20.

32. Nomikos, N., Makris, P., Vouyioukas, D. et al., Distributed joint relay-pair selection for buffer-aided successive opportunistic relaying, *Proceedings of the IEEE International Workshop on Computer Aided Modeling Analysis and Design (CAMAD)*, (2013), pp. 318–322.

33. Michalopoulos, D. S. and Karagiannidis, G. K., Two relay distributed switch and stay combining (DSSC), *IEEE Trans. Commun.* 56, (2008): 1790–1794.

34. Michalopoulos, D. S. and Karagiannidis, G. K., Selective cooperative relaying over time-varying channels, *IEEE Trans. Commun.* 58(8), (2010): 2402–2412.

35. Michalopoulos, D. S., Suraweera, H. A., Karagiannidis, G. K. et al., Amplify-and-forward relay selection with outdated channel estimates, *IEEE Trans. on Commun.* 60(5), (2012): 1278–1290.

Chapter 13

Radio Resources Management Optimization in Cognitive Radio Networks

Anargyros J. Roumeliotis, Marios I. Poulakis,
Stavroula Vassaki, and Athanasios D. Panagopoulos

Contents

13.1 Introduction

The rapid evolution of mobile communication networks from simple voice systems to mobile broadband multimedia systems has significantly increased the spectrum demands. The spectral resources are allocated by each government through National Regulatory Authorities (NRA), such as the Federal Communications Commission (FCC) in the United States, the Office of Communications (Ofcom) in the UK, the Electronic Communications Committee (ECC) of Europe, and the Conference of Postal and Telecommunications Administrations (CEPT) in Europe, to the authorized holders such as the network operators. It is widely known, that the allocation process is performed under a spectrum auction procedure in most cases, based on the command and control model, where the leased radio frequency bands are provided for specific usage to the telecommunication companies whose monetary offer is larger. Consequently, these companies and more specifically their customers are called the officially licensed users of the radio spectrum. However, the specific spectrum management method results in the underutilization of the frequency bands (Kolodzy et al., 2002) due to the fact that the licensed spectrum remains unoccupied for large periods of time or large geographical regions as only the authorized users can exploit it. While the aforementioned problem is existent, the advent of new communication standards, emerging from the 5th-generation (5G) cellular networks, should provide much greater amounts of handheld devices, such as laptops, tablets, smartphones, and so on. In addition, 5G is expected to be a vertical-driven technology, which will have to satisfy strict requirements for a vast range of different use cases. In order to make this possible, much more bandwidth is going to be required.

To overcome the contradiction between the low utilization of the spectral resources, which is the result of the fixed radio spectrum allocation and the increasing demand for spectral bandwidth, the wireless paradigm of cognitive radio network (CRN) has been proposed as a promising solution (Mitola et al., 1999). The CRN concept is an efficient way to take advantage of the underutilized frequency bands and aims at increasing the spectrum utilization compared to the static spectrum-management policies. It consists of the licensed owners of the spectrum, which are called as primary users (PUs) or incumbent users, and the unlicensed or cognitive owners, which are called as secondary users (SUs), and coexist with the former ones under flexible spectrum allocation approaches. It is noticeable that, in general, the applied spectrum sharing techniques are focused on avoiding the harmful interference of SUs to the PUs' transmission, and toward this direction some level of cooperation between the PUs and the SUs may be often required.

Combining the aforementioned remarks, the role of cognitive radio technology to the emerging 5G wireless communication systems becomes critical. This can be explained considering that the main 5G requirements are to support ubiquitous mobility under unprecedented users' traffic demands and also to provide better quality of users' experience under broadband connectivity at any time, in any place, and on any device. The 5G technology is investigated as an evolution of the existing 2G, 3G, and 4G cellular communication systems to meet the plethora of the communication challenges, and its employment will start after 2020. Simultaneously, the mandatory need for sustainable future wireless networks contains the notions of spectral, energy, and cost efficiency from the operators' side and the scope of 5G systems compared to the current networks is a reduction to the aforementioned factors of about 3–5 times, and 100+ times, respectively (Liu and Jiang, 2016).

Furthermore, the demanding targets toward the evolution of next-generation communication systems are combined with the resource constraints of wireless networks. One major constraint is the spectrum unavailability based on the inefficient spectrum utilization rather than the scarcity of usable natural frequencies (FCC Report, 2002). Consequently, the investigation and implementation of improved and efficient resource allocation mechanisms to cope with the aforementioned technical communication challenges become crucial. Toward this direction, the main objective of this chapter is a thorough analysis on the important scientific aspects related to the radio resource management (RRM) optimization in CRNs based on recent literature and also presenting significant RRM optimization results for specific architectures and communication systems.

In addition, this chapter presents the main characteristics of cognitive radio technology and the basic key enablers for the 5G communications indicating that the former is a promising technology for the challenging 5G era. In addition, the description of the RRM's importance in synchronous wireless networks, such as the CRNs, to achieve the strict technical requirements is provided. Moreover, specific research studies for providing a comprehensive classification of the RRM issue in the CRNs and a critical discussion are presented in detail. For completeness of this chapter, other open or ongoing research issues of RRM optimization in CRNs are briefly reviewed in the Section 13.4. Finally, the chapter concludes with general comments and directions.

13.2 Cognitive Radio Networks

The spectrum access of the unlicensed users is feasible through the exploitation of spectrum holes or as called spectrum white spaces, which are licensed frequency bands that are allocated but occasionally utilized by the primary users. Moreover, a first step toward the practical functionality of CRN would be a possible existence of a CR platform to assist the coexistence between the PUs and the SUs. Due to the continuously time varying environment of the wireless networks, this CR platform must be able to sense its environment, by estimating the channel state information (CSI) and reconfigure, rapidly, its operating parameters, such as transmission power, frequency band, and modulation scheme, through a cognition process (Haykin, 2005). In particular, the cognitive radio is based on the software defined radio (SDR) concept (Jondral, 2005), which is a system that has the ability to operate under different

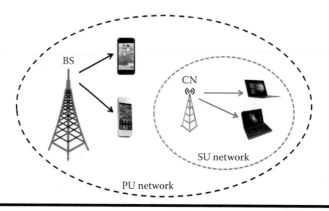

Figure 13.1   Basic cognitive radio topology. CN–Cognitive node, BS–Base station.

frequency bands, different standards, and various radio access networks and also provide diversified services with the same RF equipment. To supplement the previously mentioned analysis, a fundamental configuration of a cognitive radio system, where primary and secondary networks coexist, is presented in Figure 13.1.

Particularly, the SUs are cognitive devices that opportunistically access the spectral resources, depending on the access technique that is employed. To support efficient dynamic spectrum access (DSA), the cognitive terminals must be able to inspect (spectrum sensing) and analyze the spectrum status (spectrum management) with a view to finding and accessing the spectrum holes. At the same time, they have to take into consideration not only the coexistence with other cognitive devices (spectrum sharing), but also the necessity to change their operating frequency band in case of PU's transmission (spectrum mobility).

More specifically, in the phase of spectrum sensing the CR systems detect the spectrum usage and the transmission of the PUs and obtain crucial information for various parameters such as location, frequency band, transmission time, and power. Furthermore, the SUs can sense the spectrum on their selves (without any cooperation with the other SUs), using three main detectors that are the matched filter detector, the energy detector, and the cyclostationary feature detector. These detectors are based on the PU's signal-to-noise ratio (SNR) that is detected by a SU or under cooperation with other SUs. In particular, the matched filter detector, which is complicated, requires knowledge of primary's signal characteristics to compare the SU's received signal with the corresponding PU's known signal. On the other hand, the energy detector, whose detection is based on the energy of primary signal, is more simple and useful when the knowledge of PU's signal is absent. Finally, the cyclostationary feature detector detects PU by exploiting the statistics and mainly the periodicity in the pattern of the PU's transmission signals.

The case that SU receives the PU's signal with a SNR under an acceptable threshold, based on deep fading phenomena, makes the detection of the primary transmission by the SU infeasible. Hence, the SUs' cooperation is proposed as an ideal solution. The cooperative sensing is applied either in a centralized manner where a central controller, like a secondary base station, collects and exchanges the information with the SUs, or in a distributed manner where the SUs coexist in a without-infrastructure network and share their information with other SUs. Finally, even though the SUs acquire spectrum to transmit, it is crucial not to cause harmful interference to the PUs, which is guaranteed by the interference temperature limit that is established by the FCC (FCC Report, 2003). This metric reflects the total power that is received by the PU and generated by the noise sources and other emitters of the surrounding radio environment. In most cases, SU is obliged to change its transmission characteristics in order not to exceed this interference temperature limit.

During the spectrum-management phase, the obtained information by the spectrum sensing phase is analyzed and the SUs choose the appropriate spectrum for their transmission. Then, in the spectrum sharing phase, in contrast to the static spectrum allocation mechanism, the SUs exploit the provided frequency bands under a dynamic spectrum access. Following this dynamic process the SUs must continuously reconfigure their operation parameters in response to the changing environment and this mechanism ameliorates the system spectral efficiency. The DSA approach, as presented in Figure 13.2, is classified to the dynamic exclusive use model, open sharing model or spectrum commons model, and the hierarchical

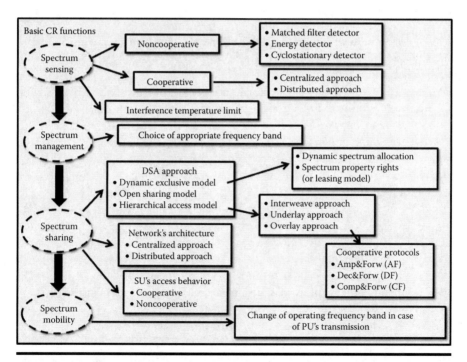

Figure 13.2 CR spectral functionalities.

access model (Zhao et al., 2007). The dynamic exclusive use model is separated in the dynamic spectrum allocation and the property rights approaches. The former ameliorates the spectrum efficiency through the exploitation of the spatial and temporal traffic statistics of different services (Xu et al., 2000), while in the latter, the PUs are considered to own the spectral resources and to cooperate with the SUs to ameliorate their performance by leasing a part of their licensed spectrum to the cognitive users.

While in the open sharing model, which is applied in the unlicensed industrial, scientific and medical (ISM) band, all the users have equal rights in the exploitation of the spectrum, in the hierarchical access model, the licensed spectrum is shared with the unlicensed users while guaranteeing that the latter do not harm the primary transmission. Moreover, the aforementioned spectrum sharing is classified in three techniques, called as interweave, underlay, or overlay (Goldsmith et al., 2009). More specifically, in the interweave technique the SU can transmit its data only with the absence of PU's transmission whereas in the underlay technique, the SU can simultaneously transmit with the PU as long as the caused interference to the primary receiver is below an acceptable threshold. Finally, according to the overlay technique, the SU acts as a relay for the PU and assists the latter's transmission in order to acquire time for its own (secondary) transmission. In this case the cooperative protocols that are mostly employed are the amplify-and-forward (AF) protocol, the decode-and-forward (DF) protocol, and the compress-and-forward (CF) protocol (Laneman et al., 2004).

Furthermore, except from the DSA approach, there are two other methods that supplement the analysis of the spectrum sharing phase and are based on the network's architecture and the SUs' access behavior (Akyildiz et al., 2006). On one hand, according to the network's architecture, the SUs access the spectrum under a centralized or a distributed manner. In the first approach the centralized entity controls and manages the spectrum allocation and the cognitive users' access, while in the distributed manner, which is deployed in a network without infrastructure (ad-hoc), each SU decides alone for the access technique depending on local information. On the other hand, the SUs' access behavior includes either the cooperation with the other SUs or the selfishness. The cooperative spectrum sharing deals with the joint communication of the unlicensed users, which possibly belong to the same service provider and maximize their system's performance under a common target. On the other side, in the noncooperative spectrum sharing each SU wants to satisfy its own benefits which are possibly in contradiction with the other users' benefits. In Figure 13.2, a brief block diagram about the basic concepts of CR spectral functionalities is presented, summarizing all that was previously discussed about the operation of cognitive networks.

The CR's standardization process is based on the actions of the IEEE 802.22 and IEEE Standard Coordinating Committee 41 (SCC41), which are the major counterparts of the CR standard (Hossain et al., 2015). The IEEE 802.22 working group on wireless regional area networks (WRANs) was formed in October 2004 and investigated the operation in the unused spectrum allocated to the TV broadcast service using the CR technology. Exploiting the CR's characteristics, the IEEE 802.22 WRANs standard targeted to guaranteeing nonharmful interference to the incumbent operation, such as digital TV and analog TV broadcasting and low power licensed devices like wireless microphones (Stevenson et al., 2009). In parallel, the IEEE SCC 41, whose ancestor was the IEEE P1900 established in March 2005, was renamed to IEEE DYSPAN-SC in 2010 and occupied with developing standards related to dynamic spectrum access networks for CR standardization (Murroni et al., 2011).

The CR protocol stack configuration is inevitable for the implementation of cognitive radio theory in the wireless networks. This protocol stack consists of the physical layer (PHY) component, the medium access control (MAC), network, transport, and application layers (Hossain et al., 2015). The first one includes the SDR module for the transmission and the reception of the signals while the other layers are implemented by adaptive protocols that take into consideration the changes in the environment. These layers are interconnected under a CR control named CR orchestration level that exploits appropriate algorithms to manage, efficiently, the information among these layers.

The security issue in the CRNs' functionality is crucial. The characteristics of the surrounding environment are subjective to various security attacks in multiple ways. Some of them are the denial-of-service (DoS) attacks, selfish (without considering the others SUs' demands and the utilized MAC protocol) and malicious users, which are either eavesdroppers (overhearing in the private transmitted data by other SUs or PUs), or jammers (causing intentional interference to other users and degrading their signals), preventing the normal operation among the PUs and SUs. Methods for diminishing these attacks consist of punishment mechanisms, such as bad reputation and authentication techniques by identifying the users. Moreover, due to the nature of CR spectrum access, they are also vulnerable to CNR-specific attacks like PU emulation attacks and spectrum sensing data falsification.

The advantages, if the CR concept starts to function, are obvious based on the scientific literature, which presents their wide application to many fields such as public safety, broadband cellular, and medical applications (Wang et al., 2011). On the one hand the police, fire, emergency medical, and public safety services must have reliable communications for every emergency scenario and on the other hand the quick access of the citizens to aforementioned services is essential. Due to the congested radio frequencies attached to public safety, the CRNs' usage becomes a promising tool for ensuring the communication reliability to a broader range of the radio spectrum. Furthermore, the CR technology expands the cellular networks not only by exploiting the television white spaces (TVWS) which provides more available spectrum to cellular operators for broadband applications, but also with offloading the traffic data generated in overloaded hotspots to the additional spectrum. Moreover, the unlicensed use of white spaces decreases the cellular operators' backhaul connection cost and increases the ability for nationwide coverage by ameliorating the rural areas' coverage. Finally, the CR's spectral flexibility is significant to the wireless medical networks due to their requirement for high quality of service (QoS), clean and less crowded spectrum band.

13.3 Cognitive Radio Technology in 5G Networks

13.3.1 Basic Requirements and Characteristics of 5G Networks

5G technology should support demanding communication characteristics as Table 13.1 describes in detail. A key feature for the 5G communications is the integration of the existing radio access techniques (RATs) with new technologies. The recent literature, such as in Boccardi et al. (2014), studies the transition from the cell-centric architectures to device-centric architectures as a possible technology direction toward the 5G implementation. Especially, in the cell-centric architectures the cell and the corresponding base station (BS) are the fundamental units of

Table 13.1 5G Performance Indicators

Performance Indicators	Definition	IMT-Advanced Enhanced Mobile Broadband	IMT-2020 Enhanced Mobile Broadband
Peak data rate (in Gbits/s)	Maximum achievable data rate under ideal conditions per user/device	1	10 or 20 (for certain scenarios)
User experienced data rate (in Mbit/s or Gbits/s)	Achievable data rate that is available across the considered target coverage area to a mobile user/device	10 Mbit/s	100 Mbit/s (for wide area coverage cases) or 1 Gbits/s (for hot spot cases)
Latency (in ms)	The contribution by the radio network to the time from when the source sends a packet to when the destination receives it	10	1

(Continued)

Table 13.1 (*Continued*)   5G Performance Indicators

Performance Indicators	Definition	IMT-Advanced Enhanced Mobile Broadband	IMT-2020 Enhanced Mobile Broadband
Connection density (devices/km^2)	Total number of connected and/or accessible devices per unit area	10^5	10^6
Mobility (in km/h)	Maximum speed at which a defined QoS (Quality of Service) and seamless transfer between radio nodes which may belong to different layers and/or radio access technologies (multilayer/-RAT) can be achieved	350	500
Energy efficiency (EE) (in bit/Joule)	(a) on the network side, EE refers to the quantity of information bits transmitted to or received from users, per unit of energy consumption of the radio access network (RAN) (b) on the device side, EE refers to quantity of information bits per unit of energy consumption of the communication module	×1 (as reference)	×100
Spectrum efficiency (bit/s/Hz)	Average data throughput per unit of spectrum resource and per cell	×1 (as reference)	×3 or ×5 (s.t. further research)
Area traffic capacity (in Mbit/s/m^2)	Total traffic throughput served per geographic area	0.1	10 (for hot spots)

Source: ITU-R Recommendation M.2083 (2015). *Framework and Overall Objectives of the Future Development of IMT for 2020 and Beyond*, Geneva, 2015.

the network, while in the device-centric architectures the smart devices communicate, exchange the information of their surrounding environment, and adjust their communication parameters properly based on their interconnection. Thus, the functionalities of the network come closer to the end-users. To have a much more realistic vision for the technical challenges of 5G networks, Table 13.1 summarizes the performance indicators and the corresponding international mobile telecommunications (IMT) requirements for the IMT-Advanced (4G) and IMT-2020 as modified by scientific research in (ITU-R M.2083, 2015). While these values may further be developed in new versions of ITU-R recommendations based on future investigation, they reveal the real targets of the performance indicators in future wireless networks.

13.3.2 Key Technologies for 5G Networks

The 5G cellular networks become realistic and promising through some key technologies, which are briefly described in the analysis here. These can be applied to the physical, MAC, network, and application layers, and they are also connected to the CR concept. More specifically, in the physical layer domain the major technology enablers are the employment of small cells and their densification, the flexible spectrum management including the millimeter wave (mm-Wave) wireless channels, the large scale antenna systems or Massive Multiple-Input Multiple-Output (Massive MIMO) and the full duplex radio technology, which is capable to double the spectrum efficiency of existing communication systems due to the nodes' ability to simultaneously transmit and receive in the same frequency band. Moreover, in MAC layer, new radio multiplexing techniques are proposed such as Filtered Bank Multicarrier (FBMC), generalized frequency division multiplexing (GFDM), Sparse Code Multiple Access (SCMA), and Interleave Division Multiple Access (IDMA) (Agiwal et al., 2016). Furthermore, in the core network the cloud radio access network (C-RAN), software defined network (SDN), and network function virtualization (NFV) are significant enablers for the upcoming 5G cellular systems and proposed as capital and operating expenditures reduction techniques for the operators. Finally, many applications that have been investigated under the 4G framework will be supported by the 5G wireless standard and the most important of them are the device-to-device (D2D) systems, vehicular communications, including vehicle-to-vehicle (V2V) and vehicle-to-infrastructure (V2I), machine-to-machine (M2M) systems, wearable devices, and Internet of things (IoT).

Due to the general consensus that a large increase in area traffic capacity is inevitable for the future communication systems, many recent works, such as Andrews et al. (2014), demonstrate that the extreme densification of the networks, called as ultra-dense networks (UDNs), consists an efficient way to achieve this technical challenge (Gotsis et al., 2017). In UDNs, the cell density is increased and becomes similar to the corresponding users' density. Moreover, the UDNs, including scenarios such as dense urban areas, stadium, campus, and apartments, are based on the heterogeneous networks (HetNets). The HetNets contain small cells of different size, like micro-, pico- and femto-cells, with different low transmission power levels coexisting under the macro-cell deployment. There are many benefits under the UDN configuration such as the coverage extension to the coverage holes, improvement of the spectral reuse, traffic offloading in the hotspot areas, and amelioration of the energy efficiency by bringing the network closer to the users. Besides the aforementioned advantages many challenges like the interference, the users' mobility, and the signaling overhead of CSI information exchanged among the nodes require further investigation.

Another promising technique toward the implementation of 5G mobile networks is the usage of the mm-Wave frequencies. The operation of the 5G systems is investigated under a unified air-interface framework employing frequency bands at lower and higher frequencies (Liu and Jiang, 2016). Especially, the necessity for more bandwidth that exists in the band of 20–300 GHz in combination with the congestion of the below 3 GHz or *beachfront* spectrum (Andrews et al., 2014) pushes the research to the mm-Wave frequency regime. In the mm-Wave bands the propagation environment is really challenging in terms of strong path and foliage loss attenuation, blockage vulnerability by moving obstacles, atmospheric attenuation (especially rain attenuation), and high penetration loss through the buildings (Rappaport et al., 2013; Kourogiorgas et al., 2015). The joint consideration of spectral and propagation characteristics of the high frequency bands makes mm-Waves reliable for high-data rates and line-of-sight (LOS) propagation in small cells, while the low frequency bands provide seamless coverage. Thus, it is easily concluded that the flexible usage of low and high frequency bands based on the communication environment is of utmost importance in future networks. Finally, another benefit of the mm-Wave frequencies is the consideration that these frequencies are reliable for self-backhauling through the highly directed beams transmitted by the corresponding array antennas.

The exploitation of the mm-Wave frequency bands paves the way to the Massive MIMO systems through the decrease of antennas' size, the decrease in their separation and the increase in the mm-Wave beams' directivity. The aforementioned characteristics enable each BS, or even the users, to be equipped with array antennas including tens or even hundreds of antenna elements, while the today's state-of-the-art systems, like LTE-A and 802.11ac, support up to 8 antennas per access node. The array antennas provide significant coverage, capacity, and resources reuse benefits due to the usage of the available spatial degrees of freedom.

As it has been already mentioned, many applications are considered as candidate technologies for the implementation of 5G communications. More specifically, the D2D systems contain mobile devices that communicate directly in short range without the assistance of the BSs and are deployed either on cellular spectrum, called as in-band D2D communication, or on unlicensed spectrum, called as out-band D2D communication (Asadi et al., 2014). The D2D concept targets to ameliorate the area traffic capacity, the spectrum efficiency, the coverage and offloading the backhaul. In a larger scale, M2M systems include huge amount of connected devices like smart grid components, metering, and sensors, whose communication requires high-data rate, very high link reliability, low latency, and real-time operation along with wide-coverage areas (Boccardi et al., 2014). The aforementioned characteristics are crucial in vehicular communications where the data for the traffic safety must be transmitted within a given time interval. Another important application within the future 5G infrastructure is the systems of wearable devices where the patients' health is monitored based on a body area network (BAN) and the recorded signals are transmitted in real time by sensors via gateways to a central server making available a more personal health care. However, collecting data in real time is bandwidth limited and the 5G wireless concept is a promising solution to this constraint (Oleshchuk et al., 2011). Finally, across the evolution to the 5G systems, the IoT concept gains large scientific and industrial interest intending to massive, ubiquitous, simultaneous wired and wireless internet connections of heterogeneous *things* with different intelligence, complexity, transmission power, energy and latency requirements (ITU-R M.2083, 2015), either in small scale like smartphones, sensors, cameras, and vehicles, or in large scale as smart grids, homes, industries, and agriculture.

13.3.3 Cognitive Radio's Role in 5G Systems

The importance of CR in the emerging 5G technology is apparent through the recent literature like Zhang et al. (2015b), Agiwal et al. (2016), Yang et al. (2016), and Panwar et al. (2016). More specifically, based on the major CRN's remarks of the aforementioned literature, the intelligent features of the CR devices such as the interaction with the environment, the cognition property, and the self-reconfiguration make them possible appropriate candidate equipment for the 5G terminals. Moreover, the tight relation between the 5G and CR technologies is strengthened by a plethora of similarities such as the ability to support different systems and networks, to contain new and flexible protocols, very advanced physical and MAC technologies, and an end-to-end integrated resource management that should include all the networks involved in the data-transmission process (Badoi et al., 2011).

Furthermore, according to Hong et al. (2014) the application of CR technology in a cellular system aids the latter to cope with the exploding and diverse mobile data traffic by expanding its operation from the licensed band with fixed, limited bandwidth to underutilized frequency bands without causing harmful interference to the incumbents, the licensed owners of these frequency bands. The low cost of leasing this opportunistic and unreliable spectrum, in comparison with purchasing a licensed band, renders the cognitive cellular networks a promising communication framework for the 5G mobile communication systems.

The importance of CR concept, primarily based on flexible spectrum management, has been widely investigated for many key technologies that are crucial for the 5G networks as the following indicative recent works prove. In Yang et al. (2016), the advanced spectrum sharing technique under technical and economical perspectives is investigated in 5G cognitive HetNets and a spectrum flow scheme, which implements both cooperative capacity offload and spectrum leasing, is proposed resulting in spectral and energy efficiency. The basic drawbacks of the current CR spectrum sensing mechanism through its application to 5G networks are given to Zhang et al. (2015b). Particularly, in the 5G framework the spectrum sensing equipment results in SUs' high cost and design complexity as well as in its additional energy consumption and waste of resources. Consequently, a new spectrum sharing scheme for the application of CR in 5G concept is proposed, whose main idea is based on removing the cognitive capability from SUs and let a new communication entity (a spectrum agent) perform spectrum sensing. This spectrum sharing mechanism targets to reduce the energy consumption and wasting of resources at the user terminal and ultimately improves the overall spectrum efficiency.

The CR technology has also been investigated under the UDN framework in Tseng et al. (2015) and Obregon et al. (2014). While in Tseng et al. (2015), graph- and genetic-based approaches are proposed in order to maximize users' throughput by minimizing the crucial for UDNs communication interference, in Obregon et al. (2014), the spectrum sharing between UDNs and radar systems, which have the role of SU and PU, respectively, is examined. More specifically, the scope is to analyze regulatory policies to ameliorate sharing conditions for indoor and outdoor UDNs operating at the S- and Ku-radar bands. Furthermore, CR applications to D2D systems are examined in Liu et al. (2016) and Cheng et al. (2012). In former, the secure D2D communication in energy harvesting large-scale cognitive cellular networks is investigated, where the energy-constrained D2D transmitter harvests energy from multiantenna equipped power beacons and communicates with the corresponding receiver using the spectrum of the primary base stations. Moreover, in Cheng et al. (2012), the CR and D2D communications are combined in cellular networks. Specifically, PUs can normally transmit via BS, while SUs can transmit either via BS or via D2D communication and the SUs' behaviors are analyzed using evolutionary game theory.

Another significant role of CR toward the implementation of 5G vision is its employment to vehicular communications that are an important part of the intelligent transportation systems (ITS). Especially, in Atallah et al. (2015) the usage of CR is investigated for further spectrum allocation to cope with the plethora of vehicular networking applications, such as traffic monitoring and collision-avoidance and in Mumtaz et al. (2015), a CR-based resource allocation policy is proposed to exploit white spaces as a mean of further offloading vehicular users. Finally, applications of CR have been also investigated in M2M communications. Specifically, in Aijaz et al. (2015), a cognitive M2M protocol stack is described for employment to IoT, while in Tragos and Angelakis (2013) the *energy efficiency* aspect is introduced to the CR's function and the CR-inspired smart objects (CRSOs) are proposed as high-energy efficiency solutions to the M2M systems due to their ability to adapt to environmental conditions.

As previously discussed, Massive MIMO technique is one of the key enablers for 5G technology. However, there are very recent works, such as Wang et al. (2015) and Barnes et al. (2015), that treat with the combination of Massive MIMO and CR systems. Particularly, in Wang et al. (2015) an underlay CR system, which includes a multiuser Massive MIMO primary network and a multiple-input single-output (MISO) secondary network, is studied and a tight lower bound of the average achievable rate, which can be used to measure the performance for any finite numbers of antennas, is derived. Moreover, in Barnes et al. (2015) the authors investigate a CRN whose secondary base station (SBS) is equipped with Massive MIMO systems and there is cooperation between this SBS located in a microcell with a corresponding relay station in a femtocell, under the deployment of a primary base station. The scope of this work is the maximization of the energy efficiency of the overall CRN while satisfying the QoS requirement of each SU given specific interference constraint for each PU.

Across the evolution to 5G, the tremendous increase of traffic load, arising from the greater number of both active users and access nodes, makes inevitable the appropriate redesign of the current backhaul links due to the fact that their static design is useless based on the mutable operation dynamics. Except from the technologies like copper, optical fiber, satellite links, micro, and mm-Waves, the CR consists an equivalent powerful technology proposed for backhaul connection providing low energy consumption. More specifically, in Lun et al. (2014) a cognitive energy efficient backhaul deployment scheme, implementing the reinforcement learning theory, is investigated based on the backhaul link diversity where the system's base stations are able to connect to different available backhaul links that are selected according to different selection schemes and QoS constraints. Finally, toward the direction to the software-driven networks in Ziegler et al. (2015), the *Cognitive and Cloud-Optimized Network Evolution* (CONE) end-to-end architecture is deployed including both 5G access domain and 5G era horizontal layers under cloud, software-defined, and cognitive networks framework. The CR characteristics, such as real-time collection of networks and usage data, associated analysis and automated proactive action, enable the automated, optimized network operation and the automated management of the resources in 5G systems.

The previous brief survey in the recent literature reveals the critical role of CR to the future wireless communication ecosystems, like the 5G. It is obvious that CR significantly benefits the 5G configuration. The high degree of interconnection among 5G and CR is demonstrated in the following Figure 13.3, which summarizes the relationship of these innovative concepts toward the direction of the promising IoT.

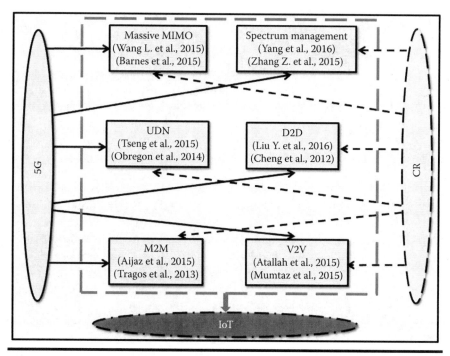

Figure 13.3   Role of cognitive radio in 5G.

13.4 Radio Resources Management in Cognitive Radio Networks

Due to the crucial impact on the performance of the wireless systems, the appropriate exploitation of the radio resources has been investigated in CRNs under the framework of various mathematical theories. In this section, a promising research field known as RRM optimization is discussed. First, to understand the general optimization framework, details of the optimization objective functions are given presenting basic characteristics of the utility theory. The utility functions are the core of the problems' optimization structure on which the RRM techniques are deployed. Subsequently, some of the most basic concepts of RRM in the CRNs, widely met in scientific literature, are described. The issues connected to the RRM field include problems referring to the power control, the channel allocation (i.e., a bandwidth portion around a central frequency), the time allocation to each user of the CRN called as scheduling or partitioning, and the efficient users' pairing whose careful examination is inevitable to ensure the amelioration of PUs' and SUs' performance under multiple requirements and different incentives.

13.4.1 Basic Utility Theory

The cognitive RRM (CRRM) field targets to the appropriate allocation of CRN's radio resources based on the users' preferences modeled by functions that come from the utility theory, which has been greatly applied in economics. In particular, the utility quantifies the users' objectives, acquiring greater value as the latter approach their goals and including either one or more performance metrics. Under the CR framework, the utility function can capture some of the most important network performance metrics (figures of merits) such as the capacity (e.g., ergodic, outage, effective, and secrecy capacity), the energy efficiency, and the spectral efficiency. In the following analysis, some significant scientific works are described to highlight the methodologies that are applied to the CRRM framework.

Especially, ergodic capacity is determined as the average Shannon capacity over all the fading wireless channels (block fading channels), which are mostly characterized by fast fading phenomena. On the contrary, in case the channels face slow fading phenomena, the outage capacity is employed (Ozarow et al., 1994), denoting the maximum achievable constant rate that can be maintained over fading blocks under a specific outage probability. Both ergodic and outage capacity concepts are investigated, from the perspective of the SUs, under the spectrum sharing process in Kang et al. (2009). Furthermore, optimal power allocation schemes are proposed to maximize the corresponding capacity, for different fading channels statistics subject to peak/average transmit and/or peak/average interference power constraints. In this case, a capacity gain for the SU is noted considering the average interference power constraints compared to the corresponding peak constraints.

The theory of effective capacity has been developed to deal with the challenge of modern communication systems to provide higher QoS level for the wireless applications. This performance metric is defined as the maximum constant arrival rate that can be supported by the channel to guarantee a QoS requirement and is particularly convenient for analyzing the statistical QoS performance of wireless transmissions where the service process is driven by the time varying wireless channel (Wu et al., 2003). Two paradigms of the effective capacity application in the CRN framework are presented in Vassaki et al. (2014) and Roumeliotis et al. (2016). In particular, the former work investigates the optimization of effective capacity, using the convex optimization theory, from the SU's perspective in an underlay cognitive network where the power allocation mechanism allows the SU's transmission with maximum power when the PU experiences an outage event due to the randomness in its channel state. Moreover, except from the traditional interference power constraint, an interference constraint, which is based on the inverse signal-to-interference-plus-noise ratio (ISINR) of the primary receiver, is investigated and the simulation results reveal that this specific constraint benefits the SU and leads to lower values of PU's outage capacity. Regarding the work in Roumeliotis et al. (2016), the authors study the maximization of PU's effective capacity under the overlay approach in a Gaussian orthogonal relay model and a

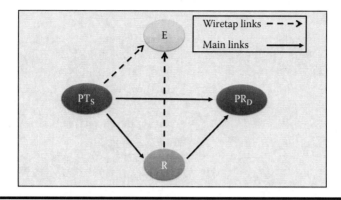

Figure 13.4 Basic configuration of a primary network with a relay and an eavesdropper.

time allocation mechanism is proposed, using the convex optimization theory, considering, jointly, the satisfaction of the minimum SU's effective capacity and the maximum SU's instantaneous power. The simulation results show the amelioration of PU's effective capacity as the SU's effective capacity or SU's QoS requirement take lower values.

Furthermore, due to the nature of CRNs, the physical layer security plays a critical role for the protection of PUs and SUs, which are vulnerable to various attackers such as eavesdroppers (E) and jammers. The secrecy capacity concept (Wyner, 1975) is defined as the maximum rate at which the information can be securely transmitted from the source (S) to the corresponding destination (D). Formally, for a Gaussian channel, secrecy capacity is expressed as the difference of the Shannon capacity of the SD link, called as the main link, and the Shannon capacity of the SE link, called as the wiretap link. In Figure 13.4, the basic configuration of a primary network (PT–PR pair) with a helper relay (R), which enhances the PT's secure transmission and an eavesdropper, is depicted according to the aforementioned notions.

During the past years, secrecy capacity has been applied extensively in wireless networks (Poulakis et al., 2016) and particularly in CRNs (Zou et al., 2014). In the latter work, a multiuser multieavesdropper CR system with multiple cognitive users (CUs) transmitting to a common cognitive base station (CBS) is considered, while multiple eavesdroppers act cooperatively or independently in intercepting the CUs–CBS transmissions, called as coordinated and uncoordinated eavesdroppers, respectively. Furthermore, the secrecy outage and diversity performance of this system are studied and closed-form expressions of the secrecy outage probability, for both coordinated and uncoordinated cases, are derived according to various scheduling schemes, such as the round robin scheme, the optimal user scheme (i.e., the CSI of all links is available), and the suboptimal user scheme (i.e., only the CSI of

CUs–CBS links is known). Another noticeable remark is the secrecy diversity order of M, where M is the number of CUs, for both optimal and suboptimal scheduling, while the round robin scheduling achieves order of only one. To conclude, all the scheduling schemes are improved as the maximum allowable interference level increases and the round robin scheduling has the worst performance with respect to the secrecy outage probability.

Finally, the technical challenges of wireless communications render substantial optimization of efficiency metrics, such as the EE and the spectral efficiency (SE), which are defined as the ratio of the rate over the power. More specifically, the maximization of EE constitutes a problem with significant impact on CR systems, which is apparent through Haider et al. (2015) and Naeem et al. (2013). Especially, in Haider et al. (2015), an interference-tolerant CRN is investigated across the spectral and energy efficiency trade-off, given specific power and interference constraints. While in the low SNR regime, the EE is better when the signal is transmitted with average power constraint, in the high SNR regime peak or average power constraints lead to similar EE. Moreover, in Naeem et al. (2013), an optimal power allocation policy under a fractional programming approach is derived to maximize the EE of the SUs' transmissions in a multiuser green CR, while guaranteeing the interference constraints due to the PUs' existence. The authors note that the EE increases with the number of SUs and decreases with the increase in number of PUs as the optimization problem becomes stricter.

13.4.2 Optimization Approaches

There are multiple criteria for the classification of the optimization problems associated with the type of objective and constraint functions. The widest category is the one of nonlinear problems whose subcategory is the convex problems. Moreover, the optimization problems are separated into continuous and discrete based on the kind of described problem, while a critical mathematic tool for discrete problems' solution is the integer programming. All the aforementioned categories are related with the construction of the optimization problem. This problem includes the maximization or minimization of an objective function, which may be a utility function, by determining the appropriate values of the optimization variables under specific constraint functions. According to the problem's complexity, the optimal solution can be derived in a closed-form expression or approximately through a heuristic approach. Especially, the linear optimization problems require polynomial time (P) to find the solution, while the integer optimization problems have larger complexity and belong to the nondeterministic polynomial time (NP) problems requiring various relaxation techniques to be approximately solved.

More specifically, convex optimization theory is a greatly applied theory guaranteeing that if a local minimum exists, this is also the global minimum. Considering that bold indexing in all equations provided next refers to vertices or matrices, the problems' structure is defined as (Boyd et al., 2004):

$$\text{minimize } f_o(\mathbf{x})$$
$$\scriptstyle \mathbf{x}$$

$$\text{subject to } f_i(\mathbf{x}) \leq 0, i = 1, ..., m \tag{13.1}$$

$$h_i(\mathbf{x}) = 0, i = 1, ..., p$$

The characterization of an optimization problem as convex includes not only the satisfaction of convexity properties for the objective function (i.e., $f_o : \mathbf{R}^n \to \mathbf{R}$), the functions of inequality constraint (i.e., $f_i : \mathbf{R}^n \to \mathbf{R}$), and the set of points, (i.e., $\mathbf{x} \in \mathbf{R}^n$), for which the objective and all constraint functions are defined, but also the affinity of the equality constraint functions, (i.e., $h_i : \mathbf{R}^n \to \mathbf{R}$). The scope of convex optimization is to find the values of the optimization variable $\mathbf{x} \in \mathbf{R}^n$ according to the problem's formulation. The most important method, which results in closed-form solution, is the Lagrangian method whose function is expressed as $L = f_o(\mathbf{x}) + \sum_{i=1}^{m} \mu_i f_i(\mathbf{x}) + \sum_{i=1}^{p} \lambda_i h_i(\mathbf{x})$, and μ_i, λ_i are the Lagrangian multipliers whose expressions are derived by $\partial L/\partial \mathbf{x} = 0$. A recent work of convex optimization theory in CRNs is described in Wang and Liu (2015) where the time allocation problem to optimize the SU's effective capacity is studied under a specific PU's effective capacity constraint and constant SU's power level. Furthermore, the comparison between an optimal time-slot allocation scheme, considering both the channel conditions and PU's delay QoS requirement, and a fixed time-slot allocation scheme, which only varies with PU's delay QoS requirement, reveals the superiority of the first scheme through the simulation results.

Another class of convex optimization problems is the second order cone programs (SOCP) that involves conic expressions under continuous optimization variables and its formal structure is (Boyd et al., 2004):

$$\text{minimize } \mathbf{f}^T \mathbf{x}$$
$$\scriptstyle \mathbf{x}$$

$$\text{subject to } \left\| \mathbf{A}_i \mathbf{x} + \mathbf{b}_i \right\|_2 \leq \mathbf{c}_i^T \mathbf{x} + d_i, i = 1, ..., m \tag{13.2}$$

$$\mathbf{F}\mathbf{x} = \mathbf{g}$$

where $\mathbf{x} \in \mathbf{R}^n$, $\mathbf{f} \in \mathbf{R}^n$, $\mathbf{A}_i \in \mathbf{R}^{n_i \times n}$, $\mathbf{b}_i \in \mathbf{R}^{n_i}$, $\mathbf{c}_i \in \mathbf{R}^n$, $d_i \in \mathbf{R}$, $\mathbf{F} \in \mathbf{R}^{p \times n}$, $\mathbf{g} \in \mathbf{R}^p$ and the inequality constraint $\left\| \mathbf{A}_i \mathbf{x} + \mathbf{b}_i \right\|_2 \leq \mathbf{c}_i^T \mathbf{x} + d_i$ is called as second-order cone constraint due to the inequality between a norm and an affine expression. An example for the implementation of SOCP in CR framework is described in Liu et al. (2011). A robust downlink beam forming method with power control under a multiuser MISO CRN is investigated for minimizing the SUs' transmit power while targeting to a lower bound on the SUs' received SINR and imposing an upper limit on the interference power at the PUs. The simulation results show that the proposed method has better performance than a nonrobust method with respect to channel changes.

To extend the applications of optimization theory, some fundamentals of non-convex optimization theory based on the difference of two convex functions (DC) are presented. This objective function formulation aims at coping with the RRM problems in nonconvex wireless networks generally, for example, in Poulakis et al. (2016) and more specifically in CRNs such as in Alvarado et al. (2014). The latter work involves an indicative usage of DC programming to find the SUs' transmission power levels for their sum-rate maximization over an underlay CR MIMO network subject to interference constraints for PUs' protection. The proposed distributed algorithm is comparable with a centralized scheme according to their performance. Furthermore, another optimization method with many applications in CRRM, such as in Naeem et al. (2013) that has been described in Section 13.4.1, is the fractional programming where the objective function is formulated as the ratio of two functions that contain the optimization variable.

In the CRRM field, there are many optimization problems whose optimization variables take integer values. If all the optimization variables require integer values then the optimization model is called as integer programming (IP) model. On the other hand, the optimization model is known as mixed integer programming (MIP) model when certain optimization variables take integer values while other take real values. Subsequently, if the objective and constraint functions are both linear then the optimization problem is known as ILP or MILP, depending on which category (IP or MIP, respectively) it belongs. Furthermore, the integer programming is strongly related with combinatorial optimization problems, which study all the possible combinations of network's radio resources to select the best combination in order to ameliorate the network performance. A paradigm of MILP in CRRM framework is examined by Hoang et al. (2006) where a CRN with multiple cells that include a base station supporting cognitive users is considered. The authors investigate channel-allocation/power-control schemes that maximize the spectrum utilization of the CRN, under opportunistic spectrum access, while minimizing the interference to PUs. Due to the NP-hard complexity of MILP problem, a heuristic solution is proposed based on the dynamic interference graph theory and the superiority of the proposed scheme in terms of the overall number of network's served users is obvious through the comparison with other literature's algorithms.

The optimization techniques in CRRM also include the dynamic programming (DP), which treats with the solution of a complex problem through the solution of a collection of simpler subproblems appropriately combining these series of solutions to give the best one for the given problem. This optimization method has a great usage in the Markov Decision Process (MDP), which provides a mathematical framework for modeling decision making in situations where the decision maker can partly control the outcome due to stochastic conditions. The MDP is characterized by the Markov property, which means that the conditional probability distribution of future states of the process depends only on the current state, not on the sequence of events that preexisted. A CRN's dynamic programming scenario is investigated in Yu et al. (2016) under a QoS differential transmission scheduling problem in the CR-based smart grid communications networks. The different QoS requirements of smart grid users (SGUs) are associated with the different priorities of SGUs in the smart grid. The scheduler allocates the channels, based on the proposed priority policy, considering the minimum transmission delay of SGUs under a semi-MDP approach with infinite stages and solved by the methodology of adaptive DP. In conclusion, the proposed priority mechanism guarantees low transmission delay for high-priority SGSUs and in the case of emergency data the transmission delay decreases to a satisfying low level for all users ensuring the differential QoS provision in the smart grid.

Finally, the stochastic programming consists of another optimization theory involving the uncertainty, which is characterized by a probability distribution on the parameters into the objective and the constraint functions. This uncertainty leads to the randomness of the solutions and the optimal objective value of the optimization problem. A widely known approach to cope with the stochastic programming is called as two-stage recourse model. In the first stage, the decision entity makes a decision which is influenced by some uncertainty and in the second stage this entity makes another decision, called as recourse decision, in response to the random event preceded. The recourse decisions target to the best correction of first stage decisions through the minimization of the expected costs of all decisions taken with respect to each random outcome. In work of Almalfouh et al. (2012) a problem of joint spectrum-sensing-duration design and power control in a point-to-point CR link under the stochastic programming is investigated. This problem is formulated as a two-stage stochastic program with recourse in order to maximize the CR achievable throughput subject to specific PU's interference constraint. The simulation results reveal that, under perfectly known CSI, the throughput achieves the highest values and shorter value of sensing duration is required for a given PU's interference constraint, while longer sensing duration is required as the PU's allowable interference becomes stricter.

13.4.3 Game Theory

It is widely known that users' decisions on their function parameters and spectrum usage are influenced not only by the environment, but also by the other users' actions, due to the coexistence between the PUs and SUs under the CRN framework. These users act cooperatively or selfishly according to their incentives, and their multiple interactions can be analyzed in a game structure, through the mathematical tool of game theory. The formal game's definition includes the set of players, that is, PUs and SUs, the action space, that is, the set of strategies and users' utility/payoff, which describes the result of the game for each user. Applications of game theory in various OSI layers in wireless networks can be found in Charilas and Panagopoulos (2010a). The multifold application of game theory in CRNs (Wang et al., 2010) contains games based on multiple categories such as the users' noncooperation, for example, potential games; users' cooperation, for example, bargain and coalitional games; economic games, for example, Cournot, Bertrand, and Stackelberg games discussed in Wang et al. (2010) and Yu (2013); auction games and games based on mechanism design.

In noncooperative games, where each user acts selfishly in order to achieve its goals, the Nash equilibrium (NE) is the solution concept. The notion of NE contains the actions' and utilities' sets of each user at a specific point where no user can benefit by changing them, while the other users keep their corresponding sets unchanged. Due to the difficulty in proving game's convergence to NE, different special structures of noncooperative games, for example, potential games, have been investigated that guarantee the existence of NE. Potential games are met in CRRM field such as in Zhong et al. (2014). The authors study the joint relay selection and discrete power control problem in a relay CRN for maximizing the rate of the secondary system given the interference power constraint at the primary receivers and the total available power constraint for the secondary relays. The latter are regarded as the game players and the power levels are the pure strategies for each SU. In addition, the utilities, based on SUs' rate, power, and penalty functions, ensure the feasibility of the network without advance knowledge of feasible power strategy profiles. Furthermore, it is proven that the proposed game not only has at least one feasible pure strategy NE, but also an optimal solution to the rate maximization problem constitutes a feasible pure strategy NE. Finally, a centralized iterative algorithm, based on best response dynamic, and decentralized algorithm, based on learning automata, are proposed, and the simulation results reveal that both algorithms have less complexity than an exhaustive search algorithm and can achieve optimal or near-optimal performance.

Besides the users' selfishness in noncooperative games, in cooperative games, largely applied in CRRM such as in Zhang et al. (2015a) and Li et al. (2010), the users' cooperation is investigated. In the first work the cooperative Nash bargaining game theory is analyzed in cognitive small cells through a joint uplink sub-channel and power allocation problem considering the cross-tier interference mitigation, minimum outage probability requirement, imperfect CSI and minimum rate requirement. The existence, uniqueness, and fairness of the solution to the aforementioned game are proved and the proposed Nash bargaining resource allocation mechanism converges to a Pareto-optimal equilibrium. Moreover, the evaluation of the proposed algorithm through simulations presents its superiority based on a better trade-off between capacity and fairness compared with a centralized maximal rate and round robin approaches. In Li et al. (2010), SUs act as relays and assist the transmissions of operators' PUs to gain access for their own transmission in the spare operators' spectrum. The fact that multiple SUs compete with each other to acquire spectral access from multiple operators, while the latter compete with each other to get assisted from multiple SUs, results in modeling this scenario as coalitional game with a nonempty core. To conclude, the authors investigate the behavior of SUs' access rates and operators' data rates as solution to an optimization problem with a concave objective function and linear constraints leading to a transferable coalitional game.

Moreover, other game theory approaches met in CRN framework are the auction theory and mechanism design. More specifically, in Khaledi et al. (2013) the spectrum allocation problem is studied under a scenario of auction theory where a primary spectrum owner (PO), which acts as the auctioneer, sells idle spectrum bands to make a profit and the SUs bid to buy spectrum bands from the PO. The CRN's channels are nonidentical, that is, have different qualities, the SUs express their preferences for each channel separately and each SU submits one vector of bids for each channel. Furthermore, numerical results show improvement in social welfare, SUs' channel capacities and PO's revenue performances in comparison with the case of identical channels. In terms of mechanism design, two algorithms are proposed in Wang et al. (2008) to suppress cheating and collusion behavior of selfish users for the improvement of spectrum efficiency. In the first algorithm, under the Bayesian mechanism design, the selfish users' incentive to cheat is eliminated and the cooperative unlicensed spectrum sharing is achievable under the threat of punishment. In the second one, under a collusion-resistant dynamic spectrum pricing auction game, licensed spectral resources are shared among PUs and SUs and users' collusion is decreased by setting up the optimal reserve price in the auction.

13.4.4 Matching Theory

The large heterogeneous deployment of synchronous wireless communication systems leads to extreme signaling overhead and high complexity. These characteristics make the turn inevitable in the RRM field from the investigation of centralized approaches to the distributed and self-optimizing mechanisms. The necessity for distributed radio resource allocation paves the way to the application of matching theory (Gale et al., 1962; Roth et al., 1992), which describes the creation of mutually beneficial relations between different sets of agents, as an efficient mathematical tool for the design and analysis of wireless networks. The great impact of matching theory is apparent by its adoption from the plethora of works in the scientific literature and especially the great application in CRNs that will be described next, as well as from its adoption in many real-world systems such as the U.S. National Resident Matching Program.

More specifically, under the framework of matching theory, the CRNs can be modeled as matching markets, which are separated as one-sided, that is, competitive markets and two-sided markets. In two-sided markets, most of the applied algorithms are based on a stable matching algorithm called as deferred acceptance (DA) algorithm (Gale et al., 1962). For the matching process, according to the DA mechanism, each user (PU/SU) makes a ranking list from the most to the least preferable users of the other set (SU/PU) according to its preference relation, which is based on its utility function. After the lists' construction, each user of one set (PUs/SUs) makes proposal for matching to the users of the other set (SUs/PUs) that belong to its preference list and the latter accept or reject the proposal. This procedure is repeated until each of the rejected users (PUs/SUs) makes proposal to all preferable users (SUs/PUs) and results either in stable matching between PUs and SUs or in unmatched users. A matching can be regarded as stable only if it leaves no pair of users on opposite sets, which were not matched to each other but would both prefer to be.

There are various classifications of matching models depending either on the users' quotas or the construction of the users' preferences (Gu et al., 2015). According to the users' quotas, the one-to-one matching (O2O) model (e.g., stable marriage problem) is related with the pairing of each PU with at most one SU, while the one-to-many (O2M) (e.g., college admissions problem) and many-to-many (M2M) matching models are associated with the matching between many SUs and one or many PUs, respectively. Based on the users' preferences, that is, utility functions, there are two mostly used subcategories, the canonical matching, where the preferences depend solely on the information available at each specific user, and the matching with externalities. In the latter subcategory, there are interdependencies between the users' preferences, that is, individual users' preferences are affected by the other users' preferences and also by the current matching.

Matching theory has an extensive deployment in CRRM field, mostly in the management of spectrum sharing and users' cooperation, as demonstrated by the following brief discussion on different works of recent literature. In Leshem et al. (2012), a multiuser CRN's spectrum allocation problem is studied under an O2O matching model, where the wireless channels play the role of PUs and the utility functions are the ergodic capacities of each cognitive user on each channel. The authors prove not only the existence of a unique stable matching of their proposed matching algorithm, which is based on DA scheme, but also the algorithm's time efficiency to lead to the stable allocation. Moreover, the authors (Bayat et al., 2013) consider an overlay CRN including multiple PUs and SUs and propose a distributed O2O users' pairing mechanism, based on auction and matching theory, whose utility functions consist of the rate and monetary factors that are applied in complete and partial received SNR scenarios. This mechanism ensures that both the PUs' and the SUs' minimum rate requirements are satisfied and contains the negotiation between PUs and SUs on the amount of monetary compensation and the time slot allocation that PUs propose to SUs for spectrum access or to relay the PUs' signal. To conclude, it is proven that the algorithm results in the best possible stable matching, which is weak Pareto-optimal and its performance is comparable to an optimal centralized solution. The O2O matching model is also considered in Feng et al. (2014) where the cooperative spectrum sharing is studied in a CR relay network between multiple PUs and SUs. The SUs act as relays for the PUs and gain time for their own transmission under complete, partially incomplete, and incomplete information scenarios. These information scenarios are related with the PUs' knowledge about the SUs' private information, which includes the SU's power transmission and channel gain of the link among secondary transmitter and primary receiver. In terms of PUs, the utility function incorporates the average data rate during the entire time period by employing SUs, while the SU's utility function is the difference between the benefit of its own transmission and the cost of its total energy consumption. In addition, the distributed proposed algorithm, which is based on the generalized Gale-Shapley algorithm and the English auction, leads the PUs to the unique Pareto-optimal equilibrium in the partially incomplete information scenario, while it converges to a stable equilibrium, in the incomplete information scenario. Finally, the authors notice that the PUs' total utilities increase with the number of SUs, and when the SUs' number is far larger than the PUs' number, the utility losses caused by partially incomplete and incomplete information are negligible.

Besides the O2O matching models, there are many CRRM works occupied with the application of O2M matching models such as Vassaki et al. (2015) and Roumeliotis et al. (2015). Both works consider the leasing process under a multiuser cooperative overlay CRN, where SUs devote a part of their transmission power to relay the PUs' message in exchange of their spectrum access. In the first work, the PU's utility function is related with the rate, while the SU's utility function is defined as a function of its rate minus a power cost due to the relay of the PU's signal. In addition, the authors compare the O2O and O2M matching algorithms and result in the conclusion that O2M matching scheme leads to better performance for the PUs, whereas the SUs are more benefited by the O2O matching scheme. In Roumeliotis et al. (2015), the appropriate pairing of PUs and SUs is examined in a matching market under an O2M matching scheme with *externalities* which refer to the timeslot duration that is assigned to each SU and also the virtual payment that is employed as a cooperation incentive among the users. Furthermore, the utility functions of both the PUs and SUs incorporate the users' rates and monetary factors and the SUs' utilities also include the SUs' energy consumption. The simulation results reveal that in terms of PUs the performance of the proposed distributed algorithm is close to the performance of a centralized scheme with less complexity.

13.4.5 Multicriteria Decision Making Theory

The diversification, not only in the characteristics of radio environment, but also in the wireless technologies, renders the decision making for RRM a complex problem based on various criteria with conflicting requirements. This problem includes many objectives with multiple parameters, while two mathematical approaches that have been investigated for its solution are the fuzzy logic and the multiattribute decision making (MADM). Besides the deployment of both theories in the widely known network selection problem (Charilas and Panagopoulos, 2010b), there are many applications of fuzzy logic and MADM in CRNs some of which will be described next. In literature, due to the tight connection of CR technology with the spectrum management, many paradigms of the aforementioned methods target to the amelioration of spectrum sensing process under appropriate SUs' channel allocation including the spectrum handoff (SH). The SH process depicts the change of SU's current operating frequency channel when a PU arrives at the channel occupied by the SU due to the PU's protection from harmful interference.

The fuzzy decision-making process includes the fuzzification module, the fuzzy reasoning, and the defuzzification module. The input parameters to the fuzzy logic system are fuzzified between 0 and 1, based on predefined membership functions (MBF). Then, in the fuzzy reasoning phase a list of IF-THEN rules, which may be based on prior knowledge or questionnaires, determine the relations of input and output variables and in the last phase the fuzzy system's output is reconfigured to a nonfuzzy number acquiring its actual value. In the terms of the MADM approach, there are various algorithms applied in CR systems such as analytic hierarchy process (AHP), simple additive weighting (SAW), multiplicative exponent weighting (MEW), technique for order of preference by similarity to ideal solution (TOPSIS), and the compromise ranking method VIKOR, whose analytical overview can be found in Zavadskas et al. (2011).

The implementation of fuzzy logic in CRRM field is presented in Matinmikko et al. (2009), where the authors strengthen the connection between the CR and fuzzy logic due to the learning property of both mechanisms, and toward this direction, the proposed fuzzy-logic-based cooperative spectrum sensing mechanism provides not only amelioration to the detection probability versus the false alarm probability, but also additional flexibility compared with *AND* and *OR* combining methods. Furthermore, in Kaur et al. (2010), the authors investigate the SUs' opportunistic spectrum access based on a fuzzy logic decision-making model including three descriptive factors, the SU's velocity, the spectrum to be utilized by the SU, and the distance between PU and SU. The scope is to find each SU's probability to access the spectrum depending on the aforementioned descriptive factors. The simulation results reveal that as SU comes closer to PU or SU moves with higher velocity, the SU's probability to have access to the spectrum increases, resulting in the improvement of spectrum utilization. The spectrum handoff problem and how the available information from the spectrum sensing leads the SU to make this decision are studied in Giupponi et al. (2008). The proposed algorithm, which is implemented at the SU's side, is based on two fuzzy logic controllers (FLCs). The first FLC is used for the estimation of the PU and SU distance and the SU's power transmission without causing harmful interference, while in the second FLC, SU's handoff decisions or adjustments to SU's transmission power are made based on the determined transmission power and SU's bit rate. Finally, the authors compare the proposed fuzzy solution with a solution based on fixed thresholds with respect to spectrum handoff rate and interference temperature measured at the PU receiver and the former's superiority is proven through simulation results.

The wide application of MADM theory in CRNs is indicatively presented next. In Rodríguez-Colina (2011), the SH problem is investigated and the selection of the available frequency bands from a finite set is derived by an algorithm based on the AHP method. The proposed algorithm incorporates two different SUs' classes of service, that is, real time (RT) and best effort (BE) and the dynamic radio environment for extracting the optimal decision. The latter includes the change of instantaneous bandwidth, SUs' SINR, QoS, and bands' occupancy. Consequently, the spectrum bands are classified from the best to the worst based on the RT, whose most significant demands are low band occupancy and high QoS and the BE, whose priority is the bandwidth and SINR. Moreover, the simulation results show that the proposed algorithm accurately selects the combination of the aforementioned parameters focusing on the satisfaction of class of service requirement and the choice of the most appropriate band is done with low latency. Finally, the SH problem is also examined in Hernández et al. (2015), where the VIKOR, SAW, and MEW algorithms are compared in the selection of best channel under the criteria of the probability of channel availability, estimated channel, time availability, SINR, and bandwidth. The performance metrics, whose evaluation is based on real spectrum occupancy taken from the GSM mobile band, are focused on handover, failed handover, bandwidth, and delay. To conclude, the simulation results depict that the VIKOR algorithm provides an efficient and effective process for channels' selection because it has low average number of handover, high rate of bandwidth utilization, and low average transmission delay.

13.4.6 Machine Learning

As the bedrock of CR framework is the dynamic adaptation of cognitive users to the fast changing conditions of the radio environment, the incorporation of learning characteristics to the cognitive engine is crucial and the interest for the application of machine learning algorithms in CRNs increases continuously, (Bkassiny et al., 2013; Alshawaqfeh et al., 2015). In particular, the machine learning field investigates algorithms that learn from sample data or past knowledge of a system and target to the optimization of system's performance by building appropriate effective models. The classification of machine learning includes three categories: the supervised, unsupervised, and reinforcement learning. In first category, the algorithms analyze the training data, that is, known inputs with known outputs, extract knowledge from them, and predict the outcome of new samples, while in second category, the algorithms examine and attempt to identify the hidden structure of the training data focusing mainly on appropriate data's clustering. The third category focuses on the behavior of users to take actions in an environment so as to maximize/minimize a specific cumulative reward/cost function. In reinforcement learning, the agent determines a sequence of actions composing a policy that maps the state of an unknown environment with an optimal action. Due to this environment's uncertainty, the agent results in the optimal action after the exploration of all current feasible actions and their consequences. Consequently, the combination among the exploitation of the existing system's past knowledge and the exploration of the current system's state leads to the most beneficial action for the agent.

In the CRRM field, machine learning techniques have been deployed in spectrum sensing process to find the appropriate channels, which will be occupied by SUs in order to eliminate collisions among PUs and SUs, and keep the interference to PUs under an acceptable threshold. In Thilina et al. (2013), the cooperative spectrum sensing is investigated under various pattern classification mechanisms for both unsupervised and supervised learning. Especially, each SU collects data for the energy levels of surrounding environment in a feature vector, called as *energy vector*, and feeds a classifier to decide the availability of the channel, categorizing the feature vector into *channel available*, where no PU is active, and *channel unavailable* classes, where at least one active PU exists. According to the training phase, which precedes the online classification, the classifier learns from the training feature vectors under different algorithms, that is, the K-means clustering and Gaussian mixture model (GMM) for the unsupervised learning and the support vector machine (SVM) and weighted K-nearest neighbor (KNN) for the supervised learning. More specifically, in K-means clustering algorithm the training feature vectors are separated in K clusters corresponding to a combined state of PUs and the classifier decides the mapping of each cluster with one of the two classes, while in GMM, the feature vectors are described by a mixture of Gaussian density functions each of which correspond to a cluster. Moreover, in SVM approach, the maximization of the margin among separating hyperplanes and feature vectors results in the support vectors, which are subsets of training vectors that fully specify the decision function. In addition, in terms of the weighted KNN technique each feature's weight is obtained by evaluating the area under the receiver operating characteristic (ROC) curve of that training feature vector. After the examination of the aforementioned classifiers under the training duration, the classification delay and the ROC curves, the simulation results reveal that the SVM classifier achieves the highest detection performance, the weighted KNN's train duration is very small, and the performance of K-means clustering is close to the SVM-linear classifier in terms of the ROC performance.

Besides the supervised and unsupervised learning, the concept of reinforcement learning is also presented in CRRM such as in van den Biggelaar et al. (2012). A system with a large circular primary cell and many secondary cells, each one with a central SBS and many SUs, which operate on the same frequency band with the PU, is deployed. Moreover, a decentralized Q-learning power allocation algorithm is investigated to maximize the SINR at the secondary receivers subject to a constraint for the PU's allowed aggregate interference. According to the Q-learning technique each SBS is an agent that aims to learn an optimal power allocation policy for its cell by interacting with the environment, which is modeled as a finite-state discrete-time stochastic system. Consequently, SBSs sense the environment state, select an action according to a policy, perform this action, sense the resulting new environment state, compute the induced cost, and update the state-action Q-value according to a rule. To conclude, the numerical results show that the use of a cost function with the property of penalizing the SUs' actions that lead to a higher than required secondary SINR is related with better results than a cost function without such penalty and the Q-learning convergence is faster as the frequency of execution increases until a value after which the convergence time remains invariable.

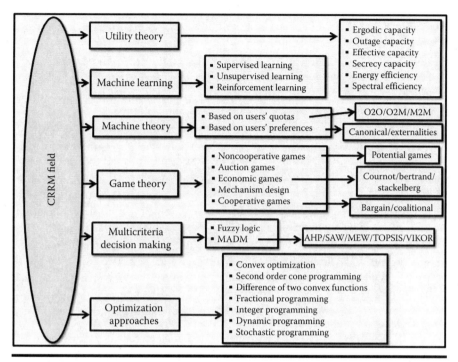

Figure 13.5 Basic RRM techniques in CR.

The wide range of different mathematical perspectives that approaches the CR resource management, as presented in this chapter, assures the importance of the RRM field in CRNs and their flexible extension using various new techniques in next-generation wireless communication systems. To summarize and obtain a comprehensive overview of the CRRM field, the aforementioned mathematical formulations and the utility functions are presented in Figure 13.5.

13.5 Open Research Issues and Concluding Remarks

Summing up, this chapter provides a thorough analysis about the concept of CRNs, which, according to recent literature, consists of a promising candidate technology toward the implementation of 5G networks. The importance of RRM in the wireless systems is depicted and a plethora of applied mathematical approaches in the CRRM framework are rigorously discussed through an extensive and comprehensive survey in significant research works. One of the most important technical challenges is the design of optimized cognitive nodes exploiting the advances in the field of software radio technologies and software-driven 5G networks. These nodes will be optimized based on a cross-layer optimization strategy in order to be able to get inputs from and provide outputs to other nodes operating at each layer in the device.

Moreover, on one hand, new RRM algorithms in CRN tailored for specific networks should be developed that will consider the information that is exchanged at a local level. On the other hand, network-wide RRM mechanisms should be optiuzed in order to have all the benefits of a cognitive approach.

Many challenges also exist in the coexistence of high frequency microwave fixed point-to-point terrestrial links and fixed satellite services operating at Ka band (27.5–29.5 GHz). There are many issues as these both links can be employed as backhaul solutions, but there are also many issues rising with the next-generation satellite networks operating at Ka band and above that, consisting of non-GEO satellite constellations and Earth Stations on Mobile Platforms (ESOMPs). Therefore, new optimized RRM schemes should be proposed to address these issues in order to dynamically control the intersystem interference that is generated between the fixed service (terrestrial) and fixed satellite service (satellite) networks.

Many technical challenges of the cognitive radio networks such as, routing, mobility, and dynamic spectrum management should be handled along with the employment of 5G networks, taking into considerations the complex radio environments and the user preferences.

References

Agiwal, M., Roy, A., and Saxena, N. (2016). Next generation 5G wireless networks: A comprehensive survey. *IEEE Communications Surveys and Tutorials*, 18(3), 1617–1655.

Aijaz, A. and Aghvami, A. H. (2015). Cognitive machine-to-machine communications for internet-of-things: A protocol stack perspective. *IEEE Internet of Things Journal*, 2(2), 103–112.

Akyildiz, I. F., Lee, W. Y., Vuran, M. C., and Mohanty, S. (2006). NeXt generation/ dynamic spectrum access/cognitive radio wireless networks: A survey. *Computer Networks*, 50(13), 2127–2159.

Almalfouh, S. M. and Stuber, G. L. (2012). Joint spectrum-sensing design and power control in cognitive radio networks: A stochastic approach. *IEEE Transactions on Wireless Communications*, 11(12), 4372–4380.

Alshawaqfeh, M., Wang, X., Ekti, A. R., Shakir, M. Z., Qaraqe, K., and Serpedin, E. (2015). A survey of machine learning algorithms and their applications in cognitive radio. In Weichold M., Hamdi M., Shakir M., Abdallah M., Karagiannidis G., Ismail M. (Eds.), *Cognitive Radio Oriented Wireless Networks*, Lecture Notes of the Institute for Computer Sciences, Social Informatics and Telecommunications Engineering, vol. 156, pp. 790–801, Springer, Cham, Switzerland.

Alvarado, A., Scutari, G., and Pang, J. S. (2014). A new decomposition method for multiuser DC-programming and its applications. *IEEE Transactions on Signal Processing*, 62(11), 2984–2998.

Andrews, J. G., Buzzi, S., Choi, W. et al. (2014). What will 5G be? *IEEE Journal on Selected Areas in Communications*, 32(6), 1065–1082.

Asadi, A., Wang, Q., and Mancuso, V. (2014). A survey on device-to-device communication in cellular networks. *IEEE Communications Surveys & Tutorials*, 16(4), 1801–1819.

Atallah, R. F., Khabbaz, M. J., and Assi, C. M. (2015). Vehicular networking: A survey on spectrum access technologies and persisting challenges. *Vehicular Communications*, 2(3), 125–149.

Badoi, C. I., Prasad, N., Croitoru, V., and Prasad, R. (2011). 5G based on cognitive radio. *Wireless Personal Communications*, 57(3), 441–464.

Barnes, S. D., Joshi, S., Maharaj, B. T., and Alfa, A. S. (2015, April). Massive MIMO and femto cells for energy efficient cognitive radio networks. In *International Conference on Cognitive Radio Oriented Wireless Networks*, (pp. 511–522). Springer International Publishing.

Bayat, S., Louie, R. H., Vucetic, B., and Li, Y. (2013). Dynamic decentralised algorithms for cognitive radio relay networks with multiple primary and secondary users utilising matching theory. *Transactions on Emerging Telecommunications Technologies*, 24(5), 486–502.

Bkassiny, M., Li, Y., and Jayaweera, S. K. (2013). A survey on machine-learning techniques in cognitive radios. *IEEE Communications Surveys & Tutorials*, 15(3), 1136–1159.

Boccardi, F., Heath, R. W., Lozano, A., Marzetta, T. L., and Popovski, P. (2014). Five disruptive technology directions for 5G. *IEEE Communications Magazine*, 52(2), 74–80.

Boyd, S. and Vandenberghe, L. (2004). *Convex Optimization*. Cambridge University Press, New York.

Charilas, D. E. and Panagopoulos, A. D. (2010a). A survey on game theory applications in wireless networks. *Computer Networks*, 54(18), 3421–3430.

Charilas, D. E. and Panagopoulos, A. D. (2010b). Multiaccess radio network enviroments. *IEEE Vehicular Technology Magazine*, 5(4), 40–49.

Cheng, P., Deng, L., Yu, H., Xu, Y. and Wang, H. (2012, April). Resource allocation for cognitive networks with D2D communication: An evolutionary approach. In *2012 IEEE Wireless Communications and Networking Conference (WCNC)*, (pp. 2671–2676). IEEE.

Federal Communications Commission. (2003). *Establishment of Interference Temperature Metric to Quantify and Manage Interference and to Expand Available Unlicensed Operation in Certain Fixed Mobile and Satellite Frequency Bands*. FCC 03-289, pp. 1–31, Washington, D.C.

Feng, X., Sun, G., Gan, X. et al. (2014). Cooperative spectrum sharing in cognitive radio networks: A distributed matching approach. *IEEE Transactions on Communications*, 62(8), 2651–2664.

Gale, D. and Shapley, L. S. (1962). College admissions and the stability of marriage. *The American Mathematical Monthly*, 69(1), 9–15.

Giupponi, L. and Pérez-Neira, A. I. (2008, May). Fuzzy-based spectrum handoff in cognitive radio networks. In *2008 3rd International Conference on Cognitive Radio Oriented Wireless Networks and Communications (CrownCom 2008)*, (pp. 1–6). IEEE.

Goldsmith, A., Jafar, S. A., Maric, I., and Srinivasa, S. (2009). Breaking spectrum gridlock with cognitive radios: An information theoretic perspective. *Proceedings of the IEEE*, 97(5), 894–914.

Gotsis, A. G. and Panagopoulos, A. D. (2017). On user association and multiple access optimisation in 5G massive MIMO empowered ultra dense networks. *Transactions on Emerging Telecommunications Technologies*, 28(4), e3037. doi:10.1002/ett.3037.

Gu, Y., Saad, W., Bennis, M., Debbah, M., and Han, Z. (2015). Matching theory for future wireless networks: Fundamentals and applications. *IEEE Communications Magazine*, 53(5), 52–59.

Haider, F., Wang, C. X., Haas, H., Hepsaydir, E., Ge, X., and Yuan, D. (2015). Spectral and energy efficiency analysis for cognitive radio networks. *IEEE Transactions on Wireless Communications*, 14(6), 2969–2980.

Haykin, S. (2005). Cognitive radio: Brain-empowered wireless communications. *IEEE Journal on Selected Areas in Communications*, 23(2), 201–220.

Hernández, C., Giral, D., and Santa, F. (2015). MCDM spectrum handover models for cognitive wireless networks. *World Academy of Science, Engineering and Technology*, 9(10), 679–682.

Hoang, A. T. and Liang, Y. C. (2006, September). Maximizing spectrum utilization of cognitive radio networks using channel allocation and power control. In *IEEE Vehicular Technology Conference*, (pp. 1–5). IEEE.

Hong, X., Wang, J., Wang, C. X., and Shi, J. (2014). Cognitive radio in 5G: A perspective on energy-spectral efficiency trade-off. *IEEE Communications Magazine*, 52(7), 46–53.

Hossain, E., Niyato, D., and Kim, D. I. (2015). Evolution and future trends of research in cognitive radio: A contemporary survey. *Wireless Communications and Mobile Computing*, 15(11), 1530–1564.

ITU-R Recommendation M.2083 (2015). *Framework and Overall Objectives of the Future Development of IMT for 2020 and Beyond*, ITU, Geneva.

Jondral, F. K. (2005). Software-defined radio: Basics and evolution to cognitive radio. *EURASIP Journal on Wireless Communications and Networking*, 2005(3), 275–283.

Kang, X., Liang, Y. C., Nallanathan, A., Garg, H. K., and Zhang, R. (2009). Optimal power allocation for fading channels in cognitive radio networks: Ergodic capacity and outage capacity. *IEEE Transactions on Wireless Communications*, 8(2), 940–950.

Kaur, M. J., Uddin, M., and Verma, H. K. (2010). Analysis of decision making operation in cognitive radio using fuzzy logic system. *International Journal of Computer Applications*, 4(10), 35–39.

Khaledi, M. and Abouzeid, A. A. (2013, February). Auction-based spectrum sharing in cognitive radio networks with heterogeneous channels. In *Information Theory and Applications Workshop (ITA)*, (pp. 1–8). IEEE.

Kolodzy, P. and Avoidance, I. (2002). *Spectrum Policy Task Force*. Federal Commun. Comm., Washington, DC, Rep. ET Docket, (02-135).

Kourogiorgas, C., Sagkriotis, S., and Panagopoulos, A. D. (2015, May). Coverage and outage capacity evaluation in 5G millimeter wave cellular systems: Impact of rain attenuation. In *2015 9th European Conference on Antennas and Propagation (EuCAP)*, (pp. 1–5). IEEE.

Laneman, J. N., Tse, D. N., and Wornell, G. W. (2004). Cooperative diversity in wireless networks: Efficient protocols and outage behavior. *IEEE Transactions on Information Theory*, 50(12), 3062–3080.

Leshem, A., Zehavi, E., and Yaffe, Y. (2012). Multichannel opportunistic carrier sensing for stable channel access control in cognitive radio systems. *IEEE Journal on Selected Areas in Communications*, 30(1), 82–95.

Li, D., Xu, Y., Liu, J., Wang, X., and Wang, X. (2010, June). A coalitional game model for cooperative cognitive radio networks. In *Proceedings of the 6th International Wireless Communications and Mobile Computing Conference*, (pp. 1006–1010). ACM, New York.

Liu, F., Wang, J., Du, R., Peng, L., and Chen, P. (2011). A second-order cone programming approach for robust downlink beamforming with power control in cognitive radio networks. *Progress in Electromagnetics Research M*, 18, 221–231.

Liu, G. and Jiang, D. (2016). 5G: Vision and requirements for mobile communication system towards year 2020. *Chinese Journal of Engineering*, 2016(2016), 8.

Liu, Y., Wang, L., Zaidi, S. A. R., Elkashlan, M., and Duong, T. Q. (2016). Secure D2D communication in large-scale cognitive cellular networks: A wireless power transfer model. *IEEE Transactions on Communications*, 64(1), 329–342.

Lun, J. and Grace, D. (2014, September). Cognitive green backhaul deployments for future 5G networks. In *Cognitive Cellular Systems (CCS)*, 2014 1st International Workshop on (pp. 1–5). IEEE.

Matinmikko, M., Rauma, T., Mustonen, M., Harjula, I., Sarvanko, H., and Mammela, A. (2009). Application of fuzzy logic to cognitive radio systems. *IEICE Transactions on Communications*, 92(12), 3572–3580.

Mitola, J. and Maguire, G. Q. (1999). Cognitive radio: Making software radios more personal. *IEEE Personal Communications*, 6(4), 13–18.

Mumtaz, S., Huq, K. M. S., Ashraf, M. I., Rodriguez, J., Monteiro, V., and Politis, C. (2015). Cognitive vehicular communication for 5G. *IEEE Communications Magazine*, 53(7), 109–117.

Murroni, M. et al. (2011). IEEE 1900.6: Spectrum sensing interfaces and data structures for dynamic spectrum access and other advanced radio communication systems standard: Technical aspects and future outlook. *IEEE Communications Magazine*, 49(12), 118–127.

Naeem, M., Illanko, K., Karmokar, A., Anpalagan, A., and Jaseemuddin, M. (2013). Optimal power allocation for green cognitive radio: Fractional programming approach. *IET Communications*, 7(12), 1279–1286.

Obregon, E., Sung, K. W., and Zander, J. (2014, April). On the sharing opportunities for ultra-dense networks in the radar bands. In *Dynamic Spectrum Access Networks (DYSPAN)*, 2014 IEEE International Symposium on (pp. 215–223). IEEE.

Oleshchuk, V. and Fensli, R. (2011). Remote patient monitoring within a future 5G infrastructure. *Wireless Personal Communications*, 57(3), 431–439.

Ozarow, L. H., Shamai, S., and Wyner, A. D. (1994). Information theoretic considerations for cellular mobile radio. *IEEE Transactions on Vehicular Technology*, 43(2), 359–378.

Panwar, N., Sharma, S., and Singh, A. K. (2016). A survey on 5G: The next generation of mobile communication. *Physical Communication*, 18, 64–84.

Poulakis, M. I., Vassaki, S., and Panagopoulos, A. D. (2016). Secure cooperative communications under secrecy outage constraint: A DC programming approach. *IEEE Wireless Communications Letters*, 5(3), 332–335.

Rappaport, T. S., Sun, S., Mayzus, R. et al. (2013). Millimeter wave mobile communications for 5G cellular: It will work!. *IEEE Access*, 1, 335–349.

Rodríguez-Colina, E. (2011, May). Multiple attribute dynamic spectrum decision making for cognitive radio networks. In *2011 Eighth International Conference on Wireless and Optical Communications Networks*, (pp. 1–5). IEEE.

Roth, A. E. and Sotomayor, M. (1992). Two-sided matching. In *Handbook of Game Theory with Economic Applications*, R. J. Aumann and S. Hart (eds.), Vol. 1, pp. 485–541, Elsevier, Amsterdam, the Netherlands.

Roumeliotis, A. J., Vassaki, S., and Panagopoulos, A. D. (2015, August). Overlay cognitive radio networks: A distributed matching scheme for user pairing. In *2015 International Wireless Communications and Mobile Computing Conference (IWCMC)*, (pp. 172–177). IEEE.

Roumeliotis, A. J., Vassaki, S., and Panagopoulos, A. D. (2016, May). Time allocation mechanism with QoS constraints in a spectrum leasing environment. In *2016 23rd International Conference on Telecommunications (ICT)*, (pp. 1–5). IEEE.

Spectrum Efficiency Working Group. (2002). *Report of the Spectrum Efficiency Working Group*. Federal Communications Commission, Tech. Rep, FCC, Washington, DC.

Stevenson, C. R., Chouinard, G., Lei, Z., Hu, W., Shellhammer, S. J., and Caldwell, W. (2009). IEEE 802.22: The first cognitive radio wireless regional area network standard. *IEEE Communications Magazine*, 47(1), 130–138.

Thilina, K. M., Choi, K. W., Saquib, N., and Hossain, E. (2013). Machine learning techniques for cooperative spectrum sensing in cognitive radio networks. *IEEE Journal on Selected Areas in Communications*, 31(11), 2209–2221.

Tragos, E. Z. and Angelakis, V. (2013, June). Cognitive radio inspired M2M communications. In *Wireless Personal Multimedia Communications (WPMC)*, 2013 16th International Symposium on (pp. 1–5). IEEE.

Tseng, F. H., Chao, H. C., and Wang, J. (2015). Ultra-dense small cell planning using cognitive radio network toward 5G. *IEEE Wireless Communications*, 22(6), 76–83.

van den Biggelaar, O., Dricot, J. M., De Doncker, P., and Horlin, F. (2012, September). Power allocation in cognitive radio networks using distributed machine learning. In *2012 IEEE 23rd International Symposium on Personal, Indoor and Mobile Radio Communications-(PIMRC)*, (pp. 826–831). IEEE.

Vassaki, S., Poulakis, M. I., and Panagopoulos, A. D. (2015, May). Spectrum leasing in cognitive radio networks: A matching theory approach. In *2015 IEEE 81st Vehicular Technology Conference (VTC Spring)*, (pp. 1–5). IEEE.

Vassaki, S., Poulakis, M. I., Panagopoulos, A. D., and Constantinou, P. (2014). QoS-driven power allocation under peak and average interference constraints in cognitive radio networks. *Wireless Personal Communications*, 78(1), 449–474.

Wang, B., Wu, Y., Ji, Z., Liu, K. R., and Clancy, T. C. (2008). Game theoretical mechanism design methods. *IEEE Signal Processing Magazine*, 25(6), 74–84.

Wang, B., Wu, Y., and Liu, K. R. (2010). Game theory for cognitive radio networks: An overview. *Computer Networks*, 54(14), 2537–2561.

Wang, J., Ghosh, M., and Challapali, K. (2011). Emerging cognitive radio applications: A survey. *IEEE Communications Magazine*, 49(3), 74–81.

Wang, L., Ngo, H. Q., Elkashlan, M., Duong, T. Q., and Wong, K. K. (2015). Massive MIMO in spectrum sharing networks: Achievable rate and power efficiency. *IEEE Systems Journal*, 11(1), 20–31.

Wang, Y. and Liu, K. R. (2015). Statistical delay QoS protection for primary users in cooperative cognitive radio networks. *IEEE Communications Letters*, 19(5), 835–838.

Wu, D. and Negi, R. (2003). Effective capacity: A wireless link model for support of quality of service. *IEEE Transactions on Wireless Communications*, 2(4), 630–643.

Wyner, A. D. (1975). The wire-tap channel. *The Bell System Technical Journal*, 54(8), 1355–1387.

Xu, L., Tonjes, R., Paila, T., Hansmann, W., Frank, M., and Albrecht, M. (2000). DRiVEing to the Internet: Dynamic radio for IP services in vehicular environments. In *Local Computer Networks, 2000. LCN 2000. Proceedings*. 25th Annual IEEE Conference on (pp. 281–289). IEEE.

Yang, C., Li, J., Guizani, M., Anpalagan, A., and Elkashlan, M. (2016). Advanced spectrum sharing in 5G cognitive heterogeneous networks. *IEEE Wireless Communications*, 23(2), 94–101.

Yu, Q. (2013). A survey of cooperative games for cognitive radio networks. *Wireless Personal Communications*, 73(3), 949–966.

Yu, R., Zhong, W., Xie, S., Zhang, Y., and Zhang, Y. (2016). QoS differential scheduling in cognitive-radio-based smart grid networks: An adaptive dynamic programming approach. *IEEE Transactions on Neural Networks and Learning Systems*, 27(2), 435–443.

Zavadskas, E. K. and Turskis, Z. (2011). Multiple criteria decision making (MCDM) methods in economics: An overview. *Technological and Economic Development of Economy*, 17(2), 397–427.

Zhang, H., Jiang, C., Beaulieu, N. C., Chu, X., Wang, X., and Quek, T. Q. (2015a). Resource allocation for cognitive small cell networks: A cooperative bargaining game theoretic approach. *IEEE Transactions on Wireless Communications*, 14(6), 3481–3493.

Zhang, Z., Zhang, W., Zeadally, S., Wang, Y., and Liu, Y. (2015b). Cognitive radio spectrum sensing framework based on multi-agent architecture for 5G networks. *IEEE Wireless Communications*, 22(6), 34–39.

Zhao, Q. and Sadler, B. M. (2007). A survey of dynamic spectrum access. *IEEE Signal Processing Magazine*, 24(3), 79–89.

Zhong, W., Chen, G., Jin, S., and Wong, K. K. (2014). Relay selection and discrete power control for cognitive relay networks via potential game. *IEEE Transactions on Signal Processing*, 62(20), 5411–5424.

Ziegler, V., Theimer, T., Sartori, C. et al. (2015, May). Architecture vision for the 5G era: Cognitive and cloud network evolution. In *2015 IEEE 81st Vehicular Technology Conference (VTC Spring)*, (pp. 1–6). IEEE.

Zou, Y., Li, X., and Liang, Y. C. (2014). Secrecy outage and diversity analysis of cognitive radio systems. *IEEE Journal on Selected Areas in Communications*, 32(11), 2222–2236.

Chapter 14

Emerging Technologies for Mobile Health

Konstantinos Karathanasis and Konstantina S. Nikita

Contents

14.1 Introduction

Over the past decade, one of the most relevant costs in the budget of developed countries is Public Health expenditure. In 2011, the Organization for Economic Cooperation and Development (OECD) defined the total health expenditure per capita (PPP) as percentage of gross domestic product (GDP) ranging from 10.2% to 11.9% for the top 10 countries (Andria et al. 2015). Regarding the United States, economists at the Centers for Medicare and Medicaid Services (CMS) estimate that health care spending will grow by 5.8% per year for the period 2010–2020, 1.1% more than the expected growth of the GDP. By 2020, national health expenditures are projected to reach $4.64 trillion, representing then almost 20% of the GDP (Keehan et al. 2011). Since rising health care costs are unsustainable, there is a growing need for finding new ways to deliver health care. At the same time life expectancy is continuously increasing and it pushes the demand for health care services upward as elderly people demand for autonomous lifestyle and high quality of life. In addition, people have higher expectations and demand more options in their health care. However, we do not take good care of ourselves to prevent diseases: obesity, heart disease, and cancer are global health issues that are worsened by the way we live, for example, poor diet, physical inactivity, and smoking. So, policy makers are crushed between two opposite forces: the need to reduce public expenses for welfare on one hand and the continuous demand for high quality health care services on the other hand. Taking into account all the above, the only solution that seems viable is a radical change in the traditional models of health care assistance toward new services in which the patient is actively involved and instructed to manage his own health (Andria et al. 2015).

Rapid advancements in wireless sensing and smart devices are creating a pervasive wireless environment that can address a wide range of major health-related challenges, such as the aging population, prevalence of chronic diseases, outbreaks of infectious diseases, support for well-being, and entertainment (Nikita et al. 2012). Innovations in human-computer interaction research have revealed effective methods for people with various disabilities to use computers or to receive computer aided medical treatment. Furthermore, the prevalence of sensor-rich smartphones, PDAs, and tablets has broadened the perspective in the roles and capacities of these mobile devices in monitoring and intervention of human health-related activities. Moreover, the demand for accurate monitoring of human behaviors and lifestyles has never been so imminent because it plays a key role in the prevention of diseases such as cancer, diabetes, coronary heart disease, and obesity.

Exploitation of recent technological advances assists in a fundamental redesign of health care processes based on the use and integration of communication technologies at all levels. Computer technologies, modern equipment and information and communication technologies (ICT) for health care solutions have given rise to the electronic health (e-health) field, while the integration of mobile computing, medical sensors, and portable devices in the health environment has enabled the subfield of mobile health (m-Health) (de Mattos and Gondim 2016). The concept of m-Health was first introduced in Istepanaian et al. (2004) and defined as *mobile computing, medical sensor, and communications technologies for health care.* Nowadays, the m-Health is bringing together major academic research and industry disciplines worldwide to achieve innovative solutions in the areas of health care delivery and technology sectors (Istepanaian and Zhang 2012). In a broad sense, m-Health is the delivery of health care services through mobile devices, which are used to capture, analyze, store, and transmit health information from multiple sources, including sensors and other biomedical acquisition systems. Mobile applications for health can address different audiences, such as doctors, nurses, patients, and even healthy people. Solutions can be used to encourage, for example, a reduction in cigarette consumption, adherence to weight loss diets and physical activity plans, or adherence to treatment plans aimed to control and prevent heart disease. As a result, m-Health solutions can improve the efficiency and effectiveness of health services, enabling cost-effective and efficient health care delivery at home and/or hospital, assisted-living and nursing home settings to reduce operating costs and promote disease management and wellness (de Mattos and Gondim 2016, Figure 14.1).

A diverse range of applications and maturing technologies are ready for translational research and practical deployment, addressing global health challenges associated with demographic, environmental, social and economic changes (Nikita 2014). The necessary steps toward large scale m-Health deployment in today's health care include setting the related devices (implantable, digestible, and wearable) standards, ensuring functionality and patient safety and defining high data rate secure protocols for biosignals. Also, to practically exploit current technology in the field of modern medical devices, the issue of energy-autonomous systems through energy scavenging applications as well as energy management and optimization issues in biomedical devices and networks must be addressed (Kim and Chu 2016). As far as sensing technologies are concerned, in-body and on-body sensors using bioelectronics, smart textiles, printable-flexible-stretchable electronics and microelectromechanical system (MEMS) are the technologies that will be used to

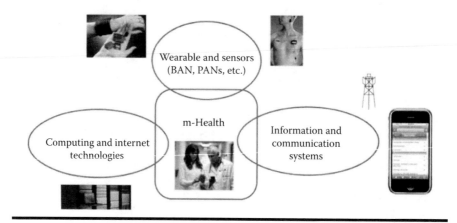

Figure 14.1   m-Health general concept. (From Istepanaian, R.S.H., Zhang, Y.T., *IEEE Trans. Inf. Technol. Biomed.*, 16(1), 1–5, 2012.)

create implantable, digestible and wearable medical devices (Nikita 2014). Finally, electronic health records through advanced big data analytics will help clinical decision support to become more efficient, better supporting evidence based medicine and health care.

Looking at m-Health applications from a different perspective, one of the biggest Internet of things (IoT) growth areas is measuring individual health metrics through self-tracking gadgets, clinical remote monitoring, wearable sensor patches, Wi-Fi scales, and a myriad of other biosensing applications (Swan 2012). As reported in Manyika et al. (2015), in 2025 remote monitoring could create as much as $1.1 trillion a year in value by improving the health of chronic-disease patients. Toward this direction, there are three ways in which the IoT is revolutionizing health care: (a) reducing device downtime through remote monitoring and support, (b) proactive fulfillment by replenishing supplies before they are needed, and (c) efficient scheduling by leveraging utilization to serve more patients. IoT devices can be used to enable remote health monitoring and emergency notification systems. These health monitoring devices can range from blood pressure and heart rate monitors to advanced devices capable of monitoring specialized implants, such as pacemakers or advanced hearing aids. Toward this direction, future 5G capabilities (5G-PPP 2015) could generate significant improvements in many health scenarios, including the management and tracking of hospital assets, robotics-assisted tele-surgery, assisted living and remote monitoring of health or wellness data, and remote application of medication (smart medication).

Within this framework, this chapter aims to provide a brief overview of the main technologies and research areas enabling m-Health realization. First, the basic sensing principles in conjunction with the types of sensors used for mobile health applications are presented. Next, propagation and communications issues for wireless biomedical applications are discussed. Wearable, implantable and ingestible medical devices are considered and aspects like spectrum regulations, patient safety and biocompatibility are analyzed along with a brief overview of commercially available products. Finally, state-of-the-art mobile health systems are presented before concluding with challenges and future opportunities in the field.

14.2 Sensing Methods and Technologies

14.2.1 Sensing Principles

During the past decades, advances in technology have had significant impact on monitoring applications in the fields of e-health, clinical diagnosis, and biomedical telemetry. The successful employment of these applications relies on the development of sensing devices called biosensors. Biosensors are the devices that provide physical, chemical, and biological data for monitoring several human functions and diseases. Over the past years, novel sophisticated tools and new materials have made it possible to construct sensing devices with increased reliability and accuracy with respect to the sensed information.

As depicted in Figure 14.2, the fundamental elements of a biosensor are the bioreceptor, an immobilized element able to recognize the target analyte (compound whose concentration is to be determined), and the transducer, which is used to convert the biochemical signal into a measurable electric signal (Monošík et al. 2012; Tamura 2014; Lioumpas et al. 2014). Based on the detection principle, biosensors can be mainly categorized into electrical, electrochemical, optical, calorimetric and piezoelectric. The basic characteristics of each category are presented next.

Electrochemical biosensors operate based on the presence of an appropriate enzyme in the bio-recognition layer (baroreceptor), which is able to provide those electroactive substances to the physicochemical transducer in order to detect a measurable signal. Native enzyme can be used as the biorecognition element or enzymes can be additionally used as labels bound to antibodies, antigens, and

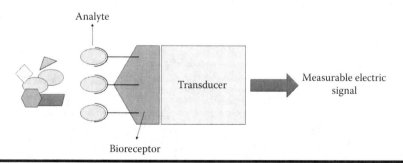

Figure 14.2 Basic components of a biosensor. (From Lioumpas et al., Sensing principles for biomedical telemetry, in K.S. Nikita [Ed.], *Handbook of Biomedical Telemetry*, Wiley-IEEE Press, pp. 56–75, 2014.)

oligonucleotides with a specific sequence, thus providing affinity-based sensors (Bakker 2004). Depending on the analyte, different enzymes are employed for its detection (Grieshaber et al. 2008). Electrochemical sensors are classified to different categories according to the transducer's operation principle, with the most common ones being the potentiometric, amperometric and impedimetric (Pohanka and Skláda 2008).

Optical methods for the detection of biomolecular interactions and their biomedical applications have attracted great research interest, since for most applications it is desirable to have a compact biosensor with high sensitivity and fast response. Furthermore, cost reduction of high quality fibers and optoelectronic components has played a major role in the development of optical biosensors (Velasco-Garcia 2009). Specific advantages are the immunity to external electromagnetic interference, increased speed of biodetection and increased bandwidth. The optical methods for detection are based on fluorescence spectroscopy, surface plasmon resonance, interferometry, and spectroscopy of guided modes in optical waveguide structures (Gauglit 2005; Velasco-Garcia 2009).

Thermal/calorimetric biosensors operation principle is based on the absorption or evolution of heat that all biological reactions taking place inside living organisms cause. Among their advantages are small size, long-term and high stability, and nonchemical contact measurement (Syam et al. 2012). Calorimetry involves measuring the heat following a biochemical reaction. Therefore, the basic part of a thermal biosensor is a sensor that measures the temperature and consequently the number of immobilized enzyme molecules, as the amount of product molecules created can be calibrated against temperature changes (Lammers and Scheper 2001). The sensor most commonly used for measuring the temperature in the reaction medium is an enzyme thermistor (Danielsson and Mosbach 1988). Thermal biosensors were recently applied as devices for detecting pesticides and pathogenic bacteria (Syam et al. 2012), and of course they are still used as detectors for the presence of particular substances or for measuring biological parameters.

The fundamental principle on which piezoelectric biosensors operation is based the piezoelectric effect in conjunction with a piezoelectric crystal. Specifically, when the substance on the surface of the piezoelectric crystal reacts with a substrate, the crystal's mass changes, causing oscillation of its resonant frequency. Choosing a crystal with proper resonant frequency is critical for the detection of the desired biomolecule. Piezoelectric biosensors have many advantages: they are small, inexpensive, and offer long-term stability and excellent temperature behavior. Also, they are capable of providing a rapid response signal. On the other hand, they cannot be used for static measurements since they only detect changes of a variable. Piezoelectric biosensors can be used as part of a flow injection analysis; they can detect substances in aqueous solutions and continuously monitor their concentrations. They have also been used for point mutation detection in human DNA (Dell'Atti et al. 2006).

Other types of biosensors are magnetic, pyroelectric and ion channel biosensors. Recently, magnetic biosensors technology overcame the problems related with size and power consumption by using giant mangetoresistance materials; detection of biomolecules in small samples with low concentration is now possible with great sensitivity and speed (Wang and Li 2008). Pyroelectric biosensors use pyroelectric materials which have the ability to generate voltage or create current whenever their temperature changes. Finally, ion channel biosensors are suitable for detecting molecules of interest such as drugs with low molecular weight, large proteins or microorganisms (Krishnamurthy et al. 2010).

Based on the aforementioned sensing principles, sensors applied to humans are divided into two major groups: noninvasive and invasive/implantable sensors. Next, the most commonly used sensor types from both categories are presented.

14.2.2 Noninvasive (On-Body) Sensors and Interfaces

A broad category of noninvasive sensors are the ones that monitor electrophysiological signals. To begin with, electrocardiography (ECG) is widely used in the diagnosis and management of many cardiac related diseases. The main components of a typical ECG monitoring system are the electrodes, amplifier, and transmitter. Recently, wireless, mobile and remote technologies have been applied to enhance the functionalities and usability of ECG, including the transition from traditional 12-lead to 3-lead ECG (Bsoul et al. 2011) and a carbon nanotube (CNT)/polydimethylsilxane (PDMS) composite-based dry ECD electrode (Jung et al. 2012). Also, a belt-type electrode has been developed to enable long-term monitoring since head movement or sweating can cause traditional electrodes to shift or fall off (Polar 2009). Electroencephalography (EEG) has long been used to record and study the electrical activity of the cerebral cortex. The international 10–20 system is a method describing the location of scalp electrodes monitoring brain signals from the underlying area of cerebral cortex. The development of wireless ambulatory EEG is crucial for achieving long-term monitoring of a patient in their everyday environment. Toward this direction, an EEG-based brain-machine interface (BMI) was proposed for assisting or repairing human cognitive functions (Xu et al. 2011). Also, wireless EEG devices, headsets and headbands have been developed for BMI, enabling vision restoration patients, control of prosthetic limbs and robotics (Sechang et al. 2012), speech generation (Guenther et al. 2012) and cognitive imaging. Another electrodiagnostic medicine technique based on electrophysiological signal monitoring is electromyography (EMG). EMG signals generated during muscle contraction are often used in rehabilitation devices because of their distinct output characteristics compared to other biosignals (Wang et al. 2015). Typical EMG systems consist of surface and reference electrodes, an amplifier, a transmitter and batteries and can be used to control smartphones, notebooks and other gadgets. To solve issues related with wires and battery maintenance problems, EMG using conductive fabric for power supply and electrical shield for noise reduction has been developed, enabling precise EMG measurement with a wearable system. Toward this direction, an EMG system has been proposed to control an electric powered-wheelchair using combinations of left, right and shoulder elevation gestures (Moon et al. 2005).

Another commonly used noninvasive sensing technique is the photoplethysmogram (PPG). In essence, PPG is an optically obtained plethysmogram, a volumetric measurement of an organ. It includes a light emitting diode (LED) and a photodetector (Allen 2007). PPG can be used to monitor breathing, hypovolemia and other respiratory conditions (Reisner et al. 2008). Heart rate monitor in conjunction with respiratory rate has also been developed for wheelchair users (Postolache et al. 2009).

Pulse oximetry is a widespread noninvasive method for monitoring a person's oxygen saturation using a sensor placed on a fingertip, earlobe or foot. Its operation principle is quite simple: light of two different wavelengths is transmitted through the body to a photodetector and the varying absorption at each wavelength is measured. This way, it is possible to measure both oxygenated and deoxygenated hemoglobin on a peripheral scale. Recently, a wearable device platform utilizing motion tracking and measuring oxygen saturation was reported (Dong et al. 2016). Wireless pulse oximeters are also, commercially available using either a fingertip sensor (iHealth 2010) or even a watch-like enclosure (Oxitone 2012). Most current wireless pulse oximeters use the Bluetooth protocol to communicate sensor readings. However, Bluetooth usage poses an issue as far as battery life is concerned, thus alternative technologies like ZigBee are also considered (Watthanawisuth et al. 2010, Figure 14.3).

Wearable pressure monitors are another type of on-body medical sensors that can be used for measuring blood pressure, intraocular pressure, and so on. Modern systems allow home blood pressure monitoring and can serve as a substitute for regular visits to the physician. Moreover, home monitoring can be used to determine whether treatments for people with high blood pressure are actually working.

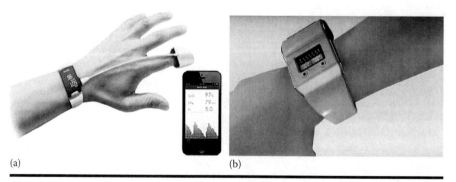

(a) (b)

Figure 14.3 (a) Fingertip. (From iHealth, Accessed June 29, 2016. https://ihealthlabs.com/fitness-devices/wireless-pulse-oximeter/, 2010.) (b) Watch-like wireless pulse-oximeter commercial devices. (From Oxitone., Accessed June 29, 2016. http://oxitone.com/products/, 2012.)

In wireless blood pressure monitoring, the system operates wirelessly to inflate the cuff and display the values. Recently, a cuffless approach to estimate blood pressure using pulse transit time (PPT) has become popular, potentially permitting continuous, noninvasive, BP monitoring after initial calibration (Mukkamala et al. 2015). Furthermore, a body worn digital MEMS barometer that determines the altitude to conserve oxygen and energy in ventilator systems has been developed (Roger 2011). It can be used as smart bandage for negative pressure wound therapy, employing differential pressure measurements. A system using a contact lens with an embedded MEMS strain gauge that measures eye curvature over a timer period has been developed. It can be used to monitor and control intraocular pressure levels for patients suffering from glaucoma as well as detect early pathological cases (Sensimed 2004).

A further category of noninvasive sensors are motion sensors. Motion analysis uses several types of motion sensors and systems for various gait analyses, physical and daily activities with different diseases. There are different types of motion sensors and some of them are briefly discussed in the following. Accelerometers and gyrosensors can be used for a variety of applications. With the development of MEMS, accelerometer sizes and cost decreased and several rehabilitation studies were performed using them. In recent studies, acceleration signals at the low back, which is close to the center of gravity, were monitored, and different signals were found in patients with different diseases (Bonato 2005; Tao et al. 2012). Similar sensors have been designed to support high-resolution motion studies of patients being treated for neuromotor conditions such as Parkinson's disease (Patel et al. 2009) and stroke (Hester et al. 2006). Another type of system is e-AR, a low-power, miniaturized ear-worn activity recognition sensor. The e-AR sensor is equipped with a MEMS three-axis accelerometer. This device allows detecting the gait cycle, seat locomotion, and acceleration (Atallah et al. 2012). Wireless gyrosensors have been used for evaluating the motion and posture of human segments while walking, by measuring the angular velocity and angle (Gouwanda and Senanayake 2010). Moreover, goniometers are sensors that measure the changes in the angle. An electronic optical fiber and flexible goniometer can be used to measure relative rotation between two body segments (Echo Wireless Goniometry 2017).

As individual health care is a growing global interest, development of textile-based electrodes and motion sensors is one of the main issues of recent smart textile research. Smart clothing offers functions such as sensing, displaying, transmission of information and energy collection. The current demand for implementing these functionalities is huge; therefore research on smart clothing is actively pursuing the development of electronic, photonic and photovoltaic textiles (Cho et al. 2011). Textile-based electrodes and motion sensors are being exploited focusing on increasing the friendliness and quality of smart clothing. In addition, over the last few years, nanotechnology has seen an increased acceptance as well as widespread use of its applications in medicine (Alzaidi et al. 2012). Wearing-unconscious devices are required for biomedical and health care applications because bulky devices closely contacted with the human skin irritate users. Organic electronics enable large-area and distributed sensor and/or actuator array and are suitable for wearing-unconscious devices. A variety of flexible, large-area applications using organic transistors has been developed, including an insole pedometer with piezoelectric energy harvester and a surface electromyogram measurement sheet for prosthetic hand control (Takamiya et al. 2014). In a recently reported work (Kulkarni et al. 2014), a graphene based wearable sensor capable of detecting airborne chemicals that serve as indicators of medical conditions for rapid and sensitive vapor detection was presented.

Although wireless temperature monitoring is common in environmental applications, not many human related applications have been developed. One example is a small electronic device, attached to the human body via a disposable patch (Feversmart 2014). The device monitors body temperature and uses a nano-bluetooth chip to transmit data in real-time to a relay unit, which then sends the temperature data to a server. Using a smartphone one can constantly monitor the body temperature in real time. Also, a temperature monitoring system using an array of wireless thermometers has been recently reported (Javadpour et al. 2015). The main modules of each thermometer are an accurate semiconductor temperature sensor, a transceiver operating at 2.4 GHz and a microcontroller (Figure 14.4).

Lastly, flexible chemical sensors have been developed applying soft-MEMS techniques onto functional polymers. Recent research efforts on wearable oxygen and glucose sensors and their applications are reported in Yao et al. (2011). Also, thick-film technology has been applied to monitor transcutaneous blood oxygen, using amperometry, incorporating integral heating element to enhance transcutaneous diffusion of oxygen at 44°C (Lam and Atkinson 2007).

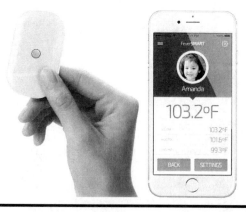

Figure 14.4 Disposable patch sensor and mobile application for remote temperature monitoring. (From Feversmart, Accessed June 30, 2016. http:// feversmart.com/, 2014.)

14.2.3 Invasive/Implantable (In-Body) Sensors

Recently, invasive and implantable wireless technology has been widely used in biomedical applications. Implantable sensors can be used to monitor pressure, motility, pH and temperature (Cumming et al. 2006). Moreover, integration of ECG devices into a chip has been implemented in implantable pacemakers and defibrillators. The most common categories of invasive and implanted sensors are discussed next.

Pressure sensors are widely employed in a variety of applications. Passive wireless pressure sensors that were originally developed for use in harsh environments have been adapted for use in the human body. A wireless, batteryless implantable, real-time blood pressure monitoring microsystem for small laboratory animals has been recently reported (Cong et al. 2010). It employs an instrumental elastic circular cuff (made of soft biocompatible material) which is wrapped around a blood vessel, avoiding vessel penetration and minimizing vessel restriction. Diaphragm and strain gauge pressure sensors are also available. As an example, the CardioMEM Champion is a small size implantable device for monitoring and treating aneurisms using a passive MEMS LC resonator. The device consists of the biocompatible sensor implanted in the wall of a patient's cardiac major artery and an external reader that patients hold in their front. The sensor sends a RF signal to the reader, indicating real-time blood pressure data. The reader will wirelessly transmit the data to the physician, enabling blood pressure monitoring after the patient goes home. Also, an implantable intracranial pressure (ICP) senor providing strong indication of brain health and monitoring of conditions affecting the central nervous system has been recently reported. The ICP sensor is based on MEMS and liquid crystal polymer is chosen as the biocompatible and flexible structure (Sattayasoonthorn et al. 2013).

Chemical sensors are mainly used to monitor pH and glucose based on an electrochemical reaction. A remotely powered implantable microsystem for continuous blood glucose monitoring has been developed. It consists of a MEMS glucose biosensor bonded to a transponder chip which transmits the measured data to an external reader (Ahmadi and Jullien 2009). In addition, a wireless label-free detection of disease related C-reactive proteins (CRPs) was reported in Chen et al. (2009a). The sensor uses a MEMS microcantilever which is deflected due to specific CRP- anti-CRP binding. The deflection of the microcantilever is detected using a position-sensitive detector and the converted biosignal is transmitted by a custom designed wireless amplitude shift keying transceiver.

In-body sensors have been also used in electroencephalography. A continuous, in vivo EEG system using an inductively powered implantable wireless neural recording device has been developed. The device uses an intergrted circuit (IC) to amplify, modulate and transmit neural signals to an external receiver (Sodagar et al. 2007). Moreover, deep brain stimulation (DBS) using brain pacemakers has become a reality in the last 20 years. Brain pacemakers, give the doctors an equal chance of defeating neurological diseases causing movement disorders like Parkinson, epilepsy, and dystonia. They are equipped with an implantable pulse generator, a battery-powered micro-electronic device which generates mild electrical signals to reactivate the almost dead cells of different organs (Goswam et al. 2015).

Magnetoelastic sensors have been frequently used for implantable sensors. They are typically composed of a piezoelectric sensor and amorphous metallic glass ribbons or wires with resonant frequency inversely proportional to their length. The resonant frequency of such a sensor shifts in response to different physical parameters such as stress, pressure, temperature, flow and liquid velocity, magnetic field, and mass loading. The fabrication and application of a miniaturized array of four magetoelastic sensors that enable simultaneous remote measurement of pH, temperature and pressure are presented in Bouropoulos et al. (2005). The platform is ideal for applications where the sensors have to be placed inside sealed, optically opaque containers and when disposable use is necessary.

Finally, microfluidic technology is becoming increasingly popular in implantable devices and for lab-on-a-chip technology. It can also be applied in implantable drug delivery systems for colonic disease (Tng et al. 2012). One of the most notable microfluidic-based drug delivery mechanisms is the Jewel insulin pump from Debiotech (JewelPUMP 2012), codeveloped with STMicroelectronics using Debiotech's microfluidic MEMS technology. The pump can be mounted on a disposable skin patch to provide continuous insulin infusion. It promises substantial improvement in the treatment efficiency and quality of life of diabetic patients.

14.3 Wireless Technologies for m-Health

In this section the propagation and communications issues for m-Health are discussed. Basic concepts as well as recent advances are examined in the areas of body area electromagnetics, (Kiourti and Nikita 2014c), antennas and RF communications (Kiourti and Nikita 2014a), inductive coupling techniques and applications (Ghovanloo and Kiani 2014), intra-body communications (Roa et al. 2014), and wireless biosensor communication standards development (Schmitt et al. 2014).

14.3.1 Body Area Electromagnetics

The field of body area electromagnetics deals with the numerical and experimental modeling tools and methods used for wireless biomedical applications. Numerical modeling provides an effective way of assessing and predicting the electromagnetic performance of such systems in terms of radiation, propagation, and interaction with human tissues. It is considered of utmost importance for the computation of electromagnetic field values in human tissues as well as the evaluation of implantable biomedical devices in cases where it is impossible to conduct measurements in real operating scenarios. When simplified canonical geometries are used to model the human body, analytical methods can be implemented whereas numerical methods are applied when millimeter resolution anatomic models are used for increased accuracy of the findings.

14.3.1.1 Numerical Phantoms

Numerical phantoms are commonly used in studies regarding wireless biomedical applications. They are divided in two categories: canonical and anatomical ones. Canonical models are simple models, typically in the shape of a sphere, cylinder or cube, which can be solved either numerically or analytically. They provide computational efficiency with standard simulation resources and they can be used to obtain preliminary results. Also, experimental phantoms complying with the numerical ones are easy to construct, a crucial step in the validation procedure. Whole body canonical models have largely been used in the literature. A planar, three-layer model consisting of a low-water-content tissue layer embedded between two high-water-content tissues has been proposed in Curto and Ammann (2007). Canonical models have also been used to model specific parts of the human body. For example, a rectangular structure filled with tissue-emulating material has been used to model the human torso in Kuhn et al. (2009). Also, the human arm has been approximated as a tapered cone of different radii and axis ratio as well as tissue compositions, as reported in Lim et al. (2011).

Anatomical models are used to obtain more accurate results. In this case, the human body is modeled by cubic cells (voxels) whose electrical properties are considered constant. By assigning the corresponding electrical properties to each voxel, the anatomical tissues and organs are easily modeled. As computing power increases and computer resources get less expensive, there is a trend to move to more detailed anatomical structures. In contemporary models, the highest complexity used for modeling the human body is about 50 tissue types, and the finest resolution is about 1 mm. In the majority of studies, the data for designing the anatomical body models are obtained from magnetic resonance imaging (MRI) or computed tomography (CT) scans. Several anatomical models have been developed for use in a wide range of applications. Data acquired from CT scans of a cancer patient have been used to create a model consisting of approximately 35,000 10-mm-edge cubic cells in Sullivan (1990). Two 2-mm-resolution whole-body Japanese models with hands placed at the side of the body have been developed using MRI. They classified over 50 types of tissues based on images of a 22 year old male and a same age female (Nagaoka et al. 2004). Moreover, an anatomical human head model comprising 24 tissue types has been used in Chen et al. (2009b). The most advanced and complete set of computational anatomical models representing a wide range of the population is the Virtual Population from IT'IS foundation. Recently, Virtual Population 3.0 was released, including significantly enhanced models for reliable effectiveness and safety evaluations of diagnostic and therapeutic applications, including medical implants safety (Gosselin et al. 2014, Figure 14.5).

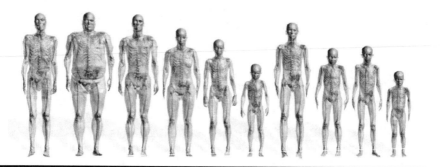

Figure 14.5 **The complete Virtual Population representing humans of different age groups and body shapes. (From Gosselin, M.-C. et al., *Phys. Med. Biol.*, 59, 5287–5303, 2014.)**

14.3.1.2 Computational Methods

Computational methods used to solve complex electromagnetic problems are divided into analytical and numerical ones. Analytical methods include solving Maxwell's equations without a direct numerical solution and inversion of large matrices. They are particularly useful to provide an insight into the physical mechanisms of electromagnetic (EM) propagation an interaction with biological tissues. The main restriction of analytical methods is the use of simplified canonical geometries for modeling the human body, which is necessary to obtain a closed form solution of the wave equation. Analytical methods have been traditionally used to treat specific problems in the literature (Weil 1975). Moreover, they can serve as valuable tools toward verifying the accuracy of numerical simulation results (Nikita et al. 2000; Gupta and Abhayapala 2008). On the other hand, numerical methods involve numerical solutions to Maxwell's equations subjected to a set of initial or boundary values and are generally implemented on powerful computing systems. They can handle complex geometries and provide some physical insight into the electromagnetic performance of simulated systems (Kiourti and Nikita 2014b). The five most common numerical techniques used in the field of bioelectromagnetics are: the method of moments (MoM) (Harrington 1968), the finite element method (FEM) (Silvester and Ferrari 1996), the finite-difference time-domain (FDTD) method (Yee 1966), the transmission line method (TLM) (Christopoulos 1996) and the multiple-multipole (MMP) method (Hafner 1990). Hybrid methods have also been derived from the combination of these and other EM propagation methods (Uzunoglu et al. 2000). These methods have been widely used to study a large variety of EM problems concerning biomedical applications in the past three decades (Zhao et al. 2006; Traille et al. 2008).

14.3.1.3 Experimental Verification

The following step after analytical or numerical evaluation of the corresponding EM problem is the experimental verification of simulation results. Therefore, physical modeling of the configuration under study is necessary. Experimental verifications can be performed with either real human subjects or physical phantoms (Kiourti and Nikita 2014b). Physical phantoms are surrogates of the human body that have electrical properties (electrical conductivity and relative permittivity) equivalent to those of biological tissues. The use of physical phantoms provides a stable and controllable EM environment which cannot be easily realized with human subjects. Moreover, safety issues make them an essential tool in the procedure of experimental testing and verification. Physical phantoms consist of materials in liquid, gel, or solid state. Several recipes have been proposed to produce such materials accounting for different types of tissue and operating frequencies. Liquid phantoms consist of an outer thin shell that must comply with specific guidelines (Kanda et al. 2004). They are easy to prepare and adjust but suffer from the limited frequency range over which they exhibit the desired electrical properties. Liquid phantoms have been used in implantable antenna testing among other applications (Liu et al. 2008). Gel phantoms cover a wider frequency range and allow more accurate modeling of the human body by stacking layers mimicking different tissue types on top of each other. They are formed by adding coagulants to the liquid solution. Such phantoms have been presented and analyzed in Karacolak et al. (2008) and Sani et al. (2010). Their main disadvantage is that they are suitable only for high-water-content tissue modeling and their properties highly degrade over time. Solid phantoms are made from materials that keep their shape over time, and they have the advantages of high accuracy in modeling the heterogeneous human body, fine mechanical stability, and minimized degradation over time. On the other hand, solid phantoms are usually expensive and complicated to fabricate. Several recipes have been proposed in literature (Onishi et al. 2005).

Canonically shaped physical phantoms have been used for biomedical applications. Rectangular and cylindrical containers filled with tissue mimicking materials have been used to test prototypes of planar inverted F and 3D spiral antennas in MICS (402–405 MHz) and ISM (2.4–2.8 GHz) bands (Karacolak et al. 2009; Abadia et al. 2009). For a more realistic approach, anatomically shaped models have also been used in many studies. The most widely used homogeneous head phantom is the Specific Anthropomorphic Mannequin (SAM) which has been proposed by IEEE (2003) and IEC (2005). Furthermore, a homogeneous anatomical model of the human head has been used in Karathanasis et al. (2012) and a human phantom of approximately 1.7 m height and 0.35 m average width has also been used as a radio propagation setup in Alomainy and Hao (2009). For all types of phantoms mentioned earlier, agreement of the electrical properties of fabricated materials with the intended theoretical values must be ensured. Toward this direction, several measurement techniques have been developed (e.g., coaxial probe, slotted line, transverse electromagnetic [TEM] line) with each applying to specific materials, frequency ranges, and applications (IEEE 2001, 1528).

Finally, the biological effects of EM radiation must be taken into account during the analysis of wireless biomedical applications, ensuring the compliance with statutory guidelines set form international organizations. The effects of EM radiation in the human body depend not only on the field level but also on its operation frequency. Wireless biomedical systems produce non-ionizing radiation and cause thermal effects through transformation of the energy of the photons into kinetic energy of the absorbing molecules. All regulations and recommendations regarding the limits on allowable absorbed power in the body are based on quantitative short-term evaluation of the thermal effects caused by EM fields. The two major standards have been set by IEEE (IEEE 1999, 2003), and the International Commission on Non-Ionizing Radiation Protection (ICNIRP 1998) while several other standards exist as well. The dosimetric quantity most commonly used to determine the interaction of EM fields with human tissues is the specific absorption rate (SAR). SAR can be experimentally measured by means of high-precision, multichannel exposition acquisition systems (EASY) and reliable high-precision dosimetric assessment systems (DASY) (Speag 1998). Acceptable levels of radiation are typically expressed in terms of maximum permission exposures and SAR values averaged over 1-g or 10-g volumes of tissue.

14.3.2 Antenna Design Considerations

The major drawback of historically reported biomedical systems has been the wired communication between medical devices and an exterior monitoring/control equipment, significantly limiting patient comfort and convenience. Therefore, devices with wireless communication functionalities appear as a highly promising option toward improving patients' quality of life and providing medical systems with constant availability, context awareness, reconfigurability, and unobrusiveness (Jovanov et al. 2003). Nowadays, advances in wireless communications further facilitate the deployment of wireless biomedical systems. In this context, it becomes obvious that, like in every other wireless application, antenna design plays a key role in the development of wireless biomedical systems. Antennas for such systems can be divided into three categories: on-body (wearable), implantable, and ingestible. On-body antennas are placed on the human body or are worn as part of a garment. Implantable antennas are integrated into medical devices that are implanted inside the human body by means of a surgical operation and ingestible antennas are used with medical devices that have the form of a capsule to be swallowed by the patient. Several challenges have to be addressed to make the wide use of the aforementioned antennas, and thus wireless medical devices, an actuality. For example, such antennas are required to be small, lightweight, robust, conformal to the body surface, and biocompatible, maintaining an improved patient safety performance. An overview of the challenges and proposed solutions for each category is discussed in the following (Nikita 2014).

14.3.2.1 On-Body Antennas and Safety Assessment

Research on on-body antennas has attracted significant interest toward the development of antennas for communication with exterior systems or other on-body antennas (Kiourti et al. 2014a). The first step in the design procedure is the selection of the operation frequency. In the United States, the Federal Communications Commission (FCC) has allocated the bands of 608–614, 1395–1400, and 1427–1432 MHz for wireless medical telemetry service (WMTS), as well as the bands of 902–928 and 2400.0–2483.5 MHz for industrial, scientific and medical (ISM) applications. In Europe, the frequency bands of 433.1–434.8 and 868.0–868.6 MHz are used for ISM applications. Finally, the Ultrawide Band (UWB) of 3.1–10.6 GHz, which has been authorized by the FCC, receives considerable attention because of the wide bandwidth offered. Microstrip and loop designs are generally applied for on-body antennas because of their conformability and light weight, but for many body postures, quarter-wavelength monopole antennas placed on a small ground plane have been shown to perform even better due to their omnidirectional pattern. In other cases, directive planar inverted-F antennas (PIFAs) are preferred. Miniature coplanar waveguide (CPW)–fed tapered slot antennas (TSA), as well as planar inverted cone antennas (PICA) have also been reported for UWB biomedical telemetry (Alomainy and Hao 2009). On-body antennas need to be flexible with low profile, easy to attach to the body or clothing. Therefore, textile antennas are also studied. Conductive materials (electrotextiles) can function as electronics while physically behaving as textiles and can thus be used to enable fabrication of textile antennas (Kiourti and Volakis 2016). Moreover, an UWB antenna made from textile materials has been reported in Osman et al. (2011). The antenna exhibited 17 GHz of bandwidth, omnidirectional patterns and adequate gain and efficiency.

Performance of on-body antennas is considerably affected by their proximity to human tissues. Commonly reported issues include antenna detuning, distortion of the radiation pattern, and degradation of the radiation efficiency. To better understand those issues, experimental investigations inside a reverberation chamber for five compact on-body antennas operating at 2.45 GHz and worn by nine individual test subjects have shown standard deviations of the antenna radiation efficiency of less than 0.6 dB and resonance frequency shifts of less than 1% (Conway et al. 2008). Furthermore, the technique of designing on-body antennas on electromagnetic band gap (EBG) substrates has been reported for overcoming radiation absorption inside the human body and achieving higher values of antenna gain (Zhu and Langley 2009; Wang et al. 2015). Finally, to achieve the necessary radiating characteristics but also reduce peak SAR level for 2.4 GHz ISM band, an artificial magnetic conductor (AMC) backed wearable antenna on latex substrate has been recently presented in Agrawal et al. (2016), Figure 14.6.

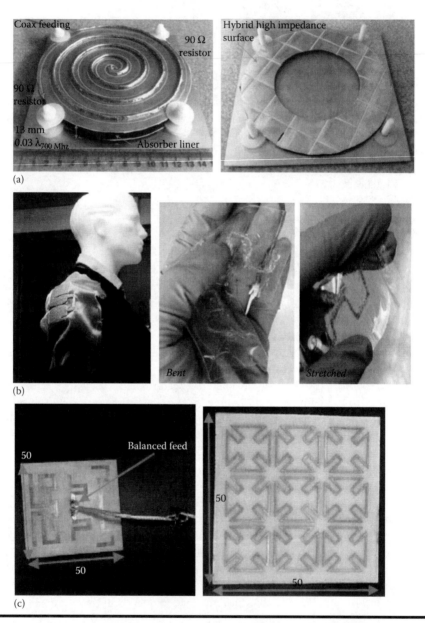

Figure 14.6 (a) Slot spiral antenna on EBG substrate. (From Wang, Z. et al., *J. Electromagnet. Wave*, 29(2), 143–153, 2015.) (b) Wearable antennas for RF communications. (From Kiourti, A., Volakis, J.L., Wearable antennas using electronic textiles for RF communications and medical monitoring. *10th European Conference on Antennas and Propagation (EuCAP)*, pp. 1–2, 2016.) (c) AMC backed wearable antenna. (From Agrawal, K. et al., *IEEE Trans. Compon. Packag. Manuf. Technol.*, 6(3), 346–358, 2016.)

Another factor that has to be taken into account when designing on-body antennas is the structure of the medical device into which the antenna is to be integrated. This involves the surrounding circuitry, including integrated circuits (ICs) and lumped elements. Therefore, antenna operation and performance are expected to vary according to the parameters that govern the operation of the medical device as a whole. To address this issue, a study that illustrated the importance of including the details of the medical device in determining and analyzing the performance of an on-body antenna has been presented in Alomainy et al. (2007). Results showed the proposed sensor matched monopole antenna design exhibited increased gain and efficiency (2.8 dB and a 29% respectively) as well as 25% enhancement in the coverage area compared to a stand-alone monopole design.

14.3.2.2 Implantable Antennas

Integrated implantable antennas are a key component of wireless implantable medical devices (Kiourti et al. 2014b; Kiourti and Nikita 2015). Operation frequencies for implantable antennas include the already mentioned ISM bands as well as the lately established Medical Implant Communications Systems (MICS) band at 402–405 MHz. Patch antennas are mostly preferred as implantable antennas because they are highly flexible in design and conformability (Kim and Rahmat-Samii 2004; Kiourti and Nikita 2012a, 2012d). Implantable antennas must also be biocompatible, in order to preserve patient safety and prevent rejection of the implant. Furthermore, they must be covered by a superstrate dielectric layer to avoid antenna short-circuit through contact with conductive human tissues. Commonly used biocompatible materials include Teflon, MACOR, and ceramic alumina (Soontornpipit et al. 2004). Insulating the antenna with a thin layer of low-loss biocompatible coating is another reported approach. Among others, materials proposed for biocompatible encapsulation include zirconia, (Skrivervik and Merli 2011) and Silastic MDX–4210 Biomedical–Grade Base Elastomer (Karacolak et al. 2010, Figure 14.7).

One of the greatest challenges in implantable antenna design is miniaturization, especially at the low-frequency MICS band. To achieve this goal, several techniques

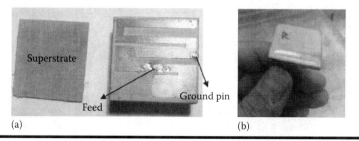

Figure 14.7 **(a) Implantable antenna and (b) antenna encased in silicone. (From Karacolak, T. et al.,** *IEEE Antennas Wireless Propag. Lett.,* **9, 334–337, 2010.)**

can be used (Kiourti et al. 2012; Kiourti and Nikita 2012b). First of all, human tissue, in which implantable antennas are intended to operate, exhibits relatively high permittivity which is exploited to miniaturize the size of the antenna. Additional miniaturization techniques include the use of high-permittivity dielectric materials (Kiourti et al. 2011), lengthening the current flow path on the patch surface (Liu et al. 2008), addition of shorting pins (Soontornpipit et al. 2004) and patch stacking (Kiourti and Nikita 2012c).

Another significant issue related to implantable antenna design is the maximum allowable power incident to the antenna, which is limited by safety guidelines. As already mentioned, SAR is generally accepted as the most appropriate dosimetric measure, and compliance with international guidelines is assessed (ICNIRP 1998; IEEE 1999). An attempt to modify the design of an implantable antenna for reducing the spatial-averaged SAR in human tissue has been presented in Kim and Rahmat-Samii (2006). Replacing the uniform-width spiral radiator of an implantable MICS PIFA with a non-uniform-width radiator was found to decrease the electric field intensity and, in turn, SAR.

Moreover, dual band operation is usually a legitimate requirement for implantable antenna design, enabling one frequency band for *wake-up* and one for data transmission (Kiourti and Nikita 2012c). This way, the power consumption of the implantable medical device is significantly reduced, increasing its lifetime. A dual-band (MICS and ISM) implantable antenna for continuous glucose monitoring has been proposed in Karacolak et al. (2008). In addition, an advanced antenna design was suggested using a π–shaped radiator with stacked and spiral structure, to support triple-band operation with data telemetry (402 MHz), wireless power transmission (433 MHz), and wake-up controller (2450 MHz) in Huang et al. (2011).

Prototype fabrication of implantable antennas meets all classical difficulties of miniature antennas: additional glue layers used to affix all components together strongly affect antenna performance by shifting its resonance frequency and degrading its matching characteristics, whereas the coaxial cable feed used to connect the antenna with the network analyzer may give rise to radiating currents on the outer part of the cable (Kiourti and Nikita 2014c). On the other hand, testing inside phantoms is easy to implement. The fabricated prototype is typically immersed inside a liquid tissue phantom and measured. Furthermore, use of animal tissue samples provides an easy approach to mimicking the frequency dependency characteristics of the electrical properties of tissues. For example, a dual-band skin-implantable patch antenna operating in the MICS and 2450 MHz ISM bands has been tested in real animal skin in Karacolak et al. (2009). Moreover, a triple-band implantable patch antenna has been tested inside a minced front leg of a pig (Huang et al. 2011). In vivo testing inside living animals is another experimental verification method. In this case, an in vivo testing protocol needs to be developed before the experimental investigations, which will deal with the choice and number of animals, presurgical preparation, anesthesia, surgical procedure, measurements, and postsurgical treatment (Kiourti et al. 2013). In vivo studies reported in the literature are thus very limited. However, the return loss frequency response of a skin-implantable antenna has been measured using rats as model animals in Karacolak et al. (2010).

14.3.2.3 Ingestible Antennas

Ingestible antennas have received a lot of research interest in the past years. The conventional methods used for diagnosing disorders of the human gastrointestinal (GI) tract cause significant patient discomfort. As a result, there is considerable ongoing work in developing ingestible antennas that can be integrated into a capsule and swallowed for examination of the entire digestive tract (Kiourti et al. 2014b). Selection of operation frequency for ingestible antennas has received a significant attention from the scientific community as it is affected by a number of competing effects. For example, antenna efficiency can improve with higher operation frequency. On the other hand, higher frequencies may cause increased radiation absorption because of the high water content of body tissues, requiring increased levels of power supply and posing questions regarding patient safety. The presence of the human body is another factor that must be taken into account when designing an ingestible antenna. The effects of the human body on the performance of an ingestible antenna inside the frequency range of 150 MHz–1.2 GHz have been investigated in Chirwa et al. (2003). A peak in the power transmitted by an ingestible medical device at approximately 650 MHz was found, and adequate communication performance was demonstrated between 600 MHz and 1 GHz. On the other hand, video transmission in the 2.45 GHz band is better developed for WLAN and Bluetooth applications and ingestible medical devices operating at this frequency can be directly connected to relevant networks. Also, higher transmission frequencies allow the use of smaller antennas and electronic components, a prerequisite for ingestible medical devices, rendering the 2.45 GHz band a promising solution. For example, an IC design for wireless capsule endoscopy at 2.4 GHz has been proposed in Xie et al. (2004).

Besides the aforementioned requirements, an ingestible antenna needs to possess omnidirectional radiation pattern and exhibit circular polarization in order to transmit signals independently of its position and orientation. Taking into account the above, normal mode helical antennas are good candidates for such applications (Liu et al. 2014). Furthermore, wireless capsule endoscopy typically includes real-time transmission of high-resolution data, therefore, antennas with miniature size but wide bandwidth are required. For example, a wideband spiral antenna for ingestible capsule endoscope systems at 500 MHz was presented in Lee et al. (2011). Two more requirements that have to be met during implantable antenna design are that of biocompatibility and durability. These requirements entail packaging of the ingestible antennas inside a shell. Capsule casings and circuitry have been shown to have a negligible effect on the performance of the ingestible antenna (Liu et al. 2014); therefore, it is not compulsory to take them into account when modeling the antenna.

Safety performance of ingestible antennas has also attracted significant scientific interest. Usually, low frequency wireless devices cause less significant biological effects than higher frequency devices as attributed to decreased levels of tissue absorption. The SAR and temperature rise performance of ingestible antennas has been analyzed at frequencies from 430 MHz to 3 GHz in Xu et al. (2008). Results showed that high values of SAR and temperature were localized at the area near the ingestible device. The antenna was safe and could be used in ingestible medical devices at input power levels of less than 25 mW in order to conform to safety regulations. Performance of ingestible antennas for various locations inside the body and antenna orientations has also been investigated. In Liu et al. (2014), three different implant positions (stomach, small intestine, and colon) were examined using a three-dimensional voxel human body. Results showed that radiation characteristics (gain and axial ratio) are affected by different implantation depths.

14.3.3 Inductive Coupling

Near-field power transmission is a viable technique to wirelessly power up devices or recharge their batteries from a short range without direct electrical contact between the energy source and the device. Moreover, the same short-range wireless link can be used to establish wide-band bidirectional data communication. As a result, wireless implantable medical devices (IMDs) can benefit from the use of near-field power and data transmission links (Ghovanloo and Kiani 2014). These devices need to transmit and receive information wirelessly across the skin barrier since breaching the skin with interconnect wires would be a source of morbidity for the patient and would increase the risk of infection. To make things more complicated, in sensory prosthetic devices which interface with the central nervous system, the quality of perception has been reported to increase with the number of stimulating sites and rate (Fernandes et al. 2012). Generally, these devices require more power and bandwidth than autonomous devices, like pacemakers. State-of-the-art visual prostheses are currently targeting beyond one thousand sites to improve the quality of the visual functions, such as mobility without a cane, face recognition, and reading large fonts (Mathieson et al. 2012). Also, inductive links can be utilized to power up wireless neural recording systems for freely behaving small animal subjects developed for neuroscience applications (Fernandes et al. 2012), surmounting the limitation of a large payload of batteries carried by the animal. In the following paragraphs, the main aspects of inductive coupling for m-Health applications are briefly discussed.

14.3.3.1 Operating Principle and Key Design Parameters

The main physical principle of inductive coupling is Faraday's law, which states that when the total magnetic flux through a conductive loop varies with time, a current is induced in the loop itself. This, in turn, results in an electromotive force

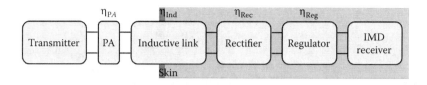

Figure 14.8 **Simplified power flow diagram in transcutaneous inductive power transmission link. (From Ghovanloo, M., Kiani, M., Inductive coupling, in K.S. Nikita (Ed.),** *Handbook of Biomedical Telemetry,* **Wiley-IEEE Press, pp. 174–207, 2014.)**

(EMF) induced in the loop (Sadiku 2007). Thus, a primary loop will generate the varying magnetic field, which concatenates with the secondary loop, resulting in an induced current in the secondary loop. A key design requirement in IMDs is to provide sufficient power delivered to the load (PDL) while maintaining high power transfer efficiency (PTE). The inductive PTE is defined as the ratio of the PDL to the power drained by the energy source. High PTE is required to reduce heat dissipation within the coils, tissue exposure to EM fields, size and weight of the external energy source and interference with nearby electronics. The typical power flow from external power source to IMD electronics indicating the various loss parameters along the way is depicted in Figure 14.8.

In an inductive power transmission link, active rectifiers can be used to maintain a high PTE. A 13.56-MHz active rectifier, which has achieved 80.2% power conversion efficiency when delivering 20 mW to the load, has been reported in Lee and Ghovanloo (2011). Furthermore, a closed-loop power transmission system can be used to maintain the received voltage constant in cases where the relative distance between the coils changes (Kiani et al. 2011). This way, voltage fluctuations occurring due to changes in coils relative distance or misalignment can be mitigated.

Another critical parameter of the induction link is coil geometry. One of the decisions that should be made early on when designing such a link is whether the coils should be wire wound or lithographically defined on a planar conductive surface. Wire-wound coils are made of filament wires twisted in a circular bundle, also known as Litz wires. The optimal number of turns for a coil made of Litz wire depends on the desired volume of the coil and parameters such as the number and diameter of individual strands (Tourkhani and Viarouge 2001). Wire-wound coils major drawback, however, is the fact that they cannot be batch fabricated or miniaturized without the use of sophisticated machinery. On the other hand, printed spiral coils (PSCs), which are lithographically defined in one or multiple layers on rigid or flexible substrates, such as PCB, can be batch fabricated. Also, they offer more flexibility in optimizing their geometries while occupying less space, deeming them attractive for IMD applications. An iterative design procedure has been reported, which starts with a set of design constraints and initial values and can result in the optimal PSC pair geometries (Jow and Ghovanloo 2007).

Finally, ensuring that the electromagnetic power absorbed in the human body meets international safety standards is one more crucial aspect that coil designs for IMD applications should take into account. Full-wave electromagnetic computational tools and experimental methods with phantoms filled with tissue simulants can be used to determine the field induced by induction devices in the human body and compare them with standard safety limits (ICNIRP 1998; IEEE 1999).

14.3.3.2 Data Transmission

Bidirectional wireless data transmission is essential for IMD systems to establish short-range wireless communication between the transmitting and receiving parts of the system. The majority of advanced IMDs have several adjustable parameters that can be fine-tuned after implantation for every individual patient according to his or her specific needs. In addition, research is conducted to equip sensory devices with a flow of stimulation commands from the external artificial sensors and signal processing units to build closed-loop neuroprosthetic devices (Schwartz et al. 2006). Sending adjustment and control commands wirelessly from the external unit to the implanted unit is known as the forward telemetry or downlink. Moreover, the same devices often need to inform the external processing components about the IMD operating status, possible faults, and in some cases the neuronal response immediately after stimulation for proper adjustment of the stimulation parameters (Venkatraman et al. 2009; Lee et al. 2010). This data flow direction is often referred to as backward telemetry or uplink.

When designing an induction link, there are two options: using a single carrier for both power and data transmission or using multiple (two or three) carriers for power, downlink and uplink. The choice of a single carrier has the advantage of robust coupling between power coils, which can lead to more reliable data transfer. Also, space is saved by reusing power coils for multiple purposes. However, achieving high PTE and large bandwidth simultaneously is challenging as they set conflicting requirements. As a result, the use of two or three carrier signals for power, downlink, and uplink with each carrier having its own pair of coils in order to decouple the data transfer link bandwidth from the PTE has been proposed in (Zhou et al. 2008; Shire et al. 2009). The use of multiple carriers in a small space introduces new challenges though, with the most important being the interference between strong power carrier with much weaker data carriers. Several research efforts have been reported offering solutions such as using orthogonal symmetrical coils (Jow and Ghovanloo 2010), coaxial coils with differential phase shift keying (DPSK) (Shire et al. 2009), and shifted coplanar coils with offset quadrature phase shift keying (OQPSK) (Zhou et al. 2008; Shire et al. 2009). Even though modulating a carrier signal provides a robust means to transfer data, generation of the carrier signal at a power level that ensures sufficient signal-to-noise ratio (SNR) at the receiver involves consuming a considerable amount of power at the transmitter, which is scarce on the IMD side. One solution to this problem is to substitute the carrier signal with a series of sharp and narrow pulses, which require much less power to generate them (Inanlou and Ghovanloo 2011).

14.3.4 Intra-Body Communications

A different approach for wireless communications in the area of the human body is becoming popular nowadays. It is based on signal transmission through the human body, called intra-body communication (IBC) and has led to the first definition of body area networks (BANs) (Roa et al. 2014). This approach was proposed by Zimmerman (1996): the architecture included a transmitter placed on the body surface that modulates an electric field, which is conducted by means of small currents through the body toward the receiver that demodulates the received signal, thus recovering the encoded information. Low carrier frequencies associated with low-power waves can be used, favoring lower consumption of the terminals and less interference with nearby devices. The IBC propagation channel is mainly established through the human body, but depending on the application and the coupling type, there are different signal pathways. Three types of body-centric communications types have been distinguished, namely off body, on body, and in body, depending on the degree of interaction of the human body with the surrounding space. For example, galvanic coupling should always be used when sensors are implanted inside the human body (in body). When all sensors are deployed on the surface of the skin, we refer to body surface sensor networks (on-body and off-body types), and both IBC galvanic and capacitive coupling schemes can be used. Some relevant examples of IBC are facilitated by the deployment of sensors in direct contact with human skin (like EMGs or ECG) for the welfare enhancement of disabled people (Liolios et al. 2010), or the estimation of muscular fatigue trough EMG (Lucev et al. 2010).

During the last decade, an important boost in IBC research has taken place by exploiting modeling approaches and techniques that have taken advantage of the rapid development of MEMS and nanotechnologies for the implementation of several laboratory prototypes with better performance. The results of attenuation on experimental measurements performed on a human cadaver obtaining a range of –10 to –40 dB in the frequency range of 1–20 MHz are reported in Tang and Bashirullah (2011). In Zedong et al. (2012), the 24–30 MHz frequency range is suggested as optimal for IBC communications, which was obtained through experimentation in a shielded chamber. On the other hand, a frequency band between 280 and 500 MHz for capacitive body-coupled communications was determined in Attard and Zammit (2012). The maximum rate reported for IBC has been 10 Mbps, thanks to the introduction of high-impedance electro-optical sensors (NTT 2015). Unfortunately, these sensors have the disadvantage of a more complex electronic design, resulting in low capacity for integration, greater power consumption, and size. Based on the above, it is clear that there is an important variability in the reported performances, which demands research efforts focused on different techniques and approximations to measure channel propagation characteristics (Yang et al. 2011).

Despite the great performance obtained by IBC systems, there are still important remaining advances to be made. From a theoretical p, although IBC prototypes have significantly improved in performance, bit rate, and consumption, there is yet no common methodology to establish the design specifications of IBC systems, although some recent exceptions can be made (Bae et al. 2012). This is due to the fact that the electromagnetic (EM) mechanisms that govern the transmission through biological tissues still remain unknown. From a technological point of view, IBC systems must consider present and future trends in the field of micro and nano-MEMS, which can lead to unobtrusive and compact solutions, like the proposed IBC system on a chip in Yan et al. (2011), while facing the need to compromise electronic complexity with performance to obtain realistic and sustainable solutions to be translated to the health care domain.

14.3.5 Biomedical Communication Standards

This section describes some examples of recent standard-based wireless technologies that are suitable for biomedical communications (Schmitt et al. 2014). In order to create a telehealth ecosystem, interoperability is important and can be achieved by means of well-defined and standardized interfaces between the various device classes. The Continua Health Alliance, with currently more than 240 member companies (care providers, device manufacturers, software and middleware vendors, information technology suppliers, silicon device manufacturers, etc.), is working on selecting connectivity standards and defining complementary guidelines for ensuring interoperability (Continua Health Alliance 2006).

The IEEE 802.15.4 standard (IEEE 2011), specifies the physical and media access control layers for low-rate wireless personal area networks (LR-WPANs) and is the basis for the ZigBee (ZigBee Alliance 2012), ISA100.11a (Kenney 2007), WirelessHART (HART 2008), and MiWi (Microchip 2010), specifications, which further extend the standard by developing the upper layers. The standard was released in 2003, amended in 2006, and enables communication of data at a maximum rate of 250 kbps.

In February 2012, IEEE published the IEEE 802.15.6 standard (referred to as 15.6) for a short-range wireless communication inside, on, or around a human body, with support of the combination of high reliability, quality of service, low power, scalable data rates (up to 10 Mbps), and interference mitigation (IEEE 2012). The 15.6 standard specifies three Physical layers (PHYs) and a unified medium access control (MAC) and security solution to address the broad requirements of medical as well as nonmedical (e.g., entertainment and well-being) body area network (BAN) applications.

IEEE 802.15.4j (IEEE 2013) defines necessary PHY/MAC modifications for the IEEE 802.15.4 standard, making it able to operate in the 2360–2400-MHz band in compliance with the FCC MBAN rules (FCC 2012). Two considerations have been taken into account in the TG 4j standardization: (1) to keep the channelization scheme flexible to accommodate harmonized coexistence with in-band primary/MBAN services and (2) to provide MAC support to enable MBAN low-power implementations.

Mature wireless transport standards suitable for connected devices in biomedical applications have been developed, solving a significant number of problems on the lower layers. However, more work at levels closer to the application is needed. A solution to this last gap of application layer interoperability has been developed within the family of International Standards Organization (ISO)/ IEEE 11073 personal health device (PHD) communication standards (Schmitt et al. 2007). These define a common framework for making an abstract model for personal health data available in transport-independent transfer syntax to establish logical connections between personal health sensors or devices (e.g., ECG sensors, heart rate meters, thermometers, or blood pressure monitors) and in-home compute engines. Besides general requirements like patient and user safety of medical devices, minimal user interaction, and unambiguous association between devices and sensors, the protocol suite is optimized to personal health device usage requirements. Personal health devices are often battery powered, which mandates the need for very low computational complexity and low power consumption. For wireless devices the latter requirement suggests not only minimizing transmit power but also reducing transmission time by minimizing protocol overhead. The ISO/IEEE 11073 PHD standards have been adopted by the Continua Health Alliance and are leveraged by its design guidelines to enable a consistent, interoperable data layer concept across the different Continua interfaces.

14.4 Wearable, Implantable, Ingestible Devices

14.4.1 Spectrum Regulations

Demand on radio spectrum for use in wireless biomedical systems is currently on the rise. This demand is driven by a rapid increase in the use of medical devices, advancements in wireless communication technologies, and the need to improve quality, reliability, and delivery of health care. Some of the most commonly used frequency bands for biomedical telemetry systems include the MICS, WMTS, ISM, and UWB bands. Regardless of the chosen band, constant management is crucial for reducing the probability of interference from other transmitting devices (Nikita 2014, Figure 14.9).

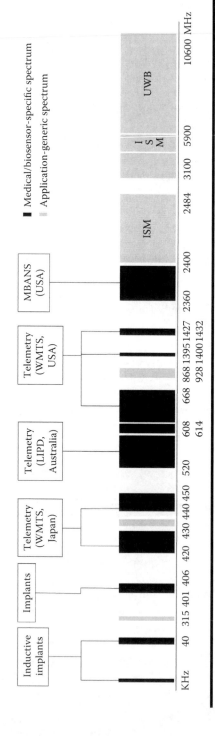

Figure 14.9 Frequency spectrum for biomedical applications depicting medical-specific and application-generic bands. (From Schmitt, L. et al., Biosensor communication technology and standards, in K.S. Nikita (Ed.), *Handbook of Biomedical Telemetry,* Wiley-IEEE Press, pp. 330–367, 2014.)

14.4.1.1 Medical Implant Communications Systems

In 1998, the International Telecommunication Union—Radiocommunication (ITU-R) outlined the use of the 402–405-MHz frequency band for Medical Implant Communications Systems (MICS) (ITU-R 1998). The MICS band is currently regulated by the U.S. Federal Communications Commission (FCC) (MICS Federal Register 1999) and the European Radiocommunications Committee (ERC 1997) and is expected to become a true global standard within several years. Two fields of application are indicated for this standard: communication between an implantable medical device and an exterior receiving station and communication between medical devices implanted within the same human body. MICS devices can use up to 300 kHz of bandwidth at a time for the complete session. Equivalently, separate transmitter and receiver bands, each with a bandwidth of 300 kHz, may be adopted as long as they are not used simultaneously. The range is typically 2 m, and the maximum power limit is set to 25 μW of equivalent radiated power (ERP).

14.4.1.2 Wireless Medical Telemetry Service

The FCC has allocated the frequency bands of 608–614, 1395–1400, and 1427–1432 MHz for wireless medical telemetry service (WMTS) in the United States (FCC 2003). These bands are very advantageous for biomedical telemetry because they allow a relatively large bandwidth for communication (e.g., four 1.5-MHz-wide channels are allowed in the 608–614-MHz WMTS band). Furthermore, WMTS bands are solely reserved for biomedical telemetry, meaning that medical devices which operate at these frequencies are protected from interference caused by other sources. However, there is currently no indication that the WMTS bands would be allotted in other parts of the world, meaning that devices cannot be marketed or used freely in countries other than the United States.

14.4.1.3 Industrial, Scientific and Medical

The industrial, scientific and medical (ISM) bands were originally reserved internationally for noncommercial use of radio frequency (RF) electromagnetic fields. They are defined by the ITU-R, but individual countries' use of the bands differs due to variations in national radio regulations. The 902–928 and 2400.0–2483.5-MHz frequency bands are used in the United States and are defined by the FCC, whereas the European countries use the 433.1–434.8 and 868.0–868.6-MHz frequency bands, which are defined by the Electronic Communications Committee (ECC). The ISM bands offer users the advantage of increased bandwidth, thus enabling video and voice transmissions. Furthermore, since government approval is not required, the ISM bands are nowadays used by a wide variety of commercial standards. However, the ISM bands are not exclusive to biomedical telemetry equipment, meaning that transmission of sensitive medical data in these bands is susceptible to interference from other devices.

14.4.1.4 Ultrawide Band

Ultrawide Band (UWB) systems are spread-spectrum communication systems or, equivalently, systems in which the bandwidth of the transmitted signal is considerably wider than the frequency content of the original information. More specifically, UWB is defined by the FCC as any communication system which has a spectral occupation of greater than 20% or occupies an instantaneous bandwidth of more than 500 MHz. The band of 3.1–10.6 GHz, which has been authorized by the FCC for unlicensed use, is nowadays receiving the most attention by standardization bodies. Extremely short pulses are transmitted, and high data rates are thus achieved. Despite the fact that UWB medical devices are currently only allowed in the United States and Singapore, regulatory efforts are already underway in Europe and Japan.

14.4.2 Patient Safety

Before medical devices can be widely accepted, the public needs to be convinced of their safety (Psathas et al. 2014). People's perception of electromagnetic radiation is generally fearful and safety standards need to be established in order to quantify biological damage and preserve patient safety. It was not until recently that research on the biological effects of medical devices with telemetry functionalities started being carried out. The approach that is currently in use for establishing such safety standards is through animal experimentation (Johnson and Guy 1972). The actual fields, current density, and absorbed energy density which cause biological damage inside the tissues of the animal are recorded and further extrapolated to human beings. The specific absorption rate (SAR), which is defined as the rate of energy deposited per unit mass of tissue, is generally accepted as the most appropriate dosimetric measure. Limits for the United States and Europe are based on recommendations from the IEEE (IEEE 1999) and the International Commission on Non-Ionizing Radiation Protection (ICNIRP 1998), respectively. Actual regulations may vary according to the scenario under study and legislation of each country. Device manufacturers must ensure that their products do not introduce higher SAR values than the specified limits. This is most commonly accomplished through in vivo experimental investigations or numerical computations. It is worth noting that the actual SAR values are not only determined by the medical device itself, but also depend on any device in the close vicinity of the human body that could influence the fields inside the human tissues. Furthermore, increasing the power incident in the medical device to improve its communication range may result in the device exceeding the regulations for maximum power absorption inside the body and must be taken into account by manufacturers.

14.4.3 Biocompatibility

Biocompatibility plays a key role in the development of implantable and ingestible devices and is an important input requirement for their design. Such devices must be biocompatible in order to preserve patient safety and prevent rejection of the implant. Furthermore, human tissues are conductive and would short circuit the implantable or ingestible components if they were allowed to be in direct contact with their metallization. Biocompatibility and prevention of undesirable short circuits are especially significant in the case of devices that are intended for long-term implantation. It is important to highlight that medical devices consist of a variety of materials. Therefore, in order to assess the biocompatibility of the device, one must consider the applied materials one by one as well as the complete medical device as a whole. Furthermore, biocompatibility of a medical device depends on the time that it is exposed to the human body as well as its specific location inside the body. In any case, the designer of the medical device is responsible for its biocompatibility and safety, rather than the physician.

14.4.4 Commercial Products

A number of commercial wireless medical devices have already been reported and indicative examples are summarized next (Nikita 2014). Ultra-lightweight, flexible and disposable skin patches, no thicker than ordinary adhesive bandages, are combined with Gentag NFC wearable sensors to simplify diagnostics, fitness, diabetes monitoring, and drug delivery. Using proprietary ASICs (application-specific integrated circuits), Gentag can custom design and license its wearable patch technology for a wide array of consumer or medical applications, which can be created to be entirely battery-free, that is, passive, powered only by NFC (Gentag 2011, Figure 14.10).

Moreover, GE Healthcare has introduced a smartphone-size imaging tool, known as the VScan, which lets physicians carry ultrasound technology in their pockets (GE Healthcare 2009). Biotronik has recently proposed a small battery-powered electrical impulse generator to be implanted in patients who are at risk of sudden cardiac death due to ventricular fibrillation and ventricular tachycardia (Biotronik 2012). Enhanced with Advanced Patient Management, Biotronik provides the cornerstones for comprehensive patient protection and care. The Medtronic

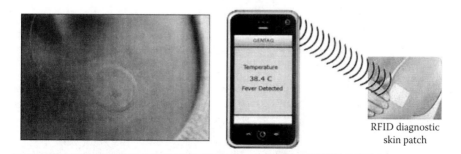

Figure 14.10   Gentag's multipurpose diagnostic skin patches. (From Gentag. 2011. Accessed July 2, 2016. http://gentag.com/nfc-skin-patches/.)

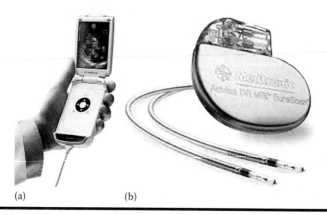

(a)   (b)

Figure 14.11 **(a) GE Healthcare VScan. (From GE Healthcare. Accessed July 2, 2016. http://www3.gehealthcare.com/en/products/categories/ultrasound/vscan_ portfolio, 2009.) (b) Medtronic Adapta with MVP. (From Medtronic, Bravo pH monitoring syestem. Accessed July 3, 2016. http://www.medtronic.com/covidien/ products/reflux-testing/bravo-reflux-testing-system, 2010.)**

Adapta with MVP pacing system offers managed ventricular pacing (MVP), atrial therapy, ventricular capture, and remote cardiac telemetry (Medtronic 2010), whereas the Medtronic Revo MRI SureScan pacing system is magnetic resonance (MR) conditional designed to allow patients to undergo MRI under the specified conditions of use (Medtronic 2011, Figure 14.11).

The Nucleus Freedom cochlear implant includes a sound processor which is worn behind the ear and a cochlear implant which is placed under the skin, behind the ear (Cochlear 2010). The sound processor captures sounds, digitizes them, and sends the digital code to the implant. The implant converts the coded sound to electrical impulses and sends them along an electrode array to further stimulate the cochlea's hearing nerve. Hearing may be managed via a remote assistant or directly from the sound processor. Furthermore, The Medtronic SynchroMed II Pump is a drug infusion system which provides precise drug delivery for chronic therapy of severe spasticity (Medtronic 2012). Another commercial product intended to provide visual perception in blind individuals is Argus II Retinal Prosthesis System. It includes a miniature video camera, a transmitter mounted on a pair of eyeglasses, a video processing unit, and a 60-electrode implanted retinal prosthesis that replaces the function of degenerated cells in the retina. Although it does not fully restore vision, it can improve a patient's ability to perceive images and movement (Second Sight 2012, Figure 14.12).

Commercial ingestible medical devices are used for gastrointestinal (GI) endoscopy and sensing of physiological parameters within the GI (pH, temperature, pressure), which allow for direct and noninvasive examination of the GI tract. Up until 2007, wireless endoscopic capsules were only developed by Given Imaging (Given Imaging 2012a). The vitamin-sized capsule (PillCam) provides a way to

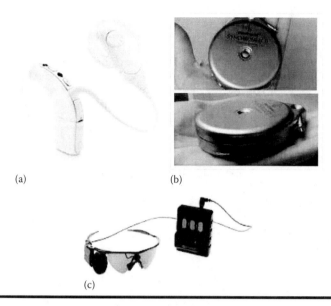

(a)   (b)

(c)

Figure 14.12 **(a) Nucleus Freedom cochlear system. (From Cochlear, Nucleus Freedom. Accessed July 3, 2016. http://www.cochlear.com/wps/wcm/connect/ us/recipients/nucleus-freedom/nucleus-freedom-basics, 2010.) (b) Medtronic SynchroMed II pump. (From Medtronic, SynchroMed II pump. Accessed July 3, 2016. https://professional.medtronic.com/pt/neuro/itb/prod/synchromed-ii/ features-specifications/#.V3jOAPl96Cg, 2012.) (c) Argus II retinal prosthesis system. (From Second Sight, Argus II retinal prosthesis system. Accessed July 3, 2016. http://www.secondsight.com/g-the-argus-ii-prosthesis-system-pf-en.html, 2012.)**

visualize, monitor, and diagnose small-bowel abnormalities including abnormalities associated with obscure GI bleeding, iron deficiency anemia, and Crohn's disease. After 2007, other companies, made significant improvements in their own endoscopic capsules. The Olympus Endocapsule is a small-bowel endoscopy system that travels through the small intestine via normal muscle contractions, taking thousands of pictures that are transmitted to a recorder worn around the waist (Olympus 2012). The IntroMedic MiroCam is a capsule endoscope based upon HBC (human body communication), a state-of-the-art patented technology utilizing the human body as a communication medium (Intromedic 2012). Apart from capsules used for visualization, monitoring and diagnosis of small-bowel abnormalities, there are also capsules with integrated sensing technologies for monitoring physiological parameters within the GI tract. For example, the Bravo pH Monitoring System is a catheter-free way to measure pH (Medtronic 2010). By using a miniature pH capsule attached to the esophagus, pH data from the esophagus can be wirelessly transmitted to a small recorder worn on a shoulder

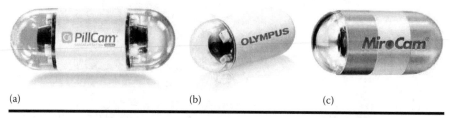

Figure 14.13 (a) PillCam by Given Imaging. (From Given Imaging, PillCam. Accessed July 3, 2016. http://www.givenimaging.com/en-int/Innovative-Solutions/Capsule-Endoscopy/Pages/default.aspx, 2012a.) (b) Olympus Endocapsule. (From Olympus, Capsule endoscopy. Accessed July 3, 2016. http://medical.olympusamerica.com/procedure/capsule-endoscopy, 2012.) (c) IntroMedic MiroCam. (From Intromedic, MiroCam. Accessed July 3, 2016. http://www.intromedic.com/eng/sub_products_2.html, 2012.)

strap or waistband. Another example is the Smartpill, an ingestible capsule that measures pressure, pH and temperature as it travels through the gastrointestinal (GI) tract to assess GI motility (Given Imaging 2012b, Figure 14.13).

14.5 Integrated m-Health Systems

A variety of integrated systems based on the use of wearable, implantable and ingestible devices have been developed for a series of applications (Kwasnicki and Yang 2014). These applications range from wireless capsule endoscopy, to gait assessment, nerve stimulation, fall detection and chronic disease management (Bourbakis and Karargyris 2014; Cancela et al. 2014; Cela et al. 2014; Fioravanti et al. 2014; Hao and Foster 2014). In this section, some characteristic examples of recently reported m-Health systems are briefly presented and discussed. By no means is this list exhaustive; the goal is to provide some insights in the directions followed by current studies toward the introduction of wireless communications, mobile computing and sensing technologies in today's health care.

An interactive m-Health system (ImHS) to establish two-way communication between long term diabetic patients and caregivers by utilizing IoT technology has been recently reported (Chang et al. 2016). The system consists of three devices: a General Packet Radio Service (GPRS) blood-glucose monitor (BGM), a telecare Android/iOS application for caregivers, and a cloud server platform that integrates the system's main functions. Via the GPRS BGM, the ImHS collects patient information—such as blood-glucose values and measurement scenarios and times (before/after meals, before/after activity, at night, and so on)—and uploads it to a cloud server in XML format via the GPRS protocol. The collected data are used to detect abnormal blood-glucose levels through a set of rules and procedures developed based on discussions with doctors and relevant publications by the World Health Organization (WHO).

As each patient's health condition is unique, the system allows patients to customize the system's detection parameters based on their doctor's suggestions. Subsequently, the ImHS will analyze this information to determine the patient's health status. According to the determined status, the system will automatically execute appropriate actions, such as sending a message to the GPRS BGM. Moreover, in a scenario in which a patient's health status is critical, the ImHS will automatically notify the patient's caregivers through the Message Queue Telemetry Transport (MQTT) protocol. Future improvements include advanced functionalities for critical health case situations, such as patient location tracking through GPS.

A novel wearable sensor system called Smart Wristlet, which can provide 24 hours fall detection service has been presented in Li et al. (2014). A fall accident is the leading cause of accidental injury and death for the elderly people (Tamura et al. 2009). For persons older than 85%, 86% of their injuries are from fall accidents. The system uses a 3-layer function architecture as depicted in Figure 14.14. Smart Wristlet is designed to sense user's activities in a noninvasive way using embedded sensors to monitor movement parameters as well as physiological data. A machine-perception based fall detection algorithm was developed to ensure reliability and practicability. Due to the high efficiency and accuracy of the algorithm, Smart Wristlet achieves fall detection precision as high as 93%, which is 3% higher than conventional methods.

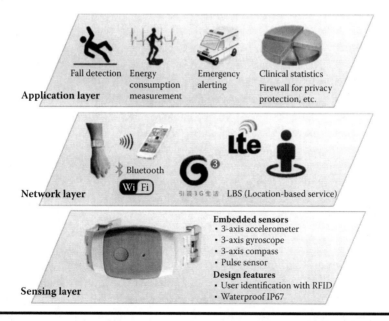

Figure 14.14   Smart Wristlet m-Health System 3-layer function architecture. (From Li, Z. et al., Fall perception for elderly care: A fall detection algorithm in smart wristlet mHealth system. *IEEE International Conference on Communications (ICC)*, pp. 4270–4274, 2014.)

More importantly, by simplifying the computational complexity and sensor data, the battery life is extended by more than 30%. This way, long-time and uninterrupted m-Health services in real applications can be ensured. For instance, with pop-up airbags on the user's body, Smart Wristlet can trace user's activity and pop-up the airbags to protect him/her when a fall event is detected. Furthermore, automated remote emergency alerting for elderly fall can be incorporated to the system since old people usually are not able to call for help in case they fall and get injured. Taking into consideration that only in the United States, elderly falls cause around 8 billion dollars direct medical costs each year (Carroll et al. 2005), Smart Wristlet targets to save more than 800 million dollars per year in a socio-economic level.

An end-to-end m-Health system for screening and management of noncommunicable diseases (NCDs) has been developed (Clifford et al. 2014). Cardiovascular diseases (CVD) account for most NCD-related deaths, followed by cancers, respiratory diseases, and diabetes (Lim 2012). Combined, NCDs are related to 80% of deaths, and studies show the risk factors include tobacco, alcohol, high blood pressure, diet, and physical inactivity (Habib and Saha 2010). The system includes simple, low-cost ($5–$20) and open-source hardware peripherals that allow a minimally trained person to collect high-quality medical data at the point-of-care through a standard smartphone. The data can be reliably transmitted even in the case of high latency network connections and stored into a web-based system that manages the Electronic Medical Records (EMRs) of the patients, making these records accessible anywhere. The novel, low-cost sensors used (semiautomatic blood pressure monitor, three lead ECG, spirometer) are connected as USB peripherals in the smartphone. Signal analysis software running on Android smartphones has been developed, guiding an untrained user through the data acquisition. Server software to allow crowd sourcing of diagnostics in order to improve accuracy and learning has been also developed. The system is fully scalable, with an open architecture to allow the addition of decision support and analysis algorithms on the back end, and peripherals and mobile phone-based software on the front end (Figure 14.15).

A smartphone based system for remote real-time tele-monitoring of physical activity has been used in a pilot study with actual patients suffering from chronic-heart-failure (CHF) (Aranki et al. 2016). Patient's physical activity and vital signs data during everyday activities are being collected via a smartphone and reported to a central server. Global position system (GPS) is also used to track outdoor activity and measure walking distance. The system is designed to assess patient activity via minute-by-minute energy expenditure estimated by accelerometry, self-reported vital signs (heart rate, blood pressure, and weight), and relevant cardiovascular symptoms (fatigue, activity, dizziness, shortness of breath, etc.). The collected data are securely transmitted to a server where they are analyzed in real-time. This way, medical staff can monitor patient status and provide medical intervention if necessary. Future plans include privacy enhancements as well as battery lifetime extension through delegation of computation procedures from the smartphone to the server. Conclusions elicited by this study can be applied to

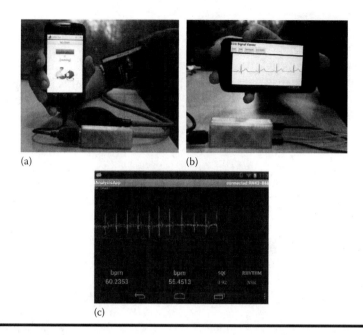

Figure 14.15 **(a) Semiautomatic blood pressure monitor; (b) ECG signal acquisition; and (c) mobile application screenshot showing two channel ECG. (From Clifford, G.D. et al., A scalable mHealth system for noncommunicable disease management.** *Global Humanitarian Technology Conference (GHTC), 2014 IEEE,* **pp. 41–48, 2014.)**

other chronic health conditions, such as diabetes and hypertension, which would also benefit from continuous monitoring through m-Health technologies.

14.6 Future Challenges and Opportunities

Current challenges in mobile health care technologies are related to several factors. To begin with, there is the issue of extended power supply lifetimes that can be achieved using energy scavenging techniques that utilize the inner workings of the body (Tentzeris et al. 2014). For example, glucose fuel cells (Zebda et al. 2013), and electrochemical gradients are some emerging techniques (Mercier et al. 2012). These systems show strong potential as new energy sources, but their large area (fuel cells) or extremely low energy density (electrochemical gradients) means these technologies need to be further developed before a complete system with a practical data acquisition rate can be built upon them. Furthermore, the majority of power consumption budget is dedicated to wireless communications (Ntouni et al. 2014); ultrawideband transceivers and energy-efficient data compression algorithms can be used to address this issue (Chin et al. 2012). Also, delegation of computation hungry procedures from mobile devices to servers can contribute in power consumption reduction (Aranki et al. 2016). In the context of implantable sensors, biocompatibility poses a significant factor. Use of biocompatible materials and careful design of the device are required to avoid traumatizing human tissues and ensuring functionality. Also, it is important to protect the privacy of data collected and disseminated by health care-related technologies. Complex security mechanisms can be used, but since they require more computational and power resources, a trade-off is crucial (Bourbakis et al. 2014). Currently, most of the health care related technology is developed on-demand. Using dynamic programming environments and cognitive surfaces we should be able to measure multiple parameters in the human body and use this data to aid in preventive medicine diagnosis. Of course there is always the computational and economic perspective, since the cost and size of health care technologies impose limitations on their use. To conclude, m-Health is a highly interdisciplinary field, therefore generating innovative ideas demands the widest possible participation and a culture that appreciates the notion that not all ideas will succeed (Modi and Mohanty 2015).

References

5G-PPP. 2015. 5G-PPP white paper on e-health vertical sector. Accessed March 17, 2017. https://5g-ppp.eu/wp-content/uploads/2014/02/5G-PPP-White-Paper-on-eHealth-Vertical-Sector.pdf.

Abadia, J., F. Merli, J.F. Zurcher, J.R. Mosig, and A.K. Skrivervik. 2009. 3D-spiral small antenna design and realization for biomedical telemetry in the MICS band. *Radioengineering* 18 (4): 359–367.

Agrawal, K., Y.-X. Guo, and B. Salam. 2016. Wearable AMC backed near-endfire antenna for on-body communications on latex substrate. *IEEE Trans. Compon. Packag. Manuf. Technol.* 6 (3): 346–358.

Ahmadi, M.M and G.A. Jullien. 2009. A wireless implantable microsystem for continuous blood glucose monitoring. *IEEE Trans. Actions on Biomed. Circuits Syst.* 3 (3): 169–180.

Allen, J. 2007. Photoplethysmography and it application in clinical physiological measurement. *Physiol. Meas.* 28 (3): R1–R39.

Alomainy, A. and Y. Hao. 2009. Modeling and characterization of biotelemetric radio channel from ingested implants considering organ contents. *IEEE Trans. Antennas Propag.* 57 (4): 999–1005.

Alomainy, A., Y. Hao, C.G. Parini, Y. Nechayev, C.C. Constantinou, and P.S. Hall. 2007. Statistical analysis and performance evaluation for on-body radio propagation with microstrip patch antennas. *IEEE Trans. Antennas Propag.* 55 (1): 245–248.

Alomainy, A., A. Sani, A. Rahman, J.G. Santas, and Y. Hao. 2009. Transient characteristics of wearable antennas and radio propagation channels for ultrawideband body-centric wireless communications. *IEEE Trans. Antennas Propag.* 57 (4): 875–884.

Alzaidi, A., L. Zhang, and H. Bajwa. 2012. Smart textiles based wireless ECG system. *Systems, Applications and Technology Conference (LISAT), 2012 IEEE*, pp. 1–5.

Andria, G., M.L. Lanzolla, G. Cavallo, G. Russo, F. Marinosci, R. Antonelli Incalzi, and M. Benvenuto. 2015. A novel approach for design and testing digital m-health applications. *Medical Measurements and Applications (MeMeA), 2015 IEEE International Symposium on*, pp. 440–444.

Aranki, D., G. Kurillo, P. Yan, D.M. Liebovitz, and R. Bajcsky. 2016. Real time telemonitoring of patients with chronic heart failure using a smartphone lessons learned. *IEEE Trans. Affect. Comput.* IEEE Early Access Articles, 7 (3): 206–219.

Atallah, L., A. Wiik, G.G. Jones, J.P. Cobb, A. Amis, and G.Z. Yang. 2012. Validation of an ear-worn sensor for gait monitoring using a force-plate instrumented treadmill. *Gait Posture* 35 (4): 674–676.

Attard, S. and S. Zammit. 2012. An empirical investigation of the capacitive body coupled communications channel for body area networks. *Proceedings of the IEEE Topical Conference Biomedical Wireless Technologies, Networks, and Sensing Systems,* pp. 85–88.

Bae, J., H. Cho, K. Song, J. Lee, and H.-J. Yoo. 2012. The signal transmission mechanism on the surface of human body for body channel communication. *IEEE Trans. Microw. Theory Tech.* 60 (3): 582–593.

Bakker, E. 2004. Electrochemical sensors. *Anal. Chem.* 76 (12): 3285–3298.

Biotronik. 2012. Lumax. Accessed July 3, 2016. https://www.biotronik.com/en-gb/products/crm/tachycardia/lumax-740-dr-t-vr-t.

Bonato, P. 2005. Advances in wearable technology and applications in physical medicine and rehabilitation. *J. NeuroEng. Rehab.* 2 (2). https://doi.org/10.1186/1743-0003-2-2.

Bourbakis, N. and A. Karargyris. 2014. Ingestible health care system paradigm for wireless capsule endoscopy. In *Handbook of Biomedical Telemetry,* by K.S. Nikita (Ed.). Wiley-IEEE Press.

Bourbakis, N., A. Pantelopoulos, and R. Kannavara. 2014. Security and privacy in biomedical telemetry: Mobile health platform for secure information exchange. In *Handbook of Biomedical Telemetry,* by K.S. Nikita (Ed.), pp. 382–418. Wiley-IEEE Press.

Bouropoulos, N., D. Kouzoudis, and C. Grimes. 2005. The real-time, in situ monitoring of calcium oxalate and brushite precipitation using magnetoelastic sensors. *Sens. Actuators B: Chem.* 109 (2): 227–232.

Bsoul, M., H. Minn, and L. Tamil. 2011. Apnea MedAssist: Real-time sleep apnea monitor using single-lead ECG. *IEEE Trans. Inf. Technol. Biomed.* 15 (3): 409–415.

Cancela, J., M. Pastorino, M.T. Arredondo, K.S. Nikita, F. Villagra, and M.A. Pastor. 2014. Feasibility study of a wearable system based on a wireless body area network for gait assessment in Parkinson's disease patients. *Sensors* 14 (3): 4618–4633.

Carroll, N.V., P.W. Slattum, and F.M. Cox. 2005. The cost of falls among the community-dwelling elderly. *J. Manag. Care Pharm.* 11 (4): 307–316.

Cela, C.J., K.C. Gosalia, A.K. RamRakhyani, G. Lazzi, S. Soora, G.J. Hayes, and M.D. Dickey. 2014. Stimulator paradigm: Artificial retina. In *Handbook of Biomedical Telemetry,* by K.S. Nikita (Ed.), pp. 593–622. Wiley-IEEE Press.

Chang, S.-H., R.-D. Chiang, S.-J. Wu, and W.-T. Chang. 2016. A context-aware, interactive M-health system for diabetics. *IEEE IT Prof.* 18 (3): 14–22.

Chen, C.-H., R.-Z. Hwang, L.-S. Huang, S.-M. Lin, H.-C. Chen, Y.-C. Yang, Y.-T. Lin, et al. 2009a. A wireless bio-MEMS sensor for C-reactive protein detection based on nanomechanics. *IEEE Trans. Biomed. Eng.* 56 (2): 462–479.

Chen, Z.N., G.C. Liu, and T.S.P. See. 2009b. Transmission of RF signals between MICS loop antennas in free space and implanted in the human head. *IEEE Trans. Antennas Propag.* 57 (6): 1850–1853.

Chin, C.A., G.V. Crosby, T. Ghosh, and R. Murimi. 2012. Advances and challenges of wireless body area networks for healthcare applications. *Computing, Networking and Communications (ICNC), 2012 International Conference on,* pp. 99–103.

Chirwa, L.C., P.A. Hammond, and D.R.S. Cumming. 2003. Electromagnetic radiation from ingested sources in the human intestine between 150 MHz and 1.2 GHz. *IEEE Trans. Biomed. Eng.* 50: 484–492.

Cho, G., K. Jeong, M. Paik, Y. Kwun, and M. Sung. 2011. Performance evaluation of textile-based electrodes and motion sensors for smart clothing. *IEEE Sensors J.* 12 (11): 3183–3193.

Christopoulos, C. 1996. *The Transmission Line Modeling Method: TLM.* Piscataway, NJ: Wiley-IEEE Press.

Clifford, G.D., C. Arteta, T. Zhu, M.A.F. Rimentel, M. Santos, J. Domingos, M. Maraci, J. Behar, and J. Oster. 2014. A scalable mHealth system for noncommunicable disease management. *Global Humanitarian Technology Conference (GHTC), 2014 IEEE,* pp. 41–48.

Cochlear. 2010. Nucleus freedom. Accessed July 3, 2016. http://www.cochlear.com/wps/wcm/connect/us/recipients/nucleus-freedom/nucleus-freedom-basics.

Cong, P., W.H. Ko, and D.J. Young. 2010. Wireless batteryless implantable blood pressure monitoring microsystem for small laboratory animals. *IEEE Sensors J.* 10 (2): 243–254.

Continua Health Alliance. 2006. Accessed July 2, 2016. http://www.continuaalliance.org/.

Conway, G.A., W.G. Scanlon, C. Orlenius, and C. Walker. 2008. In situ measurement of UHF wearable antenna radiation efficiency using a reverberation chamber. *IEEE Antennas Wireless Propag. Lett.* 7: 271–274.

Cumming, D.R.S, P.A Hammond, L. Wang, J.M Cooper, and E.A Johannessen. 2006. Wireless sensor microsystem design: A practical perspective in body sensor networks. In M. Yacoub, G.-Z. Yang (eds.), *Body Sensor Networks.* London, UK: Springer-Verlag, pp. 373–397.

Curto, S. and M.J. Ammann. 2007. Electromagnetic coupling mechanism in a layered human tissue model as reference for 434 MHz RF medical therapy applicators. *Proceedings of the IEEE Antennas and Propagation Society is an International Symposium,* pp. 3185–3188.

Danielsson, B. and B. Mosbach. 1988. Enzyme thermistors. *Meth. Enzymol.* 137: 181–197.

de Mattos, W. D. and P.R.L. Gondim. 2016. M-Health solutions using 5G networks and M2M communication. *IT Pro.* 18 (3): 24–29.

Dell'Atti, D., S. Tombelli, M. Minunni, and M. Mascini. 2006. Detection of clinically relevant point mutations by a novel piezoelectric biosensor. *Biosens Bioelectron.* 21 (10): 1876–1879.

Dong, J.C., S. Moon, and J.K. Soon. 2016. A wearable device platform for the estimation of sleep quality using simultaneously motion tracking and pulse oximetry. *IEEE International Conference on Consumer Electronics (ICCE),* pp. 49–50.

Echo Wireless Goniometry. 2017. Accessed February 07, 2017. https://www.jtechmedical.com/Echo-Wireless-Instruments/echo-goniometry.

European Radiocommunications Commission, (ERC). 1997. Recommendation 70–03 relating to the use of short range devices (SRD). *Conference of European Postal and Telecommunications in Administration (EPT),* CEPT/ERC 70–03, Annex 12, Copenhagen, Denmark.

Federal Communications Commision, (FCC). 2003. Code of federal regulations, *Title 47 Part 95, WMTS Band Plan.* http://www.fcc.gov.

Federal Communication Commission, (FCC). 2012. FCC 12-54. Amendment of the commission's rules to provide spectrum for the operation of medical body area networks. *First Report and Order and Further Notice of Proposed Rulemaking,* Washington, DC.

Fernandes, R., B. Diniz, R. Ribeiro, and M. Humayun. 2012. Artificial vision through neuronal stimulation. *Neurosci. Lett.* 519: 122–128.

Feversmart. 2014. Accessed June 30, 2016. http://feversmart.com/.

Fioravanti, A., G. Fico, A.G. Paton, J.-P. Leuteritz, A.G. Arredondo, and M.-T. Arredondo Waldmeyer. 2014. mHealth-integrated system paradigm: Diabetes management. In *Handbook of Biomedical Telemetry*, by K.S. Nikita (Ed.), pp. 623–632. Wiley-IEEE Press.

Gauglit, G. 2005. Direct optical sensors: Principles and selected applications. *Anal. Bioanal. Chem.* 381 (1): 141–155.

GE Healthcare. 2009. Accessed July 2, 2016. http://www3.gehealthcare.com/en/products/categories/ultrasound/vscan_portfolio.

Gentag. 2011. Accessed July 2, 2016. http://gentag.com/nfc-skin-patches/.

Ghovanloo, M. and M. Kiani. 2014. Inductive coupling. In *Handbook of Biomedical Telemetry*, by K.S. Nikita (Ed.), pp. 174–207. Wiley-IEEE Press.

Given Imaging. 2012a. PillCam. Accessed July 3, 2016. http://www.givenimaging.com/en-int/Innovative-Solutions/Capsule-Endoscopy/Pages/default.aspx.

Given Imaging. 2012b. SmartPill. Accessed July 3, 2016. http://www.givenimaging.com/en-int/Innovative-Solutions/Motility/SmartPill/Pages/default.aspx.

Gosselin, M.-C., E. Neufeld, H. Moser, E. Huber, S. Farcito, L. Gerbber, M. Jedensjo, et al. 2014. Development of a new generation of high-resolution anatomical models for medical device evaluation: The virtual population 3.0. *Phys. Med. Biol.* 59: 5287–5303.

Goswam, D., D. Saha, A. Saha, R. Gangul, and A. Chakraborty. 2015. A brief review of deep brain stimulation for treatment of neurological diseases. *Computing and Communication (IEMCON), 2015 International Conference and Workshop on*, pp. 1–6.

Gouwanda, D. and S.M.N.A. Senanayake. 2010. Application of gyroscopes in identifying gait symmetry in walking. *IFMBE Proceedings of the 6th World Congress Biomech*, pp. 1378–1381.

Grieshaber, D., R. MacKenzie, J. Vörös, and E. Reimhult. 2008. Electrochemical biosensors—Sensor principles and architectures. *Sensors* 8 (3): 1400–1458.

Guenther, F.H., J.S. Brumberg, E.J. Wright, A. Nieto-Castanon, J.A. Tourville, M. Panko, R. Law, et al. 2012. A wireless brain-machine interface for real-time speech synthesis. *PLoS One* 4 (12): e8218.

Gupta, A. and T.D Abhayapala. 2008. Body area networks: Radio channel modelling and propagation characteristics. In *Australian Communications Theory Workshop*. Christchurch, New Zealand: IEEE, pp. 58–63.

Habib, S.A. and S. Saha. 2010. Burden of non-communicable disease: Global overview. *Diabetes Metab. Syndr.* 4 (1): 41–47.

Hafner, C. 1990. *The Generalized Multipole Technique for Computational Electromagnetic.* Boston, MA: Artech House.

Hao, Y. and R. Foster. 2014. Wearable health care system paradigm. In *Handbook of Biomedical Telemetry*, by K.S. Nikita (Ed.), pp. 505–524. Wiley-IEEE Press.

Harrington, R. 1968. *Field Computation by Moment Method.* New York: Macmillan.

HART. 2008. *WirelessHART Device Specification, HCF_SPEC-290.*

Hester, T., R. Hughes, D.M. Sherrill, B. Knorr, M. Akay, J. Stein, and P. Bonato. 2006. Using wearable sensors to measure motor abilities following stroke. *International Workshop on Wearable and Implantable Body Sensor Networks.* Cambridge, MA: IEEE, pp. 4–8.

Huang, F.J., C.M. Lee, C.L. Chang, L.K. Chen, T.C. Yo, and C.H. Luo. 2011. Rectenna application of miniaturized implantable antenna design for triple-band biotelemetry communication. *IEEE Trans. Antennas Propag.* 59 (7): 2646–2653.

ICNIRP. 1998. Guidelines for limiting exposure to time-varying electric, magnetic, and electromagnetic fields (up to 300 GHz). *Health Phys.* 74: 494–522.

IEEE. 2012. IEEE 802.15.6. *IEEE Standard for Local and Metropolitan Area Networks— Part 15.6: Wireless Body Area Networks.*

iHealth. 2010. Accessed June 29, 2016. https://ihealthlabs.com/fitness-devices/wireless-pulse-oximeter/.

Inanlou, F. and M. Ghovanloo. 2011. Wideband near-field data transmission using pulse harmonic modulation. *IEEE Trans. Circuits Syst. I, Reg. Papers* 58 (1): 186–195.

Institute of Electrical and Electronics Engineers, (IEEE). 1999. *C95.1-1999 IEEE Standard for Safety Levels with respect to Human Exposure to Radio Frequency Electromagnetic Fields, 3 kHz to 300 GHz.*

Institute of Electrical and Electronics Engineers, (IEEE). 2001. *Draft: Recommended practice for determining the spatial-peak specific absorption rate (SAR) in the human body due to wireless communications devices: Experimental techniques.*

Institute of Electrical and Electronics Engineers, (IEEE). 2003. *Standard 1528 SCC34. IEEE recommended practice for determining the peak spatial-average specific absorption rate (SAR) in the human head from wireless communications devices: Measurement techniques.*

Institute of Electrical and Electronics Engineers, (IEEE). 2011. *IEEE 802.15.4. Part 15.4: Wireless Medium Access Control (MAC) and Physical Layer (PHY) Specifications for Low-Rate Wireless Personal Area Networks (WPANs).*

Institute of Electrical and Electronics Engineers, (IEEE). 2013. *Part 15.4: Low-Rate Wireless Personal Area Networks (LR-WPANs) Amendment 4: Alternative Physical Layer Extension to Support Medical Body Area Network (MBAN) Services Operating in the 2360 MHz– 2400 MHz Band.*

International Telecommunications Union-Radiocommunications, (ITU-R). 1998. *Recommendation ITU-R SA.1346.*

International Electrotechnical Commission, (IEC). 2005. IEC Standard 62209-1. *Human Exposure to Radio Frequency Fields from Hand-Held and Body-Mounted Wireless Communication Devices–Human Models, Instrumentations, and Procedures*, Geneva, Switzerland.

Intromedic. 2012. MiroCam. Accessed July 3, 2016. http://www.intromedic.com/eng/sub_products_2.html.

Istepanaian, R.S.H., E. Jovanov, and Y.T. Zhang. 2004. Introduction to the special section on M-Health: Beyond seamless mobility and global wireless health-care connectivity. *IEEE Trans. Inf. Technol. Biomed.* 8 (4): 405–414.

Istepanaian, R.S.H. and Y.T Zhang. 2012. Introduction to the special section: 4G Health— The long-term evolution of m-Health. *IEEE Trans. Inf. Technol. Biomed.* 16 (1): 1–5.

Javadpour, A., H. Memarzadeh-Tehran, and F. Saghafi. 2015. A temperature monitoring system incorporating an array of precision wireless thermometers. *Smart Sensors and Application (ICSSA), 2015 International Conference on,* pp. 155–160.

JewelPUMP. 2012. Accessed June 30, 2016. http://www.jewelpump.com.

Johnson, C.C. and A.W. Guy. 1972. Johnson and guy. *Proc. IEEE* 60 (6): 692–718.

Jovanov, E., A. O'Donnell-Lords, D. Raskovic, P. Cox, R. Adhami, and F. Andrasik. 2003. Stress monitoring using a distributed wireless intelligent sensor system. *IEEE Eng. Med. Biol. Mag.* 22 (3): 49–55.

Jow, U.-M. and M. Ghovanloo. 2007. Design and optimization of printed spiral coils for efficient transcutaneous inductive power transmission. *IEEE Trans. Biomed. Circuits Syst.* 1 (3): 193–202.

Jow, U.-M. and M. Ghovanloo. 2010. Optimization of data coils in a multiband wireless link for neuroprosthetic implantable devices. *IEEE Trans. Biomed. Circuits Syst.* 4 (5): 301–310.

Jung, H.C., J.H. Moon, D.H. Baek, J.H. Lee, Y.Y. Choi, J.S. Hong, and S.H. Lee. 2012. CNT/PDMS composite flexible dry electrodes for long-term ECG monitoring. *IEEE Trans. Biomed. Eng.* 59 (5): 1472–1479.

Kanda, M., M. Ballen, S. Salins, C. Chou, and Q. Balzano. 2004. Formulation and characterization of tissue equivalent liquids used for RF densitometry and dosimetry measurements. *IEEE Trans. Microw. Theory Tech.* 52 (8): 2046–2056.

Karacolak, T., R. Cooper, J. Butler, S. Fisher, and E. Topsakal. 2010. In vivo verification of implantable antennas using rats as model animals. *IEEE Antennas Wireless Propag. Lett.* 9: 334–337.

Karacolak, T., R. Cooper, and E. Topsakal. 2009. Electrical properties of rat skin and design of implantable antennas for medical wireless telemetry. *IEEE Trans. Antennas Propag.* 57 (9): 2806–2812.

Karacolak, T., A.Z. Hood, and E. Topsakal. 2008. Design of a dual-band implantable antenna and development of skin mimicking gels for continuous glucose monitoring. *IEEE Trans. Microw. Theory Techn.* 56 (4): 1001–1008.

Karathanasis, K.T., I.A. Gouzouasis, I.S. Karanasiou, and N.K. Uzunoglu. 2012. Experimental study of a hybrid microwave radiometry—Hyperthermia apparatus with the use of an anatomical head phantom. *IEEE Trans. Inf. Technol. Biomed* 16 (2): 241–247.

Keehan, S.P., A.M. Sisko, and C.J. Truffer. 2011. National health spending projections through 2020: Economic recovery and reform drive faster spending growth. *Health Aff.* 30 (8): 1594–1605.

Kenney, P. 2007. https://www.isa.org/.

Kiani, M., U. Jow, and M. Ghovanloo. 2011. Design and optimization of a 3-coil inductive link for efficient wireless power transmission. *IEEE Trans. Biomed. Circuits Syst.* 5 (6): 579–591.

Kim, J. and Y. Rahmat-Samii. 2006. SAR reduction of implanted planar inverted F antennas with non–uniform width radiator. *IEEE International Symposium Antennas Propagation.* July 2006, Alberquerque, NM, pp. 1091–1094.

Kim, J. and Y. Rahmat-Samii. 2004. Implanted antennas inside a human body: Simulations, designs, and characterizations. *IEEE Trans. Microw. Theory Tech.* 52 (8): 1934–1943.

Kim, J.-Y. and C.-H. Chu. 2016. Analysis and modeling of selected energy consumption factors for embedded ECG devices. *IEEE Sens. J.* 16 (6): 1795–1805.

Kiourti, A., M. Christopoulou, and K.S. Nikita. 2011. Performance of a novel miniature antenna implanted in the human head for wireless biotelemetry. *IEEE International Symposium Antennas Propagation.* Spokane, Washington, pp. 392–395.

Kiourti, A., J.R. Costa, C.A. Fernandes, and K.S. Nikita. 2014a. A broadband implantable and a dual-band on-body repeater antenna: Design and transmission performance. *IEEE Trans. Antennas Propag.* 62 (6): 2899–2908.

Kiourti, A., J.R. Costa, C.A. Fernandes, A.G. Santiago, and K.S. Nikita. 2012. Miniature implantable antennas for biomedical telemetry: From simulation to realization. *IEEE Trans. Biomed. Eng.* 59 (11): 3140–3147.

Kiourti, A. and K.S. Nikita. 2012a. A review on implantable patch antennas for biomedical telemetry: Challenges and solutions. *IEEE Antennas Propag. Mag.* 54 (3): 210–228.

Kiourti, A. and K.S. Nikita. 2012b. Accelerated design of optimized implantable antennas for medical telemetr. *IEEE Antennas Wireless Propag. Lett.* 11: 1655–1658.

Kiourti, A. and K.S. Nikita. 2012c. Miniature scalp-implantable antennas for telemetry in the MICS and ISM bands: Design, safety considerations and link budget analysis. *IEEE Trans. Antennas Propag.* 60 (6): 3568–3575.

Kiourti, A. and K.S. Nikita. 2012d. Recent advances in implantable antennas for medical telemetry. *IEEE Antennas Propag. Mag.* 54 (6): 190–199.

Kiourti, A. and K.S. Nikita. 2014a. Antennas and RF communication. In *Handbook of Biomedical Telemetry*, by K.S. Nikita (Ed.), pp. 209–251. Wiley-IEEE Press.

Kiourti, A. and K.S. Nikita. 2014b. Implantable antennas: A tutorial on design, fabrication, and in vitro/in vivo testing. *IEEE Microw. Mag.* 15 (4): 77–91.

Kiourti, A. and K.S. Nikita. 2014c. Numerical and experimental techniques for body area electromagnetics. In *Handbook of Biomedical Telemetry*, by K.S. Nikita (Ed.), pp. 133–172. Wiley-IEEE Press.

Kiourti, A. and K.S. Nikita. 2015. Implanted antennas in biomedical telemetry. In *Handbook of Antenna Technologies*, by Z.N. Chen (Ed.), pp. 1–33. New York: Springer.

Kiourti, A., K. Psathas, P. Lelovas, N. Kostomitsopoulos, and K.S. Nikita. 2013. In vivo tests of implantable antennas in rats: Antenna size and inter-subject considerations. *IEEE Antennas Wireless Propag. Lett.* 12: 1396–1399.

Kiourti, A., K.A. Psathas, and K.S. Nikita. 2014b. Implantable and ingestible medical devices with wireless telemetry functionalities: A review of current status and challenges. *Bioelectromagnetics* 33 (1): 1–15.

Kiourti, A. and J.L. Volakis. 2016. Wearable antennas using electronic textiles for RF communications and medical monitoring. *10th European Conference on Antennas and Propagation (EuCAP)*, pp. 1–2.

Krishnamurthy, V., S.M. Monfared, and B. Cornell. 2010. Ion-channel biosensors—Part I: Construction, operation, and clinical studies. *IEEE Trans. Nanotechn.* 9 (3): 303–312.

Kuhn, S., E. Cabot, A. Christ, M. Capstick, and N. Kuster. 2009. Assessment of the radio-frequency electromagnetic fields induced in the human body from mobile phones used with hands-free kits. *Phys. Med. Biol.* 54: 5493–5508.

Kulkarni, G.S., K. Reddy, Z. Zhong, and X. Fan. 2014. Graphene nanoelectronic heterodyne sensor for rapid and sensitive vapour detection. *Nat. Commun.* 5 (4376). doi:10.1038/ncomms5376.

Kwasnicki, R.M. and G.-Z. Yang. 2014. Clinical applications of body sensor networks. In *Handbook of Biomedical Telemetry*, by K.S. Nikita (Ed.), pp. 481–504. Wiley-IEEE Press.

Lam, Y.-Z. and J.K. Atkinson. 2007. Biomedical sensor using thick film technology for transcutaneous oxygen measurement. *Med. Eng. Phys.* 29: 291–297.

Lammers, F. and T. Scheper. 2001. Thermal biosensors in biotechnology. *Adv. Biochem. Eng./Biotechnol.* 64: 35–67.

Lee, H.M. and M. Ghovanloo. 2011. An integrated power-efficient active rectifier with offset-controlled high speed comparators for inductively powered applications. *IEEE Trans. Circuits Syst. I, Reg. Papers* 58 (8): 1749–1760.

Lee, S., K. Song, J. Yoo, and H.J. Yoo. 2010. A low-energy inductive coupling transceiver with cm-range 50-Mbps data communication in mobile device applications. *IEEE J. Solid-State Circuits* 45 (11): 2366–2374.

Lee, S.H., J. Lee, Y.J. Yoon, S. Park, C. Cheon, K. Kim, and S. Nam. 2011. A wideband spiral antenna for ingestible capsule endoscope systems: Experimental results in a human phantom and a pig. *IEEE Trans. Biomed. Eng.* 58 (6): 1734–1741.

Li, Z., A. Huang, W. Xu, W. Hu, and L. Xie. 2014. Fall perception for elderly care: A fall detection algorithm in smart wristlet mHealth system. *IEEE International Conference on Communications (ICC)*, pp. 4270–4274.

Lim, H.B., D. Baumann, and E.P. Li. 2011. A human body model for efficient numerical characterization of UWB signal propagation in wireless body area networks. *IEEE Trans. Biomed. Eng.* 58 (3): 689–69.

Lim, S.S. 2012. A comparative risk assessment of burden of disease and injury attributable to 67 risk factors and risk factor clusters in 21 regions, 1990–2010: A systematic analysis for the global burden of disease study 2010. *Lancet* 380 (9859): 2224–2260.

Lin, J. and K. S. Nikita. 2010. *Wireless Mobile Communication and Healthcare.* Berlin, Germany: Springer LNICST.

Liolios, C., C. Doukas, G. Forulas, and I. Maglogiannis. 2010. An overview of body sensor networks in enabling pervasive healthcare and assistive environments. *Proceedings of the 3rd International Conference Pervasive Technologies Related to Assistive Environments.* New York: ACM.

Lioumpas, A., G. Ntouni, and K.S. Nikita. 2014. Sensing principles for biomedical telemetry. In *Handbook of Biomedical Telemetry*, by K.S. Nikita (Ed.), pp. 56–75. Wiley-IEEE Press.

Liu, C., Y.-X. Guo, and S. Xiao. 2014. Circularly polarized helical antenna for ISM-Band ingestible capsule endoscope systems. *IEEE Trans. Antennas Propag.* 62 (12): 6027–6039.

Liu, W.C., S.H. Chen, and C.M. Wu. 2008. Implantable broadband circular stacked PIFA antenna for biotelemetry communication. *J. Electromagn. Waves Appl.* 22: 1791–1800.

Lucev, Z., I. Krois, and M. Cifrek. 2010. Application of wireless intrabody communication system to muscle fatigue monitoring. *IEEE Instrumentation and Measurement Technology Confererence*, pp. 1624–1627.

Manyika, J., M. Chui, P. Bisson, J. Woetzel, R. Dobbs, J. Bughin, and D. Aharon. 2015. *The Internet of Things: Mapping the Value Beyond the Hype.* San Francisco, CA: McKinsey & Company.

Mathieson, K., J. Loudin, G. Goetz, P. Huie, L. Wang, L. Kamins, L. Galambos, et al. 2012. Photovoltaic retinal prosthesis with high pixel density. *Nat. Photon.* 6: 391–397.

Medtronic. 2010. Adapta with MVP pacing system. Accessed July 3, 2016. http://www.medtronic.com/for-healthcare-professionals/products-therapies/cardiac-rhythm/pacemakers/adapta-with-mvp-pacing-system/index.htm.

Medtronic. 2010. Bravo pH monitoring system. Accessed July 3, 2016. http://www.medtronic.com/covidien/products/reflux-testing/bravo-reflux-testing-system.

Medtronic. 2011. RevoMRI sureScan. Accessed July 3, 2016. http://www.medtronic.com/us-en/healthcare-professionals/products/cardiac-rhythm/pacemakers/revo-mri-pacing-system.html.

Medtronic. 2012. SynchroMed II pump. Accessed July 3, 2016. https://professional.medtronic.com/pt/neuro/itb/prod/synchromed-ii/features-specifications/#.V3jOAPl96Cg.

Mercier, P.P., A.C. Lysaght, S. Bandyopadhyay, A. Chandrakasan, and K. Stankovic. 2012. Energy extraction from the biologic battery in the inner ear. *Nat. Biotechnol.* 1240–1243. doi:10.1038/nbt.2394.

Microchip. 2010. *MiWi Wireless Networking Protocol Stack, AN1006.*

MICS Federal Register. 1999. Medical implant communications service (MICS) rules reg. *Fed. Reg.* 64: 69926–69934.

Modi, K. and R.B. Mohanty. 2015. *M-Health: Challenges, Benefits, and Keys to Successfull Implementation.* Bangalore, India: Infosys Limited.

Monošík, R., M. Streďanský and E. Šturdík. 2012. Biosensors—Classification, characterization. *Acta Chim. Slov.* 5 (1): 109–120.

Moon, I., M. Lee, J. Chu, and M. Mun. 2005. Wearable EMG-based HCI for electric-powered wheelchair users with motor disabilities. *Proceedings of the 2005 IEEE International Conference on Robotics and Automation, ICRA,* pp. 2649–2654.

Mukkamala, R., J. Hahn, O.T. Inan, L.K. Mestha, C.S.K. Kim, H. Töreyin, and S. Kyal. 2015. Toward ubiquitous blood pressure monitoring via pulse transit time: Theory and practice. *IEEE Trans. Biomed. Eng.* 62 (8): 1879–1901.

Nagaoka, T., S. Watanabe, K. Sakurai, E. Kunieda, W. Wanatabe, M. Taki, and Y. Yamanaka. 2004. Development of realistic high-resolution whole-body voxel models of Japanese adult males and females of average height and weight, and application of models to radio-frequency electromagnetic-field dosimetry. *Phys. Med. Biol.* 49 (4): 1–15.

Nikita, K.S. 2014. Introduction to biomedical telemetry. In *Handbook of Biomedical Telemetry*, by K.S. Nikita (Ed.), pp. 1–24. Wiley-IEEE Press.

Nikita, K.S., P. Cavagnaro, P. Bernardi, N.K. Uzunoglu, S. Pisa, E. Piuzzi, J.N. Sahalos, et al. 2000. A study of uncertainties in modeling antenna performance and power absorption in the head of a cellular phone user. *IEEE Trans. Microw. Theory Techn.* 48 (12): 2676–2685.

Nikita, K.S., J.C. Lin, D.I. Fotiadis, and M.T. Arredondo. 2012. Editorial: Special issue on mobile and wireless technologies for healthcare delivery. *IEEE Trans. Biomed. Eng.* 59 (11): 3083–3089.

Ntouni, G., A. Lioumpas, and K.S. Nikita. 2014. Reliable and energy efficient communications for wireless biomedical implant systems. *IEEE J. Biomed. Health Inform* 18 (6): 1848–1856.

NTT. 2015. RedTacton: An innovative human area networking technology that uses the surface of the human body as a transmission path. Accessed July 6, 2016. http://www.ntt.co.jp/news/news05e/0502/050218.html.

Olympus. 2012. Capsule endoscopy. Accessed July 3, 2016. http://medical.olympusamerica.com/procedure/capsule-endoscopy.

Onishi, T., R. Ishido, and T. Takimoto. 2005. Biological tissue-equivalent agar-based solid phantoms and SAR estimation using the thermographic method in the range of 3–6 GHz. *IEICE Trans. Commun.* E88-B (9): 3733–3741.

Osman, M.R., M.K.A. Rahim, N.A. Samsuri, H.A.M Sahim, and M.F. Ali. 2011. Embroidered fully textile wearable antenna for medical monitoring applications. *Prog. Electrom. Res.* 117: 321–327.

Oxitone. 2012. Accessed June 29, 2016. http://oxitone.com/products/.

Patel, S., K. Lorincz, R. Hughes, N. Huggins, J. Growden, D. Standaert, M. Akay, J. Dy, M. Welsh, and P. Bonato. 2009. Monitoring motor fluctuations in patients with Parkinson's disease using wearable sensors. *IEEE Trans. Inf. Tech. Biomed.* 13 (6): 864–873.

Pohanka, M. and P. Skláda. 2008. Electrochemical biosensors—Sensor principles and applications. *J. Appl. Biomed* 6 (2): 57–64.

Polar. 2009. Accessed June 29, 2016. http://www.polar.com/en.

Postolache, O., P.S. Girao, J. Mendes, and G. Postolache. 2009. Unobstrusive heart rate and respiratory rate monitor embedded on a wheelchair. *IEEE International Workshop on Medical Measurements and Applications, MeMeA,* pp. 83–88.

Psathas, K., A. Kiourti, and K.S. Nikita. 2014. Safety issues in biomedical telemetry. In *Handbook of Biomedical Telemetry*, by K.S. Nikita (Ed.), pp. 445–478. Wiley-IEEE Press.

Reisner, A.T., P.A. Shaltis, D. McCombie, and H.H. Asada. 2008. Utility of the photoplethysmogram in circulatory monitoring. *Anesthesiology* 108 (5): 950–958.

Roa, L.M., J. Reina-Tosina, A. Callejon-Leblic, D. Naranjo, and M.A. Estudillo-Valderrama. 2014. Intrabody communications. In *Handbook of Biomedical Telemetr*, K.S. Nikita (Ed.), pp. 252–299. Wiley-IEEE Press.

Roger, A. 2011. Microelectronics: The medical industry's mini marvels. *Electronic Design* 6: 36–38.

Sadiku, M.N.O. 2007. *Elements of Electromagnetics, 4th ed.* New York: Oxford University Press.

Sani, A., M. Rajab, R. Foster, and Y. Hao. 2010. Antennas and propagation of implanted RFIDs for pervasive healthcare applications. *Proc. IEEE* 98 (9): 1648–1655.

Sattayasoonthorn, P., J. Suthakorn, S. Chamnanvej, J. Miao, and A.G.P. Kottapalli. 2013. LCP MEMS implantable pressure sensor for intracranial pressure measurement. *The 7th IEEE International Conference on Nano/Molecular Medicine and Engineering*, pp. 63–67.

Schmitt, L., J. Espina, T. Falck, and D. Wang. 2014. Biosensor communication technology and standards. In *Handbook of Biomedical Telemetry*, by K.S. Nikita (Ed.), pp. 330–367. Wiley-IEEE Press.

Schmitt, L., T. Falck, F. Wartena, and D. Simons. 2007. Novel ISO/IEEE 11073 standards for personal telehealth systems interoperability. *2007 Joint Workshop on High Confidence Medical Devices, Software, and Systems and Medical Device Plug-and-Play Interoperability*, pp. 146–148.

Schwartz, A., T. Cui, D. Weber, and D. Moran. 2006. Brain-controlled interfaces: Movement restoration with neural prosthetics. *Neuron* 52 (1): 205–220.

Sechang, O.H., P. S. Kumar, H. Kwno, and V. Varadan. 2012. Wireless brain-machine interface using EEG and EOG: Brain wave classification and robot control. *Proceedings of the SPIE 8344, Nanosensors, Biosensors, and Info-Tech Sensors and Systems*. doi:1117/12/918159.

Second Sight. 2012. Argus II retinal prosthesis system. Accessed July 3, 2016. http://www.secondsight.com/g-the-argus-ii-prosthesis-system-pf-en.html.

Sensimed. 2004. Sensimed triggerfish. Accessed July 3, 2016. http://www.sensimed.ch/en/sensimed-triggerfish/sensimed-triggerfish.html.

Shire, D.B., S.K. Kelly, C. Jinghua, P. Doyle, M.D. Gingerich, S.F. Cogan, W.A. Drohan, et al. 2009. Development and implantation of a minimally invasive wireless subretinal neurostimulator. *IEEE Trans. Biomed. Eng.* 56 (10): 2502–2511.

Silvester, P. and R. Ferrari. 1996. *Finite Elements for Electrical Engineers*. Cambridge: Cambridge University Press.

Skrivervik, A.K. and F. Merli. 2011. Design strategies for implantable antennas. *Antennas and Propagation Conference (LAPC), 2011 Loughborough*, pp. 1–5.

Sodagar, A.M., K.D. Wise, and K. Najafi. 2007. A fully integrated mixed-signal neural processor. *IEEE Trans. Biomed. Eng.* 54 (69): 1075–1088.

Soontornpipit, P., C.M. Furse, and Y.C. Chung. 2004. Design of implantable microstrip antenna for communication with medical implants. *IEEE Trans. Microw. Theory Tech.* 52: 1944–1951.

Speag. 1998. Accessed July 2, 2016. http://www.speag.com/.

Sullivan, D. 1990. Three-dimensional computer simulation in deep regional hyperthermia using the finite-difference time-domain method. *IEEE Trans. Microw. Theory Tech.* 38: 204–211.

Swan, M. 2012. Sensor mania! The internet of things, wearable computing. *J. Sens. Actuator* 1 (3): 217–253.

Syam, R., K.J. Davis, M.D. Pratheesh, R. Anoopraj, and S.J. Bunglavan. 2012. Biosensors: A novel approach for pathogen detection. *Vetscan* 7 (1): 14–18.

Takamiya, M., H. Fuketa, K. Ischida, T. Yokota, T. Sekitani, T. Someya, and T. Sakurai. 2014. Flexible, large-area, and distributed organic electronics closely contacted with skin for healthcare applications. *2014 IEEE 57th International Midwest Symposium on Circuits and Systems*, 829–832.

Tamura, T. 2014. Sensing technologies for biomedical telemetry. In *Handbook of Biomedical Telemetry*, by K.S. Nikita (Ed.), pp. 76–107. Wiley-IEEE Press.

Tamura, T., T. Yoshimura, M. Sekine, M. Uchida, and O. Tanaka. 2009. A wearable airbag to prevent fall injuries. *IEEE Trans. Inf. Technol. Biomed.* 13 (6): 910–914.

Tang, C.-M. and R. Bashirullah. 2011. Channel characterization for galvanic coupled in vivo biomedical devices. *Proceedings of the IEEE International Symposium Circuits and Systems*, pp. 921–92.

Tao, W., T. Liu, R. Zheng, and H. Feng. 2012. Gait analysis using wearable sensors. *Sensors* 12 (2): 2255–2283.

Tentzeris, M.M., R. Vyas, W. Wei, Y. Kawahara, L. Yang, S. Georgakopoulos, V. Lakafosis, et al. 2014. Power issues in biomedical telemetry. In *Handbook of Biomedical Telemetry*, by K.S. Nikita (Ed.), pp. 108–130. Wiley-IEEE Press.

Tng, D.J.H., R. Hu, P. Song, I. Roy, and K.-T. Yong. 2012. Approaches and challenges of engineering implantable microelectromechanical systems (MEMS) drug delivery. *Micromachines.* 3: 615–631.

Tourkhani, F. and P. Viarouge. 2001. Accurate analytical model of winding losses in round Litz wire windings. *IEEE Trans. Magn.* 37 (1): 538–543.

Traille, A., L. Yang, A. Rida, and M. Tentzeris. 2008. A novel liquid antenna for wearable bio-monitoring applications. *Microwave Symposium Digest, 2008 IEEE MTT-S International.* Atlanta, pp. 923–926.

Uzunoglu, N., K.S. Nikita, and D.I. Kaklamani. 2000. *Applied Computational Electromagnetics: State-of-the-Art and Future Trends.* NATO-ASI Series, Berlin, Germany: Springer.

Velasco-Garcia, M.N. 2009. Optical biosensors for probing at the cellular level: A review of recent progress and future prospects. *Semin. Cell Dev. Biol.* 20 (1): 27–33.

Venkatraman, S., K. Elkabany, J.D. Long, Y. Yao, and J.M. Carmena. 2009. A system for neural recording and closed-loop intracortical microstimulation in awake rodents. *IEEE Trans. Biomed. Eng.* 56 (1): 15–22.

Wang, S.X., and G. Li. 2008. Advances in giant magnetoresistance biosensors with magnetic nanoparticle tags: Review and outlook. *IEEE Trans. Magn.* 44 (7): 1687–1702.

Wang, Y.-L., A.W.Y. Su, T.-Y. Han, C.-L. Lin, and L.-C. Hsu. 2015. EMG based rehabilitation systems—Approaches for ALS patients in different stages. *2015 IEEE International Conference on Multimedia and Expo (ICME)*, pp. 1–6.

Wang, Z., K. Karathanasis, and J.L. Volakis. 2015. Axial ratio reduced ultra wideband slot spiral on hybrid impedance surfaces. *J. Electromagnet. Wave* 29 (2): 143–153.

Watthanawisuth, T., T. Lomas, A. Wisitsoraat, and A. Tuantranont. 2010. Wireless wearable pulse oximeter for health monitoring using ZigBee wireless sensor network. *Electrical Engineering/Electronics Computer Telecommunications and Information Technology (ECTI-CON), 2010 International Conference on*, pp. 575–579.

Weil, C.M. 1975. Absorption characteristics of multilayered sphere models exposed to UHF/microwave radiation. *IEEE Trans. Biomed. Eng.* 6: 468–476.

Xie, X., G. Li, X.K. Chen, X.W. Li, B.Y. Chi, and S.G. Han. 2004. A novel low power IC design for bi-directional digital wireless endoscopy capsule system. *IEEE International Workshop on Biomedical Circuits and System* (BioCAS04), Singapore, pp. S1.8.5–S1.8.8.

Xu, J., R.F. Yazicioglu, B. Grundlehner, P. Harpe, K.A.A. Makinwa, and C. Van Hoof. 2011. A160uW 8-channel active electrode system for EEG monitoring. *IEEE Trans. Biomed. Circuits Syst.* 5 (6): 555–567.

Xu, L., M.Q.H. Maz, H. Ren, and Y. Chan. 2008. Radiation characteristics of ingested wireless device at frequencies from 430 MHz to 3 GHz. *IEEE Conference of the Engineering In Medical and Biology Society,* pp. 1250–1253.

Yan, L., J. Bae, S. Lee, T. Roh, K. Song, and H.-J. Yoo. 2011. A 3.9 mW 25-electrode reconfigured sensor for wearable cardiac monitoring system. *IEEE J. Solid-State Circuits* 46 (1): 353–364.

Yang, L., H. Jingming, X. Xiaobo, and J. Ling. 2011. Measurement for channel characteristics of intra-body communication. *Cross Strait Quad-Regional Radio Science and Wireless Technology Conference,* pp. 1138–1140.

Yao, H., A.J. Shum, M. Cowan, I. Lahdesmaki, and B.A. Parviz. 2011. A contact lens with embedded sensor for monitoring tear glucose level. *Biosens. Bioelectron.* 26 (7): 3290–3296.

Yee, K.S. 1966. Numerical solution of initial boundary value problems involving Maxwell's equations in isotropic media. *IEEE Trans. Antennas Propag.* 14 (3): 302–307.

Zebda, A., S. Cosnier, J.-P. Alcaraz, M. Holzinger, A. Le Goff, C. Gondran, F. Boucher, et al. 2013. Single glucose biofuel cells implanted in rats power electronic devices. *Sci. Rep.* doi:10.1038/srep01516.

Zedong, N., L. Tengfei, W. Wenchen, G. Feng, and W. Lei. 2012. Experimental characterization of human body communication in shield chamber. *Proceedings of the IEEE-EMBS International Conference Biomedical and Health Informatics,* pp. 759–762.

Zhao, Y., Y. Hao, A. Alomainy, and C. Parini. 2006. UWB on-body radio channel modeling using ray theory and subband FDTD method. *IEEE Trans. Microw. Theory Tech.* 54 (4): 1827–1835.

Zhou, M., M.R. Yuce, and W. Liu. 2008. A non-coherent DPSK data receiver with interference cancellation for dual-band transcutaneous telemetries. *IEEE J. Solid-State Circuits* 43 (9): 2003–2012.

Zhu, S. and R. Langley. 2009. Dual-band wearable textile antenna on an EBG substrate. *IEEE Trans. Antennas Propag.* 57 (4): 926–935.

ZigBee Alliance. 2012. Accessed July 2, 2015. http://www.zigbee.org/zigbeealliance/governing-documents/.

Zimmerman, T.G. 1996. Personal area networks: Near-field intrabody communication. *IBM Syst. J.* 35 (3–4): 609–617.

Index

Note: Page numbers followed by f and t refer to figures and tables, respectively.